SHORELINES AND ISOSTASY

Thomas F. Jamieson, F.G.S., LL.D., 1829–1913, who first proposed the hypothesis of glacio-isostasy. From the George Washington Wilson Collection, Aberdeen University Library, with kind permission of Aberdeen University Library.

SHORELINES
AND
ISOSTASY

INSTITUTE OF BRITISH GEOGRAPHERS
SPECIAL PUBLICATION, NO. 16

Edited by

D. E. SMITH

and

A. G. DAWSON

Department of Geography,
Coventry (Lanchester) Polytechnic

For the British Geomorphological Research Group

1983

ACADEMIC PRESS

A Subsidiary of Harcourt Brace Jovanovich, Publishers

London · New York · Paris · San Diego · San Francisco
São Paulo · Sydney · Tokyo · Toronto

ACADEMIC PRESS INC. (LONDON) LTD.
24/28 Oval Road,
London NW1

United States Edition published by
ACADEMIC PRESS INC.
111 Fifth Avenue
New York, New York 10003

British Library Cataloguing in Publication Data

Shorelines and isostasy.—(Special publications/
Institute of British Geographers, ISSN 0073-9006;
no. 16)
1. Shore-lines 2. Isostasy
I. Smith, D. E. II. Dawson, A. G. III. Series
551.4'58 GB451.2

ISBN 0-12-652960-4

LCCCN 83-71981

Text set in 11/13 Plantin
by Dobbie Typesetting Service, Plymouth
Printed in Great Britain by
Galliard (Printers) Ltd, Great Yarmouth

CONTRIBUTORS

J. T. ANDREWS, Institute of Arctic and Alpine Research, University of Colorado, Boulder, Colorado 80309, USA

J. J. CLAGUE, Geological Survey of Canada, 100 West Pender, Vancouver, British Columbia V6B 1R8, Canada

A. G. DAWSON, Department of Geography, Coventry (Lanchester) Polytechnic, Coventry CV1 5FB, UK

R. J. DEVOY, Department of Geography, University College, Cork, Ireland

M. ERONEN, Department of Geology, University of Oulu, Linnanmaa, SF-90570, Oulu 57, Finland

R. W. FAIRBRIDGE, Department of Geological Sciences, Columbia University, New York, New York 10027, USA

J. M. GRAY, Department of Geography, Queen Mary College, University of London, Mile End Road, London E1 4NS, UK

U. HAFSTEN, Department of Botany, University of Trondheim-NLHT, N-7055 Dragvoll, Norway

J. R. HAILS, Environmental Services, Coal Division, CSR Limited, 1 O'Connell Street, Sydney, New South Wales 2001, Australia

D. HOPLEY, Department of Geography, James Cook University of North Queensland, Queensland 4811, Australia

S. B. MCCANN, Department of Geography, McMaster University, Hamilton, Ontario L85 4KJ, Canada

W. R. PELTIER, Department of Physics, University of Toronto, Toronto, Ontario M55 1A7, Canada

I. SHENNAN, Department of Geography, University of Durham, Science Laboratories, South Road, Durham DH1 3LE, UK

J. B. SISSONS, Department of Geography, University of Edinburgh, High School Yards, Edinburgh EH1 1NR, UK

D. E. Smith, Department of Geography, Coventry (Lanchester) Polytechnic, Coventry CV1 5FB, UK

D. G. Sutherland, Department of Geography, University of Edinburgh, High School Yards, Edinburgh EH1 1NR, UK

R. B. Taylor, Geological Survey of Canada, Atlantic Geoscience Centre, Bedford Institute of Oceanography, Dartmouth, Nova Scotia B2Y 4A2, Canada

A. S. Trenhaile, Department of Geography, University of Windsor, Ontario N98 3P4, Canada

PREFACE

This volume is concerned with the isostatic response in marine coastal regions to loading and unloading of the earth's crust by ice and water. It was T. F. Jamieson who first suggested, in 1865, that the growth and decay of ice sheets might be accompanied by depression and subsequent elevation of the crust, and since that time the effects of both glacio-isostasy and hydro-isostasy have been increasingly studied. Although several other forms of crustal loading and unloading are now recognized, studies of glacio-isostasy and hydro-isostasy have provided most information on crustal and sub-crustal responses. This is largely because of the detail with which the evidence for Late Quaternary relict shorelines can be studied and changes of level inferred.

The study of shorelines and isostasy is very much an interdisciplinary effort, with geographers, geologists, oceanographers, biologists and physicists amongst others contributing their own approaches. Increasing interest in recent years has led to a number of advances in knowledge being made, and this volume attempts to reflect these developments within a structured approach, which considers the context of the study, the techniques of investigation employed, and the results obtained in particular areas. The chapters comprise papers from research workers in several disciplines. Chapter one examines aspects of the relationships between isostatic and eustatic processes. Chapters two, three and four are concerned with modern coastal processes in middle and high latitudes and their relevance to the formation of relict shorelines. Chapters five and six examine several techniques of investigation concerned with the measurement and dating of relict shorelines. Chapters seven to thirteen consider shorelines and isostasy in selected areas and from different viewpoints. Chapter fourteen examines hydro-isostasy in a continental shelf environment, and chapter fifteen concludes the volume with a short historical perspective and consideration of future prospects.

The production of this volume has involved the work of many people, to whom we are most grateful. We should particularly like to thank the authors for promptly submitting and revising their manuscripts, and the referees for their constructive and most helpful advice. The Department of Geography, Coventry (Lanchester) Polytechnic, provided much help, and thanks are due to Mrs Shirley Addleton, who drew many of the maps and diagrams. Mr Callum Firth and Miss Diana Vere helped in preparing the final manuscript, and amongst several people at the Polytechnic who helped with typing we should particularly like to thank Mrs Valerie Emery, Mrs Brenda Gilbert, and Miss Joy Haddon. Finally thanks are due to Academic Press, for their help and forebearance.

September 1983
Coventry

D. E. Smith
A. G. Dawson

ACKNOWLEDGEMENTS

We should like to acknowledge the following for permission to reproduce copyright material.

Frontispiece: Aberdeen University Library. Chapter 1: Fig. 3, Geological Society of America: Fig. 5, Houston Geological Society and US Army Corps of Engineers. Chapter 2: Fig. 2, The Director, Bureau of Mineral Resources, Geology and Geophysics and Dr P. J. Cook. Chapter 3: Fig. 2, *Arctic* and Dr J. B. Maxwell; Fig. 3, The Minister, Supply and Services, Canada and Dr E. H. Owens. Chapter 5: Fig. 2, Dr J. Andrews; Figs 3 and 10, Dr J. B. Sissons; Table 4, *Norsk Geografisk Tidsskrift*, Universitetsforlaget and Mr J. Rose. Chapter 7: Fig. 2, *Boreas*; Fig. 4, *Boreas* and Drs I. Aarseth and J. Mangerud; Fig. 5, Geographical Survey of Norway and Dr S. Bakkelid. Chapter 8: Fig. 1, Elsevier Publishers, Amsterdam; Figs 3, 6, 7, 8 and 11, Editorial Board, *Atlas of Finland*. Chapter 10: Fig. 1, Uppsala University and Drs P. A. Ziegler and C. J. Louwerens. Chapter 11: Figs 3, 7, 8 and 10, Geologist's Association; Fig. 5, Royal Society and Dr R. J. Devoy; Fig. 6, *Quarterly Journal of Engineering Geology* and Professor C. Kidson; Figs 7 and 8, Professor C. Kidson; Fig. 9, *Geographical Journal*; Figs 9 and 10, Dr M. J. Tooley. Chapter 13: Fig. 1, John Wiley and Sons. Chapter 14: Figs 2–7, John Wiley and Sons.

Details of sources are quoted in the reference list for each chapter.

CONTENTS

PART FOUR: SHORELINES AND GLACIO-ISOSTASY

PART FIVE: SHORELINES AND HYDRO-ISOSTASY

PART SIX: CONCLUDING REMARKS

PART ONE

Introduction and context

ONE

Isostasy and eustasy

R. W. FAIRBRIDGE

INTRODUCTION

"Stasis" is the Greek-based word for the act of standing. "Isostasy" in geology and geophysics thus refers to the condition of equilibrium established by the earth's lithosphere, which essentially "floats" on the asthenosphere. "Eustasy" applies to sea level that tends correspondingly to seek an equilibrium with the geoid which in turn is an equipotential surface with reference to terrestrial gravity; a eustatic change in sea level may be due either to change in water volume (e.g. glacio-eustatic), or to the shape of the basin (tectonoeustatic). A regional, non-universal change in sea level (e.g. due to change in the planetary spin rate) is not eustatic, but geoidal. According to the A.G.I. Glossary the spellings "isostacy" and "eustacy" are not recommended. In Greek there is no equivalent to the letter "c".

Isostasy and eustasy are two intimately related geodynamic processes that can scarcely be understood or measured apart from each other, and yet, strangely enough, they have in the past been studied by specialists of two rather disparate disciplines, geophysics and geomorphology, the practitioners of each rarely possessing any training whatever in the other. This paradoxical situation may explain to some extent the many still unresolved problems that persist in both fields.

In this chapter I do not propose to review the vast field and its literature, but select a few small areas that may help to illuminate some of the manifold problems. The general background and some personal interpretations of Quaternary eustasy were presented by the writer two decades ago (Fairbridge, 1961), and variously criticized (Mörner, 1980). Nevertheless it may be helpful to outline briefly some of the main categories of current research and to note some of the national or international groups that are helping to direct and coordinate those studies. This information will be found in the Appendix.

TIME-SCALES

To begin with, we must recognize three very distinctive time-scales that constrain in many ways the types of investigation.

SHORELINES AND ISOSTASY
ISBN 0 12 652960 4

(a) *The long-term framework* (processes in the time-range of 10^6 to 10^9 years), studies of which currently consider such questions as eustatic cycles a few million years in length and their interactions with the isostatic behaviour of continental interiors and rifted continental margins. The latter undergo an interesting historical evolution relating to heat flow and uplift associated with the rifting phase of plate tectonics, to be followed by down-warping of the oceanic margin in consequence of removal of lateral support and progressive sedimentary loading through time (Pitman, 1978; Watts, 1982). Evolving continental margins disclose a pronounced eustatic cyclicity (Vail, 1977). Different types of mature rifted margins have undergone distinctive sedimentational and geomorphic histories (e.g. Fairbridge and Finkl, 1978; Fairbridge, 1979); in the continental interiors, isolated domes (? mantle hot spots) and quaquaversal basins (autogeosynclines) are produced, with associated unloading of uplifts by erosion and loading by sedimentation in the depressions; causative processes in the intratonic regions are still very poorly understood (Sleep and Sloss, 1980; Chen and Molnar, 1983). Convergent plate boundaries also have their eustatic/isostatic problems, but because of the tendency of subduction or orogeny to efface much of the evidence, these problems have been largely ignored.

(b) *The medium-term framework* (processes in the time-range of 10^3 to 10^6 years ago). This category involves primarily the crustal loading and unloading caused by Quaternary glaciation and deglaciation, not only on the scale of continental ice caps, but also local glaciation at various stages during each 100 000-year glacial cycle; the latter (e.g. Scottish ice, Newfoundland ice) are becoming more and more widely recognized to have played important roles during either the early stages of continental ice advances or late stages following general retreat (e.g. Sissons and Cornish, 1982; see also Sissons, this volume). The resultant complex isostatic loading effects are only just beginning to be studied.

In addition to the direct results of ice loading, widespread influence of the marginal bulge is being measured (Clague, this volume); at first it was merely postulated (e.g. Daly, 1934), and later theoretically modelled (Walcott, 1972, 1975). The bulge apparently reached to about 1000 km south of the North American (Laurentide) ice margin and about 800 km south of the Scandinavian (i.e. to the northern Mediterranean). Neotectonic, geodetic and eustatic studies suggest that the compensation (collapse) of the Weichsel/Wisconsin marginal bulge was largely completed by about 3000 years ago, but that subsidence continues in regions of high sediment loading (Mississippi delta, Chesapeake Bay, Rhine delta, southern North Sea, southern Baltic), so that discrimination is difficult.

A minor but universal factor introduced by glaciation and deglaciation is hydro-isostasy (Hopley, this volume). The removal of a layer of water 130–140 m thick from the oceans causes isostatic upwarping of the ocean floors and its reimposition during deglaciation is likely to cause up to 4 m of crustal lowering in a quasi-equatorial littoral belt (Clark *et al.*, 1978). It has been claimed by these authors that the widely observed 3–4 m mid-Holocene emerged shoreline of the tropics (Daly,

1934; Fairbridge, 1961) is a record of that hydro-isostatic reloading. However, they ignore the fact that this shoreline is only one of a series of strongly fluctuating Holocene levels (to be discussed below). They also ignore the well-known palynological evidence of the mid-Holocene thermal optimum, which indicates ∿ 2·5°C higher mid-latitude mean temperature, which would probably lead to a high sea level at that time. In as much as an oceanic reloading would theoretically diminish landwards, the 3-4 m terrace should, in any one transect, be highest near the shelf margin and decrease in altitude inland. A test of this hypothesis can be made between the Abrolhos Islands and the mainland of Western Australia, but the terraces show no difference in elevation (Fairbridge, 1948). Examination of an even better transect in West Africa, along the estuary of the Senegal River, similarly showed no difference in terrace elevation ("Project Rhéomarge": Faure *et al.*, 1980).

(c) *The short-term framework* (more or less contemporary processes in the time-range of 10^{-1} to 10^3 years). Within these more constrained time-frames, the precision of the observations rises dramatically. Whereas in a 10^7 year setting one is satisfied by elevation measurements to an accuracy of 1-10 m, in a 5000-10 000 year setting we aim for a 1-10 cm precision (Gray, this volume). In the contemporary framework the scope of observations is extended beyond the fields of stratigraphy and geomorphology, into areas covered by geodetic relevelling, satellite gravimetry, mareography (tide-gauge analyses), physical oceanography, long-term climate study and historico-archaeological documentation (Eronen, this volume). Clearly a vastly expanded field of interdisciplinary studies is thus opening up, the very surface of which has hardly been scratched at the present time.

THREE EXAMPLES

We will now consider some examples to illustrate various loading processes on various crustal types in varoius time-frames. Three "case histories" will be discussed: (a) medium- to short-term glacio-isostatic loading (Hudson Bay); (b) long- to short-term sediment loading (Mississippi Delta); and (c) long- to medium-term loading (an oceanic atoll). Each will also be considered in their relationships to the eustatic record. The short-term is treated in more detail elsewhere in this volume.

Medium-term glacio-isostasy and eustasy: Hudson Bay example

Perhaps the most striking example of the interaction between isostasy and eustasy is to be seen and measured today on the eastern Hudson Bay around the area that was the site of maximum ice loading some 15 000-10 000 years ago. This area, Great Whale River to Richmond Gulf, has been meticulously surveyed by teams from the Université de Québec (Montreal) and has been the site of one of the combined interdisciplinary field trips of the IGCP 61 and INQUA commissions (July, 1979). In other words, the data have been exhaustively checked out in the field. Interpretations

Figure 1. Photograph of a fine beach-ridge "staircase" on the west side of Richmond Gulf, Hudson Bay (Courtesy of Claude Hillaire-Marcel). Beach No. 55 (making shadow, lower right) is 1·87 m high and dated at 2420 B.P. (sidereal). It was associated with an unusually abrupt fall of sea level and accompanied by the extensive growth of coastal dunes (Filion, 1983). In classical history this cycle corresponds to the times of the Peloponnesian wars (460-404 B.C.), that were characterized by environmental stress such as drought, a similar phase in North Africa (pre-Ptolemaic aridity), and Neoglacial III in mountain regions. It correlates also with the "Greek Minimum" of sunspot proxies. (This figure may be compared with a vertical air photograph, Fig. 2, in Hillaire-Marcel, 1980, p. 217.)

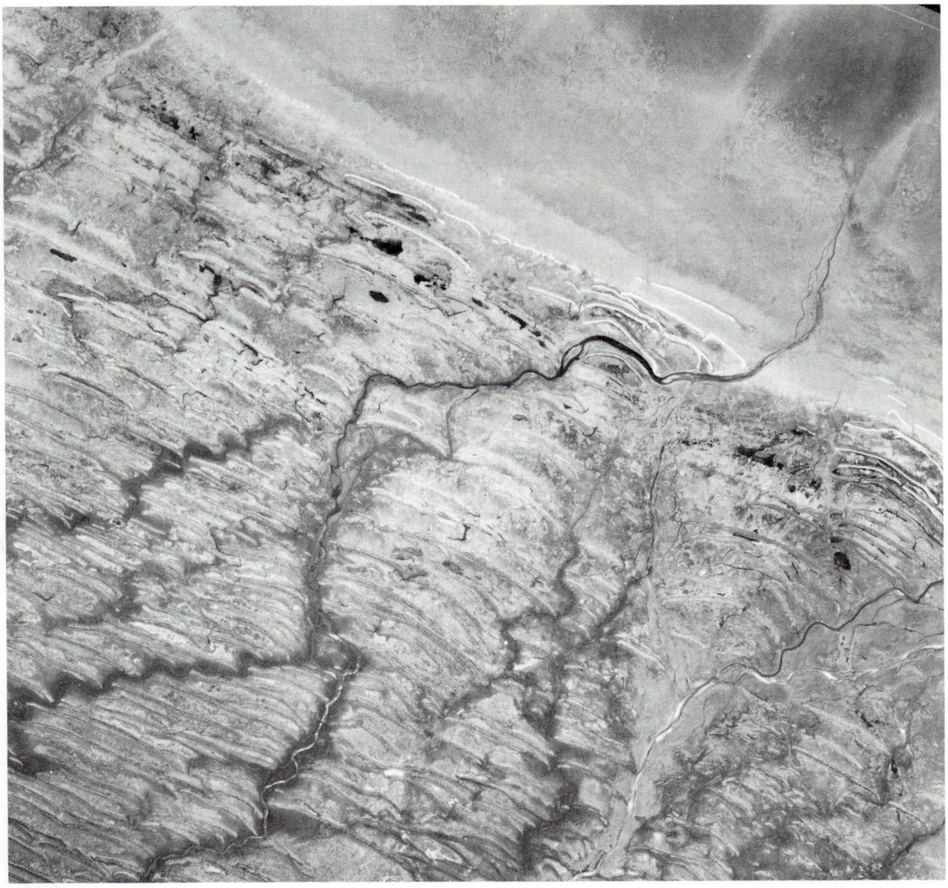

Figure 2. Vertical air photograph of the south-western coast of Hudson Bay (south-east of Fort Severn, Ontario), showing appearance of beach ridges in an area of reduced isostatic uplift (4-5 mm yr^{-1}). The beach ridge accumulation appears to reflect storminess pulsations corresponding to the same seiche effect triggered by a regular astronomic-climatic cycle as postulated for the beach-ridge "staircases" of Richmond Gulf. (Photograph: Canadian Government, Department of Energy, Mines and Resources, NAPL, A 24472-79-085.)

may still present some difficulties but the main conclusions are fairly clear (Fairbridge and Hillaire-Marcel, 1977).

The field evidence is in the form of dramatic "staircases" of raised beaches that grade upwards from present sea level to a maximum elevation of 315 m, which represents the world's highest documented post-glacial uplift (Figs 1 and 2). Above that level, the marine limit, the ground moraine is not wave-washed; below it, the erratics are largely washed clear and boulder fields winnowed. Comparable examples of glacio-isostatically uplifted beach ridges are known in many areas (e.g. Spitzbergen: Glaser, 1968; the Baltic: Hellberg, 1971; W. Scotland: Dawson, 1982) but none are so complete or well-dated.

The dating of the first post-glacial marine incursion into Hudson Bay is well established. It is believed that when the rising Flandrian transgression reached a certain critical level, the sea water reached in, under the grounded Laurentian (New Quebec) and Keewatin ice sheets, to float them and to convert the area into a "calving bay" (Peltier and Andrews, this volume). Apparently, within a few hundred years the entire Hudson Bay was clear of glacier ice and the transgressive "Tyrrell Sea" deposits were laid down over the whole area, including well inland of the present shores which at that time were still isostatically depressed. Numerous ^{14}C dates from the oldest Tyrrell Sea shells establish the initial transgression as around 8300 B.P. In this age range and with the "old carbon effect" from glacial meltwaters there are calibration problems (Sutherland, this volume), but fortunately the two principal "errors" (solar flux rate and isotopic contamination) tend to cancel each other out. Thus the 8300 year "age" may not deviate significantly from the true sidereal date (so-called "calendar years").

The "staircase" measurements permit a mean curve of water level against time to be constructed from the data. Additional ^{14}C dates confirm a straight-line time curve against a residual emergence curve. A small correction for eustatic changes then furnishes us with a reasonably accurate isostatic curve (Fig. 3). The tide-gauge

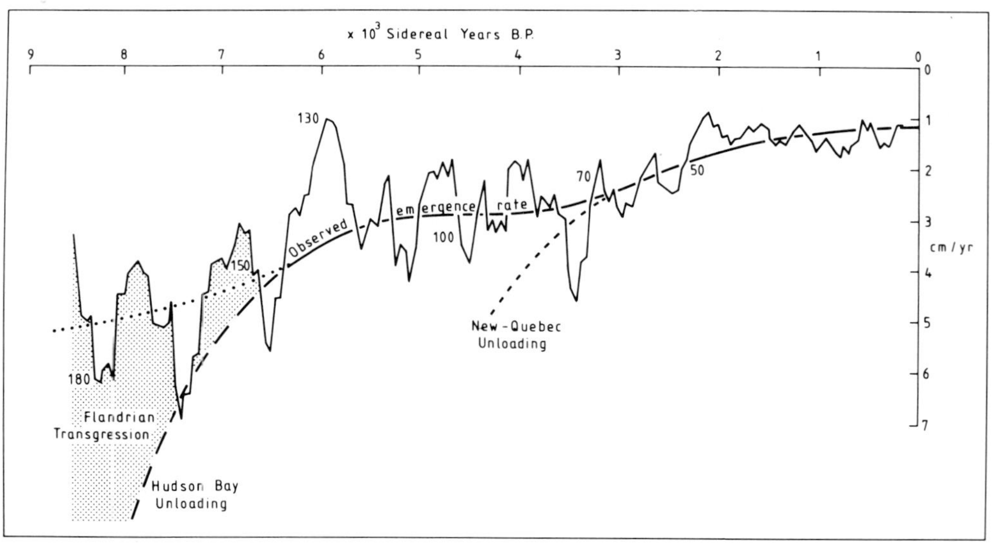

Figure 3. Graph showing variations in the coastal emergence rate at Richmond Gulf, Quebec, on the east side of Hudson Bay (about 56°N). The rate is calculated on the basis of a cycle unit (assumed to be a 45-year cycle), being the height between one ridge and the next. Assuming that an average mean sea level has been approximately constant since about 6000 B.P. (Fairbridge, 1961), then the fluctuations in rate should chronologically correspond (in mathematical sign) to eustatic and neoglacial evidence from other parts of the world (Fairbridge and Hillaire-Marcel, 1977). Prior to 6000 B.P. an adjustment is needed corresponding to the Flandrian trangression. An undulation in the emergence rate is believed to be due to the change in locus of isostatic rebound, from the Hudson Bay ice sheet (which melted earlier), to the New-Quebec (Labrador) ice, which followed with a lag of about 2000-3000 years. (Simplified from Hillaire-Marcel and Fairbridge, 1978.)

record for Churchill and other historically dated sea-level indicators show that this part of Hudson Bay is still rising isostatically at about 8 mm per year (after deduction of a global eustatic value: Fairbridge and Krebs, 1962; Gornitz *et al.*, 1982).

The Hudson Bay emergence curve may also be recalculated with the isostatic component = 0. In this way a water-level rate of change curve in cm yr^{-1} was calculated (Hillaire-Marcel and Fairbridge, 1978; Hillaire-Marcel, 1980).

The process of the detailed timing of this curve requires one key assumption. The astonishing uniformity of the staircase beach ridges suggests a regular storminess (wave-energy) cycle that is superimposed on the regular isostatic crustal uplift. It appears to be only slightly modified by eustatic factors. If we divide the age of the oldest beach (No. 185, i.e. 8300 years) by 185 it comes out very close to 45 years. Now, 45 years is an established climatic cycle in the longest time-series of North American summer temperatures (Landsberg, 1976). Hudson Bay is closed by ice for the other nine months of the year. A 45-year periodicity is also disclosed by the water levels on the Great Lakes (that reflect pressure anomalies), by tree-ring analyses near the timber line in northern Canada (Rajewicz, 1979; Hillaire-Marcel *et al.*, 1981), by geomagnetic moment, and by mean sea-level analyses (Mosetti, 1963).

If this assumption of a *c.* 45-year storminess cycle is taken as a working hypothesis, the next step is to compare it with the historical record of climate in Canada. Here we are extraordinarily lucky in having a very long proxy record based upon the careful records of fur-trading "factories" that have been preserved at the Hudson Bay Company's headquarters in London (Catchpole and Ball, 1981). The storminess record can be calibrated against mean sea level (m.s.l.) back to 1927 by means of the Churchill tide-gauge (on the western side of the bay). A last high storminess peak was reached in 1962 (Barnett, 1966, 1970), when Churchill m.s.l. was lowest. In a broad shallow basin like Hudson Bay, a strong seiche effect develops during the passage of westerly or southerly frontal systems. Mean sea level rises in the east and falls in the west. Naturally, these storms are only relevant in the three summer months—July, August, September—although in some warm cycles the open water may extend from June to October (Taylor and McCann, this volume). On our 1980 field trip, ice was only just beginning to open up near the shore on 1 July. By comparing the times of ice-opening in the Hudson Bay Company journals, and the duration of ships' passages across the bay (often beset by ice), one can derive a pattern of climatic control back for nearly 300 years. It seems that maximum beach building (i.e. storminess) corresponds to warmest summers (i.e. longest open-water, ice-free conditions).

It can be estimated that it would take at least 3–4 years to build a beach ridge of the type in question (on the average: 1 m high and 5 m wide). Thus, a unique storm would not suffice. Reviewing the climatic records for eastern Canada suggests that when the path of the Arctic summer high pressure cells, moving from north-west to south-east, lies across Quebec, there are gentle easterly or northerly winds over Hudson Bay, but the air is cold and the bay remains closed for the longest time. As the anticyclonic circulation shifts eastwards, it will be followed up by southerly and westerly air flow (but still cold and dry). This will be augmented near the shore in summer by an afternoon sea-breeze effect, helping to activate littoral sand dunes.

When the summer high pressure belt shifts northwards into Arctic Canada and Greenland, the bay is traversed by frequent frontal storms, bringing warm air and heavy rains. This is the ideal set-up for beach ridge building, and matches the times calculated for the ridges using a formula based on a calibration of twentieth century climate/sea-level data. From tide-gauge and climatic data, beach No. 1 is taken to be 1962, beach No. 2 would be 1917, and so on, back to No. 185 about 8300 B.P. Using A.D. 1950 as "B.P." datum, we allow for an imaginary beach No. 0 at A.D. 2007, which is 57 "A.P." (after present). Then the date for any beach number (n) can be calculated thus: $(n \times 45) - 57$ = date (years B.P.).

The subarctic frontal systems are further amplified when exceptional solar flares modulate the geomagnetic field near the north magnetic pole (Bucha, 1979). The solar flares, for their part, are known to fluctuate in response to the solar cycle and to longer-term planetary periodicities (see below).

Having established a working chronology, one can now examine two further parameters: (a) the anomalous episodes, when beaches were slightly higher or lower than average, and (b) long-term cyclicity. We will assume that the isostatic crustal response is a smooth one, and not a chatter or rapidly fluctuating one. To justify this assumption, we know from long-term human observations, notably in Scandinavia and Scotland, that the uplift was extremely steady. Indeed, the land upheaval is gradual enough to have triggered the "Desiccation Hypothesis" in seventeenth century Sweden (Charlesworth, 1957, p. 1344) which postulated that the ocean was slowly drying up (following the Flood of Noah). The steady uplift there has subsequently been abundantly demonstrated by multiple tide-gauge records and by repeated geodetic levelling. Conversely in the southern Baltic (Eronen, this volume) and adjacent parts of the USSR repeated levelling shows systematic or steady subsidence that in part, at least, represents the collapse of marginal bulge (Gudelis and Königsson, 1979; Kaplin *et al.*, 1982; Kliewe and Janke, 1982). In contrast to the steadiness of the general crustal adjustment, the initial phase, immediately following ice withdrawal, was marked by vigorous, but short-lived neotectonics (Mörner, 1978; Sissons and Cornish, 1982).

The coastal emergence, if it were taking place in an oceanographic-climatic environment of low energy, under essentially equilibrium conditions, should leave a uniform veneer of coastal sands and gravels. However, they are not only most distinctively concentrated in beach ridges (averaging around 45 years from peak to peak), but also arranged in secondary sequences of lower/wider ridges for a few hundred years, interrupted by episodes with higher and narrower ridges. Respectively, these correspond to epochs of slow emergence, alternating with times of rapid emergence. In the former, it seems that the eustatic component (Δ_E) is in the same mathematical sign as the isostatic component (Δ_I); in the second case Δ_E is appreciably less than Δ_I, and probably negative.

In a recent study of the Hudson Bay littoral and lake-shore dunes, with the aid of more than 100 ^{14}C dates and supported by palynology, Filion (1983) discovered that the dune building corresponded to cold, dry epochs, marked by fires near the tundra/forest border. Each of these epochs correponds to times when the eustatic sign was probably negative, and to global episodes of neoglaciation.

Study of the dating of the secondary trends in the Hudson Bay beach series discloses three significant periodicities in the beach levels that seem to be compatible with working hypothesis of planetary periodicity and similar in length to others that have been picked up in separate terrestrial time-series: *c.* 180, 556 and 1112 years. The 180-year series (see climate notes by Lamb, 1975) is remarkably close to a 177·92-year periodicity that is the precession period of the sun's looped orbital course around the barycentre of the solar system (Fairbridge, 1981b). The 1112-year periodicity is the "Quadrature Alignments" (QA-OP) recurrence of the outer planets (Stacey, 1963; Fairbridge, 1981b). The 556-year cycle is generally the most prominent, and, in addition to being the hemicycle of the QA-OP, it is the precession period of the lunar apsides. The first two cycles influence the energy level of solar radiation activity and the 556-year one, through the lunar cycle (progression of lunar perigee), influences the geomagnetic field; all three should presumably modulate the terrestrial climate but the quantitative values have never been investigated. The 45-year periodicity of the storm-beach levels is, of course, an approximation, and it could represent either the "Double Hale" (heliomagnetic) cycle as proposed by Fairbridge and Hillaire-Marcel (1977), or possibly the result of an energy pulse delivered gravitationally to the sun by the Saturn/Uranus Lap (of 45·387 years) and then transmitted as a radiation effect to the earth. The Hudson Bay emergence-rate curve (Fig. 3) discloses a distinct decrease in cycle length after the mid-Holocene, a variation that might be related to the fact that the geomagnetic field strength in the earlier stages was only one-fifth that of the late Holocene. This factor must certainly have led to an increased amplitude of solar-generated climate fluctuations in the mid-Holocene (Damon *et al.*, 1978).

Sunspot activity (and hence solar radiation) appear to be related to the planetary torque applied to the solar surface and to the sun's course around the systemic barycentre which has a mean periodicity of 19·86 years (the lap period of Jupiter and Saturn). That solar path is quasi-elliptical and precesses 360° three times during nine orbits in about 178 years, which with its harmonics, appears in many terrestrial cycles, notably the Greenland ice cores (see discussion in Fairbridge, 1981b). Furthermore there are 16 sunspot cycles of 11·12 years in the 178-year (177·92) period.

A secondary fluctuation is observed in the 19·86-year cycle, often at 4-cycle intervals (i.e. 79·44 years), which seems to control Gleissberg's "80-year" magnetic period. This appears to interact with the 178-year stress to generate an oscillatory dynamic motion within the sun which results in "grand minima in solar activity" (Landscheidt, 1981) that are measured by the inverse ^{14}C tree-ring method (Damon, 1977; Eddy, 1977). In the terrestrial proxies, these "grand minima" show up in eustatic records, as well as ice cores, tree rings and neoglaciation.

Using the same method of chronology there are numerous other beach-ridge sequences in the Arctic regions that could (eventually) be used to quantify isostatic uplift rates. In extra-glacial regions, sheaves of subparallel, almost uniformly spaced beach ridges appear to represent comparably (not necessarily synchronous) climate cycles (Hails, this volume). A well-dated series in Mexico (Curray *et al.*, 1969) may contain evidence of an 11-year sunspot period, as well as the 18·6-year nodal period, but systematic analysis is still needed.

In relatively stable, prograding coastal regions, more than 100 ridges can often be counted. Their approximately uniform elevations can be used as an indication of crustal stability, while the cyclicity of the minor variations in elevation seems to belong to the same planetary periodicities that have been postulated in the Arctic. Archaeological data and ^{14}C ages of midden shells provide added chronological time controls (Fairbridge, 1976). It should be appreciated, however, that the time-series represented by the sheaves of ridges are separated by some short gaps (of up to a few centuries) that represent times when mean sea level was some metres below the present level, following which the corresponding beaches were buried or destroyed by wave action.

Along the non-glacial coasts where sedimentary dynamics lead to prograding beaches, both isostatically induced subsidence and emergence develop contrasting effects on beach-ridge development and preservation. They are considered separately (below).

Sediment loading isostasy and eustasy: Mississippi delta example

Deltas represent the accumulation sites of major rivers where they debouch (as a rule) into the ocean. The geotectonic sites of deltas, however, are not by any means random. As pointed out by Burke (1977), the discovery of plate tectonics also disclosed that most large deltas coincided with the triple points of rifted plates that have drifted apart, leaving a "failed arm" of severely fractured crust or a wedge-like aulacogen leading into the continental interior. For physiographical reasons this re-entrant usually becomes an important fluvial confluence and deposition centre. Seismic and drilling work beneath the great deltas around the world discloses the presence of complex block-faulted basement structures. The crustal thickness below the sedimentary fill is, however, exceptionally thin as a rule, and usually interpreted as "stretched crust".

Accompanying the stretching/rifting phases, the crystalline basement also becomes hot and rises. Cooling will occupy some 20 million years or more, and during this phase, a uniform, but exponentially decreasing subsidence can be predicted (Rona, 1974; Sleep and Snell, 1976; Scrutten, 1982; Watts, 1982). However, this subsidence rate will be amplified by fluvial and deltaic sedimentary loading. Furthermore, the accumulation and progradation rates will not be uniform. They will be modulated by two cyclic trends that are not by any means always synchronous, although at times they are. The two modulating processes are (a) eustatic (which can be tectonic or climatic) and (b) climatic (which can be modulated by the eustatic component because oceanicity is increased by high eustatic stands and continentality developed by low stands).

The Mississippi River system began to operate following the opening of the Gulf

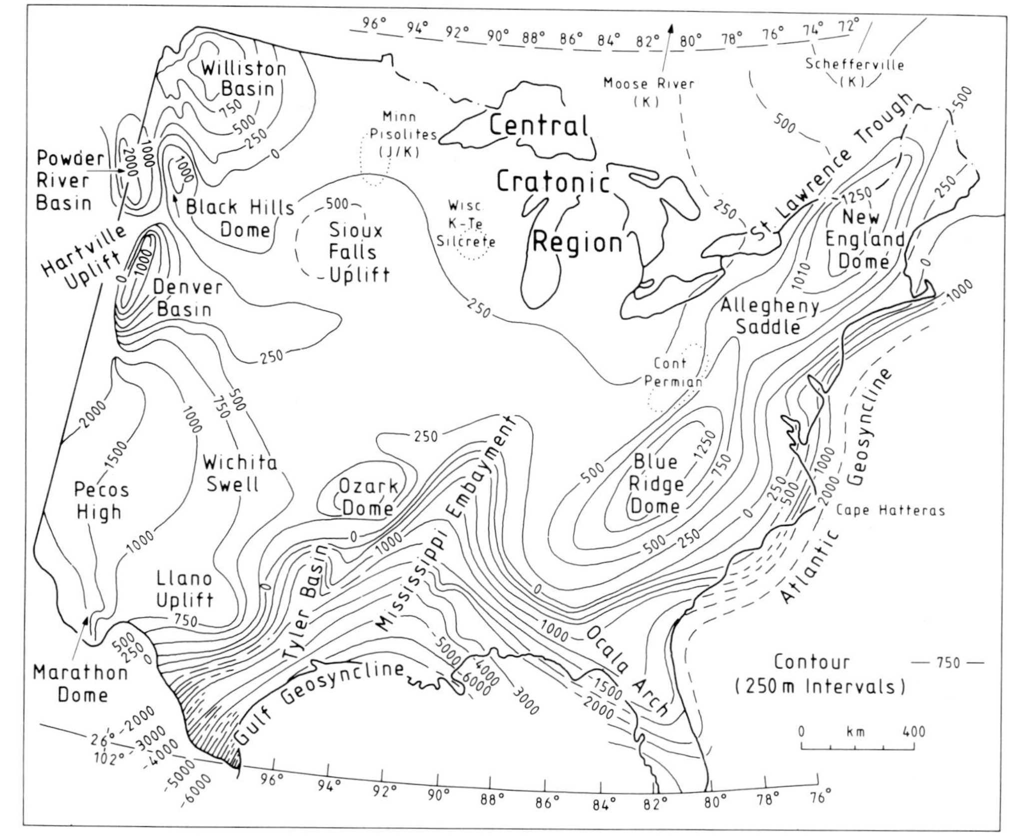

Figure 4. Structural contour map of the North American craton, exclusive of Canada and the Rocky Mountain and western belts, constructed on the basis of an assumed mid-Cretaceous peneplaned surface, dated approximately 100 m years. Traces of that surface are known in the Mid-West and on the Canadian Shield associated with crusts, paleosols and continental deposits with plant fossils. Towards the cratonic margins the surface is overlapped by marine Tertiary deposits. Parts that have been removed by erosion (e.g. in the Appalachians) have been reconstructed by geomorphological techniques. Seaward of the zero isobase the continental crust is downwarped. This isobase may be traced from the area of the continental shelf near Newfoundland, but shifts inland to the south and forms a deep re-entrant in the Mississippi Embayment, emphasizing the probable role of the latter as a triple point during the crustal opening of the Gulf of Mexico in early Jurassic times. The mid-Cretaceous datum surface is at 6000 m depth below the Mississippi delta and neotectonic subsidence continues today.

of Mexico in Mesozoic times. At first the latter was a partially closed basin with extensive evaporite deposition, but by the late Jurassic and the Cretaceous assumed a normal marine character. At that time, the hinterland (the North American craton) was largely a peneplaned land surface graded to sea level, which at the maximal stages stood about 250–300 m above m.s.l. today, as did all cratons in late Cretaceous time (Fairbridge and Finkl, 1980). In the late Cretaceous of North America, however, both the Laramide uplift in the present Rocky Mountains and the Appalachian upwarping in the east, began to furnish considerable volumes of clastic materials to the Mississippi drainage basin, and thence to the delta. Isopach maps of the Mississippi embayment and uplift projections of the Appalachians (prepared by the writer, see map Fig. 25 in Sykes, 1978; Fig. 4 here) illustrate the amount of vertical crustal displacement in approximately 1×10^8 years (since the mid-Cretaceous). One may note that while the central area of the craton has remained for the most part in the range of 250–300 m, with only minor undulations for this extended period (which has included the Quaternary glacial loading and unloading), the margins have been extensively deformed. This is nowhere more dramatic than in the Mississippi delta (floor 6000–12 000 m). Here the fracture and lineament trend that runs north-east towards the St. Lawrence River forms a broad area of subsidence that has been filling with sediment and prograding towards the Gulf of Mexico for over 150 million years. Today the area of principal sedimentation is close to the edge of the continental shelf, so that the sedimentary pattern is changing from a shallow fluvial or estuarine facies (flat-lying) to a neritic-bathyal (clinothem) style, marked by slumps and hiatuses. From the number (nine) of turbidite sequences emplaced on the adjacent floor of the Gulf of Mexico since the beginning of the Holocene (10 000 B.P.), one may estimate the major submarine slides as occurring about once every 1000–1200 years.

On the top of the delta, in a roughly central position, the floor of the Holocene is found commonly at a depth of about 200 m. At 10 000 B.P. the eustatic level was about −35 m, so that an anomaly of up to 165 m must be explained by tectonic subsidence and sedimentary compaction. In a mixed sandy, clayey and peaty sequence it is difficult to calculate the latter with precision, but the bulk of the compaction is accommodated in the top 10 m, so that in a 165-m Holocene sequence, something like 15 m may be explained that way, leaving 150 m as the isostatic-tectonic component, or 15·0 mm $year^{-1}$. This is somewhat more than that of some ancient geosynclinal basins (Kay, 1951), but may be taken as a reasonable figure for a rifted margin subsidence.

On the surface of the Holocene Mississippi delta there is a well-studied record of sedimentary deposition by multiple distributaries, creating terminal lobes which have been displaced from side to side of the deltaic fan since sea level reached its present-day stand about 6000 B.P. (Kolb and Van Lopik, 1966). These authors, and almost all other sedimentological authorities in the Gulf area (e.g. Coleman, 1977), agree that the six sedimentary lobes created by the Mississippi since the mid-Holocene have been produced by random fluctuations in the site of the principal mouth.

An alternative model for the Mississippi Holocene was presented by the writer (Fairbridge, 1968, p. 863), following a recognition that the six radiocarbon-dated lobes corresponded in time to the negative fluctuations of sea level recognized in the tropical latitudes. These are shown in Table 1.

The model for the mechanism envisaged is based on flume experiments. When the water level is lowered, the equilibrium stream debouching into standing water will maintain its course cutting down its bed to conform to the new (lowered) base level. Thus, during regressions the river course does not change, but the deposition

TABLE 1
Stages of renewed or accelerated sedimentary accumulation in the Mississippi Delta

Delta stage	^{14}C date (B.P.)	Eustatic fluctuation
Stage 1: *Sale-Cypremort* (SAC-14; HED-1; HBB-120; GBR-7)	5000–4700	Swedish (PR-5B) Emergence
Stage 2: *Cocodrie* (SAC-13; HBE-2; HBB-109; PR-6; GBR-8)	4300–4000	Bahama Emergence
Stage 3: *Teche* (SAC-9; HBD-5 and-6; HBB-75 and-65; PR -6/7; GBR-9 Neoglacial II)	3300–2800	Crane Key Emergence
Stage 4: *St. Bernard* (SAC-7; HBD-8 and-9; HBB-41; PR-9; GBR-10; Neoglacial V and VI)	2100–1700	Roman Emergence
Stage 5: *Lafourche* (SAC-5; HBD-10 and-11; HBB-31/29; PR-10; GBR-11)	1500–1200	Florida Emergence
Stage 6: *Plaquemine* (SAC-3; HBD-12; HBB-15/17; PR-11)	700–550	Paria Emergence
Stage 7: *Balize* (SAC-2; HBD-13; HBB-7; PR-12)	350–200	Little Ice Age Emergence

Notes: The delta stages and their dates are based mainly on Kolb and Van Lopik (1966), and more recent data, interpreted in terms of global eustasy (Fairbridge, 1961, 1968, p. 863; with updating). Regressions are accompanied by widespread growth of coastal peats, that are drowned by the ensuing transgression. The peats are well preserved in core samples and easy to date by radiocarbon. Equivalent events include the Solar Activity Cycle (SAC), negative fluctuations based on tree-ring ^{14}C (Eddy, 1977); the Hudson Bay dune-building cycles (HBD) of Filion (1983) that represent cold cycles and negative eustasy (HBB, 45-year storm-beach cycle: Fairbridge and Hillaire-Marcel, 1977); Swedish West Coast eustatic emergences (Mörner, 1976b; identified as "Post-Thermal Maximum Regression" PR-1, etc.); regression stages during the isostatic uplift in north-west England are also indicated (GBR-1 etc: data by Tooley, 1978). It may be noted that several of the well dated time-series are rather more refined than that the Mississippi Delta example. Nevertheless the same number of stages (and dates) with lobe or distributary shifts are indicated by reports about the Yangtze and Yellow River deltas in China, and the Ebro, Rhone and Nile deltas in the Mediterranean. One may note that these examples span a wide range of tidal and climate types, so that the distributary fluctuations should not be attributed to local weather conditions, wind changes, discharge variables, tidal characteristics or geoidal phenomena.

centre shifts seawards. During a subsequent rise of sea level, displacing the shoreline towards the inner part of the delta, the former course is drowned, and filled by longshore drift, being destined to play no further part in the history of the delta's sedimentary distribution pattern. When the next regression starts a new

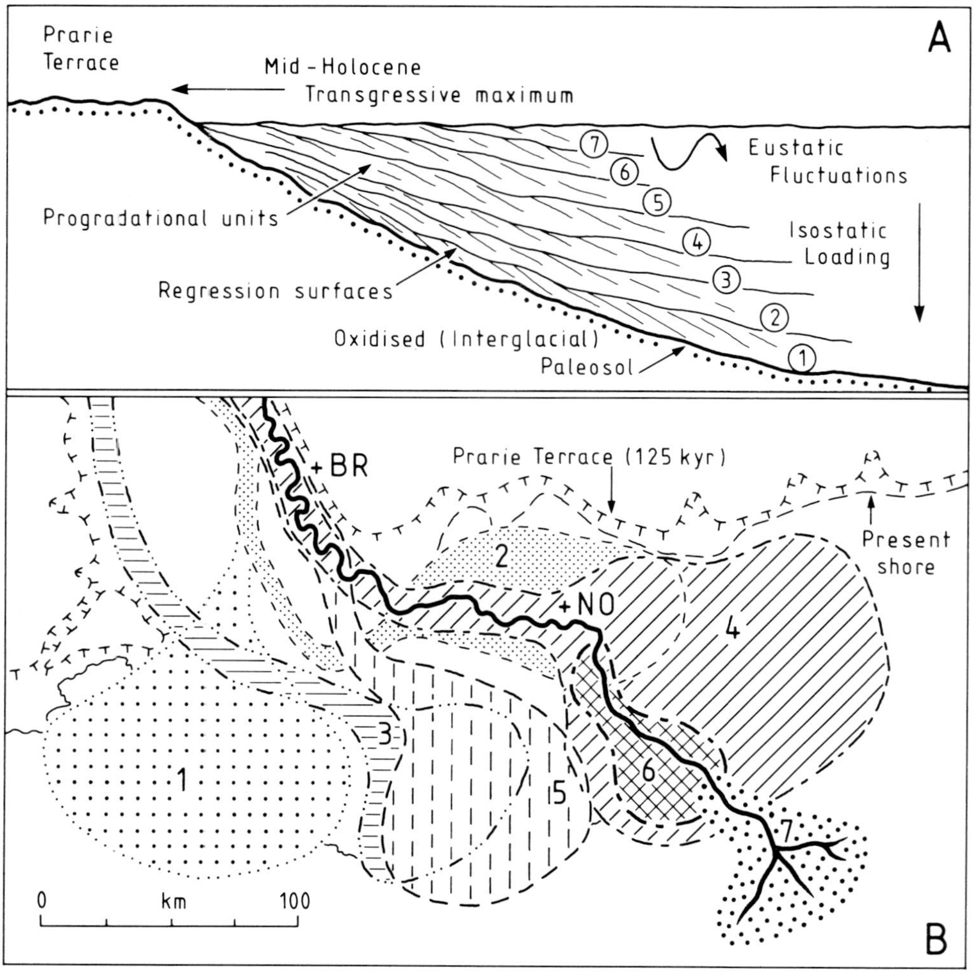

Figure 5. Interpretation of Mississippi delta. (A) Conceptualization of the Holocene sequence since 6000 B.P., the point in time when the Flandrian transgression reached present datum (m.s.l.) according to the writer (Fairbridge, 1961). A series of eustatic fluctuations (six regressive ones, and seven transgressions) has left a sequence of regressive surfaces, with peats, fossil soils and archaeological living sites (see dates by Kolb and Van Lopik, 1966), separated by progradational units that developed at high sea-level stands. Thanks to sediment loading isostasy and sediment compaction, each unit has subsided and has been capped by another. The sketch is idealized because no complete sequence can be drilled in any one place because of the migratory history of the distributaries. (B) Sketch map of the Mississippi delta lobes (adapted from Kolb and Van Lopik, 1966). *Legend:* BR, Baton Rouge; NO, New Orleans; 1, Sayle-Cypremort stage (5000-4700 B.P. ^{14}C); 2, Cocodrie stage (4300-4000 B.P.); 3, Teche stage (3300-2800 B.P.); 4, St. Bernard stage (2100-1700 B.P.); 5, Lafourche stage (1500-1200 B.P.); 6, Plaquemine stage (700-500 B.P.); 7, Balize stage (350-200 B.P.). With each negative eustatic displacement regression, the deposition site was shifted closer to the shelf break, and corresponding turbidite layers are found on the deep sea floor of the Gulf of Mexico within the same time-span.

distributary course is initiated, generally preferring an area of inter-distributary peat and swamp accumulations where the compaction and subsidence rate is highest (Fig. 5).

In this way, it becomes evident that the distributary/sedimentary-lobe history of the delta is modulated by eustatic fluctuation, superimposed on a regional subsidence pattern that is determined by isostatic and thermal crustal cooling trends, and to a small extent augmented by compaction, all three processes leading to continuous subsidence and encouraging long-term sedimentation. The subsidence rate could furthermore be accelerated slightly during the eustatic transgressions by hydroisostasy (Clark *et al.*, 1979), due to the loading by the additional water, which could be up to 6 m for very brief (century-long) episodes during some of the mid-Flandrian fluctuations (Fairbridge, 1961).

Short-term sedimentary units, however, are broken up over the upper delta into a sequence of brief secondary cycles. In each of these the succession frequently passes upwards from a subaerial exposure surface (locally with terrestrial indications, and during the Holocene marked by archaeological debris, pottery and midden shells), or swamp deposits, being covered by littoral sands, fine levée sands and eventually by marine silts and clays. During the 1000–1200 years of the average Mississippian Holocene cycle, some 5-25 m of sediment may accumulate, whereas the secondary cycles may only involve 35–50 cm.

On the surface of the Mississippi delta today in a recently dug pit, put down by archaeologists some 35 km WSW of New Orleans, approximately 2 m deep, five cycles (of exposed surface, levée sand and swamp ooze) were observed. The oldest was dated about 1800 B.P., thus indicating roughly 35 cm per cycle of 360 years each; the sediment was compacted very little.

With respect to the relatively stable (i.e. less active!) areas near the outer margins of peripheral bulges of the formerly glaciated regions, the longer-term Holocene eustatic fluctuations are now rather well documented (although still vigorously denied by the more conservative workers). The classic examples are found in New Zealand and the South Pacific (Schofield, 1960, 1977). In northern Europe, long time-series are demonstrated from north-west England (Tooley, 1978), western France (Ters, 1973; Morzadec-Kerfourn, 1974; Prigent *et al.*, 1983), the southern North Sea (Geyh, 1971, 1980) and southern Sweden (Digerfeldt, 1974; Mörner, 1976). Very similar results are now reported from the USSR (Kaplin *et al.*, 1982b). A preliminary series of maps containing a simplified global synthesis of sea-level data has been prepared by Colquhoun for the INQUA Shorelines Commission (see Appendix).

In moderately subsiding areas, such as the peripheral bulge belt of the eastern United States, the mid-Holocene littoral cycles are, of course, subsurface and difficult to recognize. Even meticulous studies have failed to identify them (e.g. Kraft *et al.*, 1979). However, in the southern peripheral area (South Carolina, Georgia and northern Florida) work in progress discloses convincing evidence of the predicted fluctuations (Colquhoun, 1981).

Volcanic loading isostasy and eustasy: oceanic atoll examples

As proposed one and a half centuries ago by Charles Darwin, the development of mid-oceanic coral atolls in the Pacific was to be explained by crustal subsidence. On his rapid reconnaissance voyage in H.M.S. *Beagle* he postulated a general subsidence of the Pacific basin, not having the time or opportunity to study large numbers of examples, which would have disclosed that for each subsiding reef complex there were other reefs which today are emerging. There are in fact many of intermediate examples (Chubb, 1957; Fairbridge, 1975).

Kuenen (1950) proposed that each atoll was subsiding into its own saucer-shaped depression, on account of volcanic loading isostasy. In this case the crust was oceanic, thin and underlain by an upper mantle lithosphere that would have a thickness and temperature proportion to its age and distance from the mid-oceanic ridge (Watts, 1976). Following this plate-tectonic reasoning, those atolls farthest from the ridge should rise from the greatest depths and should be the most isostatically compensated. During late Cretaceous-Cenozoic history major tectono-eustatic fluctuations such as indicated by Vail *et al.* (1977) have been recorded by terrestrial episodes recognized in atoll boring (Fig. 6) by tropical soils, travertine, karst features, terrestrial gastropods and other indicators such as displaced neritic facies (Labeyrie *et al.*, 1967; Ladd *et al.*, 1970; Matthews *et al.*, 1974; Thiede, 1980). It is therefore evident that these negative tectono-eustatic pulses were appreciably more rapid than the secular isostatic subsidence.

The distribution and variety of Pacific atolls and volcanic islands, however, denies any simple generalization. There are still many unknowns. Only a very small number have been drilled and few have been studied geophysically. Any analysis should consider the following points:

1. The Pacific atolls for the most part rise from volcanic seamounts that are situated in the western two-thirds of the main ocean, and not in marginal sea basins. Those atolls located within the marginal basins are largely related to subsidence of dismembered blocks (platforms) of continental crust (Kuenen, 1950; Gardner, 1970; Hayes, 1980, 1983).

2. The volcanism that created the atoll foundation seamounts is mostly limited to a discrete interval of geological time: late Mesozoic to early Cenozoic (Eocene). Thus most of the eastern Pacific, symmetrical to the East Pacific Rise, is almost free of seamounts.

3. A currently popular hypothesis of seamount distribution ascribes them to punctuated eruptions on an oceanic crust drifting across a series of mantle convective plumes or "hot spots". A shift of the North Pacific plate motion at $\sim$40 m years B.P. by 20° led to a 20° change in direction of the Emperor seamounts (Clague and Jarrad, 1973). In this case eruptions continued until after the Eocene, persisting south-east of Hawaii until the present day. No hypothesis appears to be offered that addresses itself to the general chronology.

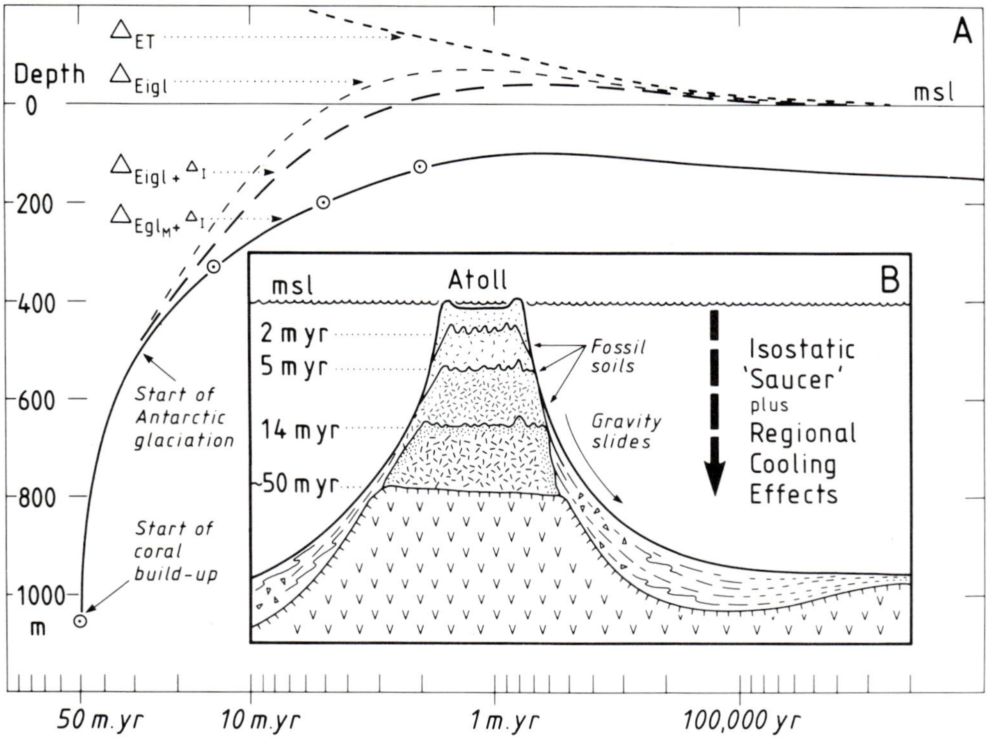

Figure 6. (A) Semi-log time curve, illustrating conceptualized subsidence of a typical mid-Pacific atoll. Data mainly from Eniwetok, Marshall Islands (lat. 11°30′N). Basalt volcanic foundation is Eocene in age (*c.* 50 m years) drilled into at 1275 m depth. Numerous subaerial exposure events are marked by leaching, unconformities, paleosol crusts, tropical forest pollen and terrestrial gastropods. These interruptions are interpreted as due to negative eustatic oscillations. The two most distinct marker horizons are the top of Tertiary "e" zone (N8/N9, mid-Langhian, 14 m years) at 330 m, and the top of "g" zone (terminal Messinan, 5 m years) at 200 m. The subsidence, i.e. change in depth (*D*) of the basalt floor with time is related to loading isostasy (Δ_I) minus the progressive Cenozoic tectono-eustatic drop (Δ_{ET} = ∿280 m), and after about 30 m year minus the oscillatory post-Miocene glacio-eustatic drop (Δ_{Egl} = ∿130 m).

The amplitude of Δ_I was ∿600 m during the first 20 m year. After 30 m year ago, approximately the beginning of general glaciation in Antarctica, Δ_{EglM} (maximal glacio-eustasy), plus Δ_I, becomes a combined factor, marked in the bores by solution unconformities, e.g. ∿14 and 5 m year. Δ_{Eigl} (interglacial eustatic trace), plus Δ_I is diverged progressively upwards. Additional factors, not shown here, postulated by Jardetzky and some writers (see reference in Fairbridge, 1961, 1975: also Marshall Islands; also 1976: *Quaternary Research* **6**, 529–566), include the geoidal rise (and fall) involved when a southern hemisphere plate moves across the equator and the geoid fluctuations influenced by core-mantle anomalies (Mörner, 1976a). (B) Profile sketch of idealized coral atoll-capped seamount (not to scale), that has isostatically subsided, and tilted a little.

4. Particularly in the Central and South Pacific the lines of atolls are disposed in belts alternating with lines of uplift, that are punctuated by mature volcanic islands, which show relatively little subsidence. These alternative "waves" of submergent and more stable crust demand some coherent theory (see, for example, Chubb, 1957; Schofield, 1977, 1980).

5. The Hawaiian chain has been utilized as a classic model for the hot spot/drift hypothesis. Thus the newest volcanism is manifest by submarine eruptions south-east of the "Big Island" of Hawaii and the isostatically subsided examples are reflected by the partly atoll-capped seamounts (Leeward Islands) of the north-west. Gravimetric studies suggest a secular trend of subsidence from south-east to north-west (Watts, 1976). But the distribution of uplifted coral reefs, especially on Lanai (Stearns, 1961), taken in conjunction with evidence of submerged shelves (Stearns, 1974), suggest a fluctuating history, or perhaps some phases of uplift punctuated by eustatic fluctuations.*

6. Northward displacement during the Neogene of the high carbonate-pelagic belt that is related to the Pacific equatorial convergence (Van Andel *et al.*, 1975), is confirmed by evidence from geometric analysis of the Pacific plate. Meridional drift would also involve a secular vertical adjustment of the equatorial bulge, which does not seem to have been considered. A question that needs to be posed is this: Is the rate of vertical isostatic adjustment of the crust perfectly in balance with the rate of geoid displacement consequent upon the meridional plate drift? If not, at any given point in time, atoll coral growth should mark sectors approaching the equator, whereas emergence and erosion should mark those departing from the equator.

CONCLUSIONS

1. Isostasy and eustasy are seen as two physical processes affecting respectively the earth's lithosphere and hydrosphere, involving vertical motions that can be monitored specifically by mean sea level, the shoreline at any given moment in time representing the trace of mutual contact between the two surfaces.

2. Gravimetric inequalities within the lithosphere, and possibly lower down (i.e. in the lower mantle or core), are known, thanks to satellite surveys, to distort the geoid, to create remarkable anomalies that in places exceed 100 m above or below the theoretical surface. The distribution of these anomalies does not relate to any known geotectonic feature or process of the lithosphere. It is not known if the sites of those anomalies migrate with respect to the crust, but if they do, it might explain some former shoreline changes (as proposed by Mörner, 1976a, 1980); no case has actually been demonstrated.

3. Of many different processes affecting mean sea level, only two are of major importance in the time-frame of one to two million years (Donovan and Jones, 1979); glacio-eustasy (involving ratio of sea water to glacier ice, i.e. the volume of liquid water in the ocean), and tectono-eustasy (involving the shape of ocean basins as controlled by plate tectonic activity and isostasy).

*Watts (personal communication) has explained that the spacing of the successive volcanic centres would cause renewed pulses of heating and uplift in preceding ones.

4. Of the many processes involved in isostasy, within the same time-frame as (3), the most important are thermal (rifted and collision plate margins; and intraplate igneous activity) and loading or unloading (glacio-eustatic; hydro-isostatic; volcanism, sedimentation and erosion).

5. An example of the interplay between glacio-eustatic unloading and eustasy is illustrated from Hudson Bay where 185 storm beaches can be used to time the crustal uplift to 315 m during 8300 years. A solar/planetary cycle is suspected in the storminess periodicity (45-year) and in longer-term trends (178-year, 556-year, 1112-year). Beach-ridge cycles of analogous but not necessarily identical periodicities are widely known in the Arctic, where isostatic uplift preserves them, whereas in stable or subsiding situations parts or all of the littoral record may be effaced or buried. In deeply indented coasts of the subtropics they are exceptionally preserved where there is an unusual sediment supply.

6. A largely buried beach and littoral complex is likely to be preserved in deltas, and the Mississippi delta contains a well-dated example. The major Holocene eustatic fluctuations recognized from the more stable regions of the world can be found reflected in the stratigraphy and distribution of the deltaic lobes. These lobes, long regarded by "the establishment" as products of energy fluctuations within a regime of dynamic equilibrium, are represented by the writer as eustatically determined, although associated climatic pulses must also modulate the fluvial discharge. Here isostatic subsidence is related to three variables: crustal cooling, sedimentary loading and hydro-isostasy (the last is minor, and its rate is variable).

APPENDIX: RESEARCH ORGANIZATIONS

Six international commissions have been contributing considerable time and effort to sea-level and coastal research in the Quaternary including the present time. They are affiliated to four international unions, as follows:

(a) The *International Union of Geodesy and Geophysics* (International Association of Geodesy): Commission on Recent Vertical Crustal Movements. Deals mainly with a geodetic relevelling, gravimetric analysis and other geophysical surveys. A quarterly bulletin is issued by its Data Centre in Prague (editor: P. Vyskočil).

(b) The *International Union of Geological Sciences* (with the collaboration of UNESCO): International Geological Correlation Project no. 61 (leader: A. L. Bloom), which terminated in 1982. Has issued an informal "Atlas" of sea level curves, covering the last 18 000 years or so. Organized (jointly with INQUA, see below) international field trips to selected study areas. Progress reports are printed in the IUGS journal *Episodes*. IGCP 61 will be succeeded by IGCP 200 in 1983.

(c) The *International Union for Quaternary Research* (an interdisciplinary union associated with ICSU and UNESCO). Three specialist commissions deal with the interlocking problems of recent chronology, relative sea-level change, and crustal motions. They are as follows: the INQUA Holocene Commission (president: L. K. Königsson, Uppsala), the INQUA Neotectonics Commission (president: N. A. Mörner, Stockholm) and the INQUA Shorelines Commission (president: D. R. Grant, Ottawa). Each of these groups also has regional subcommissions covering most of the world, so that in sum the experience of a very great number of regional experts is coordinated. The principal activity of these INQUA commissions is to organize regional field meetings and publish newsletters. A symposium volume on neotectonics has recently been issued (Fairbridge, 1981a), and a periodical bulletin of abstracts and news is edited by N. A. Mörner. A general INQUA Newsletter appears in the "house journal" *Striolae*, edited by Königsson (Uppsala). A shorelines map of the world is being prepared (and draft copies issued, 1978-1982) by the INQUA Shorelines Commission. It carries three sets of data for sites around the globe, selected preferably at about 100 km separation, covering (a) modern sea-level data (range and trend), (b) mid-Holocene (*c.* 3500 B.P.) sea level (or nearest available date), and (c) last interglacial high sea level (*c.* 120 000 B.P., i.e. Eem I). Since the Second World War the Shorelines Commission has published annotated bibliographies (Richards and Fairbridge, 1965, and subsequently by Richards).

(d) The *International Geographical Union*: Commission on the Coastal Environment. Deals strictly with contemporary effects and interactions with human needs and problems. Organizes regional field conferences and publishes a regular Newsletter (editor: H. J. Walker, Louisiana State University, Baton Rouge, Louisiana). Has published bibliographies over several decades, first in Copenhagen and the latest (1975-1978) in Japan edited by Koike.

A useful general newsletter, unconnected with any union, is *Coastal Research* (editor: W. F. Tanner, Geology Department, Florida State University, Tallahassee, Florida).

REFERENCES

Barnett, D. M. (1966). *Canadian Journal of Earth Sciences* **3**, 77-88.

Barnett, D. M. (1970). *Canadian Journal of Earth Sciences* **7**, 626-627.

Bucha, V. (1979). *Studia geophys. geod.* **23**, 55-67; 102a-102d.

Burke, K. (1977). *Ann. Rev. Earth Planet. Sci.* **v**, 5, 371-396.

Catchpole, A. J. W. and Ball, T. F. (1981). *In* "Syllogeus 33: Climatic Changes in Canada", pt. 2, pp. 48-96. National Museum of Natural Science, Ottawa.

Charlesworth, J. K. (1957). "The Quaternary Era, with Special Reference to its Glaciation". Edward Arnold, London.

Chen, W. P. and Molnar, P. (in press). *J. geophys. Res.*

Chubb, L. J. (1957). *Geol. Mag.* **94**, 221-228. (Discussion: *ibid.* **98**, 170-171; **99**, 279-283).

Clague, D. A. and Jarrad, R. D. (1973). *Bull. geol. Soc. Am.* **84**, 1135-1154.

Clark, J. A., Farrell, W. E. and Peltier, W. R. (1978). *Quaternary Research.* **9**, 265-287.

Colquhoun, D. J. *et al.* (1981). *Striae* **14**, 144-150.

Curray, J. R., Emmel, F. J. and Crampton, P. J. S. (1969). *In* "Coastal Lagoons: A Symposium" (A. A. Castanares and F. B. Phleger, eds), pp. 63-100. Universidad Autónoma, Mexico City.

Daly, R. A. (1934). "The Changing World of the Ice Age". Yale University Press, New Haven.

Damon, P. E. (1977). *In* "The Solar Output and its Variation" (O. R. White, ed.), pp. 429-448. Boulder.

Damon, P. E., Lerman, J. C. and Long, A. (1978). *Ann. Rev. Earth Planet. Sci.* **6**, 457-494.

Dawson, A. G. (1982). *Scott. J. Geol.* **18** (4), 253-265.

Donovan, D. T. and Jones, E. J. W. (1979). *Q. Jl. geol. Soc. Lond.* **136**, 187-192.

Eddy, J. A. (1977). *In* "The Solar Output and its Variation" (O. R. White, ed.), pp. 51-71. Boulder.

Fairbridge, R. W. (1948). *J. Proc. R. Soc. West. Aust.* **33** (for 1946-47), 1-43.

Fairbridge, R. W. (1961). *In* "Physics and Chemistry of the Earth, 4" (L. H. Ahrens, F. Press, Kalervo Rankama and S. K. Runcorn, eds), pp. 99-185. Pergamon Press, London.

Fairbridge, R. W. (ed.) (1968). "The Encyclopedia of Geomorphology". Reinhold (Hutchinson-Ross reprint), New York.

Fairbridge, R. W. (1975). *In* "Encyclopedia of World Regional Geology, part 1" (R. W. Fairbridge, ed.), pp. 410-415. Dowden, Hutchinson and Ross, Stroudsburg, Pennsylvania.

Fairbridge, R. W. (1976). *Science* **191**, 353-359.

Fairbridge, R. W. (1979). *Geologie Mijnb.* **58** (2), 273-276.

Fairbridge, R. W. (1981a). *Z. Geomorph.* NF, Suppl. Bd. **40**, vii-xii.

Fairbridge, R. W. (1981b). *Striae* **14**, 131-141.

Fairbridge, R. W. and Finkl, C. W. Jr (1978). *Z. Geomorph.* N.F. **22**(4), 369-389.

Fairbridge, R. W. and Finkl, C. W. Jr (1980). *J. Geol.* **88**, 69-86.

Fairbridge, R. W. and Hillaire-Marcel, C. (1977). *Nature* **268**, 413-416.

Fairbridge, R. W. and Krebs, O. A. (1962). *Geophys. J. R. astr. Soc.* **6** (4), 532-545.

Faure, H., Fontes, J. C., Hebrard, L., Monreillet, J. and Pirazzoli, P. A. (1980). *Science* **210**, 431-433.

Filion, L. (1983). Thèse Université Laval (Dép. Géographie), Québec.

Gardner, J. F. (1970). *Bull. geol. Soc. Am.* **81**, 2599-2614.

Geyh, M. A. (1971). *Geol. For. Stockh. Forh.* **93**(4), 679-690.

Geyh, M. A. (1980). *Radiocarbon* **22**(3), 695-704.

Glaser, U. (1968). *Würzb. geogr. Arb.* **22**, 1-22.

Gornitz, V., Lebedeff, S. and Hansen, J. (1982). *Science* **215**, 1611-1614.

Gudelis, V. and Königsson, L. K. (eds) (1979). "The Quaternary History of the Baltic". Uppsala.

Hayes, D. E. (ed.) (1980-83). "The Tectonic and Geologic Evolution of Southeast Asian Seas and Islands" (2 parts). American Geophysical Union, Washington.

Hellberg, K. (1971). *Lunds Univ. Naturgeogr. Inst. Rapp.* **9**, 1-124.

Hillaire-Marcel, C. (1980). *In* "Earth Rheology, Isostasy and Eustasy" (N. A. Mörner, ed.), pp. 215-230. John Wiley, Chichester.

Hillaire-Marcel, C. and Fairbridge, R. W. (1978). *Geology* **6**, 117-122.

Hillaire-Marcel, C. *et al.* (1981). *In* "Syllogeus 33: Climatic Changes in Canada", pt. 2, pp. 28-47. National Museum of Natural Science, Ottawa.

Kaplin, P. A., Klige, R. K. and Tchepaliga, A. L. (1982). "Sea Level Fluctuations". Moscow University Publishing House, Moscow.

Kay, M. (1951). *Geol. Soc. Amer., mem.* **48**.

Kliewe, H. and Janke, W. (1982). *Petermanns Geogr. Mitt.* 1982 (2), 65-74.

Kolb, C. R. and Van Lopik, J. R. (1966). *In* "Deltas" (M. L. Shirley and J. A. Ragsdale, eds), pp. 17-61. Houston Geological Society, Houston.

Kraft, J. C., Allen, E. A., Belknap, D. F., John, C. J. and Maurmeyer, E. M. (1979). *In* "Barrier Islands, from the Gulf of St. Lawrence to the Gulf of Mexico" (S. P. Leatherman, ed.), pp. 149-183. Academic Press, London and New York.

Kuenen, P. H. (1950). "Marine Geology". John Wiley, New York.

Labeyrie, J., Lalou, C. and Delibrias, G. (1969). *Cahiers du Pacifique* **13**, 59-68.

Ladd, H. S., Tracy, J. I. and Cross, M. G. (1970). *U.S. Geol. Surv. Prof. Pap.* **680-A**.

Lamb, H. H. (1972–75). "Climate: Present, Past and Future". Methuen, London.

Landsberg, H. E. (1976). *Jour. Interdiscip. Cycle Res.* **7**, 237-243.

Landscheidt, T. (1981). *Jour. Interdiscipl. Cycle Res.* **12**, 3-19.

Matthews, J. L., Heezen, B. C., Catalano, R., Coogan, A., Tharp, M., Natland, J. and Rawson, M. (1974). *Science* **184**, 462–464.

Mörner, N.-A. (1976a). *J. Geol.* **84**, 123–151.

Mörner, N.-A. (ed.) (1976b). *Boreas* **5**, 193–275.

Mörner, N.-A. (1978). *Geology* **6**, 41–45.

Mörner, N.-A. (ed.) (1980). "Earth Rheology, Isostasy and Eustasy". John Wiley, Chichester.

Morzadec-Kerfourn, M. T. (1974). *Mém. Soc. géol. miner. Bretagne* **17**, 208 pp.

Mosetti, F. (1963). *Boll. geofis. teor. appl.* **5**, 127-138.

Pitman, W. C. III (1978). *Bull. geol. Soc. Am.* **89**, 1389-1403.

Prigent, D., Visset, L., Morzadec-Kerfourn, M. T. and Lautridou, J. P. (1983). *In* "Quaternary Coastlines and Marine Archaeology" (N. C. Flemming, ed.), pp. 303-324. Academic Press, London and New York.

Rajewicz, R. (1979). "Analyse des series chronologique de l'est du Canada appliquée a l'etude des cycles climatiques". Thèse. Université de Québec, Montreal.

Richards, H. G. and Fairbridge, R. W. (1965). "Annotated Bibliography of Quaternary Shorelines (1945-1964)". Special Publication 6, Philadelphia Academy of Natural Sciences.

Rona, P. A. (1974). *Tectonophysics* **22**, 283-299.

Schofield, J. C. (1960). *N.Z. Jl. Geol. Geophys.* **3**, 467-485.

Schofield, J. C. (1977). *N.Z. Jl. Geol. Geophys.* **20**, 503-530.

Schofield, J. C. (1980). *In* "Earth Rheology, Isostasy and Eustasy" (N. A. Mörner, ed.), pp. 517-521. John Wiley, Chichester.

Scrutten, R. A. (ed.) (1982). "Dynamics of Passive Margins". American Geophysical Union, Washington.

Sissons, J. B. and Cornish, R. (1982). *Quaternary Research* **18**, 268-288.

Sleep, N. H. and Sloss, L. L. (1980). *In* "Dynamics of Plate Interiors" (A. W. Bally *et al.*, eds), pp. 93-98. Washington.

Sleep, N. H. and Snell, N. S. (1976). *Geophys. J.R. astro. Soc.* **45**, 125-154.

Stacey, C. M. (1963). *Ann. N.Y. Acad. Sci.* **105** (7), 421-460.

Stearns, H. T. (1961). *Z. Geomorph.* N.F., Suppl. 3, 1-16.

Stearns, H. T. (1974). *Bull. geol. Soc. Am.* **85**, 795-804.

Sykes, L. R. (1978). *Reviews of Geophysics and Space Physics* **16**, 621-688.

Ters, M. (1973). *Bull. Assoc. Francaise du Quaternaire* **36** (INQUA Supplement), 114-135.

Thiede, J. (1980). *In* "Géologie des Marges Continentales" (R. Blanchet and L. Montadert, eds), pp. 65-70. Oceanologica Acta, Paris.

Tooley, M. J. (1978). "Sea-level Changes: North-west England during the Flandrian Stage". Clarendon Press, Oxford.

Vail, P. R., Mitchum, R. M. and Thompson, S. (1977). *Amer. Assoc. Petrol. Geol., Mem.* **26**, 83-97.

Van Andel, T. H., Heath, G. R. and Moore, T. C. Jr (1975). *Geol. Soc. Amer. mem.* **143**, 134 pp.

Walcott, R. I. (1972). *Reviews of Geophysics and Space Physics* **10**, 849-894.

Walcott, R. I. (1975). *Eos* **56** (2), 62-72.

Watts, A. B. (1976). *J. Geophys. Res.* **81**, 1533-1553.

Watts, A. B. (1982). *Nature* **297**, 469-474.

PART TWO

Aspects of shoreline development

TWO

Coastal processes, relict shorelines and changes in sea level on selected mid- and low-latitude coasts

J. R. HAILS

INTRODUCTION

By the 1960s, a general history of relative changes in sea level and shoreline migration had been documented for the Atlantic and Gulf Shelves of the United States (see, for example, Fairbridge, 1961; Curray, 1965, Milliman and Emery, 1968). Much of the evidence had been based upon faunal assemblages, morphological features, lithologies and ^{14}C dating. By comparison, at that time, little was known about depositional shorelines and Holocene sea levels in Australia because studies were being undertaken in widely different geographical localities of the continent (Davies, 1959; Bird, 1961; Gill, 1961; Jennings, 1961; Hails, 1965; Thom, 1965). Nevertheless, and despite limited geophysical and geomorphological research, attempts were made to correlate changes of sea level in Australia with so-called "established" sequences in Europe and North America without careful attention to the accuracy of dating techniques and reliability of field methods and evidence (Hossfeld, 1950; Gill, 1956; Ward, 1971). On the other hand, Cook and Polach (1973) compared ages of beach ridges at Broad Sound, on the Queensland coast, with ages of Holocene beach ridges from other parts of the world and could find no direct correlation. Only since the early 1970s have studies been made of shallow subsurface marine deposits in order to correlate lithostratigraphic units by seismic reflection profiling and coring, and to obtain datable material from the base of Holocene sediments.

A sedimentary environment and its associated sub-environments can be either erosional or depositional. This chapter attempts to focus on the hydrodynamic conditions under which modern and relict shorelines have developed and prograded in selected areas of Australia and North America. These areas are known to the writer through his own research and that undertaken with co-workers. Coastal depositional environments, and their characteristic processes, produce sedimentary sequences which provide geologists and geomorphologists with clues to, and

SHORELINES AND ISOSTASY
ISBN 0 12 652960 4

evidence of, past oscillations in sea level. However, primary sedimentary structures, both inorganic and organic, in relict (unconsolidated) shoreline deposits are generally destroyed by weathering and leaching processes so that it is difficult to interpret precisely the hydrodynamical conditions under which deposition occurred. Furthermore, many structures may be related to more than one environment. Therefore, it is necessary to identify a particular spectrum of sedimentary structures in order to interpret ancient or relict environments with some degree of confidence. In this context, certain lithological associations are typical of specific environments so that the study of modern environments can provide information on which to construct process-response models. Thus, studies of relict shorelines in widely different geographical areas can assist the construction of a general model for that type of environment, but only as a working tool.

Erosion, progradation and the reworking of sediments are important shore and nearshore processes that modify coastal accumulations before their incorporation in the geological record. Therefore, it is also considered that Hutton's doctrine of uniformitarianism, *the present is the key to the past*, is particularly relevant to the study of depositional processes and relict shorelines.

The principal external and local factors controlling or regulating relative movements in sea level are examined briefly. These include broad-scale hydro-isostatic deformation resulting from the Holocene transgression, upwelling of water along continental margins and glacio-eustasy. Local factors pertain to meteorological changes affecting the magnitude and frequency of storms, changes in circulation and tidal levels within partly enclosed bodies of water which have been induced by sedimentation and regional flexuring and/or faulting.

Following the classification of Davies (1964), coasts have been described as microtidal (< 2 m), mesotidal (2–4 m), and macrotidal (> 4 m), according to their spring tidal range. Large tidal ranges are generally associated with wide continental shelves and accretion rather than erosion in mid- and low-latitude regions, although these processes are controlled as much by sediment supply, wave energy dissipation, and sea-level change, as by energy variation introduced by tides.

The main effect of large fluctuations in water level, though, is to extend wave action over a greater width of the beach or shoreline profile, and to dissipate energy over an enlarged surf zone so that waves break farther from the shoreline. Because tidal levels are known to be significantly affected by long-term mean sea-level variations, as well as short-term meteorological disturbances, it is to be noted that a wide range of tidal inundation, both spring and storm, can cause some ambiguity in interpreting palaeo sea levels from intertidal indicators.

Sea level is the height that the sea surface attains if it is undisturbed by tides, waves and winds. Because these disturbances are continuous, mean sea level has been adopted as the datum plane from which the height of tides and depth of water are measured. Although fluctuations in sea level at the coast are of a diurnal, seasonal and annual nature, it is the response of sea level to long-term climatic changes, particularly during the Pleistocene epoch, that is of interest to geomorphologists. Notwithstanding this comment, short-term changes that cause coastal erosion and modify depositional coasts are significant in the study of modern coastal processes, although they may never be documented in the geological record. Recent studies by Radok and Raupach (1977), for example, have shown that weather systems raise the mean sea level against the south coast of Australia through barometric pressure and wind stress operating over large areas of the Southern Ocean. In turn, tides and waves are superimposed on these mean sea levels.

BARRIERS AND CHENIERS: MODES OF ORIGIN AND SEQUENCE OF DEVELOPMENT

Relict barrier shorelines of Pleistocene age have been widely reported landward of their modern analogues, while submerged Holocene barrier remnants are fairly common features on modern continental shelves. Barrier islands, characteristic depositional features of the Gulf and Atlantic coasts of the USA, and the Queensland sector of the east Australian coast, for example, border gently shelving coastal plains adjacent to broad continental shelves. Generally, barriers are elongate bodies of sand, also containing large quantities of gravel or shell, which are bordered on the landward side by fresh water to hypersaline lagoons, or salt marshes.

The origin of barriers has captured the attention of coastal researchers for several years and the reader's attention is particularly drawn to key papers on barrier islands which have been compiled by Schwartz (1973). Suffice it to say, here, that the formation of barriers is related to several interactive processes as described in the climatic-geomorphic model described by Carr and Hails (1972). The model demonstrates the interaction between sediment supply and transport, wave climate and minor oscillations in mean sea level—an interaction actually dependent upon the paramount control effected by climatic change which, over geological time influences rates of weathering, river discharge and ultimately sediment yield to the nearshore and offshore zones.

Many mid- and low-latitude barriers in Australia and the USA are being eroded as evidenced by active transgressive dunes encroaching upon successively older beach ridges. In many instances in Australia the modern dunes trend parallel to their Pleistocene counterparts which, however, are fixed. Nevertheless, it can be shown that the climatic conditions prevailing along the coastline have not changed appreciably since the late Pleistocene. The current episode of erosion has been attributed to several processes, notably a diminishing supply of sediment being delivered to the shoreline, a renewed rise of sea level and, more locally, increased periods of storminess (Davies, 1957; Bird, 1961, 1965; Hails, 1964, 1968).

Unlike some of the recurved barriers on the south and east coasts of England, where the longshore movement of material is evident, and erosion at one end of a barrier is usually balanced by deposition at the other, the barriers on the mid- and northern New South Wales coast of Australia have been eroded practically along their entire length. Hails (1968) has demonstrated that a pattern of arcuate bays possibly promotes localized longshore currents, but the influence of these currents on barrier development is insignificant compared with the landward sweeping of sediment from the sea-floor. The arcuate pattern of the barrier supports the contention that, during progradation, barrier oceanic shorelines have been shaped by refracted south-easterly and north-easterly swell.

As Hails and Hoyt (1968) have stated, majority opinion in the 1940s and early 1950s supported the formation of depositional features on coastlines of emergence. However, agreement on the origin and progradation of barriers in various parts of the world was complicated at the time by widely differing and conflicting conclusions on Quaternary changes in sea level because of relatively few absolute dates. A shoreline can occupy an infinite number of positions during an oscillating marine transgression, but it is during the major stillstand, at maximum submergence for a particular fluctuation, that the shoreline is actually shaped and prograded. At this stage, lagoonal or estuarine-salt marsh environments develop behind barriers, as well as during phases of slow submergence. Hoyt (1967) and Hails and Hoyt (1968, 1971) have noted the absence of open marine deposits landward of many mid-latitude barriers in Australia and the USA as evidence against barrier evolution by emergence (a negative eustatic change of sea level) and from offshore bars. In most of Australia, as indicated in the Coorong case history which follows, the landward margins of lagoons behind Holocene barriers are invariably the modified shorelines of Pleistocene barriers. However, it appears that only in those areas, such as the Gippsland coast of Victoria (Bird, 1965), where Holocene barriers developed initially as longshore spits, were the older Pleistocene barriers exposed to the open sea during the Holocene transgression. On the New South Wales coast, Holocene barriers probably developed before or at the phase of maximum marine submergence slightly seaward of the modern shoreline, and then migrated to their present positions as the transgression came to a stillstand.

Barrier development from spits is not recognized to be the principal mechanism along depositional shorelines. Barriers originate from a topographical ridge along the landward side of a beach which is subsequently partly submerged. Continued submergence accompanied by transgression, though, may produce complex interfingering of barrier and lagoonal marsh sediments as the former are reworked; synchronous deposition may also occur.

In studying modern and palaeo-depositional environments, and past sea levels, it is important to differentiate, where possible, between *cheniers* and barrier systems (beaches, islands and spits) in respect of their origin and sequence of development. *Chenier plains* were originally described by Howe *et al.* (1935), who referred to the long, narrow, sandy ridges in south-western Louisiana as rising above the surrounding marshes and forming the most conspicuous topographic features of the region. Two decades later, Price (1955) described cheniers as "shallow-based, perched, sandy ridges resting on clay". However, subsequent coring studies have shown that typical cheniers have not necessarily been constructed on swamp mud, as assumed by Russell and Howe (1935) and Tanner (1961), but on a base of marine sand and mud, or silty clay (see also Byrne *et al.*, 1959; Coleman, 1966). Hoyt (1969) has described three stages in the development of a chenier: progradation of a mudflat; erosion of the seaward edge of the mudflat, reworking of the mudflat deposits and the formation of a ridge; and further progradation of the mudflat, leaving the ridge as a chenier. In brief, the chenier overlies mudflat sediments which border it on both the landward and seaward sides.

In Louisiana typical cheniers are 45-460 m wide, and several tens of kilometres long. However, the sand of a chenier ridge rarely extends more than 2 m below the associated sea level, or exceeds 6 m in thickness (Byrne *et al.*, 1959; Gould and McFarlan, 1959). Composite ridges of a chenier plain are generally no more than 1·5 km wide. In contrast, barriers commonly extend several tens of metres below sea level and, including dunes, can on average be 30-45 m thick; they may also be several kilometres wide. Low to moderate wave energy, land-sea stability, and large sediment loads from rivers in areas with broad continental shelves are necessary for the development of cheniers, although periodic variation in sediment supply is also important. Chenier plains prograde downdrift from the sediment source. If wave energy is high, fine-grained sediments are transported offshore instead of being added to the shoreline deposits.

Along chenier coasts the dimensions of individual ridges are not altered appreciably by movements of the shoreline unless, of course, there is pronounced erosion. Cheniers are characterized by gently dipping ($< 10°$) littoral, sublittoral and washover deposits, and some overlying dunes. The dip can be directed toward the ocean if formed on the foreshore or toward the land if formed by washover. Grain size increases from the base to the top of a ridge. Shell layers, several centimetres thick, are common throughout a chenier although leaching may reduce their occurrence in the upper metre or two, depending on climate and degree of exposure. Although transgressions and regressions may redistribute surficial sediments they do not produce sand sheets, as in the case of reworked barriers. Therefore, fossil cheniers should have dimensions similar to those of their modern analogues. According to Hoyt (1969), it is doubtful that longer stillstands in sea level, which were common probably in pre-Quaternary time, would increase markedly the width and thickness of individual cheniers, although the width of a chenier plain may be greater.

Cheniers and barriers, then, can be identified by the shape and extent of their respective deposits. A fundamental difference in the depositional processes is the proximity of a sizeable sediment source which can overload the distributive agencies (waves and tidal currents) along the adjacent shoreline. Thus, near such a source, progradation may alternate with erosion, and hence cheniers may develop.

HOLOCENE CHENIERS: NORTHERN AUSTRALIA

Holocene chenier plains, cheniers and tidal flats have been studied in discrete areas of tropical Queensland and Western Australia, as well as the Northern Territory, in order to assemble a number of case histories, even if these are still small in number. Because cheniers can form as storm beaches during cyclonic storms, it is important to relate the height of the underlying marine and littoral clay base on which they have been deposited, to relative sea level. This has been done at Point Stuart, 85 km east of Darwin, NT, by Clark *et al.* (1979) who dated shell samples from shelly chenier ridges, up to 60 m wide, and surveyed the height of the ridges relative to the base of the present storm beach. Over a distance of about 1·5 km there is fall of 1 m in base level, and the marine and littoral clay base of all the cheniers is between 0·3 and 1·7 m below the ridge crests. As the approximate age of the oldest ridge is within the range 6100–7100 B.P. it is inferred that sea level has not fluctuated more than 1 m since the chenier plain started to prograde seaward. If it is assumed that the ^{14}C dates, reported by Clark and associates, are the true age of shell fragments pertaining to intertidal and shallow subtidal species, the shoreline at Point Stuart has been advancing approximately at a rate between 19 and 31 cm yr^{-1}.

It is pertinent to mention here that the significance of shell dates in establishing the height of sea level along depositional shorelines must be considered carefully because marine shells can be deposited above normal high-tide level by wind, and during severe storms. The subsequent erosion, reworking and mixing of shells of different ages make interpretation of dated shell deposits uncertain and the dates may contribute little knowledge to the time of accumulation of the reworked deposits. Consequently, the interpretation of stillstands of sea level from ^{14}C dates requires both caution and a thorough knowledge of the sample environment and its sedimentological history. Most dating methods, anyway, are only applicable to specific time intervals and to particular mineralogical components of sediments (see Cullingford *et al.*, 1980; Sutherland, this volume). Also, many of the basic assumptions fundamental to various dating methods may not be upheld in their geological context.

Numerous beach and chenier ridges border the low wave energy southern and eastern margins of the Gulf of Carpentaria from the Roper River in Arnhem Land to Cape York (Smart, 1976; see also Fig. 3). The beach ridges, which are believed to be late Pleistocene in age, consist of quartzose sand and overlie a basement of fan deposits as a regressive sequence of barrier islands. According to Smart a chenier ridge, a barrier island complex and two sets of chenier-type ridges (sand ridges in a continuous sequence overlying muddy sediments), comprise a progradational sequence of ridges which post-date the Holocene transgression. Cores show that they are composed of quartzose sand, shell sand and shell, and show progressive leaching with age. The youngest ridges overlie marine sand and mud.

Available evidence indicates that the sea was several metres above present during the last interglacial high stand around *c.* 120 000 B.P. At this time, large areas of the eastern shore of the Gulf of Carpentaria were barrier islands, although the volume of sediment was smaller than that in the corresponding Holocene environment. Between 47 000 and 26 000 B.P. and from 23 000 to 11 000 B.P. the Gulf was a basin of internal drainage and exposed to subaerial processes during which time leaching, calcrete formation and lateritic weathering occurred. It should be noted that the Gulf of Carpentaria is < 70 m deep and is separated from the Arafura Sea by a sill at about -53 m; thus it was a separate basin when sea level was lower than about

−53 m. In deeper parts, below 160 m, brackish-water sediments of lacustrine deposits accumulated in shallow lakes. Sea level rose above −53 m about 11–12 000 B.P. and reached its present level between 6000 and 7000 B.P. No subsequent changes in sea level have been recorded so far, but data from the Gulf of Carpentaria are too incomplete to give a comprehensive account of fluctuations in climate and sea level in the area. Nevertheless, the general sequence of events, excluding any possible effects of hydro-isostatic flexure during the Holocene, fits in reasonably well with that for other parts of Australia.

The origin and progradational history of cheniers in the tropical coastal areas of Australia and Louisiana are basically very similar, although there are differences in the dimensions of the ridges. These differences reflect, probably, the seasonal supply of sediment from the rivers draining into the Gulf of Carpentaria which transport smaller sediment loads than the Mississippi River system in Louisiana. The role and incidence of cyclonic storms, in the formation of cheniers over time, has not been researched adequately in the remoter areas of northern Australia. Suffice it to say that erosion, and some times catastrophic destruction of mangroves during cyclones, is accompanied by a winnowing of fine intertidal muds, leaving a residue of broken and unbroken intertidal mollusc shells concentrated in the swash zone immediately upslope of the mangroves.

UNITED STATES ATLANTIC COASTAL PLAIN AND INNER SHELF

Atlantic coastal plain

The United States Atlantic Coastal Plain is a non-glaciated and non-orogenic continental coast which consists of Quaternary transgressive-regressive barrier shoreline sediments (Oaks and DuBar, 1974). Depositional conditions during the Pleistocene can be gauged from modern shoreline features and sediment accumulation along 1200 km of the coast. Barrier islands are the most prominent coastal feature and extend in an almost unbroken sequence along the Atlantic seaboard. The size of the modern (and ancient) barriers, as well as their preservation, is dependent upon tidal range, wave and current energy, sediment supply and offshore topography. For example, the mean tidal range varies from about 1 m in North Carolina to a maximum of 2·1 m along the central Georgia coast, and decreases to $<$ 0·6 m in southern Florida. Generally, wave energy varies inversely with tidal range, being higher in North Carolina and Florida. Off North Carolina and Florida the continental shelf is narrower than off Georgia. These relationships appear to have remained relatively unchanged throughout the Quaternary. However, there are several distinct types of barrier which reflect changes in the orientation and physical conditions of the coast over geological time. These changes and their modifying effects on barrier islands along the modern coast provide a key to the Pleistocene deposits which, according to lithostratigraphic studies, were subjected to similar marine processes at their time of accumulation.

Salt marshes are commonly developed landward of the barrier islands in northern Florida, Georgia and South Carolina, but are replaced by open-water lagoons farther north and in southern Florida. The lagoons are partly filled, and are bordered by salt marsh. Lagoonal-salt marsh sediments are composed of sand, silt and clay; the finer sediments are deposited during slack water. Sedimentation is aided by the salt marshes which provide protection from waves and currents. The upper surface of the marsh is useful in determining sea level for areas of former marsh associated with high Pleistocene sea levels because the salt marshes build vertically to approximately high spring-tide level.

Erosion is more prevalent than deposition along much of the Atlantic coast of the USA as evidenced by truncated dune-beach ridges, the narrowness of many barriers and historical records. In many areas salt marsh and lagoonal sediments are exposed along the beach on account of widespread erosion.

The expression of Pleistocene coastal deposits is similar to that of the Holocene accumulations along the modern coast, and the morphology of the Pleistocene features indicates that coastal sequences have not been transgressed since their deposition. Thus, erosional scarps demarcate the landward limit of Pleistocene shorelines which are associated with brackish water, barrier island, beach and other shoreline deposits. Level to gently inclined (1-2°) palaeo-lagoonal floors, characterized by oxidized and leached sediments to a depth of several metres, and incorrectly termed terraces by earlier workers, extend seaward of the Pleistocene shorelines.

Early work on the Atlantic Coastal Plain attempted to correlate transgressions by shoreline elevation, on the assumption of tectonic stability. A study of Quaternary shorelines in coastal Georgia, northern Florida, and South and North Carolina has considered tectonic movement along the coast in determining the stability of the area (Hoyt and Hails, 1967; Hails and Hoyt, 1969; Hoyt, 1969). Although no tectonic movement has been noted for any of the six Pleistocene shorelines (Wicomico (29–31 m a.s.l.), Penholoway (21–23 m), Talbot (12–14 m), Pamlico (*c.* 7·5 m), Princess Anne (*c.* 4 m) and Silver Bluff (*c.* 1·4 m)) across the 170-km breadth of coastal Georgia, there has been significant displacement in northern Florida along the flank of the Ocala Uplift. Upwarping of 20 m over a distance of 80 km, with an average displacement of 25 cm km^{-1} has occurred in northern Florida along the Wicomico barrier island, known as Trail Ridge, the most prominent topographic feature of the Georgia–Florida State boundary line (Hoyt and Hails, 1971). Less movement has occurred along the Penholoway shoreline where there has been displacement of 6–7 m, or about one-third the displacement of the Wicomico barrier system. Tectonic movement is less along the Talbot and Pamlico shorelines and no displacement has been detected along the two youngest Pleistocene shorelines (Fig. 1).

The heights of former Pleistocene barrier shorelines along the Central and Southern Atlantic Coastal Plain, USA, have been calculated from the upper limit of fossil burrows of the marine decapod *Callianassa* (*Ophiomorpha nodosa* Lundgren) and from the upper limit of lagoonal-salt marsh sediments which accrete to slightly above high-water mark (Hails and Hoyt, 1969). The significance of *Callianassa* as an indicator of former stands of sea level pertains to studies of the modern burrows of *Callianassa major* Say which show that the upper limit of burrow development in littoral sediments, adjacent to barrier beaches, approximates mean sea level (Weimer and Hoyt, 1964). Although the Pleistocene stratigraphy of the Atlantic Coastal Plain is well documented in the literature, suffice it to say that six Pleistocene barrier shorelines, representing six formations, have been identified from *Callianassa* burrows in Georgia.

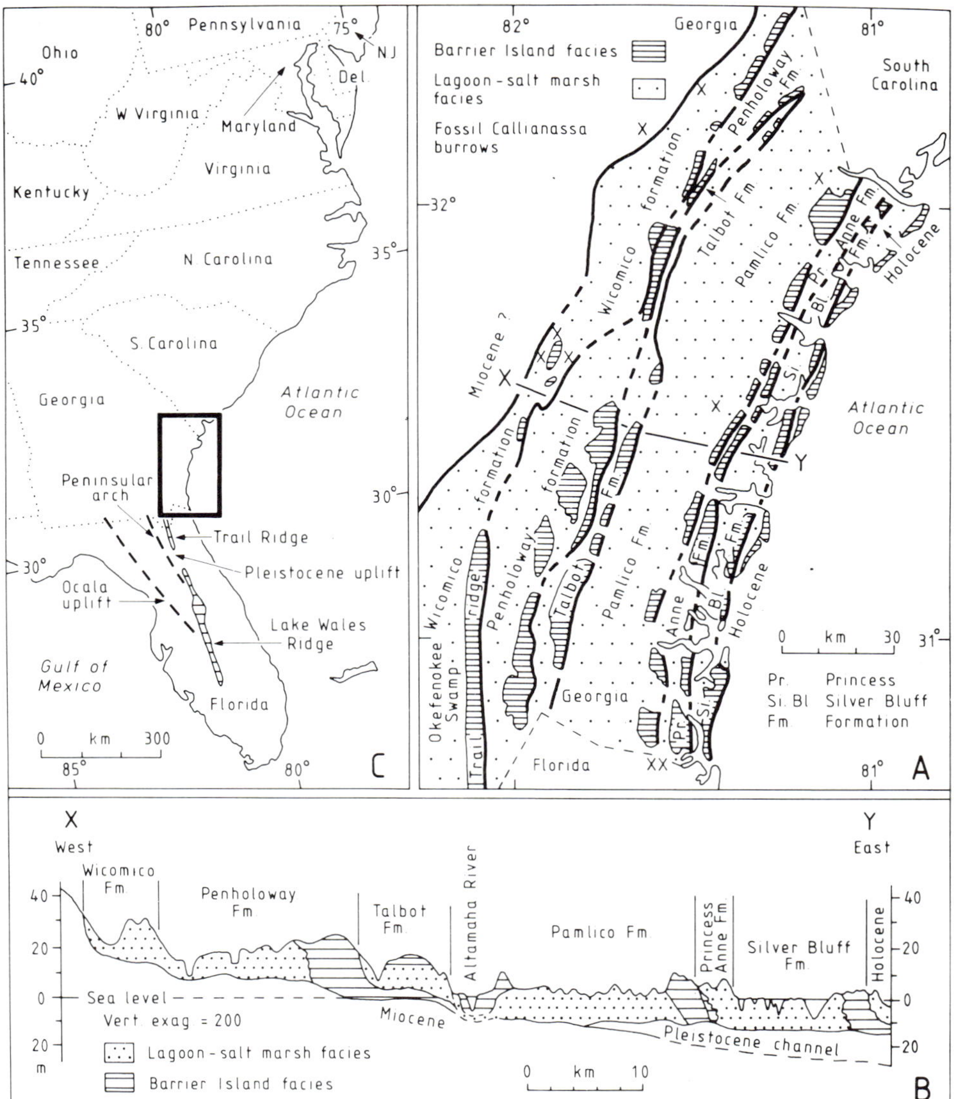

Figure 1. Relict coastlines and Holocene coast, Georgia coastal plain (A); cross-section of the Lower Georgia coastal plain (B); and structural features of northern Florida (C). After Hails and Hoyt (1969) and Hoyt and Hails (1971).

Although no coastal area is stable, *sensu stricto*, the uniform elevations of the lagoonal-salt marsh deposits comprising the Coastal Plain in Georgia, as well as *Callianassa* burrows associated with each barrier system, suggest that this sector of the Atlantic seaboard has been tectonically stable throughout the Pleistocene epoch. Of course, the dating of Pleistocene shorelines presents a difficult problem because of a lack of definitive fossils.

Uranium-disequilibrium-series dating of fossil corals, preserved in clastic sediments of inner sub-littoral depositional environments, has been undertaken by Cronin *et al.* (1981) in order to date transgressions, rather than to estimate the elevations of palaeo sea levels, along the United States Atlantic Coastal Plain. The uranium-series study, which is the first in this region, has been augmented by detailed palaeoclimatic reconstructions based on marine ostracod and pollen (continent) data, biostratigraphy and lithostratigraphy. Ostracods are benthic, shelled Crustacea, which are now used frequently in freshwater and marine continental shelf palaeoclimatic reconstruction. A variety of environments, including salt marsh, coastal and fluvial swamps, and coastal plain forests, have been represented in the pollen assemblages. Lithostratigraphic relationships between transgressive sequences have been established by extensive surface mapping and shallow subsurface cores (see Oaks and Dubar, 1974, for a detailed account of the post-Miocene of the Central and Southern Atlantic Coastal Plain).

Five relatively warm intervals during the last 500 000 years have been reported by Cronin and his associates. On the basis of multiple palaeoenvironmental criteria they determined relative sea-level positions during warm intervals to be 7 ± 5 m a.s.l. in Virginia and North Carolina 188 000 years ago; 7·5 ± 1·5 m in South Carolina 120 000 years ago; 6·5 ± 3·5 m in South Carolina 94 000 years ago; and 7 ± 3 m in Virginia-North Carolina 72 000 years ago. About 440 000 years ago maximum sea level was between 10 and 25 m a.s.l. in north-eastern South Carolina as determined from faunal palaeodepth estimates and barrier island deposits. The evidence used in the Virginia–South Carolina area for the other sea levels includes *Callianassa* burrows, nearshore faunal assemblages, beach and dune sands, and cross-bedded deltaic deposits.

However, the composite sea-level chronology for the Atlantic Coastal Plain is inconsistent with independent estimates of eustatic sea-level positions during interglacial intervals of the last 200 000 years. Cronin *et al.* advance several possible reasons in order to account for the discrepancies. These include hydro-isostatic adjustment from glacial-interglacial sea-level fluctuations, lithospheric flexure, and isostatic uplift from sediment unloading owing to erosion. Alternatively, they suggest, current eustatic sea-level estimates for the middle and late Quaternary may require revision, although their composite chronology is a first approximation!

Atlantic inner shelf

The history of sea-level stands during the last glacial and the Holocene epoch, however, is not so reliably documented, particularly seaward of the modern shoreline. Pollen analysis of Upper Quaternary peat deposits of the Atlantic inner shelf of the United States indicates that the organic material accumulated in both marine and terrestrial environments (Field *et al.*, 1979). Correlation of the stratigraphic sequences in vibrocores containing the peats with high-resolution seismic reflection profiles indicates a history of transgressive erosion on the inner shelf. Remnants of previously extensive carbonaceous sediments originally deposited in estuaries, salt-marsh lagoons and bays, which are discontinuous and usually associated with erosional unconformities, provide evidence of regional planation of the shelf by the rising Holocene sea. However, it is difficult to determine the precise position of sea level in the late Pleistocene because of the large standard deviations of ^{14}C dates, which occur near the age limit of the method and which are significantly affected by minor amounts of contamination.

The ages of Holocene peat deposits are more reliable for reconstructing former sea levels along the inner shelf but there is uncertainty, despite a similarity in age and depth of samples, about regional differences in tectonic warping. Whether it was local or affected the entire Atlantic shelf is still a subject for conjecture. Some workers have suggested that similar neotectonic effects have been produced by different processes, such as ice loading and warping along its margin, on the northern shelf, and compaction and subsidence farther south (Hicks, 1972; Dillon and Oldale, 1978). Thus, the magnitude of isostatic rebound and warping has been difficult to evaluate even if the general trend of sea level along the Atlantic inner shelf has been established for the past 12 000 years from ^{14}C dates of buried peats.

SEDIMENTARY SEQUENCE MODELS FOR HOLOCENE BARRIERS OF THE ATLANTIC COASTAL PLAIN

Morphology and vertical sedimentary sequence models for Holocene transgressive barrier systems pertaining to the Atlantic Coastal Plain have been described by Kraft *et al.* (1973) for the mid-Atlantic Delmarva Peninsula (see Table 1). There are four elements to the model, each showing a response to present wave and current conditions, as well as to the palaeotopography submerged by the ongoing transgression. These are the barrier systems, including barrier islands and baymouth barriers characterized by similar vertical transgressive sequences which, in turn, are identical to the horizontal sequence in the direction of the transgression. The simplest vertical sequence model is *the beach against highland* element. As the beach encroaches on relatively low-lying Pleistocene or pre-Pleistocene sediments, erosion provides a sediment supply to the littoral drift system. The process may be ephemeral and therefore not occur in the stratigraphic record.

Estuarine washover barrier elements are associated with large and small estuarine environments. The sands and gravels on the Delmarva Peninsula lack silt and clay probably because of high energy generated by storm waves, and the strength of longshore currents. An efficient selective mechanism of deposition and transport pertains to the formation of this type of element. Fine sediment is deposited

TABLE 1

Elements of the transgressive barrier model for the Delmarva Peninsula, Mid-Atlantic Coast, USA (after Kraft et al. *1973)*

Element	Sedimentary sequence(s)	Example of vertical sequence
Barrier systems Barrier islands Baymouth barriers Barrier island chain	Vertical transgressive sequence of sedimentary environments identical to the horizontal sequence in the direction of the transgression	Sands and gravels of the baymouth barrier Soft lagoonal mud and abundant molluscan fauna Salt marsh organic muds and peats Marsh fringe composed of sandy mud and organic debris Pleistocene sediment surface or land surface undergoing transgression
Beach against highland	Beaches constructed against 6–9 m high pre-Holocene highlands of Pleistocene or pre-Pleistocene sediments Highlands are areas of coastal erosion with a thin sequence of dune, beach and offshore bar sands encroaching upon the highlands Erosion provides a source of sediment to the littoral drift system Ephemeral situation and may not occur in the stratigraphic record	Simplest vertical sequence model Dune sands Beach sands—medium to coarse sands; low angle seaward dipping laminations Offshore bar sands Pleistocene or pre-Pleistocene sediments; areas of erosion
Estuarine washover barriers	Tendency for coarser sand and gravel to occur seaward and upward in the washover barrier Continuous over 80–120 km on Delmarva Peninsula, apart from breaks at inlets into tidal creeks *Shoestring* sands occur in long arc-like sections along western Delaware Bay coast	Dune sands Washover barrier Brackish-salt marsh with organic mud and peat Marsh fringe Pleistocene coastal sediments
Prograding spit tracts	*Normal* transgressive sedimentary sequence of coastal beach-dune sediments over lagoon and marsh OR *Regressive* vertical sequence occurs when spit migrates over the shallow marine area, and coastal spit and beach sands, and gravels, overlie estuarine and shallow marine sediments	Transgressive sedimentary sequence Dune Beach berm washover Back barrier marsh Lagoon or marsh Brackish-salt marsh with organic mud and peat Marsh fringe Beaches or other spit complex; or unconformity

dominantly in the offshore bay area, having been derived from the erosion of the dead marsh surface, and then transported by waves across the barrier system and by tidal waters across coastal marshes.

Obviously, there can be problems in differentiating between estuarine washover barrier sequences and those of the normal washover barrier-beach systems of the open Atlantic coast. Generally, though, barrier systems are larger in size and dimension than estuarine washover barriers. However, lagoonal microfaunal and marsh grass fragments in underlying muds are useful evidence for distinguishing between estuarine and ocean barrier washovers provided that lithification and post-depositional leaching do not significantly alter the original deposits.

Split complexes are somewhat anomalous elements because normal washover barrier and barrier systems are likely to be intermixed with vertical sedimentary sequences that include beach and spit sands and gravels migrating over open marine and estuarine sediments. According to Kraft *et al.* (1973), though, it is possible to identify the nature of a spit, and whether or not a transgression or a regression has occurred, if there are three points of vertical sequence control. This would be extremely difficult with a single point of control.

Other coastal geomorphic models have been described for the Mid-Atlantic Bight, USA, including the Delmarva, New Jersey, and North Carolina segments which have been divided into four compartments by Fisher (1967). These are a northern cuspate spit, and eroding headland, a southern spit, and a barrier island chain. Fisher argues that the barrier features developed from headlands by distally prograding spits because of the prevailing easterly wind and wave direction in the Bight.

SOUTH-EAST SOUTH AUSTRALIA

In south-east South Australia, relict Pleistocene barrier deposits are incorporated with a 192-km long Holocene barrier which partly encloses a lagoon (3 km wide), known as the Coorong. This, in turn, is backed by a succession of older beach-dune ridges, and intervening swampy depressions on the sites of successive emerged lagoons similar to the modern Coorong. The ridges, which are partly dissected and noted for their fixed transgressive dunes capped with calcrete, trend approximately parallel to the modern shoreline and increase in elevation inland, thus marking a sequence of Pleistocene barriers.

Various hypotheses have been advanced to explain the origin of the modern coastal barriers, as well as the stranded barrier beach ridges and associated dunes. Fenner (1931) described the modern barrier as a spit, although there is no evidence of recurves, which usually denote distinct stages in longshore spit growth. Tindale (1947) suggested that the barrier originated as an offshore bar, parallel to an older shoreline, which became emerged during a fall in sea level.

There is a general consensus among four researchers, though, that the beach-dune ridge sequence west of Naracoorte (Fig. 2) is regressive and that the barriers formed

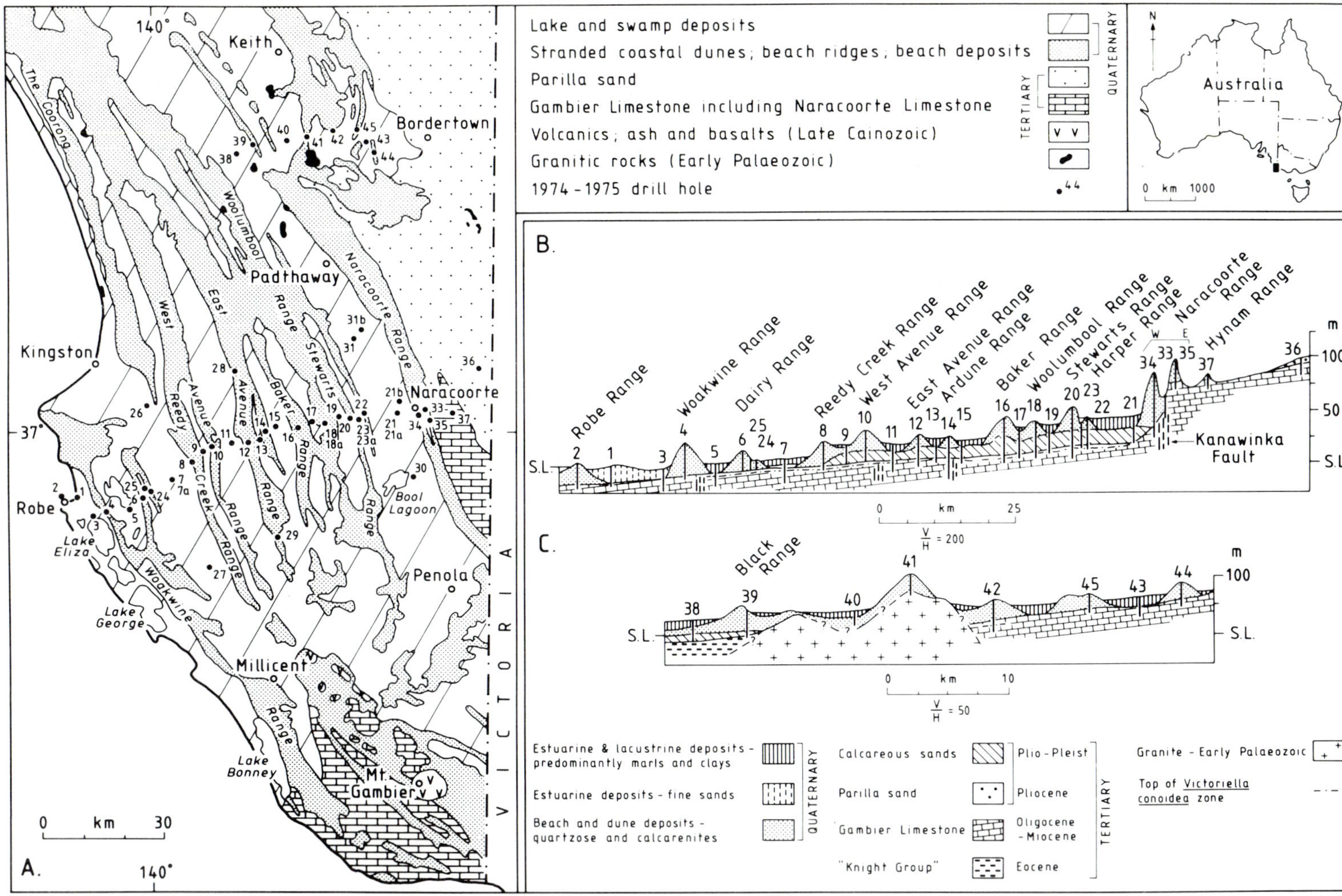

Figure 2. Geology and geomorphology of south-eastern South Australia (A), with sections through the barrier beach ridge sequence in the Robe-Naracoorte area (B) and Bordertown area (C) based upon borehole data from sites shown in (A). After Cook *et al.* (1977).

and prograded during high stands of sea level in the Pleistocene (Hossfeld, 1950; Sprigg, 1952a,b, 1958; Blackburn, 1966; Firman, 1967, 1969, 1973). Sprigg attempted to construct a sea-level curve by using the relative elevation of the barrier ridges and the Milankovitch astronomical theory despite the lack of absolute dates and limited knowledge of the Pleistocene stratigraphy.

Boreholes drilled more recently across the barrier ridge sequence between Naracoorte and Robe, some 100 km, and in the vicinity of Bordertown, have elucidated the regional stratigraphy (Cook *et al.*, 1977). It is now believed that the tectonic setting of the region has determined the essentially regressive sequence of beach-dune, estuarine-lagoonal and lacustrine facies. Uplift, associated with volcanic activity in the Mount Gambier-Mount Burr area, continued throughout the Plio-Pleistocene. This, together with numerous eustatic changes in sea level, resulted in the seaward progradation of the shoreline over a period of 690 000 years or less. Although fossils are abundant within the barrier sediments they have been abraded and therefore none is biostratigraphically diagnostic. However, magnetostratigraphy has shown that all the dune calcarenites to the west of the Naracoorte Range are normally polarized and formed during the Brunhes Normal Polarity Epoch and are therefore 690 000 years B.P., or younger (Fig. 2). Cook *et al.* report that some dunes appear to be composite beach ridges, reflecting possibly more than one change in sea level. Also, the altitudinal grouping of the ridges and intervening depressions may result from fluctuations in the rate of sea-level change, rather than from changes in the rate of uplift.

Thus, the evidence cited in this case study indicates that shoreline processes have changed very little over geological time. The regressive sequence is composed of a barrier ridge sandy facies, comprising beach and neritic (shallow marine) sands at the base, and aeolian sands at the top. These sediments are separated by a fine-grained interdunal facies deposited initially in a bay or estuarine environment, grading into lagoonal and finally saline lacustrine conditions at the top. The mechanisms, though, in this unstable region, have been both tectonism and eustasy. If the processes forming the modern Coorong barrier, which incorporates older dune calcarenites, have continued in the past, it seems unlikely that the landward margins of the lagoonal depression were exposed to the open sea during successive transgressions.

SOUTH AUSTRALIA: NORTHERN SPENCER GULF

To date, published evidence of Holocene sea levels in South Australia has come from widely different geographical localities, and it is often poorly validated on account of limited quantitative data. The writer and his colleagues have completed regional subtidal stratigraphic studies and chronological investigations of coastal sediments in northern Spencer Gulf, South Australia, in order to document the Holocene transgression more accurately. The northern part of the Gulf is virtually a land-locked body of water located some 400 km from the margin of the continental shelf and about 270 km from the Southern Ocean Coast (Fig. 3). It can be described as an inverse estuary with minimal land water input, but with high evaporation rates and the production of warm brines at its head (see Fig. 4). Water depths are < 20 m and extensive peritidal sedimentation during the late Holocene has produced wide areas of intertidal and supratidal flats.

Quaternary subsurface stratigraphy has been determined from sub-bottom profiles and vibrocores, while ^{14}C (43 samples), thermoluminescence (10 age estimates), amino-acid racemization (5 measurements) and biostratigraphic dating techniques have been used to establish a chronology. Vibrocores from supratidal flats and sabkhas, samphire-algal and mangrove-algal communities, bare or Zostera-covered sandflats and accretionary beach ridges have been analysed. Isopach and feather-edge depths of sedimentary strata have been used to determine the maximum extent of Pleistocene sea levels. Four Pleistocene marine transgressions and subsequent regressions have been identified in the subsurface cores (Belperio *et al.*, 1983; Hails *et al.*, 1983; see also Table 2).

The second oldest of the four marine transgressions is represented by the Mambray Formation which has been altered markedly by palaeosol development (pedogenesis) and is often capped by a lime-cemented crust of pseudo-calcrete, or by calcareous nodules. The palaeosols exhibit well structured clay-rich B horizons.

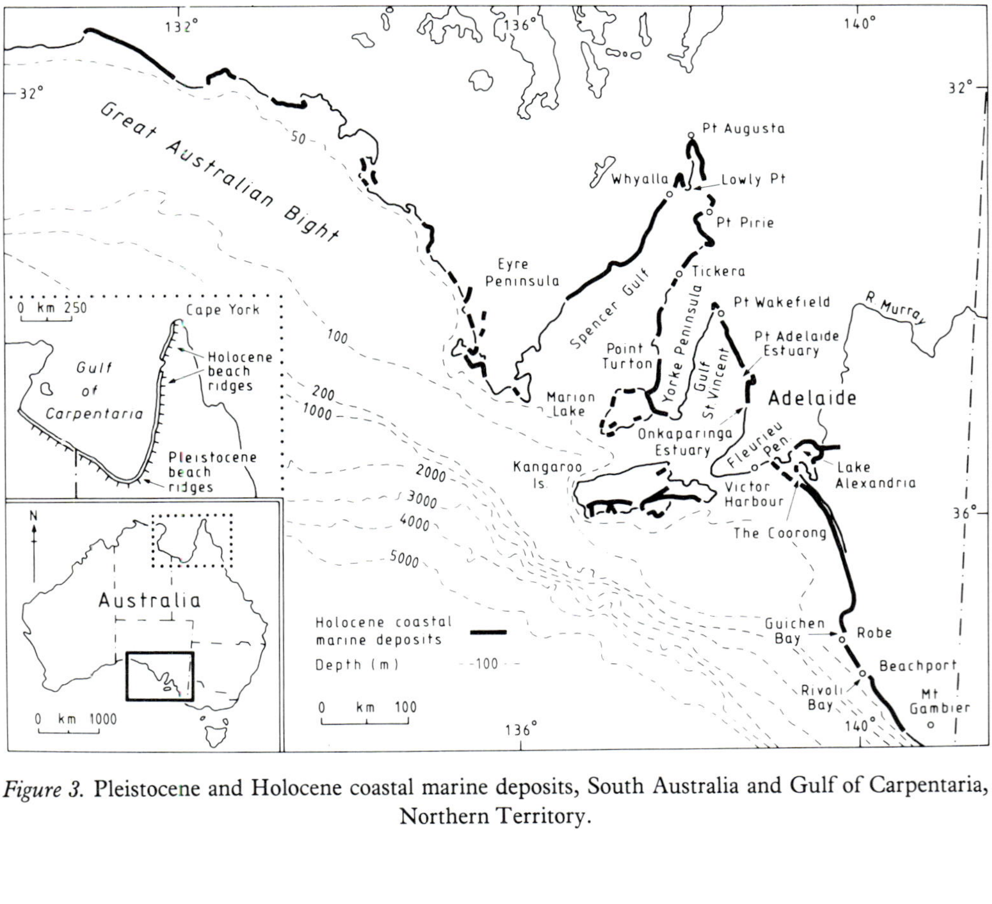

Figure 3. Pleistocene and Holocene coastal marine deposits, South Australia and Gulf of Carpentaria, Northern Territory.

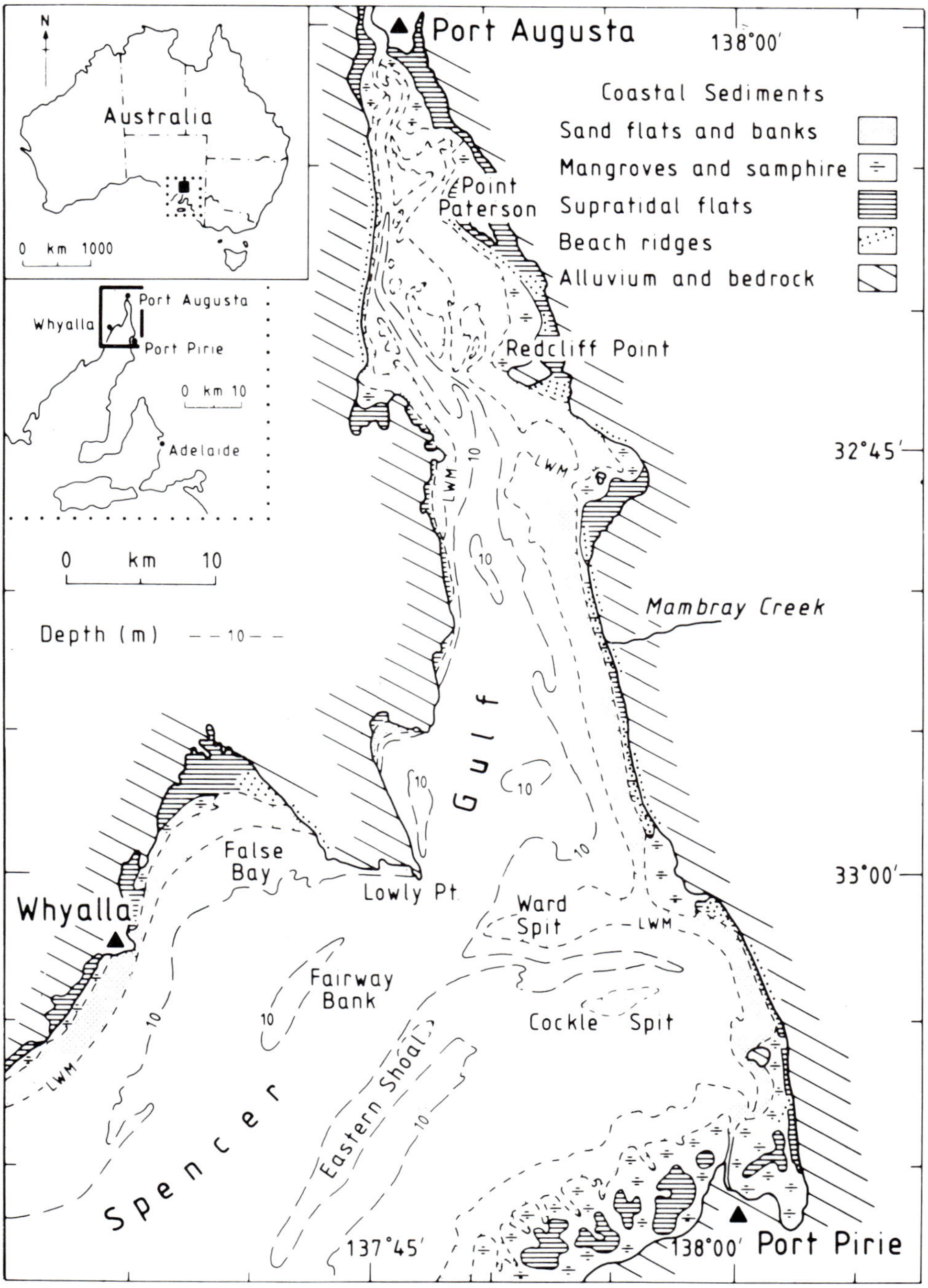

Figure 4. Coastal sediments of northern Spencer Gulf, South Australia.

TABLE 2

Stratigraphic table of strata in Northern Spencer Gulf, South Australia

		Formation	Environment	Probable Age (Ky)
QUATERNARY	Holocene	Germein Bay	Marine	0-8
	Late Pleistocene	Pooraka	Alluvial	*c.* 20
		Lowly Point	Lagoonal marine	84
		False Bay	Marine	105
		Mambray	Marine	125
		Older pleistocene marine beds	Marine	*c.* 220
		Hindmarsh Clay	Alluvial	> 220

Vibrocoring has shown that the Pleistocene sediments extend beneath the modern intertidal and supratidal zones and, in the Redcliff area, landward for a distance of 8 km. The Mambray Formation contains abundant remains of the subtropical bivalve *Anadara trapezia* and the formaminifera *Marginopora vertebralis* both of which are recognized as Pleistocene fossil indicators although they are extinct in South Australia today. Amino acid racemization (D/L enantiomer ratio of *A. trapezia*) from this formation gave an age estimate of 110 000 ± 19 000 B.P., while thermoluminescene age estimates from the same level (quartz fraction) produced ages of 90 000 ± 15 000 and 90 000 ± 24 000 B.P. Work by Belperio *et al.* (1983) shows that the age of the sedimentary components ranges from 90 000 to 160 000 B.P.

The Mambray Formation can be correlated with other Pleistocene depositional shorelines in South Australia on faunal evidence and elevation, and by leucine D/L ratios (Von der Borch *et al.*, 1980). The Glanville and equivalent formations in the Adelaide region and adjacent areas outcrop similarly at, or slightly above, low-water datum as the Mambray Formation. The Woakwine Barrier of south-east South Australia has given a uranium series age estimate of 100 000 ± 30 000 B.P. (Schwebel, 1980). Thus the Mambray Formation of northern Spencer Gulf can be tentatively correlated with isotope stage 5e and the last interglacial maxima. Equivalent deposits in eastern Australia record a sea level of 4-6 m (Marshall and Thom, 1976). The Woakwine Barrier is elevated some 12-17 m above sea level in one of the relatively few areas of tectonic upwarp in Australia (see previous section and Fig. 2). Therefore, it is suggested that the Pleistocene deposits of northern Spencer Gulf and the Adelaide region record a downwarp of about 3-4 m relative to stage 5e equivalents on "stable" coastlines, despite late Holocene uplift of at least 2·5 m (Hails *et al.*, 1983; Belperio *et al.*, 1983).

Marine transgressions equivalent to the False Bay and Lowly Point Formations can be correlated with isotope stages 5c (of Shackleton, 1969) and 5a respectively (Fig. 5). The maximum levels to which these events inundated Spencer Gulf, namely −8 m and −14 m respectively, agree broadly with other estimates reported in the literature.

Radiocarbon dates of basal Holocene sediment fractions from submarine vibrocores indicate that sea level rose rapidly from −15 m at 8000 B.P. to its present level in Spencer Gulf about 6600-6400B.P. In this context, too, the sea-level history of South Australia is similar to that of the east coast of the continent. However, there is substantial evidence for high stands of Holocene sea level at 2·5 m, and up to 3·8 m, above modern datum in the northern part of the Gulf, but how far this evidence can be extrapolated is still a matter of conjecture (Burne, 1982; Belperio *et al.*, 1983; Hails *et al.*, 1983). It can be argued that hydro-isostatic upwarping may be responsible because northern Spencer Gulf is that part of the South Australian coast farthest from the edge of the continental shelf.

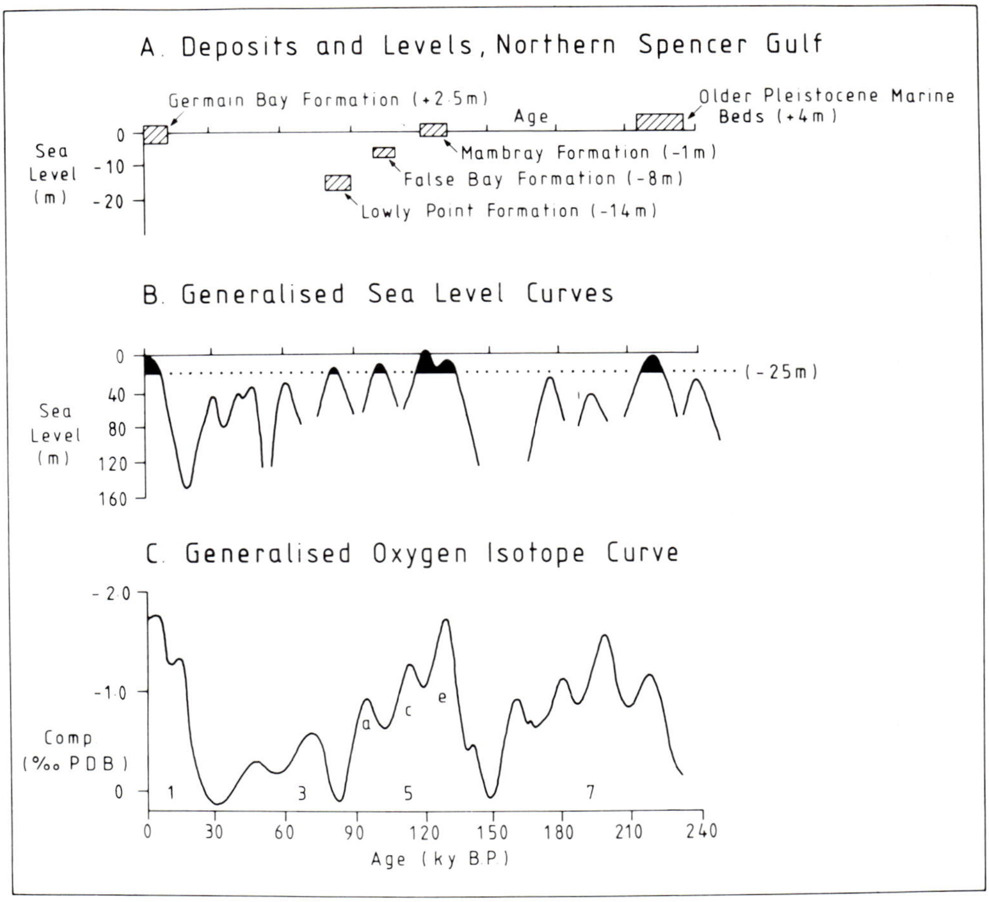

Figure 5. Levels and possible age relationships of Quaternary marine sediments in northern Spencer Gulf (A) compared with the generalized sea-level curve (B) of Bloom *et al.* (1974), Chappell (1976) and Chappell and Thom (1977), and with a "generalized oxygen isotope curve" (C) after Emiliani (1972) and Fairbanks and Matthews (1978).

Evidence presented by Belperio *et al.* (in press), points to local seismotectonism as the most likely cause of the higher stand of sea level. Shallow marine sedimentation of northern Spencer Gulf has been accompanied over geological time by the growth of seagrass, and accumulation of seagrass facies sediment, at the rate of about 1-2 m yr^{-1}, resulting in extensive carbonate banks around the shoreline. A relict bank is buried beneath the present-day intertidal and supratidal zones of Redcliff with its surface elevated about 2·5 m above present low water (Fig. 4). The age of the bank varies from > 5000 years at its most landward margin to about 1700 years at its seaward margin. Thus, a relative sea level of at least 2·5 m above present datum is suggested for this time interval. Mangrove and samphire colonization and deposition on the surface of the seagrass bank, which is related to the present-day tidal range, occurred post-1700 B.P., following regional tectonic movement. The marine regression at the time resulted in a seaward shift of the coastline and the formation of extensive intertidal areas over what were previously marginal seagrass banks.

If a time-depth curve for northern Spencer Gulf before 1600 years B.P. is corrected over a 2·5-m tectonic displacement, the final curve compares favourably with most others for the east coast of Australia. This means that the present level of the sea was reached by at least 6000 B.P., and has not deviated significantly (< 1 m) since about that time (Thom and Chappell, 1975, 1978; Belperio, 1979).

SUMMARY

Coastal depositional environments produce sedimentary sequences which provide researchers with evidence of past oscillations in sea level. Correlation of regional litho- and biostratigraphic sequences of transgressive and regressive sediments in mid- and low-latitude environments enable relict shorelines to be reconstructed with the use of morphology and sedimentary sequence models, as well as those process-response models showing the interaction between sediment supply and transport, wave climate and variations in mean sea level. This interaction is dependent upon the control effected by climatic change over geological time and interrelated rates of weathering, river discharge and sediment supply to the nearshore and offshore zones.

Tectonism within a region determines whether or not regressive sequences of beach-dune, estuarine-lagoonal, and lacustrine facies will occur during deposition accompanying high stands of sea level. Available evidence from the Gulf and Atlantic Coastal Plain of the USA and widely distributed coastal areas of Australia suggest that shoreline processes have not changed significantly over geological time. If the processes forming modern depositional shorelines, then, continued during the Quaternary, it appears unlikely that the landward margins of lagoonal salt marshes, or lagoons, were exposed to the open sea during successive transgressions. In fact, the landward margins of lagoons behind Holocene barriers are invariably the modified shorelines of Pleistocene barriers. The absence of open marine deposits landward of many mid-latitude barriers in Australia and the USA is evidence against barrier evolution by emergence and from offshore bars. The arcuate pattern of barriers on the east Australian coast indicates that, during progradation, oceanic depositional shorelines have been shaped by refracted swell.

Cheniers and barriers can be identified by the shape and extent of their respective deposits. A difference in the processes leading to the development of these features is, in the case of cheniers, the proximity of a sizeable sediment source which can overload the distributive agencies along the adjacent shoreline. Near such a source progradation may alternate with erosion and hence cheniers may develop.

Hydro-isostatic adjustment from glacial-interglacial sea-level fluctuations, lithospheric flexure, and isostatic uplift from sediment unloading on account of erosion have been cited as reasons why the composite sea-level chronology for the Atlantic Coastal Plain is inconsistent with independent estimates of eustatic sea-level positions during interglacial intervals of the past 200 000 years. Time-depth curves, corrected for tectonic displacement, show that present sea level was reached at least 6000 years B.P. in Australia, and has not deviated significantly (< 1 m) since about that time.

REFERENCES

Belperio, A. P. (1979). *Marine Geology* **32**, M1-M9.

Belperio, A. P., Hails, J. R., Gostin, V. A. and Polach, H. A. (in press). *Marine Geology*.

Bird, E. C. F. (1961). *Geogr. J.* **127**, 460-468.

Bird, E. C. F. (1965). *Proc. R. Soc. Vict.* **79**, 75-88.

Blackburn, G. (1966). *Aust. J. Sci.* **29**, 50-52.

Bloom, A. L., Broecker, W. S., Chappell, J., Matthews, R. S. and Mesolella, K. J. (1974). *Quaternary Research* **4**, 185-205.

Burne, R. V. (1982). *BMR Journal of Australian Geology and Geophysics* **7**, 35-45.

Byrne, J. V., Leroy, D. C. and Riley, C. M. (1959). *Trans. Gulf—Cst. Ass. geol. Socs.* **9**, 1-23.

Carr, A. P. and Hails, J. R. (1972). *International Geography* **1**, 0105, 22nd IGU, Montreal.

Chappell, J. (1976). *In* "The Origin of the Australians" (R. L. Kirk and A. G. Thorn, eds), pp. 11-22. Human Biology, Series No. 6, Australian Institute of Aboriginal Studies, Canberra.

Chappell, J. and Thom, B. G. (1977). *In* "Sunda and Sahul" (J. Allen, J. Golson and R. Jones, eds), pp. 275-291. Pre-Historic Studies in South East Asia, Melanesia and Australia. Academic Press, London and New York.

Clarke, M. F., Wasson, R. J. and Williams, M. A. J. (1979). *Search* **10**, 90-92.

Coleman, J. N. (1966). *Tech. Rep. Louisiana State University, Coastal Studies Institute* **29**.

Cook, P. J. and Polach, H. A. (1973). *Marine Geology* **14**, 253-268.

Cook, P. J., Colwell, J. B., Firman, J. B., Lindsay, J. M., Schwebel, D. A. and Von Der Borch, C. C. (1977). *Journal of Australian Geology and Geophysics* **2**, 81-88.

Cronin, T. M., Szabo, B. J., Ager, T. A., Hazel, J. E. and Owens, J. P. (1981). *Science* **211**, 233-240.

Cullingford, R. A., Davidson, D. A. and Lewin, J. (eds) (1980). "Timescales in Geomorphology". John Wiley, Chichester.
Curray, J. R. (1965). *In* "The Quaternary of the United States" (H. E. Wright, Jr and D. G. Frey, eds), pp. 723-735. Princeton University Press, Princeton.
Davies, J. L. (1957). *Aust. J. Sci.* **20**, 105-111.
Davies, J. L. (1959). *Proc. R. Soc. Tas.* **93**, 89-95.
Davies, J. L. (1964). *Z. Geomorph.* **8**, 127-142.
Dillon, W. P. and Oldale, R. N. (1978). *Geology* **6**, 56-60.
Emiliani, C. (1972). *Science* **178**, 398-401.
Fairbanks, R. G. and Matthews, R. K. (1978). *Quaternary Research* **10**, 181-196.
Fairbridge, R. W. (1961). *In* "Physics and Chemistry of the Earth, 4" (L. H. Ahrens, F. Press and R. K. Runcorn, eds), pp. 99-185. Pergamon Press, London.
Fenner, C. (1931). "South Australia: A Geographic Study". Melbourne.
Field, M. E., Meisburger, E. P., Stanley, E. A. and Williams, S. J. (1979). *Bull. geol. Soc. Am.* **90**, 618-628.
Firman, J. B. (1967). *Geol. Surv. S. Aust., Q. Geol. Notes* **22**, 4-7.
Firman, J. B. (1969). *In* "Handbook of South Australian Geology" (L. W. Parkin, ed.), pp. 204-223. *Geol. Surv. S. Aust.* Adelaide.
Firman, J. B. (1973). *Geol. Surv. S. Aust., Rep. of Investigation* **39**.
Fisher, J. J. (1967). *In "Abstracts with Programs, Annual Meeting of Geol. Soc. of Am".* Special Paper No. 115, pp. 66-67. New Orleans.
Gill, E. D. (1956). *Quaternaria* **3**, 133-138.
Gill, E. D. (1961). *Z. Geomorph. NF Suppl.* **3**, 73-79.
Gould, H. R. and McFarlan, E. Jr (1959). *Trans. Gulf-Cst. Ass. geol. Socs.* **9**, 261-270.
Hails, J. R. (1964). *Aust. Geogr.* **9**, 207-217.
Hails, J. R. (1965). *Aust. Geogr. Stud.* **3**, 63-78.
Hails, J. R. (1968). *Trans. Inst. Br. Geogr.* **44**, 133-149.
Hails, J. R. and Gostin, V. A. (1978). *Trans. R. Soc. S. Aust.* **102**, 169-173.
Hails, J. R. and Hoyt, J. H. (1968). *Z. Geomorph. Suppl.* **7**, 24-55.
Hails, J. R. and Hoyt, J. H. (1969). *Trans. Inst. Br. Geogr.* **46**, 53-68.
Hails, J. R. and Hoyt, J. H. (1971). *Quaternaria* **14**, 255-264.
Hails, J. R., Belperio, A. P. and Gostin, V. A. (in press). *Marine Geology*.
Hicks, S. D. (1972). *Shore and Beach* April, 20-23.
Hossfeld, P. S. (1950). *Trans. R. Soc. S. Aust.* **73**, 232-79.
Howe, H. V., Russell, R. J., McGuirt, J. H., Craft, B. C. and Stephenson, M. B. (1935). *Bull. Louisiana geol. Surv.* **6**.
Hoyt, J. H. (1967). *Bull. geol. Soc. Am.* **78**, 11256-1136.
Hoyt, J. H. (1969). *Bull. Am. Ass. Petrol. Geol.* **53**, 299-306.
Hoyt, J. H. and Hails, J. R. (1967). *Science* **155**, 1541-1543.
Hoyt, J. H. and Hails, J. R. (1971). *Quaternaria* **15**, 51-63.
Jennings, J. N. (1961). *Z. Geomorph. Suppl.* **3**, 80-84.
Kraft, J. C., Biggs, R. B. and Halsey, S. D. (1973). *In* "Coastal Geomorphology" (D. R. Coates, ed.), pp. 321-354. State University of New York, Binghampton.
Marshall, J. F. and Thom, B. G. (1976). *Nature* **263**, 120-121.
Milliman, J. D. and Emery, K. O. (1968). *Science* **162**, 1121-1123.
Oaks, R. Q. and Dubar, J. R. (eds) (1974). "Post-Miocene Stratigraphy Central and Southern Atlantic Coastal Plain". Utah State University Press, Logan, Utah.
Price, W. A. (1955). *Quaternaria* **2**, 75-86.

Radok, R. and Raupach, M. (1977). *Inst. Engrs. 3rd Aust. Conf. Coastal and Ocean Eng.*, pp. 103-109.
Russell, R. J. and Howe, H. F. (1935). *Geogr. Rev.* **25**, 449-461.
Schwartz, M. L. (1973). "Barrier Islands". Dowden, Hutchinson and Ross.
Schwebel, D. A. (1980). "Quaternary stratigraphy of the southeast of South Australia". Unpublished. Ph.D. Thesis, Flinders University of South Australia.
Shackleton, N. J. (1969). *Proc. R. Soc. Lond. B.* **174**, 135-154.
Smart, J. (1976). *BMR Journal of Australian Geology and Geophysics* **1**, 211-218.
Sprigg, R. C. (1952a). *Bull. geol. Surv. S. Aust.* **29**.
Sprigg, R. C. (1952b). *Rep 18th International Congr.* **13** (Sect. M), 226-237.
Sprigg, R. C. (1958). *Trans. R. Soc. S. Aust.* **82**, 183-193.
Tanner, W. F. (1961). *J. sedim. Petrol.* **31**, 87-95.
Thom, B. G. (1965). *J.R. Soc. N.S.W.* **98**, 23-36.
Thom, B. G. (1973). *Progress in Geography* **5**, 170-245.
Thom, B. G. and Chappell, J. (1975). *Search* **6**, 90-93.
Thom, B. G. and Chappell, J. (1978). *Phil. Trans. R. Soc. A.* **291**, 187-194.
Tindale, N. B. (1947). *Records S. Aust. Museum* **8**, 619-652.
Von der Borch, C. C., Bada, J. L. and Schwebel, D. L. (1980). *Trans. R. Soc. S. Aust.* **104**, 167-170.
Ward, W. T. (1965). *J. Geol.* **73**, 592-602.
Ward, W. T. (1971). *Geol. Mign.* **50**, 703-718.
Weimer, R. J. and Hoyt, J. H. (1964). *J. Palaeont.* **38**, 761-767.

THREE

Coastal depositional landforms in northern Canada

R. B. TAYLOR and S. B. MCCANN

Beaches formed in polar climates are characterized by features that generally are not found in beaches formed in non-polar climates — R. L. Nichols (1961)

INTRODUCTION

The quotation from Nichols' account of the characteristics of beaches formed in polar climates was selected for two reasons. First, his description of some of the distinctive characteristics of Arctic and Antarctic beaches (Table 1) predates the present discussion by more than 20 years, during which time it has remained a useful summary statement. Secondly, despite the latitudinal constraints implied by his use of the terms polar, arctic and antarctic, Nichols quite clearly intended his remarks to apply to all those regions "where there is or has been abundant sea ice, lake ice, water-terminating glaciers or deeply frozen ground" (1961, p. 694). There are advantages to such an inclusive definition, of cold-region rather than high-latitude coasts, particularly in the Canadian context. If the contemporary abundance of sea ice is used as the criterion for defining cold-region coasts then almost 90% of the Canadian coastline falls within this category. Most of this lies within the seasonal sea ice zone (Wadhams, 1980; Weeks, 1980) but perennial polar pack ice impinges on the northern coasts of the Queen Elizabeth Islands (Fig. 1).

In Canada research on sea ice-related processes has ranged through 30° of latitude, from the Gulf of St. Lawrence to the Arctic Ocean, encompassing a wide variety of ice environments and ice effects which range from negligible to overwhelming. Though we have been concerned with the special rôle of ice, we have always worked within the framework of general shoreline dynamics, recognizing that, in many locations, ice is merely an additional, rather than a dominant, variable in the system and that the long-term coastal evolution at most of our research sites is dominated by isostatic emergence. In many instances the suites of elevated coastal landforms are much more impressive than the modern shoreline features.

SHORELINES AND ISOSTASY
ISBN 0 12 652960 4

TABLE 1

Some characteristics of modern and elevated Arctic and Antarctic beaches (after Nichols, 1961)

1.	Beaches may rest on ice (buried glacial ice and icefoot)
2.	Beaches may be pitted due to melting of buried ice (glacial ice, icefoot, sea ice floes)
3.	Ice-pushed and ice-deposited ridges may occur
4.	Beach ridges may terminate abruptly due to presence of ice when they were formed
5.	Ice-rafted sediments may occur
6.	Beach gravel may be poorly rounded
7.	Beaches may be modified by subaerial periglacial processes (solifluction and development of patterned ground)
8.	Bedrock surfaces may be striated by drift ice action
9.	Beach ridges may be interrupted by erosional channels cut by meltwater streams
10.	Beaches may occur in association with ice-contact features and glaciomarine deposits
11.	Ventifacts may occur
12.	Beaches may contain cold water fossils
13.	Beaches may contain preserved soft parts of cold water fossils

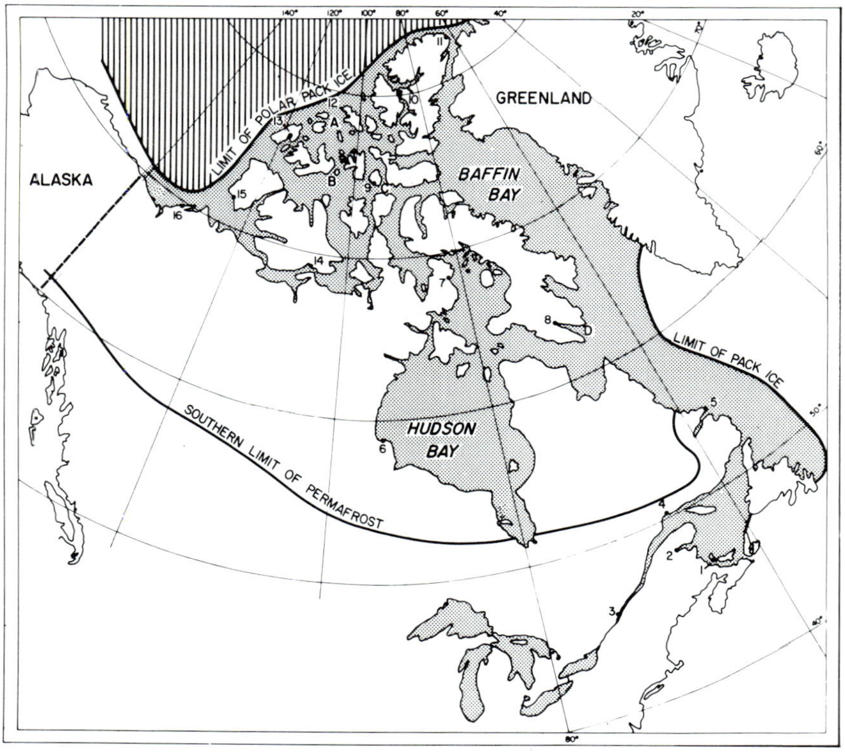

Figure 1. Distribution of sea ice and permafrost in Canada. Locations shown in Fig. 3 are indicated by numbers (1-16), whilst A,B,C,D are ice-affected environments discussed.

This chapter focuses on the geological action of drift ice in the marine littoral environment, and is organized in four parts. The first examines the spatial, temporal and physical characteristics of sea ice cover in Canada, providing a background for the second part which deals with the effects of ice in the littoral zone. Part three discusses the coastal depositional landforms in four different Canadian northern settings. Part four summarizes those characteristics of modern, ice-affected depositional shorelines which are likely to be preserved in ancient shoreline sequences and provide diagnostic evidence of the conditions of deposition.

In concluding this introduction, attention is drawn to two different, earlier reviews of cold-region coastal phenomena (John and Sugden, 1975; Weeks, 1980), and to the use of the term "glaciel" to describe drift-ice action (Hamelin, 1959, 1976; Dionne, 1972a, 1975, 1978).

SEA ICE IN CANADA

Distribution

There is a wealth of data on sea ice conditions in Canadian waters (e.g. Bilello, 1960, 1980; Swithinbank, 1960; Allen, 1964; Lindsay, 1969) but the two maps (Figs 2A,B) recently presented by Maxwell (1981), are the most useful for present purposes, as they provide an overview of winter and summer ice conditions. The mid-winter compilation (Fig. 2A) indicates ice type, defined by age, and ice motion. The age distinction is an important one; second- and multi-year ice is more resistant to abrasion and melting, and commonly much thicker, than first-year ice. There is a broad zone, including most of the Queen Elizabeth Islands, Banks and Victoria Islands and the mainland coastline south of these, in which fast ice conditions prevail almost everywhere during winter. This zone separates the mobile, second-year and multi-year, polar pack ice, to the north-west, from the mobile, first-year, pack ice to the east, in Baffin Bay and Davis Strait. A wide belt of fast ice surrounds most of Baffin Island. The mid-August compilation (Fig. 2B), for a "good" (i.e. for navigation) ice year, indicates ice cover and type and maximum extent of open water. It demonstrates very clearly that the potential for wave generation and for significant wave action at the shore is severely limited in the northern and western channels of the Queen Elizabeth Islands and in the western half of Parry Channel, where there is 70% cover of second- and multi-year ice, and somewhat restricted in McClintock Channel, Gulf of Boothia and Foxe Basin. Even in the areas designated as open water, varying amounts of residual pack ice, often in bands with 80–100% concentration, may filter out wave trains and protect the coast from wave action.

Annual regime

In terms of shoreline development at a site it is convenient to divide the "ice year" into four periods—freeze-up, complete ice cover in the littoral zone, break-up, and open water—and consider the processes operating in each period.

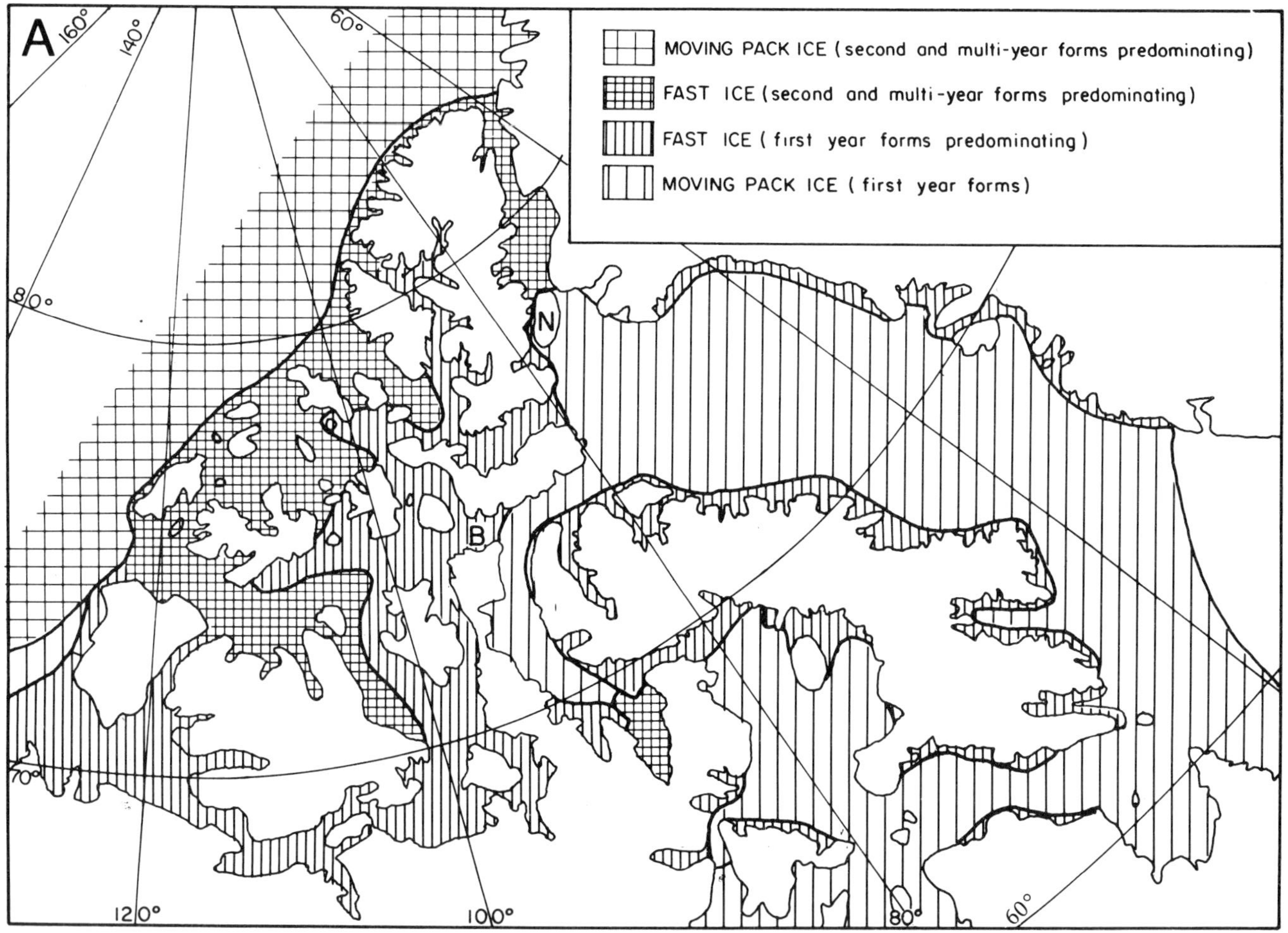

Figure 2A. Mid-winter sea ice cover in northern Canada. Polynyas except for North Water (N) are omitted; ice is in motion during some winters in Barrow Strait (B).

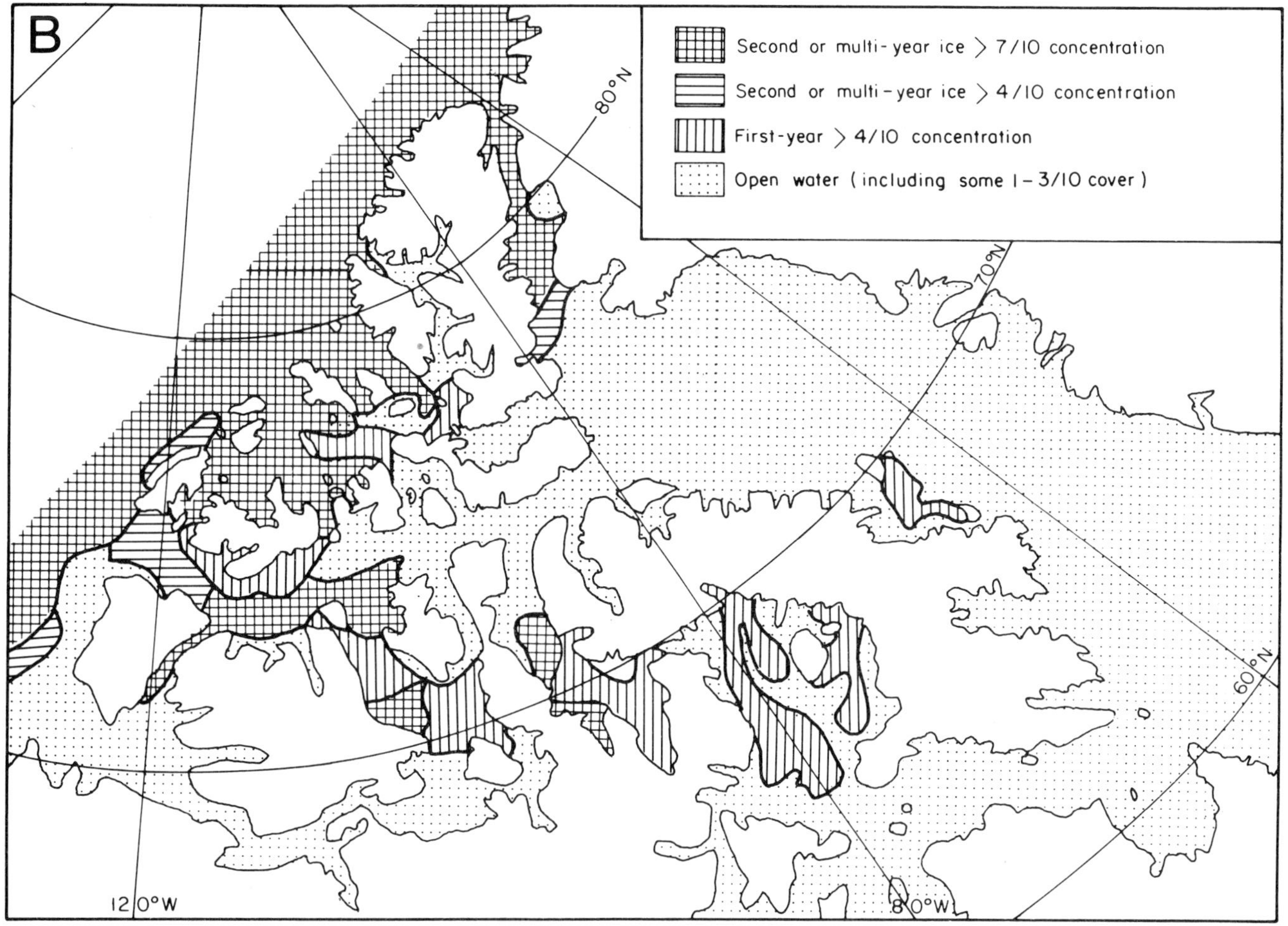

Figure 2B. Mid-August sea ice cover in a good year (both Figs simplified after Maxwell, 1981).

Freeze-up

The variables influencing freeze-up and the type of ice which develops at the shore in the autumn are temperature, wind, wave and pack ice conditions, amount of snowfall, and stage of the monthly tidal cycle. New ice begins to form in the intertidal zone when daily minimum temperatures fall below the freezing point of sea water. A thin glaze of ice forms on the foreshore and interstitial water in the sediments begins to freeze. Ice accumulation proceeds by the deposition and consolidation of ice slush, stranded floes, ice cakes and ice boulders, and by the freezing of wave swash and spray and snow. The net result is the development of an icefoot. The process may be interrupted by periods of wave action and higher temperatures. Thus, the duration of the freeze-up period, from the date on which new ice first appears on the beach to the date when the beach is completely immobilized by ice, may vary considerably from year to year (McCann and Taylor, 1975; McCann *et al.*, 1981a).

Complete ice cover in the littoral zone

The width and character of the zone of fast ice, which develops seawards from the shore and remains in place throughout the winter, varies considerably. Thus, the fast ice edge off eastern Baffin Island is commonly 100 km offshore in very deep water (Jacobs *et al.*, 1975), whereas in the Beaufort Sea it is rarely more than 30 km offshore in water depths of about 20 m, where it is stabilized at the drift limit of grounded pressure ridges (Cooper, 1974; Kovacs and Mellor, 1974; Stringer, 1974; Reimnitz *et al.*, 1978). In the constricted channels of the Queen Elizabeth Islands fast ice fills the whole width of the channels. During winter the alternate grounding and refloating of the ice cover in the intertidal zone seaward of the icefoot, particularly in areas with moderate or large tidal range, may be an important process. Layers of sediment are incorporated in the ice as it thickens by vertical accretion from the base, providing the potential for ice rafting of sediments during break-up.

Break-up

The variables influencing the timing and progression of break-up in the littoral zone are temperature, land drainage, sediment content of the ice, wind and tidal conditions, and the position of major offshore leads. The process begins with the melting of the snow cover on the ice, the enlargement of the tidal cracks in the littoral zone and break-up at river mouths due to the influx of fresh water during the nival freshet. A wide shore lead commonly develops alongshore, separating the icefoot, which tends to persist, from a still relatively solid cover of sea ice offshore. The ice in the intertidal zone breaks up earlier and melts more quickly than the offshore ice because of its fractured condition, due to repeated grounding, and its decreased albedo due to dark incorporated sediments. This effect is most apparent where tidal range is large (Rosen, 1979; McCann *et al.*, 1981a). Throughout most of the area shown as open water on Fig. 2B, the break-up period, from the inception of melt to the ablation and physical removal of most of the ice in the littoral and offshore zones, commonly lasts four or five weeks. During this period the melting of buried ice in the beach produces pitted topography and the melting of ice floes releases ice rafted sediments.

Open water

The open water period as considered here is defined as the period when the offshore ice cover is insufficient to inhibit the generation and propagation of waves large enough to cause pronounced reworking of shoreline sediments. Table 2 attempts to define the length of this period for five locations in the Arctic archipelago, using ice cover information, which may be compared with the wider ranging compilation (Fig. 3) by Owens (1977) using data from Allen (1964). Both illustrate the decrease in the length of the open water season northwards and westwards from the eastern Arctic approaches through the channels of the Queen Elizabeth Islands to the margins of the perennial polar pack. Table 2 also shows the year to year variability in the length of the open water period.

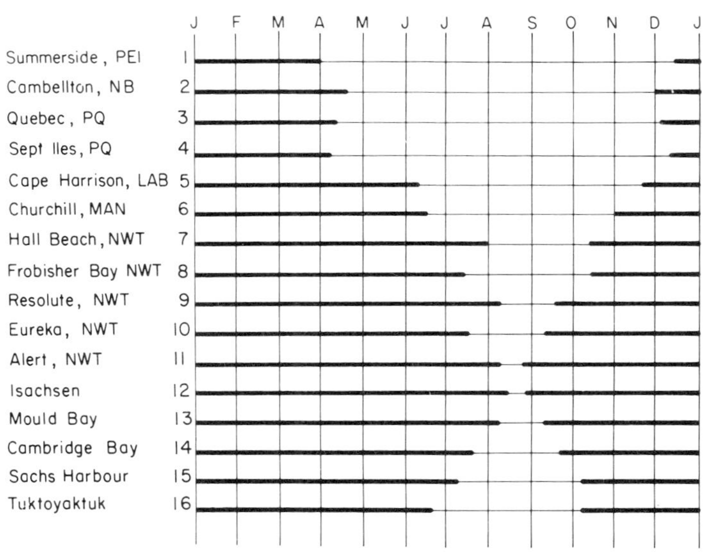

Figure 3. Average duration of ice in the littoral zone at selected Canadian locations, shown in Fig. 1 (after Owens, 1977).

TABLE 2
Open water season at selected eastern Arctic locations

		Start of open water season	End of open water season	Length of open water season (days)
1.	North coast, Somerset Island	July 18–July 23	Sept. 24–Oct. 8	67
2.	Pond Inlet, Baffin Island	Aug. 6–Aug. 13	Oct. 15–Oct. 22	60
3.	Radstock Bay, Devon Island	Aug. 7–Aug. 26	Sept. 24–Oct. 8	44
4.	Makinson Inlet Ellesmere Island	Aug. 20–Aug. 27	Sept. 24–Oct. 8	32
5.	Hooker Bay Bathurst Island	Aug. 27–Sept. 10	Sept. 24–Sept. 30	25

EFFECTS OF ICE IN THE COASTAL ZONE

The following section focuses on the direct and indirect effects of sea ice on depositional shorelines. There is no consideration of the effects of ice of terrestrial origin (i.e. tidewater glaciers and icebergs) but it is useful to begin with a brief discussion of permafrost, ground ice and ice-bonded sediments.

Permafrost and frost table effects

Many studies have documented the presence of permafrost beneath Arctic beaches, spits and barriers (e.g. Rex, 1964; McCann and Hannell, 1971; Taylor and McCann, 1974; Owens and Harper, 1977). Sequential measurements of the depth of thaw (i.e. depth of the upper surface of ice-bonded sediments) and of the depth of the frost table (depth to the 0°C isotherm), across the intertidal zone have defined the seasonal variation in the thickness of the potentially mobile unfrozen layer of surface sediments. Maximum thaw depths are usually less than 90 cm throughout the Canadian Arctic archipelago and occur towards the end of the open water season. During storms, when the beach is combed down to the frost table, rapid thawing and physical erosion of the ice-bonded sediments occurs (Taylor, 1980b), but beach profile changes are delayed. The presence of ice-bonded sediments at shallow depths also restricts the depth of beach disturbance by ice push. Ice push early in the open water season, or after freeze-back has commenced, results only in very shallow scars and ridges.

Massive ground ice, either as wedges or lenses, is an important factor in the erosion of soft sediment Arctic coastal cliffs. Thermal erosion of the exposed ice in the tundra cliffs of Alaska (Harper, 1978), the Yukon (McDonald and Lewis, 1973) and Siberia (Aré, 1972) produces block and flow slumping of unconsolidated materials, and high rates of cliff retreat.

Indirect effects of sea ice

The most important indirect effect of sea ice is to reduce overall wave energy levels at the shore by limiting wave generation offshore and by absorbing wave energy in the littoral zone. The annual beach regime in the seasonal sea ice zone includes a long period when wave-induced littoral processes, associated with swash-backwash interactions and longshore currents, are inoperative. Annual longshore transport rates are frequently much lower than on ice-free coasts with similar fetches and winds. In high-latitude settings experiencing only a short open water season with variable concentrations of mobile pack ice offshore, there is usually a large variation from year to year in the rates of longshore sediment transport and the scale of beach profile modifications. Thus, Hume and Schalk (1967) estimated that a single catastrophic storm at Barrow, Alaska in 1963 moved the equivalent of 20 years' normal sediment transport. Wave action, in storms which might be expected to occur annually or once every two or three years in the eastern Canadian Arctic (McCann, 1972, 1973; McCann and Taylor, 1975; Taylor, 1978a, 1980b, 1981) is usually accompanied by the presence of wind- and wave-driven ice in the littoral zone and the exposure of the substrate of ice-bonded sediments in the beach face. Multi-year ice floes in the littoral zone are abraded by waves into ice boulders, which become grounded, impede longshore transport and protect the upper beach. Beach profile changes are somewhat less than might be expected on ice-free beaches with similar sediments and similar waves. Post-storm beach recovery is commonly rapid, and, with the exception of beach pits, the profile character of the gravel beaches of Lancaster Sound differs little from that of mid-latitude gravel beaches. The reduced annual wave energy levels due to the short open water season are, however, reflected in the degree of roundness of the beach gravel (King and Buckley, 1968; McCann and Owens, 1969). Longshore and shore-normal variations in the size and sorting of beach sediments tend to be less defined than in mid-latitude beaches, and, locally, chaotic assemblages of all grades of sediments are related to the presence and melting of grounded ice.

Direct effects of sea ice

Direct sea ice interaction with the coast results in the erosion, deposition and reworking of sediment and the formation of distinctive morphological features by the processes of (1) ice rafting, (2) ice push and scour, and (3) ice melt. The shore icefoot (4) can either directly or indirectly affect coastal morphology.

Ice rafting

The idea that large quantities of sediment can be transported long distances by floating or "drift" ice has a long history (Lyell, 1830-33, 1867) and many modern studies describe the effects of this process, which involves the entrainment of sediment in the ice, its transport due to the movement of ice floes by wind, wave and tidal currents, and its release or deposition as the ice melts. Very large quantities of mud are recycled in this manner in meso- and macro-tidal, mid-latitude estuarine settings such as the St. Lawrence estuary (Dionne, 1981b) and Chignecto Bay, Bay of Fundy (Gordon and Desplanque, 1981), where the offshore ice cover is continually in motion throughout the winter. In most sand and gravel beach settings throughout the seasonal sea ice zone, ice rafting is a relatively minor agent of sediment transport, but the small amounts of sediment emplaced in this way are commonly lithologically and texturally different from local beach materials. In temperate and low Arctic salt marshes, rafting of vegetation frozen to the base of sea ice pans can be a locally important process (Dionne, 1968, 1972b; Ugolini, 1975).

Intertidal boulder accumulations characterize much of the eastern Canadian seaboard affected by seasonal sea ice. Boulders may occur singly, scattered across tidal flats, but are commonly arranged in mounds, ridges and barricades (Brochu, 1961; Dionne, 1962, 1968, 1972c, 1978, 1979, 1981a; Knight and Dalrymple, 1976; Rosen, 1979; Lauriol and Gray, 1980; Gilbert and Aitken, 1981; Guilcher, 1981; McCann *et al.*, 1981a,b). There is no doubt that many of the boulders, including some of the largest in all areas are transported short distances across the tidal flats by sea ice-related processes. However, there is some speculation about the exact nature of the processes involved (Drake and McCann, 1982). A limiting factor in the ice rafting of boulders is the entrainment mechanism. Rolling and sliding along the bed may be a more common mode of transport than rafting for the largest boulders.

Ice push and scour

Ice pushed ridges and associated scour marks and striations are common along most high-latitude marine and lacustrine shores (Nichols, 1953; Hume and Schalk, 1964; Bird, 1967; Dionne, 1969b, 1978; Mansikkaniemi, 1970, 1976; Kovacs and Sodhi, 1980). They occur as a result of wind-induced onshore movement of pack ice or large ice floes, and additionally due to thermal expansion of ice in the lacustrine environment (Alestalo and Häikiö, 1979; Alestalo, 1980). Onshore, ice push features are typically observed within 10 m of mean high tide level, but can be found more than 100 m inland across low lying shores (Taylor, 1978b; Kovacs and Sodhi, 1980). As sea ice moves inland, it scours the ground surface pushing sediment into low ridges, or, where ice buckles, forming shore ice pile-ups. Ice push and ice override are most frequent and severe along coastal promontories. Since wave action during storms late in the open water season usually mobilizes sediment down to the frost table (McCann, 1973; Taylor, 1981) ice push structures within the reach of waves are short-lived. Those formed farther inland are modified by subaerial processes but can persist for several centuries (Bird, 1967).

Offshore, the grounding of single ice blocks, ice pressure ridge keels, and ice island fragments, produces an irregular sea-bed topography of scour marks, pits, ridges and nearshore shoals. Ice interaction with the sea bed is best documented in the Beaufort Sea where it is most intense in the stamukhi or shear zone between the 15 and 30 m isobaths (Rex, 1955; Pelletier and Shearer, 1972; Reimnitz *et al.*, 1972, 1977, 1978; Kovacs and Mellor, 1974; Reimnitz and Barnes, 1974; Shearer and Blasco, 1975; Hnatiuk and Brown, 1977; Lewis, 1977; Lewis *et al.*, 1982). Scour track morphology depends on: water depth, ice characteristics, sea-bed slope and sediment properties. Some larger scour tracks are flanked by levées and terminate in large craters. Scour tracks are obliterated by waves in the littoral zone, modified by repeated ice groundings in the stamukhi zone and tend to infill because they are preferential sites of sediment deposition. In shallow waters the sea bed undergoes complete reworking every 50 years and possibly more frequently in the stamukhi zone (Reimnitz *et al.*, 1977). With regard to the preservation of these features, Lewis *et al.* (1982) suggest that, in the Beaufort Sea, scour marks beyond the 50-m isobath are relict. Dionne (1977) and Dredge (1982) have observed relict ice scours along the floors of former glacial lakes in northern Canada, and Hodgson (1982) has found tracks resembling ice scour on some of the north-western Queen Elizabeth Islands. The main effects of ice push and ice scouring are irregular topography, reworking of surficial sediments, transfer of sediment into and out of the littoral zone, and disruption of coastal drainage.

Ice melt

Beach pitting, mound topography and collapse structures resulting from the melting of ice are extensively documented (Rex, 1964; Greene, 1970; Dionne and Laverdiere, 1972; Short and Wiseman, 1974; Short, 1976; Taylor and McCann, 1976). In the Canadian arctic islands, beach pitting is commonly observed following the melting of sea ice, driven onshore during storms. The largest pits, greater than 2 m in diameter, are produced across the foreshore following the melting of partially buried multi-year ice boulders. Smaller pits form as buried brash ice melts at the storm swash limit and in overwash deposits. Collapse structures, dirt cones and other micro-relief structures observed following ice break-up are the remnants of melting shorefast ice, including the icefoot. The nature of the micro-relief in any one year depends very much on the characteristics (thickness, width, ice types, sediment inclusions) of the ice which was formed in the previous autumn.

Icefoot

There are several accounts of the development, decay and geomorphic effects of the icefoot in different environments (e.g. Wright and Priestly, 1922; Bentham, 1937; Joyce, 1950; Feyling-Hanssen, 1953; McCann and Carlisle, 1972; Dionne, 1973; Marsh *et al.*, 1973; Owens, 1976, in press; Evenson and Cohn, 1979; Nielsen, 1979). The form of the icefoot varies according to shore gradient, tidal range, and wave and ice conditions during its formation. On exposed coasts, it may extend well above normal high water level due to the freezing of storm wave swash and spray. On tidal coasts it may extend to low water mark, and is separated from the sea ice by tidal cracks. Moore (1968) and Greene (1970) use the term "Kaimoo" to describe what Rex (1964) called the gravel-sand-icefoot, a layered ice-sediment feature developed on the upper beach by wave action before the formation of an icefoot. Nielson (1979) has documented the erosional effects of the icefoot on rock shores, and Ellis and Wilce (1969) attributed the barren nature of the upper part of the intertidal zone to the icefoot and exposure to freezing. However, early development of an icefoot in the fall or its persistence in the spring and summer when open water conditions prevail offshore, can protect the beach from subsequent wave action.

EXAMPLES OF ICE-AFFECTED COASTAL ENVIRONMENTS

In this section, the processes, morphologies and sediments in four coastal settings in northern Canada (A,B,C,D in Fig. 1) are described to illustrate the range of conditions that exist and to provide type examples of the key coastal environments. The first three examples show the effects of the increase in the length of the open water season, and hence the importance of wave action, from west to east in the Queen Elizabeth islands; the fourth example considers the macrotidal shorelines of southern Baffin Island.

North-west Queen Elizabeth Islands

The north-western Queen Elizabeth Islands (Fig. 4) fringe the Arctic Ocean and consist of lowland terrain developed in unresistant Mesozoic-Cenozoic sedimentary rocks. The microtidal ($<0{\cdot}4$ m range) shorelines are ice-locked for at least 11 months each year and ice action and terrestrial processes dominate the coastal environment (Horn, 1967; Taylor, 1980c; Owens *et al.*, 1981). With the exception of shore leads, there are no clearly defined areas of open water within the inter-island channels. In most years water motion at the shore is restricted to wavelets and tidal oscillations which produce water-level marks, winnow fines from the foreshore and smooth features produced by other processes. Large areas of open water occasionally occur along the southern shores of the region and low (0·5 m) sand berms are common.

The extensive coastal plains are dominated by fluvial processes and a dense network of streams transports large quantities of sediment to the coasts so that deltas are a recurring feature. Material delivered to the coast remains essentially unaffected by marine processes, and shore zone sediments correspond closely to the weathered products of the underlying bedrock. The abundance of silt and clay on some shores is conducive to slope failure, even on low gradient slopes, and earth flow levées and scars persist, producing very hummocky terrain. Where these shores are also subject to ice override and ice push, shore topography becomes very complex and it is difficult to distinguish earth flow levées from ice push ridges. Massive ground ice at shallow depths often provides the basal shear plane for slope failure, and melting of the exposed ice causes thermokarst topography. Across the sandy shores, which are not so disturbed, aeolian processes are locally important

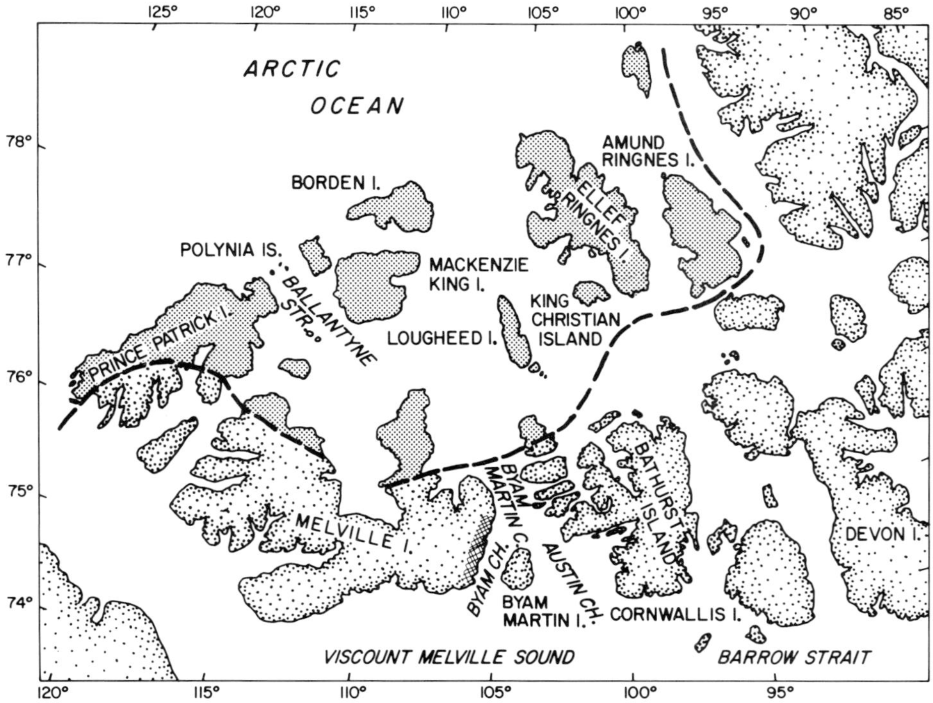

Figure 4. Ice-dominated coastal environments of the north-western Queen Elizabeth Islands (shaded area) and eastern Melville Island (cross-hatched area).

forming deflation hollows and sand drifts in the lee of ice push ridges. Ice push, ice override and ice ridging occur everywhere but are particularly important on shores fringing the Arctic Ocean and at promontories and deltas. The largest ice-built features occur offshore along the shear zone between the Arctic Ocean pack ice and the landfast ice. Sediment mounds up to 18 m high, which mark the shoals off Borden and Ellef Ringnes Islands (McClintock and Haughton, 1858), are formed on the shallow seafloor (Hudson *et al.*, 1981).

On King Christian and the Ringnes Islands relative sea level was 80–100 m above present 10 000 years ago, but the highest well-formed strandlines which occur at about 30 m were formed 6000 years ago (Hodgson, 1982). Elsewhere in the islands, discontinuous strandlines of different elevations are visible on aerial photographs but have little morphological expression on the ground. It would appear that low wave energy and ice dominant conditions similar to those of today have prevailed throughout the Holocene. Though only a few large ice-built ridges and some ice scours are preserved in the emerged coast, several characteristic features of the region emphasize the absence of waves (Table 3).

TABLE 3

Coastal characteristics of the north-west Queen Elizabeth Islands emphasizing the low wave energy environment

1.	Numerous river deltas of all sizes protrude offshore
2.	Relict river delta surfaces indicate few lateral channel deflections at delta mouths during Holocene (Hodgson, 1982)
3.	Absence of well developed wave-built or erosional features alongshore, e.g. sand bars or wave cut bluffs
4.	Terrestial or ice-built features persist across modern beach slope, e.g. sand drifts, earth flow levées, and ice push mounds
5.	Proximity of nearshore mud-sand boundary to beach, poorly compacted foreshore sediment and steep foreshore slopes of 2° to 14°
6.	Vegetation extends to high tide line along very flat ice-sheltered shores
7.	Large mounds of sediment offshore and concentric gravel berms built around small interchannel islands by sea ice pressure become permanent coastal features

Eastern Melville Island

The east coast of Melville Island, facing the Byam and Byam Martin Channels (Fig. 4), experiences less severe ice conditions than the coasts of the north-western islands. Wave action becomes more important, but ice scour and ice push, associated with grounding sea ice in the nearshore zone, are probably the critical factors influencing shoreline development. McLaren (1982) has defined three types of depositional shoreline (deltas, sand beaches and gravel beaches) and explained their genesis and distribution in terms of the interaction of fluvial, marine and ice processes during Holocene emergence. He suggested that sea ice has continually interacted with the nearshore bottom during emergence and that ice push has carried sediment landward, the type of material in transport depending on the distribution and thickness of a nearshore sand unit. Where this unit was thicker than 3 m, grounding sea ice was unable to penetrate to the underlying gravelly glaciomarine unit, and only sand was pushed shorewards to form the beaches. Contemporary sand beaches have a low-gradient (2°) foreshore, a low berm, and a flat featureless backshore: ice push features are absent, having been quickly removed by subaerial processes. Where the nearshore sand unit was absent or less than 3 m thick, ice scouring of the glaciomarine unit occurred and ice push provided a mixture of sediments for reworking by wave action. The finer fractions were winnowed out and the lag gravels built into beach ridges. At promontories, beach ridges consist of very coarse angular material derived from the erosion of bedrock exposed offshore. Contemporary gravel beaches have a steep (8·5°) foreshore slope that commonly exhibits irregular mounds of ice pushed sediment.

Raised strandlines are best exhibited in the sectors of gravel shoreline, where a series of well-defined beach ridges extend to elevations in excess of 30 m. The raised beach gravel is finer and more angular than that on the modern beach, due to frost shattering. Along the sandy shorelines there is little evidence of former sea levels, the original sand berm having been levelled off by wind and slope processes. The surfaces of raised sand deltas have been similarly modified, but exposures of foreset bedding allow these deposits to be distinguished from the former sandy beaches.

East central Arctic Islands

In the sheltered wave environment of the east central Arctic archipelago (Fig. 5) infrequent, short periods of intense wave action are chiefly responsible for the coastal depositional landforms (McCann, 1973; Taylor, 1978a). With limited fetch, a mixed, semi-diurnal tidal range of 1·2–2·8 m and a short open water season (Table 2), simple, small-scale, depositional features are developed. The most common and conspicuous features are narrow, gravel beach ridges, flights of which extend to, or near to, the marine limit, which varies from 76 to over 120 m above sea level (Blake, 1975; Dyke, 1979). A few locations exhibit more complex features (spits, barriers and lagoons), the necessary conditions are a slowly emerging coast fringed by a wide, shallow marine bench. Type examples of the depositional coastal

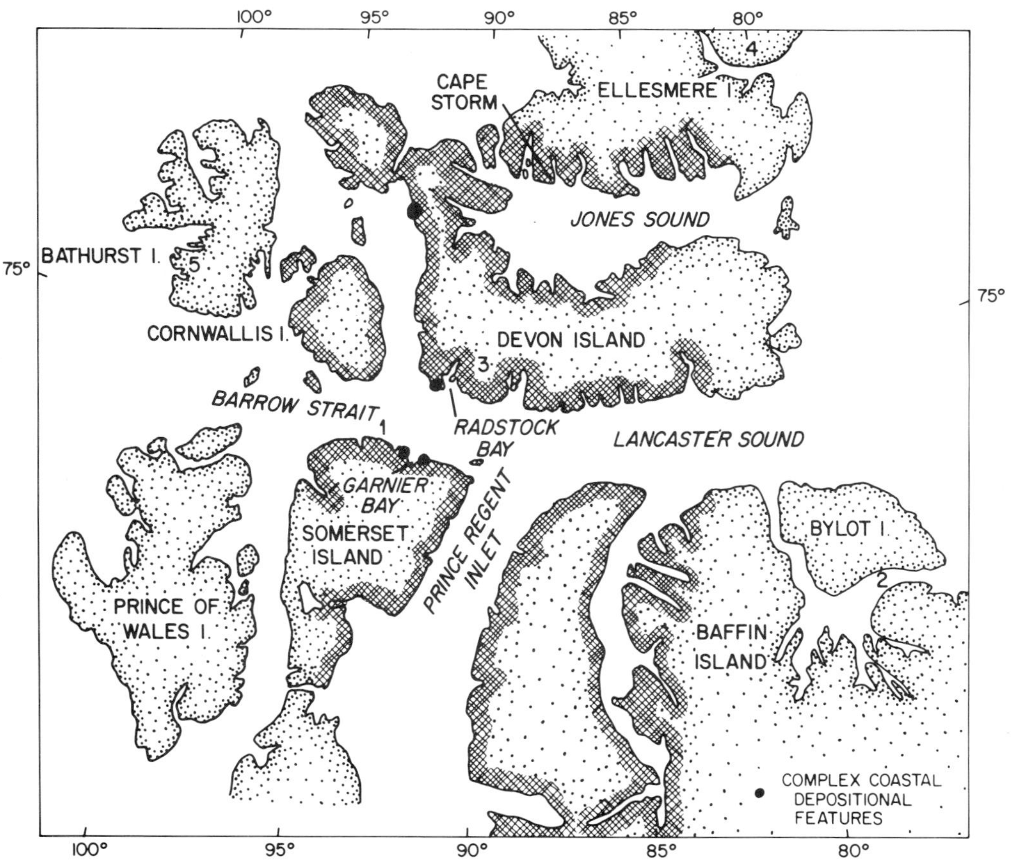

Figure 5. Sheltered wave environments of the east central arctic islands (cross-hatching). Numbers (1-5) indicate locations described in Table 2.

features of this region are described from Radstock Bay, Devon Island and northern Somerset Island where beach evolution and coastal processes have been monitored since 1969 (Owens and McCann, 1970; McCann, 1973; Taylor, 1980a).

The modern shoreline is characterized by steep (6–14°), reflective, gravel beaches with a series of swash ridges built across the upper foreshore (McCann and Owens, 1970). Except for irregularities in the profile introduced by the presence of ice (pitting and ice push ridges) the beach profile resembles the step or summer profile described by Orford (1977) along mid-latitude shorelines. Poor to moderately sorted, subangular limestone pebbles (4–32 mm) constitute the bulk of the modern beach sediment (Owens and McCann, 1970). Glacial deposits are not abundant and, since sediment must be derived from local bedrock sources, beach sediment thickness is less than a few metres. Swash ridges (0·5 m high) are transported across the foreshore by small, short period waves, and during higher wave energy events, they are either incorporated into a larger storm ridge or destroyed. The highest modern ridges are formed when strong onshore winds coincide with maximum open water fetch and spring tidal levels. Some of the beaches are modified to elevations 4·9 m above mean high tide level (Taylor, 1978a). Wave height, period and run-up increase during storms but once the waves reach a critical steepness, plunging breakers comb down the lower beach slope. If grounded multi-year sea ice lines the shore, it protects the upper beach from the full force of waves, but, by shifting the wave breaker zone to its location, it causes severe scouring of sediment around the base of the ice. Constructional waves following a storm can restore the lower beach slope to its pre-storm profile within 15 days (McCann, 1973; Taylor, 1981).

It has been suggested (King, 1969) that gravel beach ridges on the emerged coast of western Baffin Island were formed during single storms. The question arises as to what magnitude of storm would produce those ridges. At Radstock Bay, Devon Island, the largest observed storm in 1969, which had a recurrence interval of 5–7 years, did not produce a large storm ridge but rather it substantially eroded raised and modern beach deposits. Subsequent annual surveys until 1972 and in 1976 and 1981 indicate that many storm ridges have formed and been eroded but no new ridge comparable to those on the emerged coast ever developed. It is suggested that this is due to the fact that it takes a substantial period of time to accumulate sufficient sediment along a sector of coast before a ridge of comparable size to the raised beach ridges can develop. For instance, along northern Somerset Island, only four major beach ridges have formed during the last 1420 ± 50 years B.P. (^{14}C date on wood—GSC 2704). In southern Ellesmere Island, Blake (1975) dated a whalebone, frozen at the base of the second swale from the sea, at 940 ± 70 B.P. (GSC—980-2), and dated driftwood, lodged in beach deposits less than 1·5 m above present sea level, at 140–160 ± 60 B.P. (GSC—1352, GSC—1550).

Beach development has been continuous since deglaciation 9000–9500 B.P. During this period coastal emergence has decreased from 4 to 11 m/century during the first 1000 years to less than 0·5 m/century during the last 2400 years (Barr, 1971; Blake, 1975; Dyke, 1979). The highest beach ridges, at the base of upland slopes, have been buried or modified by periglacial processes, but the lower ridges are well preserved due to the coarseness of the sediment and relict depositional sequences are easily traced. Two types of beach ridge morphology have evolved: ridges of similar elevation separated by well defined swales, develop across marine benches or very gradual slopes; and stepped beach ridges, as much as 2 m below each other, develop across the steeper coastal slopes. Wave-built sediment facies across the raised beaches are similar to those observed in lower latitude gravel beaches. Occasionally, ice melt features infilled by overwash deposits, ice push mounds, and truncated ice-bonded sediments are observed in sections. Surface gravels of raised beaches are more angular than either the subsurface or modern beach sediments because of solution and frost weathering.

South-east Baffin Island

The most striking depositional features of south-east Baffin Island are the wide, boulder-strewn, intertidal flats which occur around parts of Frobisher Bay, Cumberland Sound and Hudson Strait (Gilbert and Aitken, 1981; McCann *et al.*, 1981a,b). This region is selected as an example to illustrate the interaction of sea ice and tidal processes in a macro-tidal environment. Mean tidal range approaches 5 m toward the head of Cumberland Sound and increases to 7·3 m at the head of Frobisher Bay; the ranges of spring tides are 7·7 m and 11·6 m respectively. The effects of sea ice on tidal flats have also been studied in several other areas in eastern Canada: Minas Basin, Bay of Fundy (Knight and Dalrymple, 1976); the St. Lawrence estuary (Brochu, 1961; Dionne, 1968, 1969a,b, 1972b,c, 1973, 1975, 1979; Guilcher, 1981); central Labrador (Rosen, 1979); and Ungava Bay (Lauriol and Gray, 1980). A brief review of this work is given in McCann *et al.* (1981a) and Drake and McCann (1982).

At the head of Frobisher Bay break-up commences in mid-June and ice free conditions usually prevail from mid-July to late October. The highly fractured and sediment laden ice cover in the wide intertidal zone breaks up before the floating sea ice in the centre of the inlet. This early phase of break-up, when rapidly disintegrating and melting floes of tidal flat ice are moved across the flats by tidal currents, is probably the most dynamic period of the year in terms of sea ice action. Because the ablating ice is contained within the intertidal zone, sediment incorporated within it is redeposited there, though at different locations to where it was entrained. The tidal flats are everywhere littered with cobbles and boulders, resting on and embedded in finer sediments, which range from clay to gravel but are predominantly coarse sand. Rocking, pushing, over-turning and sliding appear to be the important modes of movement for the larger boulders. In Pangnirtung Fiord on the north shore of Cumberland Sound, the most striking feature is the nearly continuous boulder barricade which fringes the outer margin of the tidal flats. Gilbert and Aitken (1981) conclude that this has formed by seaward transport of boulders across the flats to the zone where sea ice interacts with the shorefast ice, and suggest that the history of sea-level change, involving a relatively stable sea level from about 6000 B.P., is an important factor in the overall development of the tidal flats. Raised boulder barricades have been utilized as diagnostic indicators of former sea levels in northern Labrador by Løken (1962) and in Ungava Bay (Gray *et al.*, 1980; and Lauriol and Gray, 1980).

SUMMARY

Attention has been focused on the geological action of drift ice and the preservation potential of ice-induced features in four different coastal environments. Indirectly, ice, by reducing or eliminating waves, affects the development of depositional coastal features, and directly, drift ice produces unique morphological and sedimentological imprints across the coastal zone that can give information about former depositional environments. The indirect effects of sea ice are clearly reflected in the sediment characteristics, size of wave-built features and rates of coastal change. As expected, the indirect effect of ice increases in a north and west direction across the arctic islands and from south to north across eastern Canada. Along ice-locked shores (e.g. north-west Queen Elizabeth Islands), coastal sediments are a reflection of the underlying geology because of the inability of waves to rework and transport sediment alongshore. Terrestrial and sea ice processes dominate the coastal plan and profile (Table 3). Farther east (e.g. east-central arctic islands), the beaches are primarily modified by waves during infrequent, short period storm events. In this region, well defined coastal depositional features exist, but it takes considerable time for sufficient sediment to accumulate before modern beach ridges can develop, comparable to those on emerged shores. The size of wave-built beach ridges increases from the north-west Queen Elizabeth Islands to the eastern Arctic because of the greater tidal range and the greater duration and extent of open water. Permafrost can affect the coastal zone in two contrasting ways. Where massive ground ice in unconsolidated sediment is exposed and melts, it leads to rapid coastal retreat or thermokarst features. Where permafrost lies at shallow depths it restricts scouring by sea ice and, if exposed during storms, ice-bonded coarse sediment can delay beach erosion.

Not all of the ice-induced features listed by Nichols (1961) in Table 1 can be found in any given coastal region; however, similar morphological and sedimentological imprints are left by drift ice on all shores within the seasonal sea ice zone. An exception is the boulder barricade that is restricted to tidal flats where boulders are abundant (e.g. south-east Baffin Island and Labrador). A primary effect of sea ice is the reworking of surficial coastal sediments. Many of the depositional features built by ice are merely the product of local erosion. Similar ice-built features can occur across the coastal zone, but it is recognized that ice-pushed ridges and thermokarst features are best defined in the backshore, beach pits are largest across the foreshore zone, and scour marks are best exhibited across the nearshore and shallow shelf. In terms of the sedimentological imprints left by drift ice the most common are lateral and vertical discontinuities in sediment textures and sedimentary structures due to the infilling of ice-melt features and the deposition of ice-rafted sediment. Martini

(1980) has suggested the term "textural-structural inversions" for these and similar features which indicate the juxtaposition in the same environment (or deposit) of two or more energy states of different processes. Deformation structures are produced by ice push and ice loading on finer coastal sediments. An additional effect of ice is the absence of the biota beneath the icefoot and the physical damage to nearshore biota caused by moving sea ice.

The preservation potential of depositional features or the unique imprints of sea ice across emergent coasts is a function of the sediments involved and the coastal zone in which they form. Across the sandy shores of the north-west Queen Elizabeth Islands wave and ice-built features are quickly erased by subaerial processes and where earth flow levées exist in the raised coast, it is difficult to distinguish them from ice-pushed ridges. In contrast, coastal depositional features formed in gravel (e.g. beach ridges) are well defined across uplifted coasts. Ice-built features of gravel can persist for centuries but are modified by slope and periglacial processes. We confirm the observation of Nichols (1953) that ice push ridges are not found on the higher raised beach ridges. Across the coastal zone, the morphological imprints of ice are least preserved if formed within the zone of storm wave activity. Features above the reach of waves can persist longer. Ice scour marks are rarely preserved in the littoral zone but are found across the floor of relict glacial lakes. Even though the morphological features formed by ice can be erased or reworked by waves or subaerial processes, the sedimentological signatures of direct ice action, such as deformation structures, heterogeneous deposits, textural structural inversions and truncation of beds, can be preserved at depth. Thus, Reinson and Rosen (1982) have described sedimentary units within a modern barrier beach in Labrador, which they suggest represent break-up and freeze-up events in the yearly cycle of beachface accretion, concluding that one or both units could be preserved in the rock record. However, something of the frustration that might attend the search for, and interpretation of, these features in elevated shoreline sequences is captured in the following remark by Martini (1980, p. 95):

> It is a common experience, and a rather annoying one, for geologists to realise that many ice-generated features observed on recent coasts are not readily recognizable in cross-sections of ancient deposits.

REFERENCES

Adams, W. P. and Mathewson, S. A. (1976). *Revue Géogr. Montréal* **30** (1-2), 187-196.
Alestalo, J. (1980). *Fennia* **158**, 27-39.
Alestalo, J. and Haikio, J. (1979). *Fennia* **157**, 51-92.
Allen, W. T. R. (1964). *Can. Met. Branch*, CIR-4116, Ice—17, Toronto.
Aré, F. E. (1972). *International Geographical Congress Paper* 78-79.
Barr, W. (1971). *Arctic* **24**, 249-268.
Bentham, R. (1937). *In* "Arctic Journeys. The Story of the Oxford University Ellesmere Land Expedition" (E. Shackleton, ed.), pp. 328-332. Hodder and Stoughton, London.

Bilello, M. A. (1960). *Arctic* **14**, 2-24.
Bilello, M. A. (1980). *In* "Sea-Ice Processes and Models". *Proceedings of the Arctic Ice Dynamics Joint Experiment, International Commission on Snow and Ice Symposium* (R. S. Pritchard, ed.). University of Washington Press, Seattle.
Bird, J. B. (1967). "The Physiography of Arctic Canada". John Hopkins Press, Baltimore.
Blake, W. Jr (1975). *Geogr. Annlr.* **A57**, 1-71.
Brochu, M. (1961). *Min. Mines et Relevés Techn., Ét. Géogr.* **30**. Ottawa.
Cooper, P. F. Jr (1974). *In* "The Coast and Shelf of the Beaufort Sea" (J. C. Reed and J. E. Slater, eds), pp. 235-242. Arctic Institute of North America, Arlington.
Dionne, J. C. (1962). *Can. Geogr.* **7**, 69-77.
Dionne, J. C. (1968). *Am. J. Sci.* **266**, 380-388.
Dionne, J. C. (1969a). *Revue Géogr. Montréal* **23**, 5-20.
Dionne, J. C. (1969b). *J. sedim. Petrol.* **39**, 1174-1181.
Dionne, J. C. (1972a). *Centre de Recherches For Estières des Laurentides, Région de Québec, Quebec, Rapport* Q-F-X-34.
Dionne, J. C. (1972b). *Z. Geomorph. Suppl. Band* **13**, 131-162.
Dionne, J. C. (1972c). *Revue Géogr. Montréal* **26**, 125-152.
Dionne, J. C. (1973). *Cah. Géogr. Québ.* **41**, 221-250.
Dionne, J. C. (1975). *In* "Fluvial Processes" (P. G. Johnson, ed.), pp. 25-37. University of Ottawa Press, Ottawa.
Dionne, J. C. (1977). *Maritime Sediments* **13**, 79-81.
Dionne, J. C. (1978). *Géogr. Physique et Quaternaire* **32**, 3-70.
Dionne, J. C. (1979). *Maritime Sediments* **15**, 5-13.
Dionne, J. C. (1981a). *Workshop on Ice Action on Shores: Proceedings, National Research Council of Canada, ACROSES Publ.* **3**, 53-80.
Dionne, J. C. (1981b). *Workshop on Ice Action on Shores: Proceedings, National Research Council of Canada, ACROSES Publ.* **3**, 81-87.
Dionne, J. C. and Laverdière, C. (1972). *Canadian Journal of Earth Sciences* **9**, 979-990.
Drake, J. J. and McCann, S. B. (1982). *Canadian Journal of Earth Sciences* **19**, 748-754.
Dredge, L. A. (1982). *Canadian Journal of Earth Sciences* **19**, 1079-1087.
Dyke, A. S. (1979). *Current Research, Part B, Geol. Surv. Can. Paper* **79-1B**, 307-318.
Ellis, D. V. and Wilce, R. T. (1969). *Arctic* **14**, 224-235.
Evenson, E. B. and Cohn, B. P. (1979). *Z. Geomorph.* **23**, 58-75.
Feyling-Hanssen, R. W. (1953). *Norsk geogr. Tidsskr.* **14**, 45-52.
Gilbert, R. and Aiken, A. E. (1981). *Workshop on Ice Action on Shores: Proceedings, National National Research Council of Canada, ACROSES Publ.* **3**, 89-103.
Gordon, D. C. and Desplanque, C. (1981). *Workshop on Ice Action on Shores: Proceedings, National Research Council of Canada, ACROSES Publ.* **3**, 35-52.
Gray, J. T., De Boutray, B., Hillaire Marcel, C. and Lauriol, B. (1980). *Arctic and Alpine Research* **12**, 19-30.
Greene, H. B. (1970). *J. sedim. Petrol.* **40**, 419-427.
Guilcher, A. (1981). *Géographie Physique et Quaternaire* **35**, 155-169.
Hamelin, L-E. (1959). "Dictionnaire Français—Anglais des Glaces Flottantes". Québec, Université Laval; Institute de Géographie, Travaux No. 9 (unpublished).
Hamelin, L-E. (1976). *Revue Géogr. Montréal* **30**, 233-236.
Harper, J. R. (1978). "The Physical Processes Affecting the Stability of Tundra Cliff Coasts". Ph.D. Dissertation, Louisiana State University.

Hnatiuk, J. and Brown, K. O. (1977). *Proceedings 9th Annual Offshore Technology Conference, Houston.*
Hodgson, D. A. (1982). *Geol. Surv. Can. Paper* 81-89.
Horn, D. R. (1967). "Recent Marine Sediments and Submarine Topography, Sverdrup Islands, Canadian Arctic Archipelago". Ph.D. Dissertation, University of Texas.
Hudson, R. D., Pilkington, G. R. and Metge, M. (1981). *Workshop on Ice Action on shores: Proceedings, National Research Council of Canada, ACROSES Publ.* **3**, 15-34.
Hume, J. D. and Schalk, M. (1964). *Am. J. Sci.* **262**, 267-273.
Hume, J. D. and Schalk, M. (1967). *Arctic* **20**, 86-103.
Jacobs, J. D., Barry, R. G. and Weaver, R. L. (1975). *Polar Record* **17**, 521-536.
John, B. S. and Sugden, D. (1975). *Progress in Geography* **7**, 53-132.
Joyce, J. R. F. (1950). *J. Geol.* **58**, 646-649.
King, C. A. M. (1969). *Geogr. Annlr.* **A51**, 207-218.
King, C. A. M. and Buckley, J. T. (1968). *J. Sedim. Petrol.* **38**, 482-520.
Knight, R. J. and Dalrymple, R. W. (1976). *Revue Géogr. Montréal* **30**, 65-85.
Kovacs, A. and Mellor, M. (1974). *In* "The Coast and Shelf of the Beaufort Sea" (J. C. Reed and J. E. Slater, eds), pp. 113-161. Arctic Institute of North America, Arlington.
Kovacs, A. and Sodhi, D. S. (1980). *Cold Regions Sci. Technol.* **2**, 209-288.
Lauriol, B. and Gray, J. T. (1980). *In* "The Coastline of Canada" (S. B. McCann, ed.), pp. 281-292. *Geol. Surv. Can. Paper* 80-10.
Lewis, C. F. M. (1977). *Proceedings, 4th International Conference on Port and Ocean Engineering under Arctic Conditions* **I**, 568-579. Memorial University of Newfoundland.
Lewis, C. F. M., Blasco, S. M., Pelletier, B. R. and Sparkes, R. (1982). *Abstracts of Papers, Eleventh International Congress on Sedimentology*. Hamilton.
Lindsay, D. G. (1969). *Canadian Institute of Mining and Metallurgy, Spec. Vol.* **10**, 45-60.
Løken, O. H. (1962). *Geogrl. Bull., Ottawa* **17**, 23-56.
Lyell, C. (1830-1833). "Principles of Geology", 1st ed., Vol. 1, Vol. 2, 1832; Vol. 3, 1833. John Murray, London.
Lyell, C. (1867). "Principles of Geology", 10th ed., Vol. 1. John Murray, London.
Mansikkaniemi, H. (1970). *Publications Instituti Geographici Universitatis Turkuensis* **50**, 1-30.
Mansikkaniemi, H. (1976). *Fennia* **144**, 1-17.
Marsh, W. M., Marsh, B. D. and Dozier, J. (1973). *Am. J. Sci.* **273**, 48-64.
Martini, I. P. (1980). *Canadian Coastal Conference: Proceedings of the National Research Council of Canada, Ottawa*, 93-102.
Maxwell, J. B. (1981). *Arctic* **34**, 225-240.
McCann, S. B. (1972). *In* "International Geography" (W. P. Adams and F. M. Helleiner, eds), Vol. 1, pp. 41-43. University of Toronto Press, Toronto.
McCann, S. B. (1973). *In* "Coastal Geomorphology" (D. R. Coates, ed.), pp. 141-155. *Proc. Third Annual Geomorphology Symposium S.U.N.Y. Binghamton.*
McCann, S. B. and Carlisle, R. J. (1972). *Inst. Br. Geogr. Spec. Publ.* **4**, 175-186.
McCann, S. B. and Hannell, F. G. (1971). *J. Glaciol.* **10**, 155-157.
McCann, S. B. and Owens, E. H. (1969). *Arctic and Alpine Research* **1**, 267-278.
McCann, S. B. and Owens, E. H. (1970). *Shore and Beach* **38**, 26-30.
McCann, S. B. and Taylor, R. B. (1975). *Arctic and Alpine Research* **7**, 379-386.
McCann, S. B., Dale, J. E. and Hale, P. B. (1981a). *Géographie Physique et Quaternaire* **35**, 183-204.

McCann, S. B., Dale, J. E. and Hale, P. B. (1981b). *Workshop on Ice Action on shores: Proceedings, National Research Council of Canada, ACROSES Publ.* **3**, 105-115.

McClintock, Sir, F. L. and Haughton, S. (1858). "Reminiscences of Arctic Ice-travel in Search of Sir John Franklin and his Companions". John Murray, London.

McDonald, B. C. and Lewis, C. P. (1973). *Environmental-Social Committee, Northern Pipelines, Task Force on Northern Oil Development, Report* 73-39.

McLaren, P. (1982). *Bull. geol. Surv. Can.* 333.

Moore, G. W. (1968). *In* "Encyclopedia of Geomorphology" (R. W. Fairbridge, ed.), pp. 21-22. Reinhold, New York.

Nichols, R. L. (1953). *J. Glaciol.* **2**, 172-175.

Nichols, R. L. (1961) *Am. J. Sci.* **259**, 694-708.

Nielsen, N. (1979). *Z. Geomorph.* **23**, 321-331.

Orford, J. D. (1977). *Earth Surface Processes* **2**, 381-400.

Owens, E. H. (1976). *Revue Géogr. Montréal* **30**, 95-104.

Owens, E. H. (1977). "Coastal Environments of Canada: The Impact and Clean-up of Oil Spills; Fisheries and Environment Canada". *Environmental Protection Service, Economic and Technical Review Report* EPS-3-EC-77-13.

Owens, E. H. (in press). *In* "The Encyclopedia of Beaches and Coastal Environments" (M. L. Schwartz, ed.). Academic Press, New York and London.

Owens, E. H. and Harper, J. R. (1977). *Arctic* **30**, 397-414.

Owens, E. H. and McCann, S. B. (1970). *Am. J. Sci.* **268**, 397-414.

Owens, E. H., Taylor, R. B., Miles, M. and Forbes, D. L. (1981). *Current Research, Part B, Geol. Surv. Can. Paper* 81-1B, 39-48.

Pelletier, B. R. and Shearer, J. M. (1972). *24th International Geological Congress, Section 8, Marine Geology and Geophysics*, 251-261. Montreal.

Reimnitz, E. and Barnes, P. W. (1974). *In* "The Coast and Shelf of the Beaufort Sea" (J. C . Reed and J. E. Slater, eds), pp. 301-351. Arctic Institute of North America, Arlington.

Reimnitz, E., Barnes, P., Forgatsch, T. and Rodeick, C. (1972). *Marine Geology* **13**, 323-334.

Reimnitz, E., Barnes, P. W., Toimil, L. J. and Melchior, J. (1977). *Geology* **5**, 405-408.

Reimnitz, E., Toimil, L. and Barnes, P. (1978). *Marine Geology* **28**, 179-210.

Reinson, G. E. and Rosen, P. S. (1982). *J. sedim. Petrol* **52**, 463-471.

Rex, R. W. (1955). *Arctic* **8**, 177-186.

Rex, R. W. (1964). *In* "Papers in Marine Geology" (R. L. Miller, ed.), pp. 384-400. MacMillan, New York.

Rosen, P. S. (1979). *J. sedim. Petrol.* **49**, 1113-1123.

Shearer, J. and Blasco, S. (1975). *Geol. Surv. Can. Paper* **75-1**, Part A, 483-493.

Short, A. D. (1976). *Revue Géogr. Montréal* **30**, 115-122.

Short, A. D. and Wiseman, W. J. (1974). *Arctic* **27**, 215-224.

Stringer, W. J. (1974). *In* "The Coast and Shelf of the Beaufort Sea" (J. C. Reed and J. E. Slater, eds), pp. 165-172. Arctic Institute of North America, Arlington.

Swithinbank, C. W. M. (1960). "Ice Atlas of Arctic Canada". *Defense Research Board*, Queens Printer, Ottawa.

Taylor, R. B. (1978a). *Fourth International Conference on Port and Ocean Engineering Under Arctic Conditions, Vol. 2*, 904-916. Memorial University of Newfoundland.

Taylor, R. B. (1978b). *Arctic* **31**, 133-149.

Taylor, R. B. (1980a). *In* "The Coastline of Canada", (S. B. McCann, ed.), pp. 239-250. *Geol. Surv. Can. Paper* 80-10.
Taylor, R. B. (1980b). *Canadian Coastal Conference Proceedings*, 103-121. Burlington, Ontario.
Taylor, R. B. (1980c). *Geol. Surv. Can. Open File* 693.
Taylor, R. B. (1981). *Workshop on Ice Action on Shores: Proceedings, National Research Council of Canada, ACROSES Publ.* **3**, 133-149.
Taylor, R. B. and McCann, S. B. (1974). *J. Glaciol.* **13**, 321-322.
Taylor, R. B. and McCann, S. B. (1976). *Revue Géogr. Montréal* **30**, 123-132.
Ugolini, F. C. (1975). *Science, N.Y.* **188**, 51-53.
Wadhams, P. (1980). *Cold Regions Sci. Technol.* **2**, 38-89.
Weeks, W. F. (1980). *Cold Regions Sci. Technol.* **2**, 1-35.
Wright, C. S. and Priestly, R. E. (1922). "Glaciology". British (Terra Nova) Antarctic Expedition, 1910-1913. Harrison, London.

FOUR

The development of shore platforms in high latitudes

A. S. TRENHAILE

INTRODUCTION

Much has been written about the development of rock coasts in the temperate regions of the world, but we know far less about these coasts in cool and cold climates. Although many opinions have been expressed on the formation of shore platforms in the mid-latitudes (Trenhaile, 1980), there is a lack of even basic agreement on the efficiency of coastal erosional processes in high latitudes. The authors of several texts believed that coastal development is very slow in cool climates (Bird, 1967; Zenkovitch, 1967; Davies, 1972), but shore platforms appear to be forming in some areas at the present time (e.g. Adie, 1964; Nichols, 1968; Sollid *et al.*, 1973; Moign, 1974a,b; Chastain, 1976). The rate and mode of development of rock coasts in cool environments need to be understood, not only because they constitute a significant proportion of the world's coasts, but also because much of the temperate zone experienced, and often still bears the imprint of, erosion which occurred under cool climatic conditions during the Pleistocene.

Of particular concern are the questions: (a) are cliff retreat and platform formation accomplished by different processes than those in more temperate latitudes; and (b) if the answer to (a) is yes, are these differences in type or intensity reflected in the development of zonally distinct morphologies. A distinction will be made in this chapter between processes which are responsible for cliff erosion and those which modify or planate the surface at its base. Many of the processes which operate in high latitudes are the same as those in the mid- and low-latitudes. They include mechanical wave action, and chemical and bioerosional weathering. Frost, coastal ice and isostatic changes in the relative level of the land and sea, may, however, produce zonal peculiarities which serve to distinguish the coastal landforms of cool regions from those which have developed in warmer climates.

SHORELINES AND ISOSTASY
ISBN 0 12 652960 4

FROST

Coastal engineers have found that concrete and stone may weather rapidly in the intertidal and spray zones of cool coastal regions (e.g. ACI, 1980). Indeed, Mather (1969) considered that this environment is as severe as any to which concretes are normally exposed. The geomorphological implications were recognized by a number of Scandinavian workers (Rekstad, 1915; Vogt, 1917). In North America, the effect of frost was described on the shores of freshwater bodies by Taber (1950), and later on the Atlantic coast of Canada by Dionne (e.g. 1972, 1978, 1980).

In the Bay of Fundy, nearly all the major cliff erosive events occur in spring (Thomson, 1979), which supports the conclusions reached by workers in Norway, Sweden, Finland, Western Canada, and Northern Ireland. Small talus accumulations occur at the base of the cliffs in Fundy, as well as those farther north in Gaspé (Trenhaile, 1980). In the Arctic, where there is often insufficient wave action to remove the weathered debris, the cliffs may become progressively buried under a talus apron (Bird, 1967; Howarth and Bones, 1972). The presence of talus and the occurrence of rockfalls during the spring thaw, or during the periods of most frequent alternations about the freezing point, have convinced many workers that gelifraction plays a very important role in the retreat of slopes in cold areas. Wyroll (1977), however, found that the failure of rock slopes in Labrador-Ungava could not be attributed to frost action, and he challenged the validity of the frost-dominated slope development model. Other data, provided by workers who have considered the occurrence of freeze-thaw cycles, casts some doubt on the general efficacy of frost action in coastal and alpine environments (Cook and Raiche, 1962; Gardner, 1969; Thorn, 1979).

Cliffs

Tricart (1969) believed that periglacial cliffs retreat rapidly, but the Scandinavian and Spitsbergen literature, for example, are contradictory and inconclusive. Jahn (1961) measured a rate of retreat of between 2·5 and 5 cm yr^{-1} on a sea cliff in Spitsbergen, but Rudberg (1967) estimated that the cliffs of Gotland, Sweden, generally retreat at a rate of only 4–6 mm yr^{-1}. In northern Scandinavia, Rapp (1960a) calculated that valley sides have retreated by 1–2 m in the postglacial period, and by only 0·04–0·15 mm yr^{-1} between 1952 and 1960. Limestone, sandstone and chert cliffs which border a fjord in western Spitsbergen, but are presently above the level of wave action, were estimated to have retreated by 0·02–0·2 mm yr^{-1} between 1882 and 1954, and by 0·34–0·5 mm yr^{-1} over the last 10 000 years (Rapp, 1960b). In the same general area, but about 5 km from the coast, he calculated a recent rate of retreat of 0·05–0·5 mm yr^{-1}. In northern Finland, Soderman (1980) found that mean recessional rates for the Holocene are only between 0·07 and 0·18 mm yr^{-1}. Rapp emphasized that disintegration is presently slow, and it may be significant that

Rapp's, Rudberg's and Soderman's data, in contrast to Jahn's, are insufficient to account for the width of many shore platforms in high latitudes. Nevertheless, it must be questioned whether some of the data, which refer to mountain and valley slopes, are relevant to coastal cliffs which are subjected to saline splash and spray, and where weathered debris is removed by ice or wave action.

Platforms

Many workers believe that frost action plays an important role in the formation of shore platforms in high latitudes. This view was developed to explain the occurrence of strandflats, particularly where they have formed in narrow fjords and in other areas where wave action is fairly weak (Rekstad, 1915; Vogt, 1917; Nansen, 1922; Gronlie, 1924; Nordenskjold, 1928; Holtedahl, 1960). Many workers stressed the effect of frost on coastal cliffs, but Tietze (1962) suggested that strandflats are the result of frost action under shelf ice, as it rises and falls with the tide. The strandflat, however, is widest and best developed on headlands. According to Nansen (1904), the strandflat is always narrow in the outer fjords, and is absent in the inner fjords, because the sea which formed it acted less strongly here. Nansen (1922) later accorded the major role in the cutting of the strandflat to a combination of frost and wave action. He emphasized the particularly favourable conditions created by frost action on the cliffs, particularly in fjords and other sheltered areas, where narrow, rugged ledges have developed close to the high tide level. More recently, Moign (1974a,b) and Guilcher (1974) have argued that gelifraction is an essential mechanism for the formation of strandflats and contemporary shore platforms in Spitsbergen, particularly in fissile and fractured rocks. Andersen (1968) and Sollid *et al.* (1973) considered that frost action played a major role in the formation of the Norwegian Main Line platform (Younger Dryas), which has also formed in narrow fjords where wave action is feeble. The degree of development of the recent rock terraces, however, is determined by their exposure to wave action, although their occasional occurrence in sheltered locations, suggests that frost may also play a role (Sollid *et al.*, 1973). In western Scotland, low temperatures during the Loch Lomond stadial (Younger Dryas), may have expedited the development of shore platforms in sheltered areas (Sissons, 1974). Increased storminess during the Loch Lomond stadial, however, would also have encouraged marine erosion and the removal of the debris (Gray, 1978). Dawson (1980) also argued that very intense frost shattering favoured the formation of quite wide platforms during the Loch Lomond stadial. These platforms are commonly between 30 and 70 m in width, and they are found in some apparently very resistant igneous and metamorphic rocks. Dawson (1980) proposed that the shoreline was produced as a result of frost action on coastal cliffs and the seaward removal of debris by sea ice. Elsewhere, it has been claimed that the extensive limestone platforms of Anticosti Island in the Gulf of St. Lawrence are the result of solution and frost acting at the cliff base (Corbel, 1958). Frost action in the intertidal zone may also have contributed to the formation of shore platforms on the Shantarski Islands of eastern Siberia (Gubkin, 1969). Even in the fairly mild climate of southern England, chalk wave-cut platforms were damaged when they froze during a particularly severe winter (Williams and Robinson, 1981).

Mechanisms

It is not possible in this chapter to discuss in detail the various mechanisms which may contribute to the effects of frost action (a recent review has been provided by McGreevy, 1981). None the less, it is essential that workers appreciate some of the factors which govern the efficacy of frost in the coastal environment. Freeze-thaw cycles may occur in coastal areas because of fluctuations in air temperatures, and also in the intertidal zone, because of tidal wetting and drying; rocks may freeze in air during low tidal periods, but rapidly thaw when they become immersed at high tide (Nansen, 1922). In coastal Maine, concrete specimens which were placed at the mid-tidal level, thawed in water to about 2·8°C and froze in air to between −2 and −23°C (Cook, 1952). Rocks in the intertidal zone may therefore experience many more freeze-thaw cycles than those above the high tide level, where they are dependent upon fluctuating air temperatures. In Maine, more than 200 freeze-thaw cycles per year have been recorded in the intertidal zone, compared with only 30–40 inland (Kennedy and Mather, 1953). The very sudden changes in temperature caused by tidal ebb and flow further emphasize the especially severe conditions within the tidal zone.

The winter 0°C isotherms for air and surface sea water are almost coincident in the northern and southern hemispheres. The 0°C water isotherm essentially provides a polar limit to the zone in which freeze-thaw cycles are induced by tidal fluctuations. The frequency of these cycles declines towards the low latitudes, until the coasts are completely frost free. Their greatest frequency, therefore, is in the area immediately equatorwards of the 0°C water isotherm. This would include: the Alaskan Panhandle and northern British Columbia; New England and possibly southern Nova Scotia; north-western Europe; Japan, Korea, and the Yellow Sea coast of China; southern Chile and Argentina; and possibly the southern extremity of New Zealand. Fluctuations in air temperatures may also subject rocks in the intertidal and supratidal zones to frost action. In North America, the frequency of freeze-thaw cycles increases southwards in Canada and northwards in the United States (Russell, 1943; Visher, 1945; Fraser, 1959; Williams, 1964). The greatest frequency on the Atlantic coast occurs in New England, eastern Québec and the Maritime provinces. On the Pacific coast, maximum frequency is farther north, occurring along the Alaskan Panhandle (Williams, 1964). The duration and intensity of the freezing period, however, increases northwards. This general pattern is probably repeated along the margins of the other continental landmasses. There is considerable overlap between areas in which large numbers of freeze-thaw cycles are induced by tides, and those which experience frequent air temperature fluctuations about the freezing point. The latter, however, extend polewards of the 0°C water isotherm. Although the efficacy of frost action probably diminishes polewards in this zone, some frost mechanisms are facilitated by long periods of freezing. Frost may therefore accomplish some rock weathering even in very high latitudes.

The distinction is often made between frost damage caused by the formation of ice in joints, bedding planes, and other rock crevices, and disintegration produced by ice filling rock capillaries. Water freezing in crevices may, under ideal conditions, exert pressures against confining walls of 214 MPa at −22°C (Winkler, 1973). Pressures of this magnitude, however, are probably unattainable in the field. Grawe (1936) emphasized that a closed system containing only water is necessary for maximum pressures to be exerted in rock crevices, and he objected to the uncritical acceptance of frost as the dominant agent of rock weathering in cold climates. Nevertheless, stresses which are only one-tenth of the theoretical maximum, exceed the tensile strengths of even the most resistant rocks. Frost damage to porous materials has been attributed to several mechanisms. They include: the segregation of ice crystals into layers parallel to the cooling surfaces (Collins, 1944); the pressures exerted by ice crystals growing in the capillaries (Powers, 1955); the generation of hydraulic pressures (Powers, 1955); and the freezing of supercooled water which is expelled into cracks and other crevices (Litvan, 1976). These mechanisms depend upon the formation of ice, but their effects may be virtually indistinguishable in fine-grained argillaceous rocks, from temperature dependent wetting and drying, caused by the adsorption of water molecules onto internal clay surfaces (Dunn and Hudec, 1972; Hudec, 1973).

Nansen introduced an important misconception regarding the relative significance of freezing fresh and salt water. Nansen (1904, 1922), and later Guilcher (1958) and Davies (1972), believed that the freezing of fresh water is more deleterious to rocks than the freezing of salt water. Experimental work, however, has shown that frost damage is greater in saline than in freshwater solutions. Salts lower the freezing point and probably reduce the number of freeze-thaw cycles which can occur in a given period, but, they also increase the osmotic pressures which are generated, and they permit higher levels of saturation to be maintained during freezing (Powers, 1975; Litvan, 1976; MacInnes and Whiting, 1979). The greatest frost damage has repeatedly been found to occur in solutions which are about one-half to one and a half times the normal salinity of sea water (Verbeck and Klieger, 1957; Browne and Cady, 1975; Litvan, 1976; MacInnes and Whiting, 1979; Trenhaile and Rudakas, 1981).

Because of differences in experimental design and in the collection and preparation of rock samples, it is difficult to use experimental data to rank the susceptibility of rock types to frost action. Comparison of the results of numerous studies, however, suggests that the vulnerability of rocks generally decreases with increasing grain size and tensile strength, and increases with the number and prominence of lines of weakness. Grain size is important because it affects a rock's pore size and continuity, permeability, and degree of saturation (Lewis *et al.*, 1953; Walker *et al.*, 1969). The literature suggests that the most susceptible lithologies include well foliated, fine-grained metamorphic rocks such as phyllites, argillites, and slates, fine-grained quartzites, fine-grained, low rank greywackes, fissile shales, siltstones, claystones, and friable chalks, and other fissile sedimentary rocks containing mica and illite clays, and soft fibrous serpentinites. The most durable lithologies usually include unfoliated metamorphics, such as medium and coarse-grained quartzites, very fine-grained, low porosity lithographic and other crystalline limestones, and igneous rocks such as basalts, coarse-grained granites, peridotites, and pyroxenites (Trenhaile and Rudakas, 1981).

Experimental results are contradictory regarding the efficacy of various freeze-thaw cycles in the breakdown of porous materials. Rapid freezing has usually been found to be more deleterious than slow freezing (Thomas, 1938; Battle, 1960; Wiman, 1963; Arni, 1966; Potts, 1970; Lautridou, 1978). The greatest damage, however, has occasionally been found to be associated with slow freezing, although the number of cycles which can occur in a given period decreases with the wavelength of the cycle (Taber, 1950; Tricart, 1956; Williams and Robinson, 1981). The explanation for these differences may lie in the various mechanisms involved in frost action. Fast freezing facilitates the generation of hydraulic pressures, but slow freezing favours crystal growth and the segregation of ice into layers.

The effect of frost action within a rock depends upon its degree of saturation at the time of freezing. Because of the expansion which is attendant to the freezing of water, if a rock is less than about 91·7% saturated, assuming that the water is uniformly distributed throughout the rock, expansion may be accommodated without generating damaging pressures (Powers, 1955). Because water is rarely equally distributed, however, the critical saturation level may be considerably less than 91·7% in some rocks (Fagerlund, 1975). Nevertheless, critical saturation may be very difficult to attain under natural conditions (Verbeck and Landgren, 1960). Dunn and Hudec (1972), for example, found that the most resistant of the carbonate rocks which they investigated, were those which were unable to absorb or retain much water. Carbonates which were sensitive to frost or to temperature dependent wetting and drying, were those which were at least 80% saturated after being immersed for 24 hours in water (Hudec, 1973).

Position within the intertidal zone affects the characteristics of freeze-thaw cycles, the salinity of the water, and the amount of time in which a rock is immersed. Rocks may experience the greatest number of cycles just below the high tide level, where they are exposed for the greatest period of time. In this area, rocks experience most of the cycles associated with fluctuations in air temperatures, as well as those which depend upon thawing in water. Rocks which are near the high tide level, however, are frozen for much longer than those at the low tide level, where they are immersed for long periods. In most of the intertidal zone, the rocks are washed during every tidal flow. More saline conditions may occur above the mean high water level, because of the evaporation of water provided by exceptionally high tides, storm waves and spray. This factor, however, may be of little significance at the time of year when frost is most active. Probably of greater importance are the dilution of pool waters and the leaching of salts from the upper intertidal zone, by groundwater or by meltwater issuing from icefoots. Frost action cannot be important in the coastal zone unless the rocks can attain critical levels of saturation. Frost weathering of beach gravels may be promoted where they are embedded in a matrix of mud (Taber, 1950) or by an underlayer of sand (Martini, 1975), which permit high levels of saturation to be maintained during freezing. It is not clear, however, whether saturation can be attained on bare rock platforms. The immersion period is greatest at the low tide level, and because of the pressures induced by the tidal head, water is also probably absorbed most quickly here. Rocks are immersed for almost 12 hours near the low tide level in a semi-diurnal environment, and for only a brief period near the high tide level. Therefore, the slower a rock-type absorbs water, the closer it must be to the low tide level to attain a given level of saturation in the time which is available. Not only must a rock attain saturation, but it must also retain it while it is being frozen. Rocks with large capillaries absorb water quickly, but water is also rapidly exuded when they are no longer immersed (Fagerlund, 1975). It has been noted previously, however, that the presence of salts increases the ability of porous materials to absorb and to retain water. None the less, preliminary studies (Mercan (1980)—work is continuing in this area), suggest that very few rocks are able to attain critical saturation levels in the immersion periods available in the intertidal zone. Rock cores were subjected to variable periods of immersion in artificial sea water, and to periods of drying at 7 or 17°C. Tidal data were used to simulate natural wetting and drying cycles at various elevations within the intertidal zones of macro- to micro-tidal environments. Only the most porous rocks, when immersed for periods which simulated conditions near to the low tide level, attained degrees of saturation which were even close to their critical levels. Because these rocks also lose water quickly, however, it is doubtful whether high levels of saturation could be maintained for very long after emersion. Rocks which are sensitive to high levels of relative humidity, may attain critical saturation levels without immersion in water (Hudec, 1973). Because of the humid conditions of cool coastal regions, fine-grained argillaceous rocks may be saturated with largely adsorbed water throughout the intertidal zone. Adsorbed water is virtually unfreezable, but damaging pressures may be exerted by the expansion and contraction of the rocks, as water is alternately adsorbed and exuded in response to rising and falling temperatures (Dunn and Hudec, 1965; Litvan, 1972; Hudec, 1977). Temperature dependent wetting and drying may also contribute to frost action within the tidal zone. As temperatures fall, the number of adsorbed water layers must decline in the small capillaries of the rock. This water is expelled into the larger capillaries or cracks, where being no longer adsorbed, it freezes and so contributes to the breakdown of the rock by normal frost action.

It may be even more difficult for saturation to be attained on steep cliffs, than it is on gently sloping intertidal platforms. Shale beds and other suitable strata in the cliff face may become saturated with ground- or meltwater, however, and possibly by direct adsorption from humid air. Salts supplied by spray are most abundant at low elevations, but they are known to ascend to heights which are far greater than that of the highest cliffs (e.g. Woodcock, 1953). The salinity of cliff solutions, however, may be lowered by mixing with ground water, rain or snowmelt.

A preliminary attempt has been made to determine experimentally the role of processes associated with temperature fluctuations about the freezing point, in the development of a rock coast in Gaspé, Québec, Canada (Trenhaile and Rudakas, 1981). With the proviso that the rocks are able to attain saturation, it was found that physical weathering may play an important role in the erosion of the shale and argillite cliffs, and in that portion of the intertidal platform which is close to the high tide level (see also Guilcher, 1981). There was, however, no evidence to suggest that weathering is able to planate intertidal surfaces, particularly where they consist of a variety of rock types.

Nansen (1904) believed that the strandflat was cut by the combined action of frost and waves. Although his views were substantially the same, he later emphasized frost action associated with the icefoot, not only as an agent of cliff recession, but also as an agent of planation (Nansen, 1922). Gardner's (1969) data suggest that freeze-thaw is most active at the periphery of snow patches, where there is meltwater and more frequent fluctuations in temperature about the freezing point than on bare rock faces. Nansen (1922), however, appears to have partly based his belief that frost can planate rock surfaces on the erroneous assumption that the freezing of sea water is an ineffective disruptive mechanism, compared with that of fresh water. He noted that surface layers of fresh water may be provided by rivers or by melting glaciers in fjords, sounds, or enclosed parts of the sea, or by the summer melting of sea ice in coastal waters. He believed that the effect of frost may be greatest just above mean sea level because of the presence of an icefoot at the high water level, and because at lower levels the rocks are protected by water against frost action. No evidence has been presented to indicate that frost operates down to a specific datum, however, and the experimental data provided by numerous engineering and geological investigations suggest that it does not. Frost action or temperature dependent wetting and drying certainly appear to make an important contribution to the recession of coastal cliffs in sheltered, high-latitude areas, but this does not in itself explain the way in which the surfaces are planated at their base.

ICE

Ice may move loose debris in the coastal zone. Goldthwait (1933) and Knight and Dalrymple (1976), for example, have described the rafting of rocks of between 1 and 3 m in diameter in eastern Canada. The presence of erratics on the sea-floor further testifies to the ability of ice to convey loose debris from the coastal zone, although much of this material was entrained by glacial rather than by floating shore ice. The ability of shore ice to transport material is incontrovertible (Charlesworth, 1957; Bird, 1967; Dionne, 1979), but it is questionable whether it generally operates on a sufficient scale to constitute a major process in the development of high latitude rock coasts (Martini, 1981). Nansen (1922) considered that the transportive role of ice is negligible, and Thomson (1979) found that ice rafting accounts for only a small amount of the transport of large debris in the Bay of Fundy. Moign (1974a,b) believed that sea ice is essentially protective, and that its role in the development of shore platforms in Spitsbergen is negligible. In field experiments carried out in Finland, marked cobbles and larger rocks were occasionally moved short distances seawards, but shifts in position attributable to ice were generally found to be less than a few metres each year (Mansikkaniemi, 1976).

When an icefoot is attached to a platform, it affords protection to the cliff behind. In the spring, however, snow insulates its upper surface, so that ablation generally proceeds from below. Eventually, when the tidal level is low, the icefoot may be perched or suspended on the cliff face. This exerts stress on the cliff, and rocks may be dislodged when the icefoot collapses (Moign, 1974a,b; Nielsen, 1979). In Greenland, fresh rock may be broken off along cracks produced by ice blocks thrown up by waves against the cliff in summer or autumn (Nielsen, 1979). A large amount of debris may be carried by an icefoot as it floats away from the cliff (Joyce, 1950; Corbel, 1958; Howarth and Bones, 1972; Nielsen, 1979), but Nansen (1922) considered it to be unimportant because much of the icefoot melts *in situ*. The icefoot may affect the cliff in several other ways: Corbel (1958) considered that it aids solution on the limestone cliffs of Anticosti Island in the St. Lawrence Estuary; and Thomson (1979) suggested that it might promote slope failure by raising the cleft water pressures in the cliff. Moign (1974a,b) believed that the overall effect of the icefoot is minor, although she considered that it could excavate a visor and notch in the cliff base. Several early workers suggested that the icefoot could produce terraces (see, for example, discussions in Nansen, 1922, and Charlesworth, 1957), but Nielsen (1979) considered that any ledges formed in this way would be small.

It has been proposed that shore platforms may be planated by ice. Chalmers (1882) noted that floating ice cuts striations in eastern Québec, and he considered it to be capable of considerable shore erosion. According to Nansen (1922), Hansen believed that the Norwegian strandflat was cut by wave action, and by drifting ice scouring the sides of sounds and fjords. Similarly, Goldthwait (1933) suggested that the Micmac terrace and cliff in the central parts of the St. Lawrence Estuary may have resulted from the sawing action of pack or batture ice, driven up and down the estuary by strong tidal currents. Fairbridge (1977) even claimed that the wide shore platforms of temperate areas such as Western Europe and southernmost Australia, were cut by frost action and floating ice during the Pleistocene. This assumption cannot be reconciled with the accumulated evidence of more than a century's study of shore platforms in these areas (Trenhaile, 1980), and, as Dionne (1978) has noted, it certainly exaggerates the role of ice in the formation of platforms and strandflats in high latitudes. Zenkovitch (1967) doubted that ice could abrade in the coastal zone, and Nansen (1904, 1922) thought that shore platforms could not be cut by floating ice, which he considered to be an insignificant erosive agent. This view was supported by Bird (1967), who found that erosion by floating ice is rarely important in the Canadian Arctic. Although Dionne (1972) suggested that sea ice, raised and lowered by the tide, may contribute to the planation of shore platforms in eastern Québec, this view has not been repeated or developed in later publications. Dionne

(1980) found that abrasional forms may occur in weak rocks, but he considered that the abrasion of hard rocks by sea ice is very slight. Dionne's (1980) conclusion that rock platforms in James Bay are scarcely affected by the erosive action of ice, supports Moign's (1974a,b) rejection of this factor for the formation of platforms in north-western Spitsbergen. This may be because floating ice cannot keep abrasives in contact with submerged rock surfaces in the intertidal zone (Dionne, 1980). Ice is, however, able to dislodge frost-rived rock, which may facilitate its removal by the ice itself, or by wave action. Nevertheless, most workers consider that shore ice plays a largely protective role on rock coasts, and that it is essentially impotent in the erosion of intertidal platforms.

WAVE ACTION

Although we do not understand the way in which frost or temperature dependent wetting and drying operate in the coastal zone, there is abundant, though somewhat ambiguous field evidence to suggest that they assume an important role in the erosion of marine cliffs in high latitudes. They are also probably active in the intertidal zone, but only where high levels of saturation can be attained. It is difficult to understand how these mechanisms can planate intertidal platforms, however, since the experimental data do not suggest that their efficacy rapidly diminishes, or that they cease to operate, at any specific datum within the intertidal zone. They may be capable of producing irregular, narrow ramps in sheltered areas where wave action is feeble, but it is questionable whether they can account for the wide, well planated surfaces which have been reported from many areas. Ice is not an efficient planation agent, although it may contribute to the removal of the weathered debris.

It has been suggested that platforms in sheltered areas in Australasia may be formed by the weathering of the cliffs down to the level of permanent saturation, and the removal of the fine weathered debris by weak wave action. In more exposed areas, wave cut ramps may be levelled and smoothed by a combination of chemical weathering, salt crystallization, and other processes associated with alternate wetting and drying (Trenhaile, 1980). In high latitudes, however, chemical weathering is generally considered to be slow (Kelly and Zumberge, 1961), and salt weathering, although active in high latitudes, cannot account for the formation of wide shore platforms in these areas.

Nansen (1904, 1922) assumed that since waves operate over a vertical range of elevations, they cannot make a sharp incision in the land. There is, however, a considerable concentration of wave energy within portions of the tidal range (Trenhaile, 1978, 1980; Trenhaile and Layzell, 1981; Carr and Graff, 1982), and there can be little doubt that mechanical wave action is the dominant planation mechanism in the storm wave environments of the mid-latitudes. Most workers also acknowledge the importance of wave erosion on many high-latitude coasts, even though it can only be significant during the ice-free periods of the year. Erosional bedrock forms are locally spectacular in high latitudes, particularly in high energy, storm wave environments (John and Sugden, 1975). Even in low energy environments, however, catastrophic changes may occur when the immediate offshore area becomes ice free. Shore platforms, stacks, caves, cliffs and notches occur in polar environments (Adie, 1964; Nichols, 1968; Calkin and Nichols, 1972); some of these features are contemporary, but others may have developed in interglacial periods when sea ice was less of a hindrance to wave action (John and Sugden, 1975). Numerous workers thought that the Norwegian strandflat is essentially a plain of marine abrasion (for example, Reusch, 1894; Richter, 1896; Vogt, 1907; Rekstad, 1915; Johnson, 1919; Strom, 1948), and the Icelandic strandflat may also be primarily the product of abrasion (Thorarinsson *et al.*, 1959). Nansen (1904, 1922) recognized the importance of wave erosion and transportation in the formation of strandflats in a number of areas, and several workers considered that it is essential for the development of platforms and strandflats in Spitsbergen (Hoel, 1909; Peach, 1916; Werenskiold, 1953; Moign, 1974a,b; Guilcher, 1974). Furthermore, John and Sugden (1975) believe that the strandflats of the Antarctic Peninsula are undoubtedly of marine origin. Frost action, however, may be necessary for effective wave erosion where the coast is protected by ice for six or seven months of the year; although the climate of these areas may be too severe for some frost mechanisms. Nansen (1922) and Moign (1974a,b) stressed that wide strandflats can only develop where frost and waves can act together, and for this reason their best development is at capes and in other areas where wave action is strong.

In any area, wave cut shore platforms can only develop if the wave energy reaching the shore is sufficient to erode the local rock. In the mid-latitudes, wide shore platforms have often formed in sheltered embayments because the rocks are less resistant to wave action than those on the more exposed headlands. In high latitudes, frost further weakens the rock, so that fairly weak waves are able to cut shore platforms in sheltered areas. Caution is advisable before a coastal environment is classified as being sheltered. Considerable amounts of wave energy may be expended during storms on apparently sheltered stretches of coast. In the storm wave environments of the northern hemisphere, there can be few truly sheltered coasts, particularly in western Britain and Scandinavia. In an ice free coastal inlet in Alaska, which is completely sheltered from Pacific swell, narrow shore platforms have been cut by locally generated waves, despite the small fetch, and the fact that this area is particularly suitable for frost action (Chastain, 1976). Near Penarth, in South Wales, a wide wave cut platform has developed in an area which appears to be protected from the dominant southwesterly waves (Trenhaile, 1972). Numerous other examples could be given of wave cut platforms which have developed in areas which seem to be sheltered from storm waves.

If frost and ice were able to smooth and level wide shore platforms without the assistance of wave action, then one would certainly expect such platforms to be well represented in eastern Canada. There is a strong relationship between the tidal range and the gradient of platforms which are wave cut (see, for example, summaries in Trenhaile, 1980, and Trenhaile and Layzell, 1981). The enormous variation in the tidal range in eastern Canada provides an ideal laboratory in which to determine its effect on platform morphology, and consequently the role of wave action in an area which is fairly sheltered, and which experiences severe frost and ice conditions. Platforms were surveyed in Gaspé in the St. Lawrence Estuary, Gros Morne National Park in western Newfoundland, and in the Minas Basin in the Bay of Fundy. Spring tidal range is about 2 m in Gros Morne, 2·25–3·5 m in Gaspé, and 13·5 to more than 15·5 m in Fundy. Wide platforms have developed in shales, argillites and greywackes in Gaspé (Trenhaile, 1978), in limestones, sandstones, siltstones, mudstones and shales in Newfoundland, and in conglomerates, sandstones, shales and basalts in Fundy. Platforms slope between 0·1 and 0·6° in Gaspé, 2 and 4·5° in Fundy, and they are horizontal in Newfoundland. Furthermore, an independent investigation has found that within the western Bay of Fundy, platform gradients increase eastwards in accordance with an increase in the tidal range (Thomson, 1979). As in southern Britain (Trenhaile, 1974), platform gradients vary between areas according to differences in the tidal ranges. This empirical relationship may be rewritten as a relationship between tidal range and the way in which wave energy is distributed within the intertidal zone (Trenhaile and Layzell, 1981). This relationship exists in eastern Canada even though frost and probably temperature dependent wetting and drying must assume a much greater role in coastal erosion than in southern Britain. The relationship between the wave energy distribution and platform morphology also occurs despite the restrictions placed on wave activity by coastal ice for two or three months each year, and despite the sheltered nature of the Minas Basin, and to a lesser extent, the St. Lawrence Estuary. The existence of this relationship suggests that despite the important role of frost on the cliffs of eastern Canada, it is wave action which determines the form of platform surfaces. The platforms of eastern Canada provide further support for the conclusions of Moign (1974a,b) and Guilcher (1974), that wave erosion plays a fundamental role in the formation of shore platforms in cool climates.

ISOSTASY

Most of the shore platforms in high latitudes experience uplift or downwarping associated with isostatic recovery. The degree to which these platforms are able to adjust to changes in the relative sea level depends upon the rate of change, the intensity of the erosive processes, and the hardness of the rocks. Where wave action is weak and the rocks are resistant, platforms may bear the imprint of former sea levels. Trenhaile and Layzell's (1981) model suggests that because changes in the relative sea level cause wave energy to be distributed over a considerable range of elevations, platforms are steeper and wider than they would be if the sea level was stable. The steepness and width of these platforms must therefore increase with the rate at which the relative sea level changes. In sheltered areas, platforms may exhibit the effects of a rate of change which has been retarding over a long period of time.

The gradient of these platforms would be greater than those cut in similar rocks in more exposed areas. This is because platforms which are subject to vigorous wave action are able to adjust to changing sea levels more quickly than those in sheltered environments. They are therefore less likely to retain vestiges of former changes in sea level which are faster than those which are occurring today.

Platforms may adjust fairly rapidly to forces which disturb their progress towards a state of equilibrium (Trenhaile and Layzell, 1981). This is generally possible because they tend to develop in rocks which erode quite easily, such as chalks or weathered granites, and particularly in alternations of siltstones, sandstones, limestones, or greywackes, with shales, mudstones, or argillites. Isostatic rebound probably increases the rate of platform erosion, but unfortunately, there are very little data on rates of platform erosion, and almost none for platforms in high latitudes. Rates of cliff recession have been estimated and measured in many areas, but much of the data are of questionable reliability (Trenhaile, 1980). Nevertheless, the morphology of platforms in southern Britain and eastern Canada is closely related to tidal and other contemporary morphogenic factors. Assuming a contemporary rise in sea level of 1 mm yr^{-1}, this suggests that they have been able to adjust to rates of change in the relative sea level of between -1 and $+3$ mm yr^{-1} (Valentin, 1953; Andrews, 1970). Whether they are able to adjust to rates of 5-8 mm yr^{-1}, which occur in parts of northern Canada (Andrews, 1970), and, by extrapolation of data from the Baltic and the Gulf of Bothnia, to rates of up to about 6 mm yr^{-1} in northern Norway (West, 1967), cannot be determined until accurately surveyed platform profiles are available for those areas. In the Bay of Fundy, however, platforms do appear to have at least partly adjusted to a rate of subsidence of 24 mm yr^{-1} in the last 4000 years, although about three-fifths of this change may be attributed to an increase in the tidal range, rather than to the change in the mean sea level (Grant, 1970).

CONCLUSIONS

Frost action may play an important, and in some cases an essential role in the retreat of coastal cliffs in high latitudes. It is probably most significant in southern Canada and the northern United States, Iceland, Scandinavia, around the shores of the Sea of Japan and the Sea of Okhotsk in eastern Asia, and southern South America. Frost action is most effective in fine-grained rocks of low tensile strength, in situations where they can attain high levels of saturation. The role of ice is less clear, although it may contribute to the excavation of loose debris from the intertidal zone. Most workers acknowledge the important role of marine erosion and transportation on many high-latitude coasts, despite the presence of ice for a considerable portion of the year. The areas of most active frost action are also storm wave environments, where wave action is vigorous. Furthermore, rocks which are most susceptible to frost are generally those which are also most vulnerable to wave action. Frost action may be essential in particularly sheltered locations, but unless there is sufficient wave action to remove the often coarse weathered debris, it cannot operate effectively. Although very high latitudes do not provide the best conditions for frost action, some frost mechanisms operate in severe climatic conditions, and because of the nearly continuous ice cover, it is possible that frost may assume a proportionately greater role in the development of rock coasts in these areas than in lower latitudes.

REFERENCES

A.C.I. (1980). *American Concrete Institute, Detroit, Publ.* **SP-65**, 1-627.

Adie, R. J. (1964). *In* "Antarctic Geology" (R. J. Adie, ed.), pp. 29-32. Columbia University Press, Amsterdam.

Andersen, B. C. (1968). *Norges geol. Unders.* **256**, 1-160.

Andrews, J. T. (1970). "A Geomorphological Study of Post-glacial Uplift with Particular Reference to Arctic Canada". *Inst. Br. Geogr. Spec. Publ.* No. 2, 156 pp.

Arni, T. H. (1966). *A.S.T.M. Stand. Spec. Tech. Publ.* **169A**, 261-274.

Battle, W. R. B. (1960). *Royal Geographic Society Research Series* **4**, 83-95.

Bird, J. B. (1967). "The Physiography of Arctic Canada". Johns Hopkins Press, Baltimore.

Browne, R. P. and Cady, P. D. (1975). *American Concrete Institute, Detroit, Publ.* **SP-47** (Durability of concrete), 101-19.

Calkin, P. E. and Nichols, R. L. (1972). *In* "Antarctic Geology and Geophysics" (R. S. Adie, ed.), pp. 625-643. Columbia University Press, Oslo.

Carr, A. P. and Graff, J. (1982). *Trans. Inst. Br. Geogr.* N.S. **7**, 240-245.

Chalmers, R. (1882). *Trans. R. Soc. Can.* **1**, (4), 285-286.

Charlesworth, J. K. (1957). "The Quaternary Era". Edward Arnold, London.

Chastain, C. (1976). "Hydrodynamic processes and sea cliff/platform erosion, Favorite Channel, Alaska". Unpublished Ph.D. thesis, University of California, Santa Cruz.

Collins, A. R. (1944). *J. Instn. Civ. Engrs.* **23**, 29-41.

Cook, F. A. and Raiche, V. G. (1962). *Geogr. Bull.* **18**, 64-78.

Cook, H. K. (1952). *Proceedings A.S.T.M. Stand.* **52**, 1169-1180.

Corbel, J. (1958). *Cah. Géogr. Québ.* **2**, 193-216.

Davies, J. L. (1972). "Geographical Variations in Coastal Development". Oliver and Boyd, Edinburgh.

Dawson, A. G. (1980). *In* "Studies in the Late-glacial of North-west Europe" (J. J. Lowe, J. M. Gray and J. E. Robinson, eds), pp. 45-53. Pergamon Press, Oxford.

Dionne, J. C. (1972). *Association Canadienne-Française pour L'Advancement des Sciences*, 40ê Congrès, Ottawa, 1-9.

Dionne, J. C. (1978). *Geographie Physique Quaternaire* **32**, 3-70.

Dionne, J. C. (1979). *Earth Science Reviews* **15**, 185-212.

Dionne, J. C. (1980). *Proceedings Canadian Coastal Conference on the Littoral* (Burlington) 80-92.

Dunn, J. R. and Hudec, P. P. (1965). *Department of Geology, Rensselaer Polytechnic Institute Contributions* **65-8**, 1-149.

Dunn, J. R. and Hudec, P. P. (1972). *Highways Research Board, Research Record* **393**, 65-78.

Fagerlund, G. (1975). *American Concrete Institute, Detroit, Publ.* **SP-47** (Durability of concrete), 13-65.

Fairbridge, R. W. (1977). *Studia geol. pol.* **52**, 135-141.

Fraser, J. K. (1959). *Arctic* **12**, 40-53.

Gardner, J. (1969). *Geogr. Annlr.* **A51**, 114-120.

Goldthwait, J. W. (1933). *In* "Pleistocene Geology of the Central St. Lawrence", *Mem. geol. Surv. Canada* **359**, 144-145.

Grant, D. R. (1970). *Canadian Journal of Earth Sciences* **7**, 676-689.

Grawe, O. R. (1936). *J. Geol.* **44**, 173-182.

Gray, J. M. (1978). *Trans. Inst. Br. Geogr.* N.S. **3**, 151-164.

Gronlie, O. T. (1921). *In* "Report of the Scientific Expedition to Nova Zemlya 1921" (O. Holtedahl, ed.). Videnskapsselskapet, Kristiania.

Gubkin, N. M. (1969). *Vestnik Moskovskogo Universiteta Geografiya* **2**, 111-113.

Guilcher, A. (1958). "Coastal and Submarine Morphology". London.

Guilcher, A. (1974). *Annls. Geogr.* **83**, 1-33.

Guilcher, A. (1981). *Géographie Physique Quaternaire* **35**, 155-169.

Hoel, A. (1909). *Norsk geol. Tidsskr.* **1**, 1-28.

Holtedahl, H. (1960). *In* "Guide to Excursions" (J. A. Dons, ed.). 21st International Geological Congress, Oslo.

Howarth, P. J. and Bones, J. G. (1972). *In* "Polar Geomorphology" (R. J. Price and D. E. Sugden, eds). *Inst. Br. Geogr. Spec. Publ.* **4**, 139-153.

Hudec, P. P. (1973). *In* "Canadian Arctic Geology" (J. D. Aitken and D. J. Glass, eds), pp. 313-335. Saskatoon.

Hudec, P. P. (1977). *Proceedings 2nd International Symposium Water-Rock Interaction, Strasbourg* **4**, 38-45.

Jahn, A. (1961). "Quantitative analysis of some periglacial processes in Spitsbergen", *Uniwersytet Wroclawski Im Boleslawa Bieruta, Zeszyty Nauk. Nauki Przyrodnicze,* 11, Warsaw.

John, B. S. and Sugden, D. E. (1975). *Progress in Geography* **7**, 53-132.

Johnson, D. W. (1919). "Shore Processes and Shoreline Development". John Wiley, New York.

Joyce, J. R. F. (1950). *J. Geol.* **58**, 646-649.

Kelly, W. C. and Zumberge, J. H. (1961). *J. Geol.* **69**, 433-446.

Kennedy, T. B. and Mather, K. (1953). *Proceedings American Concrete Institute* **50**, 141-172.

Knight, R. J. and Dalrymple, R. W. (1976). *Revue Géogr. Montréal* **30**, 65-85.

Lautridou, J. P. (1978). *Inter-Nord* **15**, 5-13.

Lewis, D. W., Dolch, W. L. and Woods, K. P. (1953). *Proceedings A.S.T.M. Stand.* **53**, 949-958.

Litvan, G. G. (1972). *J. Am. Ceram. Soc.* **55**, 38-42.

Litvan, G. G. (1976). *Cement and Concrete Research* **6**, 351-356.

MacInnes, C. and Whiting, J. D. (1979). *Cement and Concrete Research* **9**, 325-336.

Mansikkaniemi, H. (1976). *Fennia* **148**, 1-17.

Martini, A. (1975). *Acta Universitatis Wratislaviensis (Studia Gogaficzne)* **251**, 187-193.

Martini, I. P. (1981). *Z. Geomorph.* **25**, 1-16.

Mather, B. (1969). *Proc. Am. Soc. Civ. Engrs. Conf.* (Civil engineering in the oceans, Miami Beach).

McGreevy, J. P. (1981). *Progress in Physical Geography* **5**, 56-75.

Mercan, D. W. (1980). "A preliminary investigation of tidal regime water saturation level relating to intertidal shore platforms". Unpublished B.A. thesis, University of Windsor, Ontario.

Moign, A. (1974a). "Strandflats immergés et émergés du Spitsberg central et nord-occidental", Thesis Brest, publ. Lille III.
Moign, A. (1974b). *Inter-Nord* **13-14**, 57-72.
Nansen, F. (1904). "The Norwegian North Polar Expedition 1893-1896", pp. 102-130. Longman, Green and Co.; reprinted in 1969 by Greenwood Press, New York.
Nansen, F. (1922). "The strandflat and isostasy", *Videnskapssel skapets Skrifter 1, Mat.-Naturv. Klasse 1921*, no. 11. Kristiana.
Nichols, R. L. (1968). *J. Glaciol.* **7**, 449-478.
Nielsen, N. (1979). *Z. Geomorph.* **23**, 321-331.
Nordenskjold, O. (1928). *Am. Geogr. Soc. Spec. Publ.* **8** (W. L. G. Joerg, ed.).
Peach, A. M. E. (1916). *Trans. Edinb. geol. Soc.* **10**.
Potts, A. S. (1970). *Trans. Inst. Br. Geogr.* **49**, 109-123.
Powers, T. C. (1955). *Proceedings A.S.T.M. Stand.* **55**, 1132-1155.
Powers, T. C. (1975). *American Concrete Institute, Detroit, Publ.* **SP-47** (Durability of concrete), 1-11.
Rapp, A. (1960a). *Geogr. Annlr.* **A42**, 65-200.
Rapp, A. (1960b). *Norsk Polarinstitutt Skrifter* **119**, 1-96.
Rekstad, J. (1895). *Norsk geol. Tidsskr.* **111**, 8.
Reusch, H. (1894). *J. Geol.* **11**, 347-349.
Richter, E. (1896). *Globus* **69**, 313-319.
Rudberg, S. (1967). *Geogr. Annlr.* **A49**, 283-299.
Russell, R. J. (1943). *Trans. Am. geophys. Un.* **1**, 125-133.
Sissons, J. B. (1974). *Boreas* **3**, 41-48.
Soderman, G. (1980). *Fennia* **158**, 83-152.
Sollid, J. L., Andersen, S., Harme, N., Kjeldsen, O., Salvigsen, O., Sturd, S., Tveita, T. and Wilhelmsen, A. (1973). *Norsk geogr. Tidsskr.* **27**, 233-325.
Strom, K. M. (1948). *Geogr. J.* **62**, 19-27.
Taber, S. (1950). *Am. J. Sci.* **248**, 784-793.
Thomas, W. N. (1938). *Great Br. Dept. Sci. Ind. Res., Building Res. Stat. Tech. Paper* **197**, 1-146.
Thomson, K. W. (1979). "Coastal erosion in part of the Bay of Fundy, Nova Scotia". Unpublished M.Sc. thesis, University of Alberta.
Thorarinsson, S., Einarsson, T. and Kjartansson, G. (1959). *Geogr. Annlr.* **A41**, 135-169.
Thorn, C. E. (1979). *Earth Surface Processes* **4**, 211-228.
Tietze, W. (1962). *Petermanns geogr. Mitt.* **106**, 1-20.
Trenhaile, A. S. (1972). *Trans. Inst. Br. Geogr.* **56**, 127-144.
Trenhaile, A. S. (1974). *Trans. Inst. Br. Geogr.* **62**, 129-142.
Trenhaile, A. S. (1978). *Ann Ass. Am. Geogr.* **68**, 95-114.
Trenhaile, A. S. (1980). *Progress in Physical Geography* **4**,, 1-23.
Trenhaile, A. S. and Layzell, M. G. J. (1981). *Trans. Inst. Br. Geogr.* N.S. **6**, 82-102.
Trenhaile, A. S. and Rudakas, P. A. (1981). *Géographie Physique Quaternaire* **35**, 171-181.
Tricart, J. (1956). *Biul. peryglac.* **4**, 285-318.
Tricart, J. (1969). "Geomorphology of Cold Environments". MacMillan, London.
Valentin, H. (1953). *Geogr. J.* **119**, 299-305.
Verbeck, G. J. and Klieger, P. (1957). *Highway Research Board Bull.* **150**, 1-13.
Verbeck, G. J. and Landgren, R. (1960). *Proceedings A.S.T.M. Stand.* **60**, 1063-1079.
Visher, S. S. (1945). *Geol. Soc. Am. Bull.* **56**, 713-736.
Vogt, J. H. L. (1907). *Norsk geol. Tidsskr.* **1**, 8-22.

Vogt, J. H. L. (1917). *Norsk geol. Tidsskr.* **IV**.

Walker, R. D., Pence, H. J., Hazlett, W. H. and Ong, W. J. (1969). *Highway Research Board, Progress Report* **65**, 14-21.

Werenskiold, W. (1953). *Geogr. Tidsskr. Copenhagen* **52**, 302-309.

West, R. G. (1967). "Pleistocene Geology and Biology", 2nd ed. Longman, London.

Williams, L. (1964). *Ann. Ass. Am. Geogr.* **54**, 597-611.

Williams, R. B. G. and Robinson, D. A. (1981). *Earth Surface Processes and Landforms* **6**, 1-9.

Wiman, S. (1963). *Geogr. Annlr.* **A45**, 113-121.

Winkler, E. M. (1973). "Stone: Properties, Durability in Man's Environment". New York.

Woodcock, A. H. (1953). *J. Meteorol.* **10**, 362-371.

Wyrwoll, K. H. (1977). *In* "Reviews in Engineering Geology", 3 (D. C. Coates, ed.), pp. 59-67. Geological Society of America.

Zenkovitch, V. P. (1967). "Processes of Coastal Development". Oliver and Boyd, Edinburgh.

PART THREE

Methodology and techniques

FIVE

The measurement of relict shoreline altitudes in areas affected by glacio-isostasy, with particular reference to Scotland

J. M. GRAY

INTRODUCTION

Since isostasy and eustasy involve essentially vertical movements of land and sea level, shoreline altitude is an essential component of *every* study of relict shorelines. Yet despite this fundamental importance of shoreline altitude, it is surprising that many papers concerned with sea level change omit any reference to the methods by which the altitudes cited in these papers have been obtained. Others give only partial information, omitting for example to say what datum or what instrument has been used. And even when some information is given, it is frequently vague on details. For example, use of a "level" could mean a hand-held or accurate level, while many papers quote altitudes above "mean sea level" without saying precisely how mean sea level was established. Furthermore, correlations based on altitude are frequently made without reference to possible error sources in the altitude data being used.

It is the aim of this chapter to draw attention to the importance of shoreline altitude measurement and to review the techniques by which the measurements can be made. Andrews (1967, 1970a) emphasized that there are substantial differences in the research methods used in the various glacio-isostatic areas. In large part these stem from differences in the available evidence and characteristics of these areas, e.g. spatial differences in shoreline development, availability of dateable materials, coastal accessibility. In this chapter the detailed methodology developed in Scotland will be described first, following which alternative methods will be described for use in areas where the basic techniques used in Scotland become increasingly impractical. First, however, it is necessary to discuss why a particularly detailed methodology has been developed in Scotland.

SHORELINES AND ISOSTASY
ISBN 0 12 652960 4

THE SCOTTISH GLACIO-ISOSTATIC CENTRE

It is only since 1962 that a rigorous methodology has been employed in Scotland, beginning in the south-east and subsequently being applied to other sections of the Scottish coastline. Figure 1 shows stretches of the Scottish coastline so far examined using the methods described below. Other areas are currently being studied or written up, while a number of coastal areas have been selectively examined (e.g. relict shore platforms in the Kintyre area (Gray, 1978)). It is not the intention of this chapter to discuss in detail those British studies that have used other methods. Some of these are discussed more fully elsewhere (e.g. Sissons, 1967b; Gray, 1972a,

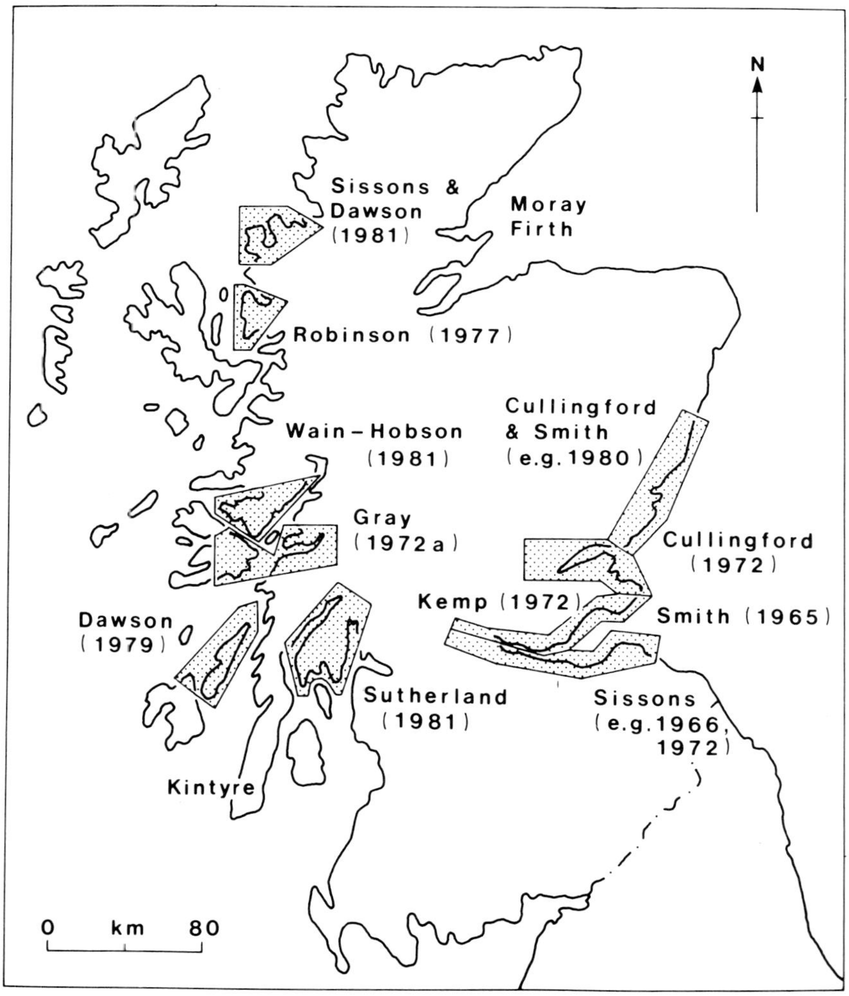

Figure 1. Map showing where detailed, systematic studies of relict shoreline altitudes have been completed in Scotland.

1974b, 1975b). The Scottish research has been directed by Dr J. B. Sissons at the Geography Department, University of Edinburgh, and apart from his foresight in developing a rigorous methodology, there are at least three major reasons for the detailed nature of shoreline research in Scotland.

First, relative to Fennoscandia and North America, the Scottish isostatic centre is very small (Nikonov, 1980). This means that a much more manageable length of coastline is involved, with a reasonable prospect that the whole length will be studied in detail over a relatively short span of time.

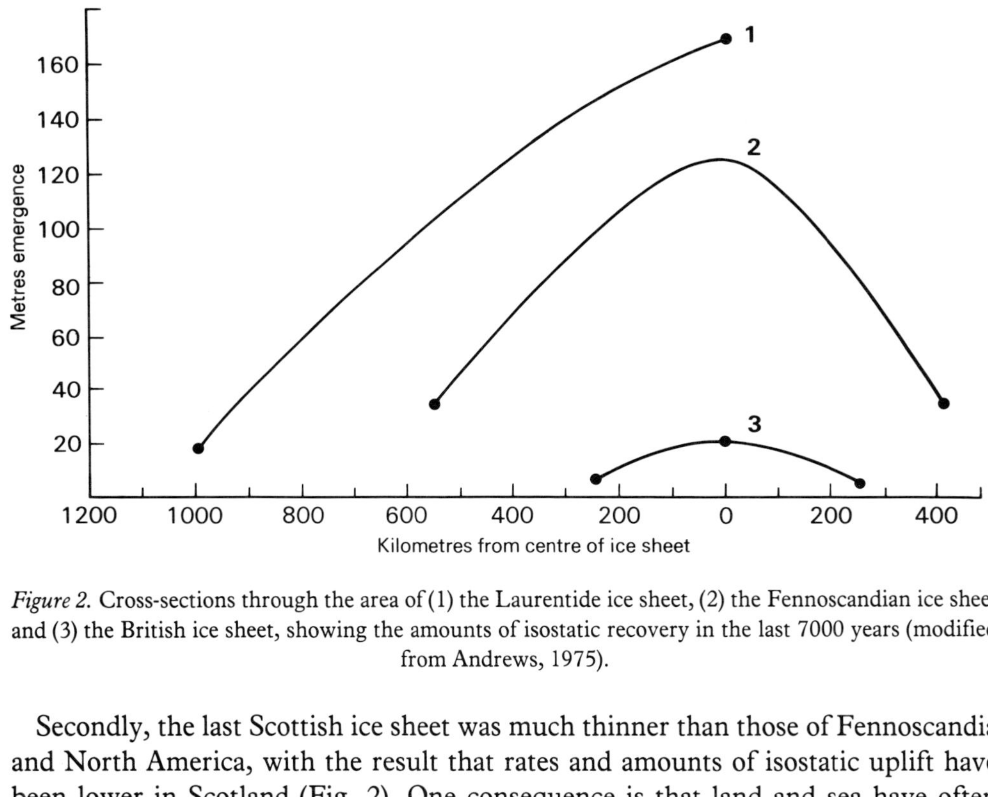

Figure 2. Cross-sections through the area of (1) the Laurentide ice sheet, (2) the Fennoscandian ice sheet and (3) the British ice sheet, showing the amounts of isostatic recovery in the last 7000 years (modified from Andrews, 1975).

Secondly, the last Scottish ice sheet was much thinner than those of Fennoscandia and North America, with the result that rates and amounts of isostatic uplift have been lower in Scotland (Fig. 2). One consequence is that land and sea have often been rising at approximately the same rates during the Lateglacial and Holocene. Although Andrews (1970a, pp. 93–95) has questioned Wright's (1914) isokinetic theory that shorelines are formed when the rate of isostatic uplift is equalled by the rate of sea level rise, it is reasonable to conclude that if the two rates do approximately match, then shorelines are likely to be better developed than in areas where isostatic rebound has outstripped sea level rise, other things being equal. There is in fact a marked contrast in shoreline development between Scotland and Canada. Andrews (1970a, p. 76), for example, states that

> Beaches or terraces are . . . poorly developed over most of Arctic Canada. . . . At the present time, no equivalents are known in terms of their geographical continuity, of the main post-glacial shoreline of Britain.

As well as this clear development, the Scottish relict shorelines are confined to a fairly narrow coastal zone. The highest raised beaches in Scotland reach only about 42 m O.D. (Sutherland, 1981) and it is rare to find a shoreline more than 10 km from the sea. This compares with shoreline altitudes of over 285 m and distances from the coast of over 450 km in parts of North America (Andrews, 1970a). Thus, as well as a much shorter length of coastline, a much smaller vertical and areal coastal fringe is involved in the study of Scottish shorelines. Furthermore, the fact that individual shorelines are often separated by small vertical intervals, means that only accurate and detailed work can hope to distinguish them and enable reliable correlation (see Fig. 3). In addition stratigraphical approaches are facilitated because of the likelihood of transgression occurring.

Thirdly, the Scottish coastline is much more accessible than most of the coastline of Fennoscandia, the Canadian Arctic or most other glacio-isostatic centres. The coastline of eastern and southern Scotland is quite densely populated and frequently

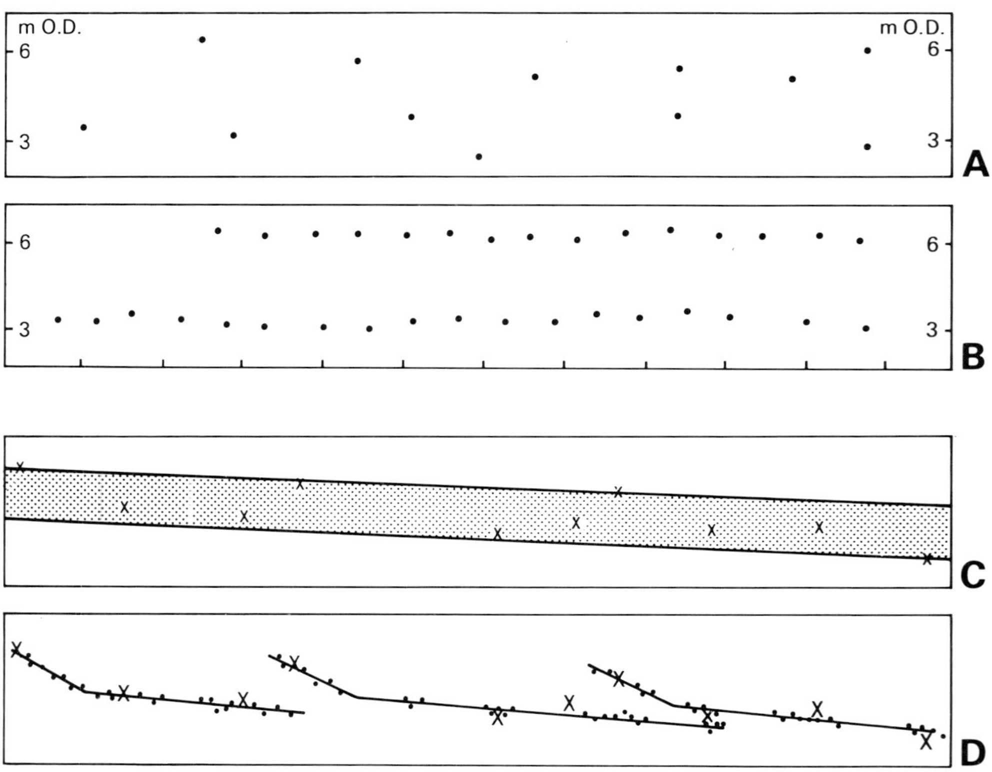

Figure 3. Diagrams to illustrate the importance of accurate and detailed measurements in areas where shorelines are closely spaced vertically (after Sissons, 1967a). (A) altitudes of two raised shorelines in East Lothian, Scotland measured by aneroid barometer; (B) altitudes of the same shorelines, measured by accurate levelling; (C) altitudes of the highest marine feature along a coast, made at widely spaced locations. A shoreline (shaded) sloping from left to right is suggested; (D) the same stretch of coast representing detailed investigation. What was formerly interpreted as a gently sloping shoreline has been resolved into three shorelines and associated more steeply sloping outwash spreads.

roads have been built along the raised beaches themselves. In western and northern Scotland, as well as in the Hebridean islands, the coastline is less accessible but it is still rare for part of the coast to be more than 25 km from a road. But as well as the accessibility and convenience brought by the road network, it is along these same roads that the Ordnance Survey has normally placed its bench marks, which thus provide an accessible common datum for the whole country. The climate too makes fieldwork much easier than in the Canadian or Fennoscandian Arctic.

Shoreline altitude measurement in Scotland is therefore considerably easier than in most other glacio-isostatic centres of the world and it should not be surprising that a particularly detailed and accurate methodology has been developed there. There is also a long history of shoreline study in Scotland dating back to the classic work of T. F. Jamieson (1865, 1882). However, one field in which Scotland is not ideal is in shoreline dating, there being a notable absence of related biogenic materials, particularly in the Lateglacial beach deposits.

BASIC METHODOLOGY USED IN SCOTLAND

Morphology and stratigraphy

A prerequisite of measurement of Scottish shoreline altitudes has been morphological mapping, usually at a scale of 1:10 000 or 1:10 560. Prior to the fieldwork, mapping has been carried out on aerial photographs, and this stage can save considerable time and effort on the more remote sections of coast, while in other areas it acts as a basic framework that can be modified in the field.

The procedure in Scotland has been to study relatively short stretches of coast at a time (i.e. between about 100 and 300 km, see Fig. 1) and to map all depositional features and erosional terraces along these coastal fringes. Since there is often a close relationship between ice-limits and raised beaches, the mapping has included fluvio-glacial and glacial features as well as marine, although it is frequently difficult to distinguish outwash terraces from raised beaches on morphology alone. What morphological mapping does achieve is to record the position and extent of all flats regardless of origin and to map the relationships of these terraces to each other and to other landforms. It also allows an assessment of the suitability of the terraces for measurement (Cullingford, 1972). Figure 4 shows an example of a morphological map for part of the Scottish coastline. Problems of publication have meant that detailed maps have rarely been produced outside theses (Smith, 1965; Cullingford, 1972; Gray, 1972a; Dawson, 1979; Sutherland, 1981), but Sissons and Smith (1965a), Cullingford (1977) and Cullingford and Smith (1980) have published reasonably detailed maps.

As well as examining natural field sections, stratigraphy has been examined by a number of augering and boring techniques. The simplest of these involves the use of a soil auger which can penetrate to a depth of about one metre. Where there are fine or soft deposits, such as unconsolidated marine clay and peat, deeper penetration

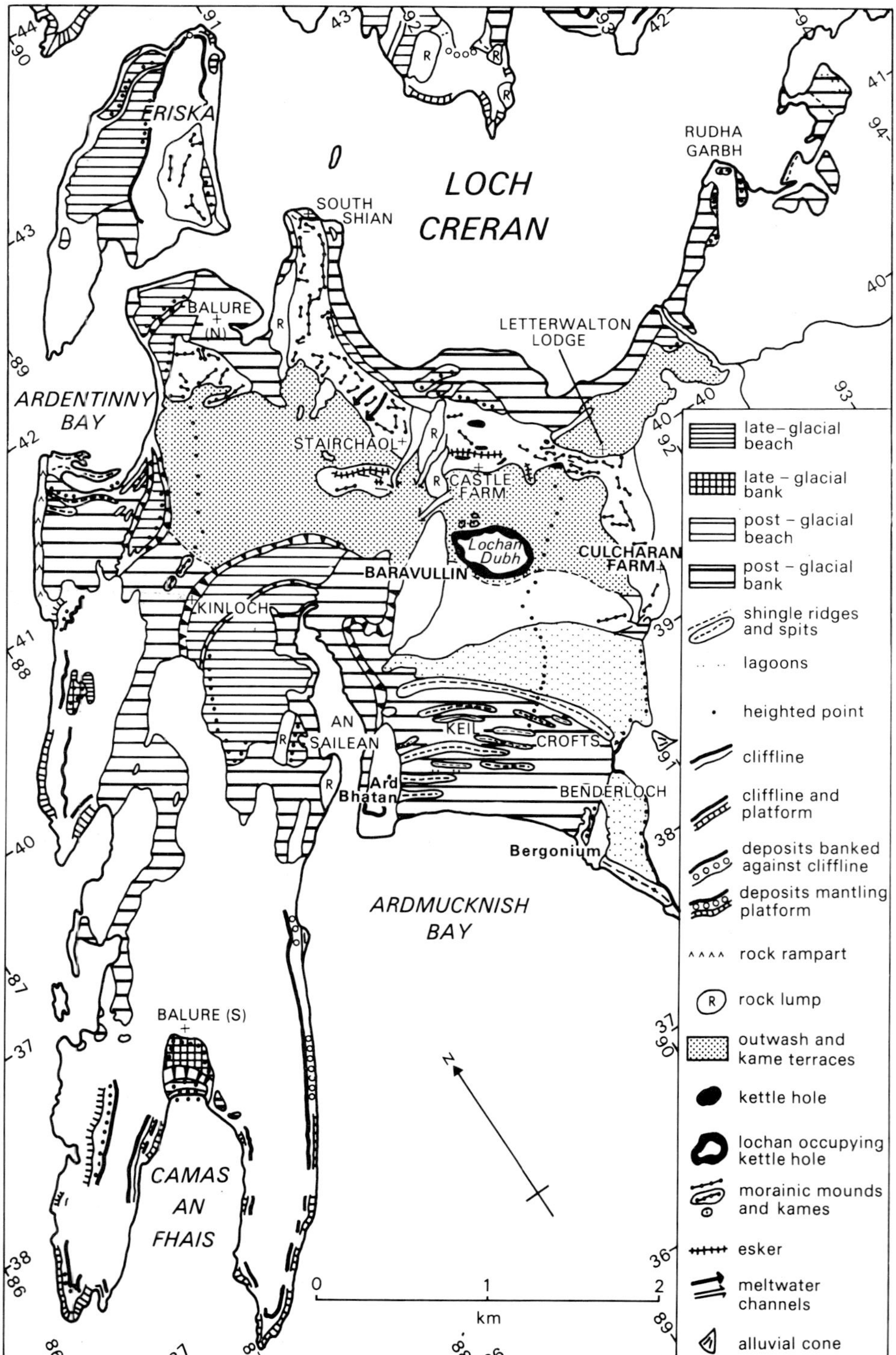

Figure 4. Example of a morphological map of part of the Scottish coast—western Benderloch, Argyll. Original mapping scale of 1:10 560.

is often possible by means of a small Hiller peat sampler strengthened at the extension rod connections, or other hand-operated borer, a technique that has proved very successful on the carselands (relict mud flats) of south-east Scotland (e.g. Sissons and Smith, 1965b; Sissons, 1966, 1969, 1972; Kemp, 1971; Cullingford, 1977). Not only has the method allowed the detailed and precise recording of the stratigraphy of these areas, but it has also led to the identification of peat-covered buried beaches. Sissons (1972, p. 155) describes the procedure for determining the position of the back edge of a buried beach and, by employing levelling and data analysis methods described in later sections of this chapter, has been able to reconstruct three buried shorelines. The hand-operated boring technique has the advantage of being rapid, cheap and precise while the apparatus is fairly easily tansported and can be operated single-handed. Unfortunately penetration may be halted by stones, compacted peat, tree fragments and tough or coarse sediments. Attempts to penetrate these may lead to "damage to either the equipment or the operator"! (Cullingford, 1972, p. 124).

Deep penetration in resistant sediments is possible with the use of power operated borers and in Britain many useful borehole records are held by the Institute of Geological Sciences, sand and gravel companies, site investigation firms, civil engineers, etc. Examples of shoreline studies that have utilized borehole data are those by Sissons (1969, 1981), Sissons and Rhind (1970) and Browne and Graham (1981). There are, of course, problems in using this data including the uneven distribution of the bores and inaccuracies in describing, recording and locating them. However, with experience and when supplemented by hand boring, these commercial bore records have proved to be very valuable in particular areas.

Altitude measurement

Datum lines

The detailed work in Scotland has involved accurate levelling of shoreline altitudes from Ordnance Survey bench marks (Newlyn datum). The use of bench marks means that shoreline altitudes are directly comparable from area to area since they are related to a common datum (Ordnance Datum (O.D.) Newlyn). Gross errors are, of course, still possible due to movement of bench marks by, for example, mining subsidence or rebuilding of a wall on which a bench mark is sited. Thus where possible the altitude of bench marks has been checked by levelling their altitude from neighbouring ones. In areas where bench marks are sparse, temporary ones have been established on boulders etc., to save always having to go back to the original bench mark. However, care has been taken to avoid the accumulation of errors through establishing further temporary bench marks from these temporary bench marks.

On remote parts of the Scottish coastline bench marks related to the original datum (Liverpool) have sometimes been used since these tend to be much more widespread. Newlyn datum bench marks are mainly located along roads, whereas those tied to the old datum are frequently found along very remote parts of the coastline. Thus, if the difference between the two datums is known, the O.D. Liverpool bench marks can be used in these remote areas. During the relevelling to establish the altitude of bench marks relative to O.D. Newlyn, the Ordnance Survey frequently simply relevelled existing bench marks so that for a number of them altitudes relative to both datums are available. The difference between the two is not constant, however. For example, around Oban in western Scotland, Newlyn altitudes are about 0·60–0·65 m above Liverpool ones as shown on the 2nd ed. 1:10 560 maps of Scotland, whereas at Port Appin, about 15 km to the north, the difference is less than 0·50 m (Gray, 1975a). A similar procedure has been carried out to correct local datum bench marks on some of the Scottish islands (e.g. Dawson, 1979).

Instruments

The levelling has been carried out using various models of surveyor's or builder's levels, mainly of automatic type. These are accurate instruments, though errors can still occur due to improper field use or misalignment of the cross-hairs in the level (collimation error). In order to check for such errors all the detailed Scottish studies have involved starting and ending traverses at the same point, and correcting the altitudes by distributing the closing error through the traverse cumulatively at each instrument position. Where closing errors have been large (e.g. over 0·15 m), traverses, or parts of them, have been repeated. Collimation error can be reduced by keeping backsights and foresights approximately equal, thus cancelling out misalignment of the cross-hairs. The standard method ("2 peg test") to check for collimation error and to realign the instrument has been employed frequently during fieldwork.

The method described above is time-consuming and requires two people, but it provides accurate altitude data and thus avoids one of the major uncertainties of trying to correlate shoreline fragments.

Measurement procedure

The choice of an accurate methodology does not, of course, guarantee meaningful results. Much depends on *what* is measured, *where* the measurements are taken, and *how many* measurements are made. Taking these in turn, in Scotland the emphasis regarding what to measure has been on determining the altitude of coastal erosional and depositional *terraces*, presumed to be relict beaches, mud flats, shore platforms etc. (Figs 5 and 6). There are of course dangers in relying on morphological recognition of former coastal features, though as described earlier, stratigraphical and sedimentological information has frequently been used. Accurate altitude measurement itself may be capable of distinguishing outwash terraces from extensive beach flats. In south-east Scotland, Sissons *et al.* (1966) have shown that relict outwash terraces have much steeper gradients than the Lateglacial shorelines

Figure 5. Depositional raised beach with backing cliff mainly in drift, Kintyre peninsula, Scotland.

Figure 6. Erosional shore platform and cliffline, Isle of Islay, Scotland.

to which they grade (see Fig. 3D). There is however the problem that distinct terraces may be absent from areas that were none the less penetrated by the sea (Paterson *et al.*, 1981), and thus care must be taken in arguing from negative evidence. Coastal rock terraces backed by steep slopes are not always marine in origin either (e.g. eroded, horizontal, Tertiary lava flows in western Scotland), and thus it is important to have evidence of marine erosion (e.g. undercut cliffs, caves, stacks, etc. (Figs 7 and 8)) and a thorough knowledge of the local geological structure. Where peat overlies a terrace, steel rods have been used to facilitate measurement of the terrace surface rather than the peat surface, which in western Scotland may be 1–2 m higher. Although some measurements have been made on features other than terraces (e.g. storm ridges in western Scotland (Gray, 1972a; Dawson, 1979; Sissons and Dawson, 1981; Sutherland, 1981)), these have not formed the primary data for shoreline recognition.

As regards where to measure, this has been towards the back edges of the terraces, though some metres away from the backing slope to avoid the worst effects of slopewash or solifluction from it. Other places avoided have included areas of obvious blown sand, alluvial fans and gullies, ground disturbed or close to detectable former field boundaries, and areas of subsidence.

In deciding how many measurements to make, there are two basic decisions to be taken: first, how many features should be measured, and secondly how many measurements should be made on each feature. These should be recognized as problems in sampling and as in other areas, erroneous results may occur if the sample is unrepresentative or too small. In the detailed Scottish work the procedure has been to measure *all* well-developed marine terraces within the limited areas studied, and thus it is the "total population" of fragments that has been measured. The second problem remains however. Smith (1965, p. 39) found that "variations from measuring similar points at intervals along the same marine terrace, serve to cast doubt upon the conclusions of those who have only measured one point". The procedure has been to take several measurements on each terrace fragment using a systematic sampling method with measurements taken every 50–100 m along the back edges of the terraces, though this distance is reduced for small fragments and is varied to avoid channels, fans, etc.

Relationship to sea level

One of the most taxing problems in the study of raised shorelines comes in trying to relate the altitude of the relict marine features to the part of the vertical tidal cycle at which they are formed. Some studies have therefore included altitude measurement of features on the modern coast in order to relate them to tidal regime. In south-east Scotland, Sissons *et al.* (1966) measured the altitude of the back edge of modern mud flats, and found that they were about 1 m below high water level as defined by Admiralty Tide Tables. Therefore, they suggested that the raised estuarine flats they had measured were similarly related to high water levels. Both modern mud flats and the high water level rise slightly up estuaries and this has been allowed for in discussing isostatic tilting of relict shorelines.

Figure 7. Raised natural arch, Isle of Kerrera, Scotland.

Figure 8. Raised sea-stack (the Dog Stone), Oban, Scotland.

In western Scotland both Dawson (1979) and Sutherland (1981) have measured altitudes of features on the modern coast. The latter author measured the altitudes of ten intertidal flats and found a close correspondence with mean high water of Ordinary Spring Tides (MHWOST) as recorded at tide-gauges close to five of these features.

The relationship between modern coastal landforms and present sea level is recorded in many other studies from various parts of the world, but the most relevant measurements for each study of relict shorelines are those taken within the study area itself. Unfortunately, it is never possible to determine whether present conditions are applicable to the past. The estuaries of south-east Scotland, for example, were both wider and deeper when the Lateglacial and Holocene shorelines were formed, and as well as topographic changes, tidal range, storminess, etc., may vary through time. For Scottish Lateglacial shorelines, polar analogues are often more appropriate but even these may not be exact (Gray, 1978).

Measurement error

Although the Scottish methodology described above clearly gives accurate results relative to other procedures used both in Scotland and elsewhere, it is only recently that a quantitative assessment of the accuracy has been made. Sutherland (1981) discusses the error sources in his study in western Scotland and quantifies them as shown in Table 1.

In this table, since category (iii) includes instrumental errors, addition of (i), (iii) and (iv) gives a figure of ±0·379 m for the degree of uncertainty in the heights of

TABLE 1
Error magnitudes in the study of Sutherland (1981)

Error		Magnitude
(i) Datum errors—based on Ordnance Survey figures for bench mark accuracy		±0·01 m
(ii) Instrument errors—based on closing errors of levelled traverses and subsequent altitude correction		±0·011 m
(iii) Measurement errors—based on resurveying terraces		±0·068 m
(iv) Landform errors—based on variability of measurements on the same feature		±0·301 m
(v) Relationship to tidal regime errors—based on variability of different types of coastal landform around the coast		
	intertidal deltas	±0·16 m
	shingle ridges	±0·46 m
	erosional terraces	±0·23 m

landforms surveyed. This may then be added to the error for individual landform types (v), giving for example $\pm 0{\cdot}54$ m for intertidal deltas. As Sutherland points out, the figures are only relevant for his study, and many of them depend on the type of area studied and the skill and judgement of the fieldworker. Also they do not include gross errors or misinterpretation of evidence, which invalidate the rest of the survey, and which are best avoided by a systematic approach to fieldwork. Sutherland argues that the accuracy is likely to be higher for south-east Scotland because of a smaller landform error, which is by far the largest influence on the uncertainty of the altitude of the surveyed landforms.

The accurate methodology used in Scotland means that correlation of relict shoreline fragments can be made with much greater confidence than in those studies where there may be an uncertainty of several metres in the altitude of individual points.

Shoreline representation and correlation

Shoreline diagrams

In order to represent and correlate shoreline altitudes, shoreline diagrams are used of which there are two main types. In Scotland it is the *equidistant* shoreline diagram that has been favoured over the shoreline *relation* diagram. As its name implies, the equidistant diagram has a linear x-axis and is produced by projecting the altitudes obtained in a limited area into a vertical plane drawn normal to the isobases. Herein lies one of the problems in this research, since it may not be possible to determine isobase trend until the measurements are correlated on the basis of the shoreline diagram (Andrews, 1967). In most of the Scottish studies there have been some particularly well developed shorelines that have acted as guides for isobase trends of other shorelines. In many studies the line of projection has been rotated and plots made every 5-15°. Those projections producing minimum gradients or the best visual or statistical alignments of points have then been used. Errors may occur due to isobase curvature so that when projected into a single plane the points plot in the wrong position relative to each other (Fig. 9). As Sutherland (1981) points out the resultant error becomes greater as isobase curvature increases (i.e. nearer the centre of uplift), as shoreline gradient increases (i.e. more important for Lateglacial than Holocene shorelines, and more important in Fennoscandia or N. America than in Scotland), and as distance from the projection plane increases.

On equidistant shoreline diagrams, shorelines have been identified as marked alignments of points, though in some cases a number of constraints have been brought to bear on the problem (Cullingford, 1972, 1977). These have been named the staircase constraint, continuity constraint and ice margin constraint. Cullingford developed a shorthand method of detailing the constraints in the field evidence, with commas indicating breaks in field evidence along a stretch of coast:—

$$\frac{A}{B} - \underset{H}{\underline{\frac{\frac{(E)}{D}}{C} - F}} - G, (I),))J - K, \overset{O}{\overline{L - M}} - \frac{N}{P}$$

where *A*, *B*, *C* etc., are shoreline fragments

$\frac{A}{B}$: fragment *A* occurs above fragment *B* in a staircase and thus cannot be correlated with it (staircase constraint).

F — *G*: fragments *F* and *G* are morphologically continuous and thus must be correlated (continuity constraint).

)) *J*: fragment *J* is contemporaneous with a specific former ice margin and thus cannot be correlated with any shorelines within the ice limit (ice margin constraint).

(*I*): the measurements on fragment *I* are of doubtful validity for altitude analysis.

The constraints are of most use in areas such as eastern Scotland where raised beach fragments are abundant (e.g. Cullingford and Smith, 1980), but could be used for example in many parts of Fennoscandia.

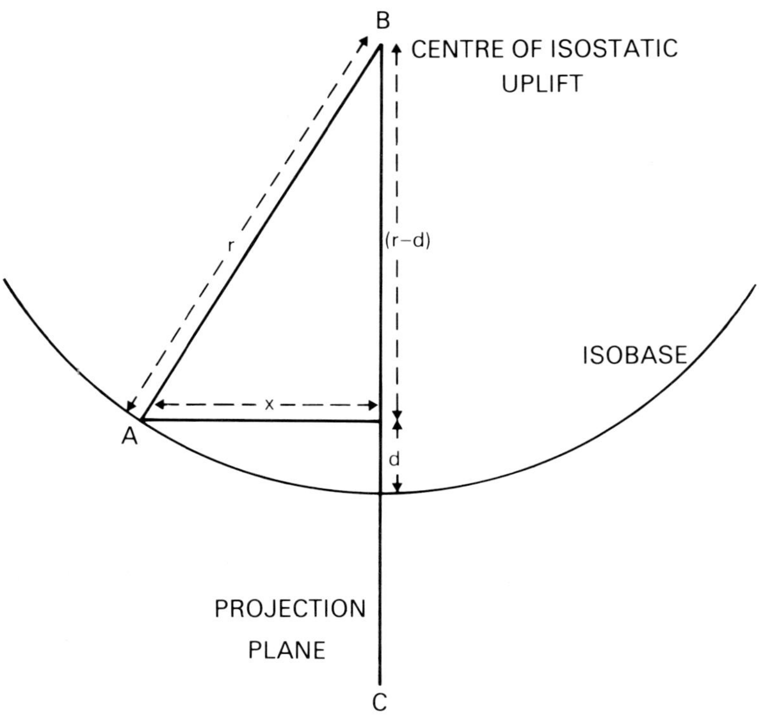

Figure 9. Diagrammatic representation of the horizontal displacement (d) caused by projecting the altitude of a raised shoreline at A into the projection plane BC (from Sutherland, 1981). $d = r - \sqrt{r^2 - x^2}$.

Statistical techniques

Lines to represent shorelines have been calculated by regression techniques. So far only linear regression has been used in Scotland since little if any curvature has been identified from equidistant shoreline diagrams. Sissons (1972) used ± 1 standard deviation limits, rather than single lines to represent his shorelines, and this is a refinement worth pursuing (Fig. 10), though note that the confidence limits should be curved rather than parallel.

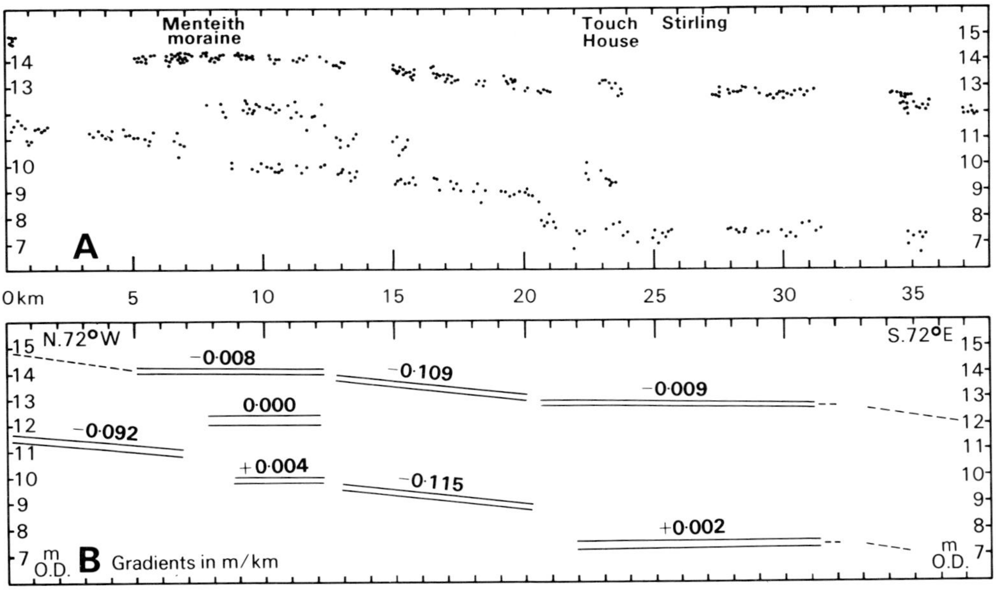

Figure 10. (A) Altitudes of three shorelines in the Carse of Stirling, Scotland. The dots in the uppermost group relate to the Main Post-glacial shoreline, those in the middle group to the High Buried Shoreline, and those in the lowest group to the Main Buried Shoreline. (B) Gradients of the shorelines. The double lines are one standard deviation on either side of the calculated line (from Sissons, 1972).

Because shoreline displacement occurs in three dimensions, trend surface analysis has been applied in order to define areal displacement patterns (e.g. Smith *et al.*, 1969; Gray, 1974a, 1978). One of the major problems in the application of trend surface analysis is in deciding at which order to stop. In some studies no reason is given as to why a specific order surface is shown; in others it appears to be due to the upper limit computed by the program used. Some operators have stopped when they have considered that there has been an insignificant increase in explanation; others have terminated analysis when, in their opinion, autocorrelation of residuals has been eliminated. All these methods introduce a large measure of subjectivity into what is a critical decision. Chayes (1970, p. 1273) stressed that what was urgently required was "some reasonably reliable rule of thumb by which the trend mapper could tell whether inclusion of terms of higher order is improving his maps or merely making them look fancy". He followed Krumbein and Graybill (1965) in suggesting the use of F-ratios to test each order in turn to determine whether it has contributed significant additional explanation to that of the previous order, stopping execution at the last order to improve the explanation significantly. The only subjectivity involved is in choosing the level of probability at which to test, but the 95% level is the usual standard (see Table 2). This technique is a very useful general indicator of the surface order at which to stop, though it must be used with caution (Baird *et al.*, 1971).

TABLE 2
The calculation of F-ratios for the contribution of successively higher-order trend surfaces (from Gray, 1974a)

	Sum of squares (m)	Degrees of freedom	Variance (m)	F	Confidence level (%)
Due to linear	241·694	2	120·847	231·95	>99·9
Deviations from linear	53·685	103	0·521		
Due to quadratic	10·466	3	3·489	8·08	>99·9
Deviations from quadratic	43·219	100	0·432		
Due to cubic	6·380	4	1·595	4·15	>99
Deviations from cubic	36·839	96	0·384		
Due to quartic	1·541	5	0·308	0·79	<95
Deviations from quartic	35·298	91	0·388		

The shoreline data collected in Scotland are not entirely suitable for trend surface analysis (Gray, 1972b) because:

(i) the data distribution consists of clusters of heights;
(ii) the residuals are not independent since all the heights on a single fragment are automatically correlated with each other;
(iii) the residuals are not normally distributed since, due to local conditions of beach formation, it is likely that in many cases residuals on each fragment will be autocorrelated.

However, all three inadequacies can be substantially overcome if each fragment is represented by a single height. The sample used by the author has been the arithmetic mean of the heights on each fragment plotted at mean eastings and northings, i.e. geographical mean sample. Trials have indicated that trend surfaces computed for total sampled populations and geographical mean samples do not differ significantly (Robinson, 1972; Gray, 1972b). Sutherland (1981) argues that the mean altitude should be plotted half way along the back edge of the shoreline fragments rather than at mean easting and northing, while Cullingford (1977) and Cullingford and Smith (1980) have favoured the use of total sampled populations.

As well as providing regional trends, the study of residuals from regression lines and trend surfaces may lead to important additional information regarding shoreline displacement and formation (Fig. 11). For example, residuals have been used to suggest shoreline dislocation and to show the effect of coastal configuration (Gray, 1974a), as well as drawing attention to the effect of geology on the altitude of a relict shore platform (Gray, 1978).

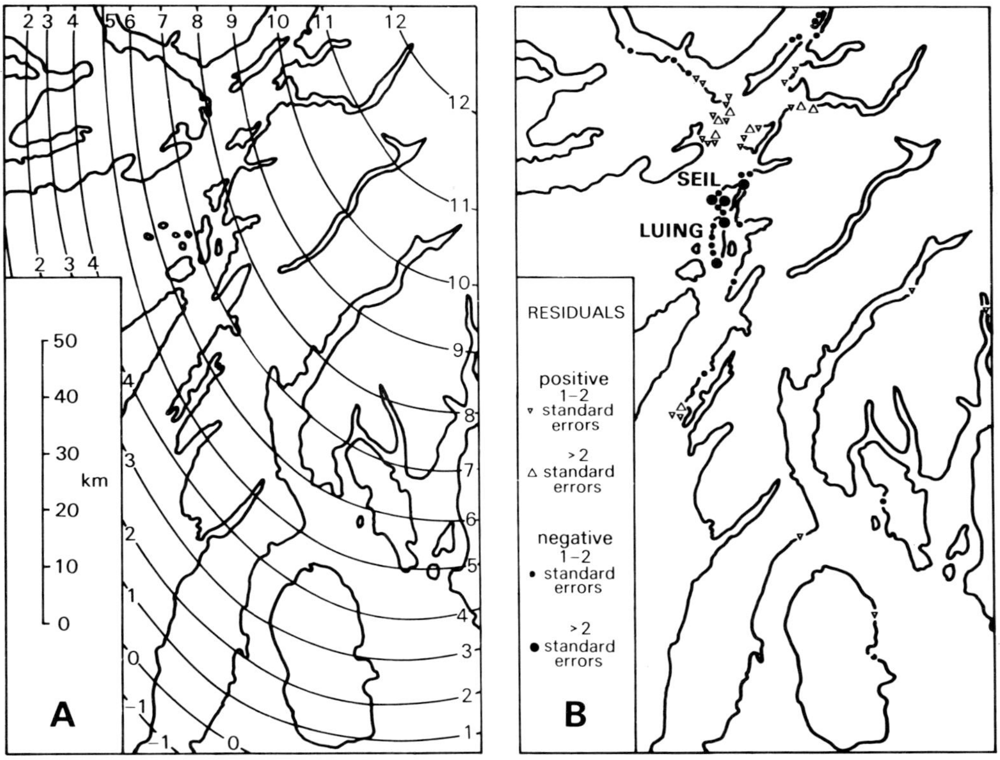

Figure 11. (A) Quadratic trend surfaces for the Main Lateglacial Shoreline in the South-west Scottish Highlands. (B) Residuals from the quadratic surface. Note the clusters of negative residuals on the islands of Seil and Luing (mainly slate bedrock) and positive residuals to the north (from Gray, 1978).

ALTERNATIVE METHODS

As has been emphasized, the use of the basic methodology described above would be impractical in many coastal areas of the world. This section examines some alternative methods, including some used in Scotland, and where possible attempts an assessment of their accuracy and potential.

Datum lines

Bench marks tied to a national datum are present in many countries, but before strict comparison of altitudes between countries can be achieved it is necessary to convert the altitudes, since national datums vary (Jardine, 1976). Republic of Ireland O.D., for example, is about 2·5–2·7 m below British O.D. Newlyn (Synge, 1977).

Alternatives to bench marks may be classified as water level or surrogate water level methods. They may be used in cases where shorelines lie close to the sea (or a lake), but are of limited use in places where relict shorelines lie many kilometres inland.

Water levels

In its simplest form the water level method assumes a water surface to be horizontal so that it provides a datum over a wide area from which to measure. In practice it has been used mainly to transfer along a coast altitudes measured from bench marks or some arbitrary datum point. Thus, for example, Goldthwait (1908), Ives (1960), Synge (1969), Andrews (1970b) and Cato and Linden (1973) used the method to measure relict shoreline altitudes around lakes, while Sutherland (1981) used the water surface of a canal. In such cases the effects of lake tides or seiches can probably be ignored, though under certain meteorological situations wave and wind conditions can introduce errors (Goldthwait, 1908) as can seasonal or longer-term variations in lake level (Robinson, 1908).

The sea surface has also been used, though unless there is a very low tidal range significant errors are introduced due to variations in the water level. Washburn and Stuiver (1962) used the method in East Greenland where the tidal range is about 1 m. Two refinements have been attempted, one involving synchronous measurement and the other using tidal corrections.

The method of synchronous sea-level marking has been used by the present author on mainland coasts of western Scotland which are devoid of bench marks, or on small remote islands. At a prearranged time *on a calm day*, the level of the sea is chiselled at two places, one close to a bench mark or other datum, the other close to the features to be measured. The altitude of the former is then levelled from the bench mark or datum, and the altitude of the latter is assumed to be the same and is used to measure the altitude of the remote features. The method has been tested by chiselling both marks close to bench marks, levelling to both and comparing the two heights. In five trials so far carried out by the author the maximum difference has been 0·02 m, and this is supported by a recent test by Sissons and Dawson (1981). The results so far obtained suggest that over distances of up to 10 km at least, the method has considerable potential and could be developed in other remote areas.

The second refinement is to correct sea level at the time of observation by subsequently correcting to national datum using tide tables listing times and altitudes of tidal positions at nearby coastal stations (e.g. Hoppe *et al.*, 1969; Synge,

1969; Rose, 1978; Ase, 1980). The accuracy of the method depends on the proximity and number of these stations, the nature of the coastline and the time of measurement (most accurate at times of slack tide). Accuracy also depends on weather conditions, which may not only create difficult wave conditions, but may also falsify what are only predicted tide levels in the tables. The method has been commonly used in remote areas but no systematic assessment of its accuracy has been carried out.

A further possibility is to use the mid-point between local high and low tides as used by John and Sugden (1971) in the South Shetland Islands, Antarctica, and Archer (1968) in E. Hudson Bay, Canada. The latter author argued that mean values, especially when averaged over a number of days, would not differ significantly from place to place, but it is not always clear how the mid-point has been obtained.

Surrogate water levels

Instead of using water levels, there are a number of datum lines that use mainly biological surrogates of sea level. One of these is the upper limit of growth of barnacles such as *Balanus balanoides* and *Chthamalus stellatus* (e.g. Donner, 1959; Donner *et al.*, 1977; Corner, 1978) or of growing seaweed such as *Fucus vesiculosus* (e.g. Andersen, 1968; Moller and Sollid, 1972; Sollid *et al.*, 1973; Løken, 1978). The methods were developed by Tanner (e.g. 1930) and have been used extensively in Scandinavia.

The most detailed tests of the level of *Fucus vesiculosus* have been carried out by Moller and Sollid (1972) and Sollid *et al.* (1973). In the Lofoten-Vesteralen-Ofoten area of northern Norway 20 measurements showed the upper limit to lie 0·22 ± 0·09 m above mean tide level in interior areas and 0·35 ± 0·10 m above mean tide level in outer coasts. In Finnmark 55 observations gave 0·3 ± 0·2 m above mean tide level. Regional variations were explained by differences in salinity between interior areas and outer coasts, and winter freezing towards the heads of fjords where "it occasionally reaches no higher than mean tide level" (Sollid *et al.*, 1973).

Sutherland (1981) measured the upper limit of barnacles in his study area (Fig. 1). A total of 85 measurements at 21 localities gave a range of 0·56–1·37 m O.D.; the mean was 1·16 ± 0·21 m O.D. Sutherland found a fall in the barnacle limit up the sea lochs and demonstrated statistically that this was due to a decrease in salinity. The method appears unreliable at salinities less than 90% but using altitudes from only those areas where salinities were above this figure, Sutherland found a reasonable consistency in his area at 1·20 ± 0·12 m. Further testing of this type is needed, particularly in areas where barnacle lines have been used as datum levels, but the method has considerable potential in remote areas. Sollid *et al.* (1973), however, found that the irregular development of *Balanus balanoides* in Finnmark, northern Norway made it unsuitable as a datum there.

A second type of biological surrogate is the lower limit of continuous land-based vegetation, though it has rarely been used as a datum. It has, however, been tested in western Scotland where Gray (1972a) and Sutherland (1981) both found it to lie at 2·4 ± 0·4 m O.D. Problems occur in defining when the vegetation is no longer "continuous" and in distinguishing land-based from salt-marsh vegetation. Sutherland found that the line approximates to mean high water of Ordinary Spring Tides in sheltered areas but deviates from it in exposed locations and areas of low salinity. Somewhat similar is the lower limit of lichen-covered boulders adjacent to the shore (Andrews, 1966).

Thirdly, the upper limit of sea-drifted materials, particularly seaweed, has been used by several authors. Dawson (1979), for example, used the limit of *Pelvetia canaliculatus* deposited during high water of ordinary spring tides as a datum for the island of Scarba in western Scotland on which bench marks are absent. The altitude used was that of *Pelvetia* on the neighbouring island of Jura where it was believed to relate to high water under calm conditions. Synge and Stephens (1966) and Synge (1969) used a similar datum based on the upper limit of fresh drifted sea salad (*Enteromorpha* sp.). Sutherland (1981) argued that this type of datum is impractical for use over a field season due to tidal variations and the occurrence of storms moving the drift limit to much higher levels. Instead he tested the variability of the upper limit of all drifted material reflecting the upper limit of constructive wave action normally during winter. It thus differs from Synge and Stephens limit but is similar to the "upper line of flotsam" datum used by Matthews (1967) and Barr (1971) in Arctic Canada. Sutherland noted that in some areas such material is sparse and tends to decompose as summer progresses. From measurement at 25 localities he found the limit to lie at 3·25 ± 0·37 m O.D. with a range of 1·44 m.

A non-biological surrogate is the flat surface of the icefoot used in the Canadian Arctic by Blake (1970, 1975) and Barr (1971). It persists for most of the summer and was used because of the absence of tide-gauge data. Blake argued that it lies close to the level of the highest tides.

Finally some authors appear to have used combinations of surrogates. Dyke (1979), for example, used the "high tide line" in Baffin Island "because it is easily recognizable. It is marked by the highest level of kelp jetsam, the highest level of presently active marine deposition, and the lowest level of lichen cover."

Conclusions

There is no lack of alternatives to bench mark datums for use in remote areas. Although all the methods require much more testing, the barnacle and growing seaweed lines appear to have potential, though there are situations where they are less accurate or absent. Lake levels and synchronous sea-level marking are also accurate, but upper levels of flotsam and lower limits of vegetation or lichens are probably less reliable. Of uncertain accuracy are the use of the icefoot and the tide table corrections of sea levels.

Instruments

Although the use of accurate levelling is increasing both in Arctic Canada (e.g. Andrews, 1970a; Blake, 1970, 1975; Pheasant and Andrews, 1973; McLaren and Barnett, 1978) and in Fennoscandia (e.g. Hoppe *et al.*, 1969; Cato and Linden, 1973; Rose, 1978; Rose and Synge, 1979; Corner, 1980), most shoreline altitudes quoted in the literature for these and other isostatic centres have been obtained with aneroid barometers/altimeters, or sometimes with hand-held levels. As early as 1908 both Goldthwait and Robinson warned of the shortcomings of these instruments and Rose (1981, p. 327) believes that "with the exception of use in reconnaissance or general studies in areas with a wide range of shoreline displacement, these implements create more problems than they solve". Their main advantage is that (although they provide more accurate results with two-person use) they can both be used by one person, whereas the surveyor's level and staff cannot. They also allow more rapid field measurement.

Aneroid barometer

The difficulties involved in the use of the aneroid barometer for measuring altitudes were ably discussed by Sparks (1953). Since it operates on the principle of pressure variation with altitude, accurate results clearly depend on either the presence of a stable background atmospheric pressure (i.e. anticyclones) or large, unidirectional changes in pressure that can be corrected reliably. Sparks advised against its use during the passage of fronts, unstable air masses, irregular development of pressure systems or high winds. Other possible sources of error include temperature differences between readings and in allowing time for the instrument to respond to ascents and descents. For the year 1948–1949 Sparks found that in north-west Scotland only 33% of summer days were suitable for this type of work for errors of ±1·5 m. Thus studies using this method must trade accuracy against field time. Table 3 shows estimated accuracies of altitudes in some studies using aneroids, though it is not always clear whether other error sources are incorporated in these figures. The most rigorous testing known to the author is in an undergraduate dissertation by Baillie (1972) in which under very favourable meteorological and terrain conditions an accuracy of only ±4·2% (2 standard deviations) was obtained. Lower accuracy is likely under more typical field conditions. Attempts to reduce the errors have involved taking frequent control or check readings, e.g. at a base point every 30 min (Moller and Sollid, 1972; Blake, 1975), repeating traverses (e.g. Løken, 1962; Ives, 1964; Dyke, 1979), checking some measurements by levelling (e.g. Sugden and John, 1965; Andersen, 1968) or correcting for changes in pressure and/or temperature as recorded at local meteorological stations (e.g. Sim, 1960; Washburn and Stuiver, 1962; Matthews, 1967; Archer, 1968; Andrews and Falconer, 1969). Traverses should be closed and closing errors distributed (as with levelling) but, like many other aspects of shoreline measurement, such detailed information is rarely given. Webber *et al.* (1970) found the aneroid unsuitable

TABLE 3
Error estimates/calculations in shoreline altitudes determined by aneroid barometer

Authors	Error	Location
Sparks (1953)	± 1·5 m (suitable weather conditions)	Britain
Løken (1962)	± 0·7 m	Labrador
Washburn and Stuiver (1962)	± 0·5 m (< 5 m altitude) ± 1 m (5-25 m) ± 2 m (> 25 m)	E. Greenland
Ives (1964)	± 1·5 m	Baffin Island
Matthews (1967)	± 1·8 m	N. Ungava (Canada)
Andrews (1966)	± 2 m	Baffin Island
Archer (1968)	± 3 m (1%) (depending on pressure ± 7·5 m (3%) conditions)	E. Hudson Bay
Andersen (1968)	± 1 m	N. Norway
Andrews and Falconer (1969)	± 3 m	Hudson Bay
Sugden (1972)	± 5 m	W. Greenland
Baillie (1972)	± 1·6 m (4·2%)	Scotland
Moller and Sollid (1972)	± 1 m	N. Norway
Sollid *et al.* (1973)	± 1 m	N. Norway
Andrews and Retherford (1978)	± 3 m	British Columbia
Dyke (1979)	± 2 m	Baffin Island

in their study in South Hudson Bay due to the flat terrain and slight altitudinal variation.

For greatest reliability two synchronized aneroid barometers can be used, one of which remains at a fixed point and is read at frequent intervals to record variations in atmospheric pressure, while a second person completes closed traverses with the other instrument, noting the time of altitude measurements made. The fixed point readings are then used to correct the altitudes surveyed with the other. Two barometers can also facilitate traverses that begin and end at different points. The key is that one of the instruments should always remain stationary and thus be able to provide a record of atmospheric pressure variation while the other is moving.

Hand-held level

Hand-held levels such as Abney levels or clinometers have been used in some studies in Fennoscandia and elsewhere (e.g. Sollid *et al.*, 1973; Donner *et al.*, 1977; Donner, 1978; Corner, 1978). The last author claimed an accuracy of < *c.* 2%. Synge and Stephens (1966) claimed an accuracy of 0·5% but Sissons (1967b) estimated that 2–4% was more likely. He stated that

> not only is the hand level unsatisfactory for accurate work but their (Synge and Stephens') method of using it, which involves setting the instrument to give a

horizontal line of sight and ascending slopes in a series of steps, permits errors of various kinds.

These include the possibility of variable and inaccurate foot-to-eye height of the user at each step depending on stance, slope angle etc.; the problem of identifying ground location for the subsequent stance, and the quantification of the part step at the end of the traverse. The method is also unsuitable over irregular or flat terrain, or for use over long distances, and it is difficult to close traverses to check for error since this involves sighting onto spots that have just been left. Used in this way the method appears to have little potential.

A modified method was described by Baillie (1972). He used an Abney level with a surveying staff in a similar way to accurate levelling, and also used a 1·5-m high, wooden rod on which to rest the level to eliminate variable stance (i.e. as a substitute for a tripod). At the 2 standard deviation level, Baillie found an error of ±0·8% in his results on a test traverse, and concluded that under more typical field conditions and without the use of the rod and staff, the errors must be much greater. He stated that Synge and Stephens' claimed error of ±0·5% must be considered "unrealistic". Although Baillie's method is quicker to use and involves lighter equipment than accurate levelling, it does involve two people and, with its lower accuracy and problems in use over long distances and some terrain conditions, has little to commend it.

Conclusions

Present alternatives to accurate levelling for measuring shoreline altitudes are inaccurate and are not suitable for all types of terrain. Accurate levelling appears to be gaining favour. Tacheometry may be used in future as a more rapid, but less accurate means of transferring altitude over long distances, and it is possible that very accurate but presently very bulky light reflection equipment will be used instead, particularly as more portable and less expensive models are introduced.

Some authors have estimated shoreline altitudes from topographic maps (e.g. Futyma, 1981) and with the development of photogrammetric contouring, this practice may increase. It also seems likely that direct photogrammetric measurement of shoreline altitude will be developed for remote areas (see, for example, Matthews, 1967, Fig. 4; Blake, 1975, p. 26). Problems will remain, however, in relating heights to a standard datum, positioning on the back edge, avoiding fans, channels etc., and allowing for overlying peat. Matthews (1967) gives the following comparative accuracies of shoreline altitudes—±3 m for the photogrammetric method, ±1·8 m for the aneroid barameter, and ±1 m for levelled heights.

Measurement procedures and sea-level corrections

Whereas in Scotland the vast majority of features measured have been raised terraces of one sort or another (raised beaches, estuarine mud flats, shore platforms etc.), these are generally much less common in the larger glacio-isostatic centres, particularly North America. As already argued, this probably relates to the slower rates of isostatic recovery in Scotland. Andrews (1970, p. 76) pointed out that "beaches or terraces are in consequence poorly developed over most of Arctic Canada" except in peripheral areas or at proglacial lake shorelines. Instead, a wide variety of forms has been measured and used in shoreline reconstruction both in North America and Fennoscandia, including shingle ridges, boulder barricades, beach notches, delta lips and washing limits. Sim (1960) presented the four main criteria that have been used to define the marine limit (upper limit of marine action) in Arctic Canada:

(i) lower limit of undisturbed ground moraine;
(ii) lower limit of perched boulders;
(iii) upper limit of beach ridges, rolled cobbles or deltaic deposits;
(iv) upper limit of marine shells, marine algae or marine mammals.

Andersen (1968), Andrews (1970a) and Agrell (1974) discussed the problems in identification and reliability of these types of field evidence. For example, perched boulders can occur below the marine limits, and marine shells may be ice transported. Similar reservations have been expressed about the "washing limits" used by Synge and Stephens (1966) in western Scotland (Gray, 1972a) though this should not be taken to imply that reliable washing limits do not exist in Scotland or elsewhere. Indeed, they have been used extensively in Scandinavia where they include Sim's categories (i) and (ii) listed above (see, e.g. Stephens and Synge, 1966; Synge, 1969).

Since these coastal features are related differently to mean tide level, attempts have been made to measure similar features on the modern coastline, and then to apply suitable altitude corrections to the relict forms. Andrews (1970a, p. 79) argued that "on a moderately exposed coast with a tidal range of 4 m the relationship of delta surfaces, storm beaches and wave-cut terraces may be such that these synchronous deposits could vary in elevation with respect to mean high tide by 10 m". Andrews notes that on western Baffin Island, delta lips are 3–4 m below the marine limit, a figure that corresponds to the present tidal range. Even though erosion of the delta face at lower water levels has probably occurred, addition of 3·5 m to relict delta lip altitudes seems reasonable to correct them to the marine limit. In northern Norway, Rose (1978) measured a variety of features on the modern shore relative to mean sea level, and applied appropriate corrections to the relict forms (see Table 4). More limited work of this nature has been carried out by Moller and Sollid (1972), Sollid *et al.* (1973), Rose and Synge (1979) and Corner (1980) (see Rose (1981) for summary). Andrews (1970a) argued that once such corrections are carried out, a standard error of ± 1 m is appropriate for the height of a shoreline. The particular problem of relating marine organisms to sea level at the time of growth is considered elsewhere in this volume (Sutherland).

With some notable exceptions, little information is given in most of the Fennoscandian and North American literature regarding how the features measured were selected for measurement, or of the spacing and number of measurements on

TABLE 4

Relationship between the altitude of actively forming shoreline landforms and contemporaneous mean sea level in the Bugoyfjord of N. Norway (after Rose, 1978)

Landform	Levelling position	Deviation ± 1 standard deviation from sea level	Correction
Drift platform	Inner break of slope	+1·8 ± 0·22 m	−2·0 m
Rock platform	Inner break of slope	+1·3 ± 0·07 m	−1·0 m
Beach ridge	Crest	+2·2 ± 0·12 m	−2·0 m
Beach ridge around delta	Crest	+0·7 ± 0·08 m	−0·5 m
Delta	Lowest position on subaerial surface	−0·7 ± 0·25 m	+0·5 m
Clay platform	Highest position	+0·7 ± 0·57 m	−0·5 m
Ice-pushed boulders	Base of boulders	0·0 ± 0·30 m	0·0 m
Ice-pushed boulders	Top of boulders	+1·5 ± 0·03 m	−1·5 m

each feature. Furthermore, the normal practice is to quote a single altitude for each feature without indicating how this figure was derived.

Shoreline representation and correlation

As already stated, different approaches to shoreline research have been adopted in North America, Fennoscandia and Britain (Andrews, 1967, 1970a). In North America the emphasis has been in establishing the age and altitude of sea level at particular localities with the intention of constructing curves of isostatic uplift against time. This is very different from the shoreline reconstruction approach used in Fennoscandia, as in Scotland, where the main aim has been to establish the spatial pattern of isostatic uplift by use of shoreline diagrams and isobase maps. In Fennoscandia, however, there has been much more emphasis on shoreline correlation by the use of time stratigraphical markers and ^{14}C dates, compared with the largely morphological correlations applied in Scotland. This is not to imply that techniques are in any way restricted by area. Uplift curves have been drawn for Scotland (e.g. Sissons and Brooks, 1971; Cullingford *et al.*, 1980), while equidistant shoreline diagrams have been constructed in North America (e.g. Løken, 1962; Andrews, 1970a, Figs 6–8), especially for proglacial lake shorelines. Non-linear regression has been used in some studies (e.g. Andrews *et al.*, 1970).

Shoreline dating and time-stratigraphic correlation are considered in Chapter 6 by Sutherland. The rest of this section will be devoted to the alternative shoreline diagram to the equidistant diagram, *viz.* the shoreline relation diagram. First developed in Fennoscandia and used there extensively in the 1930s and 1940s, a modified version of the shoreline relation diagram has recently found favour in North America.

Whereas individual equidistant diagrams are drawn for specific small sections of the dome of isostatic uplift, one aim of shoreline relation diagrams is to represent the uplifted area as a whole on a single diagram. As used in Fennoscandia, the method was to insert a well developed shoreline as a straight line reference level, and then to fix the horizontal position of shoreline sites at various localities in the isostatic uplift area from the altitude of the reference level at each point (see Fig. 12). In this way shoreline altitudes from anywhere in the uplifted area can be drawn on a single diagram, provided that the reference level is well established at each locality. If a second reference level at a higher level is present, comparison of shoreline gradients aids the correlation of shoreline altitudes at intermediate altitudes.

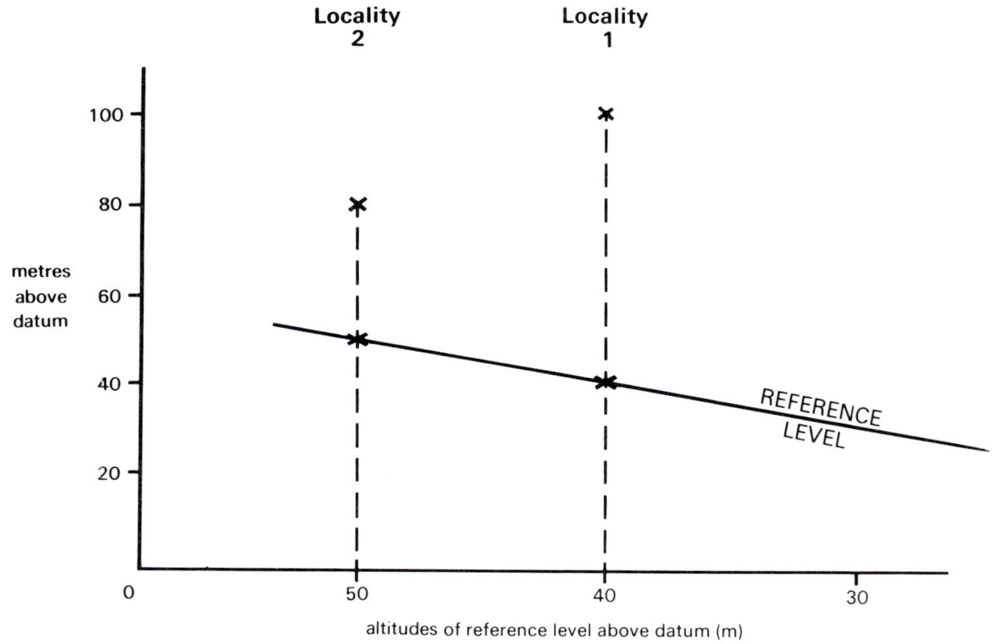

Figure 12. Diagram to illustrate the traditional use of the shoreline relation diagram. Reference level is plotted as a straight line. At locality 1, the reference level is at 40 m and hence this fixes the position along the x-axis of locality 1 where a higher shoreline at 100 m also occurs. At locality 2, which may be on the other side of the uplift dome, the reference is at 50 m and there is a higher shoreline at 80 m.

The shoreline relation (SR) diagram has been modified for use in Arctic Canada where reference levels cannot be traced over large areas. Instead the x-axis is assigned an age (e.g. 3000 B.P.) and is used as a reference level. Older age lines are then inserted above it, based on the generalized age–altitude uplift curve for Arctic Canada (Fig. 13).

Shoreline relation diagrams have been criticized by a number of Fennoscandian workers (e.g. Marthinussen, 1960; Donner, 1965; Andersen, 1965) and they have rarely been used in Fennoscandia in recent years. Criticism has focused on the assumptions made, which include:

(i) the age and altitude of the reference level must be known accurately and it must be synchronous;
(ii) shorelines must bear a constant relationship to each other (the so-called "proportionality principle"). This will not be the case if, for example, the centre of isostatic uplift has moved;
(iii) post-glacial rebound must have been a smoothly decelerating function, similar in all parts of the rebound centre.

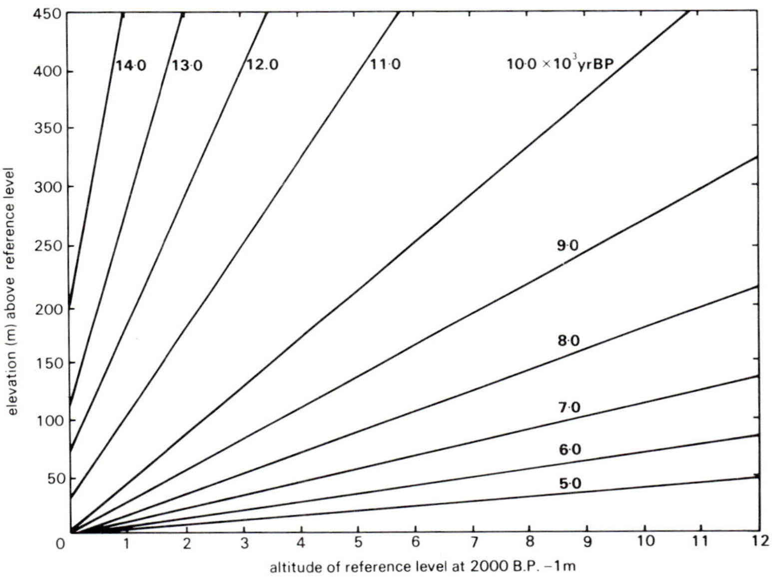

Figure 13. Shoreline relation diagram constructed for Arctic Canada. The lines are theoretical isochromes. 2000 B.P. is used as the reference level and plotted as the x-axis. A 1-m correction has been made for eustatic sea level at that time (from Andrews, 1969).

Research in Scotland gives some cause to doubt these three assumptions:

(i) Andrews' (1970a) nomogram for "Highland Britain" was applied to the Isle of Arran by Gemmell (1973), but the age and altitude of the reference level are open to question (Gray, 1975b).
(ii) Cullingford (1972) found that the isobases of two well-developed shorelines in the Tay area are not parallel. Similarly Fig. 14 indicates that if the two shorelines shown in eastern Scotland correlate with the two in the west, there was an eastward shift of over 30 km in the centre of isostatic uplift in Scotland during the early Holocene. Similar shifts are implied in other isostatic centres (e.g. Andrews and Barnett, 1972).
(iii) Sissons (1972, and this volume) and Sissons and Cornish (1982) have found clear evidence that faulting and localized uplift have occurred during rebound, e.g. in the Forth estuary (Fig. 10), and Gray (1974a) illustrates discontinuities in the gradient of a shoreline in western Scotland. More importantly there is

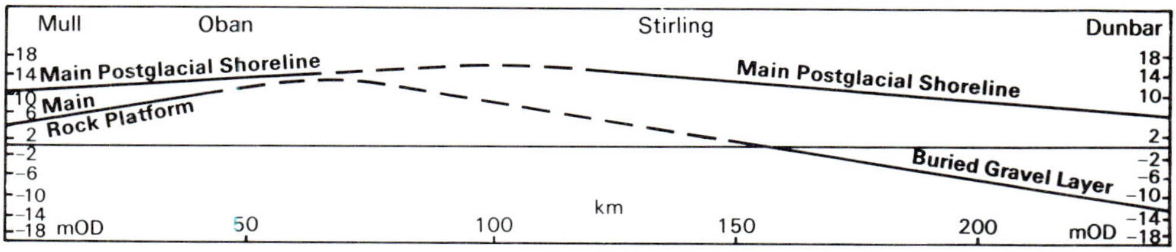

Figure 14. Cross-section through the Scottish isostatic centre showing two shorelines on both east and west coasts, and their projection across the isostatic centre. A shift in the centre of uplift is implied.

growing evidence that the Loch Lomond Advance in Scotland redepressed or retarded crustal uplift (Sutherland, 1981). These results challenge the assumptions that isostatic rebound proceeds by simple, continuous tilting and at a smoothly decelerating rate, and suggest larger irregularities in uplift in North America and Fennoscandia with their far greater isostatic movements (Sissons and Cornish, 1982).

Clearly problems remain in the use of SR diagrams but as Andrews (1970a, p. 53) says, they can still provide valuable age *estimates* of marine limits and lower strandlines.

CONCLUSIONS

The detailed methodology developed in Scotland, involving accurate levelling from bench marks of terrace altitudes, is impractical in most other centres of isostatic uplift. Other methods have been used in these areas, but there is considerable scope for testing their accuracy and developing those techniques with the greatest potential. The methods used in shoreline study vary greatly from area to area and all workers in this field can gain from studying the methods and approaches used elsewhere.

ACKNOWLEDGEMENTS

I am very grateful to Dr D. G. Sutherland, the editors of this book, and the anonymous referee for their comments.

REFERENCES

Agrell, H. (1974). *Bull. geol. Instn. Univ. Upsala* **4**, 125-188.
Andersen, B. G. (1965). *In* "The Geological Systems: The Quaternary, 1" (K. Ranakama, ed.), pp. 92-138. Wiley Interscience, New York.

Andersen, B. G. (1968). *Norges. geol. Unders.* **256**, 160 pp.

Andrews, J. T. (1966). *Geogr. Bull.* **8**, 174-193.

Andrews, J. T. (1967). *Geogr. Bull.* **9**, 71-74.

Andrews, J. T. (1969). *Arctic and Alpine Research* **1**, 221-222.

Andrews, J. T. (1970a). *Inst. Br. Geogr. Sp. Publ.* No. 2, 156 pp.

Andrews, J. T. (1970b). *Arctic and Alpine Research* **2**, 115-134.

Andrews, J. T. (1975). "Glacial Systems". Duxbury Press, Massachusetts.

Andrews, J. T. and Barnett, D. M. (1972). *Geogr. Annlr.* **A54**, 1-11.

Andrews, J. T. and Falconer, G. (1969). *Bull. geol. Soc. Am.* **6**, 263-276.

Andrews, J. T. and Retherford, R. M. (1978). *Canadian Journal of Earth Sciences* **15**, 341-350.

Andrews, J. T., Buckely, J. T. and England, J. H. (1970). *Bull. geol. Soc. Am.* **81**, 1123-1148.

Archer, D. R. (1968). *Arctic* **21**, 153-160.

Åse, L.-E. (1980). *Geogr. Annlr.* **A62**, 209-218.

Baillie, D. (1972). "Accuracies of heighting methods and contour maps". Unpublished. BSc. dissertation, Department of Geography, University of Edinburgh.

Baird, A. K., Baird, K. W. and Morton, D. M. (1971). *Bull. geol. Soc. Am.* **82**, 1219-1234.

Barr, W. (1971). *Arctic* **24**, 249-268.

Blake, W. (1970). *Canadian Journal of Earth Sciences* **7**, 634-664.

Blake, W. (1975). *Geogr. Annlr.* **A57**, 1-71.

Browne, M. A. E. and Graham, D. K. (1981). *Quaternary Newsletter* **34**, 1-7.

Cato, I. and Linden, A. (1973). *Geol. För. Stockh. Förh.* **94**, 25-35.

Chayes, F. (1970). *Bull. geol. Soc. Am.* **81**, 1273-1278.

Corner, G. D. (1978). *Norsk. geogr. Tidsskr.* **32**, 137-142.

Corner, G. D. (1980). *Boreas* **9**, 239-249.

Cullingford, R. A. (1972). "Lateglacial and postglacial shoreline displacement in the Earn-Tay area and eastern Fife". Unpublished. Ph.D. thesis, University of Edinburgh.

Cullingford, R. A. (1977). *In* "Studies in the Scottish Lateglacial Environment" (J. M. Gray and J. J. Lowe, eds), pp. 15-32. Pergamon Press, Oxford.

Cullingford, R. A. and Smith, D. E. (1980). *Boreas* **9**, 21-38.

Cullingford, R. A., Caseldine, C. J. and Gotts, P. E. (1980). *Nature* **284**, 159-161.

Dawson, A. G. (1979). "The raised shorelines of Jura, Scarba and N.E. Islay". Unpublished Ph.D. thesis, University of Edinburgh.

Dawson, A. G. (1982). *Scott. J. Geol.* **18**, 253-265.

Donner, J. J. (1959). *Ann. Acad. Sci. Fenn, AIII, Geol-Geogr.* **53**, 1-25.

Donner, J. J. (1965). *Baltica* **2**, 11-20.

Donner, J. J. (1978). *Geogr. Annlr.* **A60**, 63-72.

Donner, J. J., Eronen, M. and Jungner, H. (1977). *Norsk. geogr. Tidsskr.* **31**, 103-128.

Dyke, A. S. (1979). *Arctic and Alpine Research* **11**, 179-202.

Futyma, R. P. (1981). *Quaternary Research* **15**, 291-310.

Gemmell, A. M. D. (1973). *Trans. Inst. Br. Geogr.* **59**, 25-39.

Goldthwait, J. W. (1908). *J. Geol.* **16**, 459-476.

Gray, J. M. (1972a). "The inter-, late-, and post-glacial shorelines, and ice-limits of Lorn and eastern Mull". Unpublished Ph.D. thesis, University of Edinburgh.

Gray, J. M. (1972b). *Area* **4**, 275-279.

Gray, J. M. (1974a). *Trans. Inst. Br. Geogr.* **61**, 81-99.

Gray, J. M. (1974b). *Boreas* **3**, 129-138.
Gray, J. M. (1975a). "Measurement and analysis of Scottish raised shoreline altitudes". Department of Geography, Queen Mary College, University of London, Occasional Paper No. 2, 40 pp.
Gray, J. M. (1975b). *Trans. Inst. Br. Geogr.* **64**, 129-132.
Gray, J. M. (1978). *Trans. Inst. Br. Geogr.* N.S. **3**, 151-164.
Hoppe, G., Schytt, V., Hagglom, A. and Osterholm, H. (1969). *Geogr. Annlr.* **A51**, 185-192.
Ives, J. D. (1960). *Geogr. Bull.* **14**, 44-70.
Ives, J. D. (1964). *Geogr. Bull.* **21**, 54-65.
Jamieson, T. F. (1865). *Q. Jl. geol. Soc. Lond.* **21**, 161-203.
Jamieson, T. F. (1882). *Geol. Mag.* **9**, 400-407 and 457-466.
Jardine, W. G. (1976). *Geol. För. Stock. Förh.* **98**, 78-82.
John, B. S. and Sugden, D. E. (1971). *Brit. Ant. Surv. Bull.* **24**, 45-111.
Kemp, D. D. (1971). "The stratigraphy and sub-carse morphology of an area on the northern side of the River Forth between Lake of Menteith and Kincardine-on-Forth". Unpublished Ph.D. thesis, University of Edinburgh.
Krumbein, W. C. and Graybill, F. A. (1965). "An Introduction to Statistical Models in Geology". McGraw-Hill, New York.
Løken, O. H. (1962). *Geogr. Bull.* **17**, 23-56.
Løken, O. H. (1978). *Canadian Journal of Earth Sciences* **15**, 1547-1553.
Marthinussen, M. (1960). *In* "Geology of Norway" (O. Holtendahl, ed.), pp. 416-429. *Norges geol. Unders.*
Matthews, B. (1967). *Arctic* **20**, 176-202.
McLaren, P. and Barnett, D. M. (1978). *Arctic* **31**, 415-427.
Moller, J. J. and Sollid, J. L. (1972). *Norsk. geogr. Tidsskr.* **26**, 101-133.
Morner, N.-A. (1980). *In* "Earth Rheology, Isostasy and Eustasy" (N.-A. Mörner, ed.), pp. 251-284. John Wiley, Chichester.
Nikonov, A. A. (1980). *In* "Earth Rheology, Isostasy and Eustasy" (N.-A. Mörner, ed.), pp. 341-354. John Wiley, Chichester.
Paterson, I. B., Armstrong, M. and Browne, M. A. E. (1981). *Inst. Geol. Sci. Rep.* No. 81/7, 35 pp.
Pheasant, D. R. and Andrews, J. T. (1973). *Canadian Journal of Earth Sciences* **10**, 1621-1641.
Robinson, G. (1972). *Area* **4**, 104-113.
Robinson, H. H. (1908). *J. Geol.* **16**, 347-356.
Robinson, M. (1977). "Glacial limits, sea-level changes and vegetational development in part of Wester Ross". Unpublished Ph.D. thesis, University of Edinburgh.
Rose, J. (1978). *Norsk. geogr. Tidsskr.* **32**, 121-135.
Rose, J. (1981). *In* "Geomorphological Techniques" (A. Goudie, ed.), pp. 327-341. George Allen and Unwin, London.
Rose, J. and Synge, F. M. (1979). *Quaest. Geogr.* **5**, 125-151.
Sim, V. M. (1960). *Arctic* **13**, 178-193.
Sissons, J. B. (1966). *Trans. Inst. Br. Geogr.* **39**, 19-29.
Sissons, J. B. (1967a). "The Evolution of Scotland's Scenery". Oliver and Boyd, Edinburgh.
Sissons, J. B. (1967b). *Trans. Inst. Br. Geogr.* **42**, 163-168.
Sissons, J. B. (1969). *Trans. Inst. Br. Geogr.* **48**, 19-50.
Sissons, J. B. (1972). *Trans. Inst. Br. Geogr.* **55**, 145-159.
Sissons, J. B. (1981). *Scott. J. Geol.* **17**, 7-19.

Sissons, J. B. and Brooks, C. L. (1971). *Nature Phys. Sci.* **234**, 124-127.
Sissons, J. B. and Cornish, R. (1982). *Nature* **297**, 213-214.
Sissons, J. B. and Dawson, A. G. (1981). *Proc. Geol. Ass.* **92**, 115-124.
Sissons, J. B. and Rhind, D. W. (1970). *Scott. J. Geol.* **6**, 272-284.
Sissons, J. B. and Smith, D. E. (1965a). *Trans. R. Soc. Edinb.* **66**, 143-168.
Sissons, J. B. and Smith, D. E. (1965b). *Scott. J. Geol.* **1**, 247-255.
Sissons, J. B., Smith, D. E. and Cullingford, R. A. (1966). *Trans. Inst. Br. Geogr.* **39**, 9-18.
Smith, D. E. (1965). "Late and postglacial changes of shoreline on the north side of the Forth valley and estuary". Unpublished Ph.D. thesis, University of Edinburgh.
Smith, D. E., Sissons, J. B. and Cullingford, R. A. (1969). *Trans. Inst. Br. Geogr.* **46**, 45-52.
Sollid, J. L., Andersen, S., Harme, N., Kjeldsen, O., Salvigsen, O., Sturod, S., Tveita, T. and Wilhelmsen, A. (1973). *Norsk. geogr. Tidsskr.* **27**, 233-325.
Sparks, B. W. (1953). *Geogr. J.* **119**, 73-80.
Stephens, N. and Synge, F. M. (1966). *In* "Essays in Geomorphology" (G. H. Dury, ed.), pp. 1-51. Heinemann, London.
Sugden, D. E. and John, B. S. (1965). *Geogr. J.* **131**, 235-247.
Sugden, D. E. (1972). *Arctic and Alpine Research* **4**, 97-117.
Sutherland, D. G. (1981). "The raised shorelines and deglaciation of the Loch Long/Loch Fyne area, western Scotland". Unpublished Ph.D. thesis, University of Edinburgh.
Synge, F. M. (1969). *Geogr. Annlr.* **A51**, 193-206.
Synge, F. M. (1977). *In* "The Quaternary History of the Irish Sea" (C. Kidson and M. J. Tooley, eds), pp. 199-222. Seel House Press, Liverpool.
Synge, F. M. and Stephens, N. (1966). *Trans. Inst. Br. Geogr.* **39**, 101-125.
Tanner, V. (1930). *Bull. Commn. geol. Finl.* **88**, 600 pp.
Washburn, A. L. and Stuiver, M. (1962). *Arctic* **15**, 66-73.
Webber, P. J., Richardson, J. W. and Andrews, J. T. (1970). *Canadian Journal of Earth Sciences* **7**, 317-325.
Wain-Hobson, T. (1981). "Aspects of glacial and postglacial history of North-West Argyll". Unpublished Ph.D. thesis, University of Edinburgh.
Wright, W. B. (1914). "The Quaternary Ice Age". Macmillan, London.

SIX

The dating of former shorelines

D. G. SUTHERLAND

INTRODUCTION

The study of former shorelines is of considerable importance to a variety of disciplines. Geophysicists gain information that improves their understanding of the rheological properties of the earth. Engineers concerned with construction near the coast must have information on rates of submergence or emergence. Archaeologists have long been aware that in the past (as today) the coastline has been a focus for man's activities. Geologists and geomorphologists interested in the minutiae of the history of the earth have found knowledge of shoreline variations essential to an understanding of other phenomena such as growth and shrinkage of ice sheets or changes in the behaviour of river systems. Fundamental to all studies of shoreline change, however, is some idea of the timing of the events being examined.

The dating of former shorelines, therefore, is necessary whether the study is of relative movements of sea level at a specific locality or whether particular shorelines are being correlated over a wide area. This chapter attempts to review many of the techniques, both relative and "absolute" in nature, that have been developed to provide information as to the dating of sea-level changes. A number of general text books cover certain aspects of dating of Quaternary deposits (e.g. Bowen, 1978; Goudie, 1981), whilst various review papers also include discussion of a variety of the methods frequently applied to shoreline dating (e.g. Broecker and Bender, 1972; Thom, 1973; Andrews and Miller, 1980; Rose, 1981). There appears to be no publication, however, that attempts to assess the limitations of the various techniques both from the technical point of view and with respect to the association of the material and the shoreline being dated.

Prior to discussion of particular methods, two basic principles that are applicable to all studies may be stated:

(i) *The event or feature being dated must be adequately defined with respect to its contemporaneous sea level.* Many dated samples do not accord with this principle, often only allowing a general statement to be made as to sea level being above or below the sample at a particular time. Such samples may create more problems than they resolve and can be a waste of valuable resources.

SHORELINES AND ISOSTASY
ISBN 0 12 652960 4

(ii) *Independent methods of dating or cross-checking on the reliability of the dated sample should always be employed.* All methods of dating have an associated degree of uncertainty. The uncertainty may be capable of statistical quantification or it may only relate to relatively broad intervals of time into which the event or feature may be placed. It therefore follows that greater certainty can be accorded to those shoreline ages that have been derived by more than one method. In particular, dated samples should be located in a relative chronological framework based on stratigraphical or morphological studies, preferably both. A further matter of concern is the physical or chemical integrity of the sample and checks should be made to ascertain the presence or otherwise of contaminating agents. This last statement is particularly true for the various radiometric and biochemical dating methods.

HISTORICAL AND ARCHAEOLOGICAL EVIDENCE

There are many historical and archaeological structures that have been built in relation to sea level. If a close relationship to the contemporaneous sea level can be established then knowledge of the historical period or archaeological culture may provide either an accurate date on the sea level or a means of correlation of shorelines over a wide area. Numerous studies have made use of the above line of argument in Britain (e.g. Movius, 1953), North America (e.g. Grant, 1970), Scandinavia (e.g. Donner, 1964) as well as more widely (e.g. Flemming, 1969, 1972). Unfortunately, the two basic requirements of this approach, that of a relationship to sea level and that of an accurate independent chronology, are not often encountered together.

Historical evidence, whilst frequently appearing to be precise, inevitably relates to small absolute movements of sea level. It is therefore amenable to misinterpretation resulting from changes of tidal regime that are independent of absolute movements of sea level (Johnson and Winter, 1927). Furthermore, uncertainties in altitude of only a few decimetres that would be negligible in studies of much greater relative movements of the shoreline can be of significant magnitude in the historical context. Akeroyd (1972) has reviewed in detail much of the historical and archaeological evidence relating to sea-level change in south-east England and has effectively demonstrated the considerable degree of uncertainty inherent in this approach even in an intensively studied area.

Archaeological evidence, even more than historical evidence, has difficulties in associating the human artifacts or structures with specific sea levels. The wide application of ^{14}C dating to archaeological material has also demonstrated that distinct cultures can be found to have occupied different parts of the same region at the same time and therefore are not a suitable basis for correlating shorelines, even should it be possible to relate the archaeological evidence to sea level in an unambiguous fashion.

It seems most likely that in the future independently dated shorelines will be used to place limits on archaeological or historical interpretations rather than *vice versa*. Certain exceptions undoubtedly occur, a particular instance being the historically inscribed "water marks" that are located around the Swedish coast (Bergsten, 1954) that in a unique way provide information on the present rate of emergence at various localities.

LICHENOMETRY

In areas of continuing land emergence, as is experienced near the centres of the former northern hemisphere ice sheets, bare rock surfaces are being produced as the sea regresses. Similar rock surfaces that are now at higher altitudes have been colonized by lichens, and in polar regions where growth of competing vegetation is inhibited, the size of the lichens should relate to the lapse in time between the present and when the rock surface emerged from the sea. Donner *et al.* (1977) measured the maximum thallus diameters of *Rhizocarpon* spp. on rock surfaces that resulted from *c.* 5000 years of emergence at three separate localities around Varangerfjord in northern Norway. It was found that whilst each locality showed a systematic growth curve there was little relationship between the growth curves from the separate localities, except perhaps during the first 500 years of growth. Donner *et al.* (1977) concluded by agreeing with Beschel (1961) that conditions for lichen growth varied greatly by the shore and hence that this approach was of limited utility. It may be, however, that by paying attention to particular species of lichen this method could be applied in detailed studies of shoreline displacement in polar areas during the last 3000–4000 years. A particular relationship between the tidal cycle and the lowest position for lichen growth has been used as a surrogate datum plane (see Gray, this volume) and this would permit the lichenometrically dated rock surface to be related to a particular part of the tidal cycle.

PALYNOLOGY

Pollen analysis is one of the most important techniques applied to the study of sea-level changes. It has formed the basis of a significant body of work, particularly in north-western Europe, of both Holocene sea-level changes (e.g. Godwin, 1943; Jelgersma, 1961; Tooley, 1978) and of various interglacial sea levels (see West, 1972; Zagwijn, 1979).

There are two principal advantages to pollen analytical work in the study of sea-level changes. First, it allows transgressive or regressive contacts, as represented by seral development of marine, brackish and freshwater vegetation, to be identified to within a few centimetres. It thus also, by logical extension, indicates those sections in which there are hiatuses due to non-sedimentation or to erosion. This matter is considered further on pp. 140–143.

Secondly, palynology can be used directly in the dating of the transgressive or regressive contacts identified. Broadly, pollen analysis is used as a dating method either by comparing local pollen zones with regionally established pollen assemblage zones (PAZs) or, by identifying discrete events, such as the appearance or disappearance of a particular species and comparing this event with other dated sites (Bradshaw, in press). Both these methods have significant limitations. PAZs are of different durations, varying in north-western Europe, for example, from a few hundred years in the early Holocene to several thousand years in the late Holocene (cf. Hibbert *et al.*, 1971). Furthermore, PAZ boundaries are diachronous (Smith and Pilcher, 1973) and should only be used for the region in which they have been established. This argument applies also to the use of individual species as particular stratigraphic markers as the immigration of many species is also diachronous. The use of isochrone maps (Huntley and Birks, 1982), if available, may reduce considerably the error resulting from such diachroneity. With either approach palynology is in essence a relative dating technique unless calibrated against an independent dating method such as ^{14}C. This inevitably introduces further uncertainties. The usual approach in pollen analytical studies of Holocene sea-level change is to use ^{14}C as the prime dating method and correlation with independently ^{14}C-dated pollen diagrams as a means to check the validity of the local dates (cf. Tooley, 1978).

In studies of interglacial sea levels pollen analysis is of great value in initially identifying the particular interglacial and thereafter the particular pollen zone of the interglacial to which the sea level is referred. A potential problem exists if the complete sequence of interglacial pollen zones is not present at any one locality (Bowen, 1978) and this can lead to the erroneous correlation of diverse sites that appear to fit into a sequence of progressive vegetational change. Thus, although during the Ipswichian Interglacial in eastern and south-eastern England sea level appears to have risen from below −13 m Ordnance Datum (OD) in Ip I (Gaunt *et al.*, 1974) to above this altitude by zone Ip IIb, culminating somewhat above OD in zone Ip III (Sparks and West, 1970; West, 1972; Gaunt *et al.*, 1974), full demonstrations of these correlations await the discovery of a complete sequence of Ipswichian deposits. Similar types of evidence are available from the Netherlands (Jelgersma *et al.*, 1979) that allow the course of the Eemian Interglacial sea level to be traced from *c.* −40 m below present in pollen zone E3 to a maximum of *c.* −8 m below present in pollen zone E5 and back to *c.* −23 m below present in zone E6b.

In north-western Europe, therefore, pollen analysis has provided a major body of work on the relative chronology of interglacial sea levels (West, 1972; Zagwijn, 1979), and an important research objective of the future will be to apply certain other dating techniques described below to place these sea levels in an "absolute" chronology and allow comparison with the U-series dated sea-level curves established for elsewhere (e.g. Bloom *et al.*, 1974) as well as with the deep-sea core evidence (cf. Bowen, 1978).

PALAEOMAGNETISM

Three aspects of palaeomagnetism are of use in dating sediments: polarity reversals that have been established on time-scales of millions of years; polarity excursions in which the virtual geomagnetic pole deviates by more than 45° of latitude from the pole and then returns to the original polarity, and which have been proposed on the time-scales of tens of thousands of years; and geomagnetic secular variations that are now well established during the last 10 000 years (Thompson, 1977). Neither the reversal time-scale nor polarity excursions have been much used in dating shorelines but study of secular variation together with other geomagnetic properties of sediments such as susceptibility and intensity has considerable utility for studying sea-level changes. Like pollen analysis these palaeomagnetic measurements can fulfil the dual role of both identifying changes from freshwater to marine sedimentation (or *vice versa*) and allowing an estimation to be made of the age of such an event.

In the relevant palaeomagnetic studies measurements are made of the relative declination, inclination, susceptibility and intensity of the natural remanent magnetism of the sediment. These measurements are rapid, can be non-destructive of the sediment core and certain ones (e.g. susceptibility) can be made in the field as a control on further sampling (Oldfield, 1981). The measurements of susceptibility and intensity show marked variations that accord with marine transgressions or regressions (e.g. Dickson *et al.*, 1978; Björck *et al.*, 1982). This is probably the result of the differing particle size of the sediments in the marine or freshwater layers.

Measurements of relative declination and inclination have been shown to have a reproducible pattern in sediments deposited during the last 10 000 years (Mackereth, 1971; Thompson, 1977; Turner and Thompson, 1981). This pattern can be used to date sediments by comparison with a "master curve" of such variations such as has been established for Britain (Thompson and Turner, 1979). As with pollen assemblage zones this master curve is of regional rather than global significance (Oldfield, 1981). It is ultimately dependent upon ^{14}C for other than the relative dating of events and hence has attendant uncertainties.

In principle the curves of variation in declination and inclination should be capable of extension backwards in time through the period corresponding to the Lateglacial of north-western Europe. In certain extra-glacial sites even longer sequences are possible (e.g. Creer, 1974). In practice, however, reproducibility at these earlier periods has proved difficult, possibly because of the variable response of non-organic clays to the retention of palaeomagnetic characteristics.

The full realization of the palaeomagnetic method is dependent upon the nature of the sediment. The particle size is critical for orientation measurements because above a certain coarseness (*c.* 63 μm) particles do not freely orient in the earth's magnetic field. Similarly, susceptibility and intensity are dependent upon particle size as magnetic particles concentrate in specific size fractions. These factors together with the rôle of small amounts of organic material in "fixing" the palaeomagnetic signal are not fully understood and more research is required to quantify them. At present, empirical evidence indicates that certain basins do not give interpretable palaeomagnetic measurements. A further problem with the study of Lateglacial sediments is that the rapid climatic changes of this period can produce alterations in the sediment budget that give rise to variations in susceptibility similar to those produced by marine incursions (Bjorck *et al.*, 1982) and thus, in certain circumstances, mask the marine-freshwater contact. Other indicators of climatic change or marine conditions such as pollen or diatom analyses may be employed to differentiate these effects.

In studying marine-freshwater sedimentary changes it is advantageous to work in lake basins; thus the corresponding shoreline may not be directly dated but only those times when the lake basin has been invaded by or isolated from the sea (see pp. 143–144 for further discussion of isolation basins). Thus, altitudinal limits can only be placed on the relevant fluctuations of the sea if the threshold altitude of the basin at the time of invasion/isolation is known. If a shoreline is developed around the basin that can be associated with a particular transgression, then the age of the shoreline is clearly bracketed by the dates of the invasion and isolation sedimentary contacts.

The further extension of the geomagnetic secular variation dating master curve is to be hoped for. Added dating possibilities during the Pleistocene may come from resolution of the controversial polarity excursions that have been proposed (e.g. Bonhommet and Zahringer, 1969; Mörner *et al.*, 1971; Barbetti and McElhinny, 1972; Noel, 1975), whilst application of the reversal chronology to the early Pleistocene interglacial sequence of north-western Europe should help in placing this sequence and its associated sea-level information in an absolute chronology. Palaeomagnetic dating seems set to play a much larger rôle in the dating of sea-level changes than it has hitherto.

VARVES

Varves are rythmically deposited sediments in which each rythmite accords to a single year. In principle where varves can be identified and correlated from place to place an exact chronology may be established that can be used to date, for example, ice-sheet retreat or associated sea levels. If it were possible to relate this varve chronology to the present day then an absolute method of dating would exist. This realization led to a number of classic studies in Sweden (de Geer, 1912, 1940), Finland (Sauramo, 1923) and North America (Antevs, 1925) that related the retreats of the Scandinavian and Laurentide ice sheets to floating varve chronologies.

Only in Sweden, however, has it been possible to relate the varve chronology to the present (Liden, 1938) and only in Sweden has extensive rechecking of the original varve chronology allowed its errors to be assessed (Nilsson, 1968; Wenner, 1968; Fromm, 1970) and comparisons made with ^{14}C dating (Tauber, 1970; Lundqvist, 1975; Mörner, 1980). These studies have shown that although in principle the varve chronology could give dates of events to the nearest year, in practice the Swedish varve chronology increases in uncertainty as earlier events are considered, being assigned errors of +360, −200 years at *c.* 7000 B.P. and +475, −300 years at *c.* 12 000 B.P. (Fromm, 1970). Thus it is perhaps slightly less accurate than ^{14}C dating during the same period. The method is not subject, however, to the problems of contamination that can invalidate any one ^{14}C date, and over discrete time periods or for specific events the varve chronology should be capable of greater resolution than ^{14}C dating and hence serve as a useful check on that method.

A varve chronology can be applied to the dating of shorelines either directly or indirectly. The indirect method involves using the varve chronology to date particular phases of ice retreat that can, in turn, be related to shorelines (e.g. Sauramo, 1923; Donner, 1978). This method is of wide application as glacial retreat stages have frequently been found to be associated with specific shorelines, but major errors may occur if the shoreline is wrongly associated with the ice retreat features.

The direct method has been most convincingly demonstrated by Liden (1938) along the Angerman River in Sweden. Angermanland is characterized by strongly seasonal river regimes and a high rate of glacio-isostatic uplift such that silty clay non-glacial varves have been formed and preserved by the continuing regression of the sea. Liden was able to trace certain varve sequences into the surfaces of deltas that had formed during the emergence of the land. The varves therefore directly date the deltas which Liden associated with shorelines 2-5 m above the delta surfaces. Liden was thus able to construct the first closely dated emergence curve for a glacio-isostatically uplifted area. This method should be applicable in other areas of rapid uplift and, because of its potential for high chronological resolution, it may serve as a check on the ^{14}C dating of associated shorelines.

RADIOCARBON DATING

Since its inception in the late 1940s by Libby and co-workers (Libby, 1955), ^{14}C dating has become the prime dating method for events in the last 50 000 years or so. In general, ^{14}C dates younger than *c.* 25 000 B.P. are widely accepted as reliable whilst dates greater than this age are often considered to be so subject to minor amounts of contamination as to be unreliable unless supported by independent methods of dating. It should be emphasized, however, that all samples should be subject to cross-checks whatever their age.

In discussing the application of the ^{14}C method to the dating of shorelines it is useful to consider separately deposits of marine origin (chiefly carbonates) and terrestrial materials associated with marine events.

Deposits of marine origin

There are two outstanding problems to be satisfied for ^{14}C dates on marine substances to be of use in the dating of shorelines: (a) the relationship between ^{14}C

ages on marine organisms and those on terrestrial materials; and (b) the relationship between the samples dated and the contemporaneous sea level. Marine carbonates are also subject to post-depositional recrystallization and contamination by detritus lodged in the matrix, and the means of detecting such effects are discussed on pp. 146–148.

(a) The majority of the carbon in the active carbon cycle is dissolved in the oceans. Because the mass turn-over of the oceans is relatively slow compared to the mixing of the atmosphere, the oceans are impoverished in ^{14}C relative to the atmosphere from which photosynthesizing terrestrial vegetation derives its carbon. Marine carbonates or plants that assimilate their carbon from sea water therefore have an apparently greater age than do contemporaneous terrestrial materials (Mangerud, 1972). The magnitude of this apparent age varies according to the rate of mixing of the surface layer of the oceans, the presence of upwelling of deep "old" water and dilution effects near the coasts (Broecker *et al.*, 1960; Mangerud, 1972; Mangerud and Gulliksen, 1975). Estuaries with high freshwater inputs containing ancient carbon from bedrock may also produce anomalously high ages (Scholl and Stuiver, 1967; Broecker and Bender, 1972; Mangerud, 1972) and limestone reefs may locally have the same effect (Polach and Golson, 1966).

The apparent age of sea water may be established by ^{14}C dating marine organisms that were collected live prior to atom bomb testing. Those samples collected during the last 100 years may show the influence of the "industrial effect" caused by the burning of fossil fuel and this should be allowed for in apparent age calculations (Mangerud and Gulliksen, 1975). Older samples may be influenced by natural variations in ^{14}C production (Stuiver, 1978) and Olsson (1980) has attempted to correct for this. Numerous studies have been made on the apparent age of sea water (e.g. Mangerud, 1972; Hjort, 1973; Krog and Tauber, 1974; Mangerud and Gulliksen, 1975; Olsson, 1980; Harkness, in press) revealing that whilst ages as much as 2500 years may be achieved (Rafter and O'Brien, 1970) more typically the ages are 300-500 years. Figure 1 is a compilation of available results for the coasts of the North Atlantic, as far as the slightly differing methods of age calculation allow. The eastern Atlantic coastal areas are typified by ages of 300-500 years with a suggestion of an age gradient from south to north around the British coast (Harkness, in press) as well as along the Norwegian coast (Mangerud and Gulliksen, 1975). The Arctic coasts and the east coast of Greenland appear to have higher ages of the order of 500–700 years.

In addition to the age difference between terrestrial and marine materials due to the above carbon cycle effects, a further problem results from isotopic fractionation. This is the biological or chemical process whereby different isotopes are preferred or discriminated against because of their different atomic masses. On average marine carbonates have *c.* 5% more ^{14}C because of isotopic fractionation than do terrestrial materials such as grasses, leaves and wood (Craig, 1953). This fractionation effect is equivalent to an age difference of *c.* 410 years.

It has been the practice of a number of radiocarbon laboratories to argue that the variations in ^{14}C age resulting from the apparent age of sea water and from isotopic fractionation effectively cancel each other out and that hence it is not necessary to make adjustments for either (e.g. Nydal *et al.*, 1972; Krog and Tauber, 1974). Such an approach is unsatisfactory because, first, the apparent age of sea water may be well in excess of 400 years and is thus best considered as a separate problem from

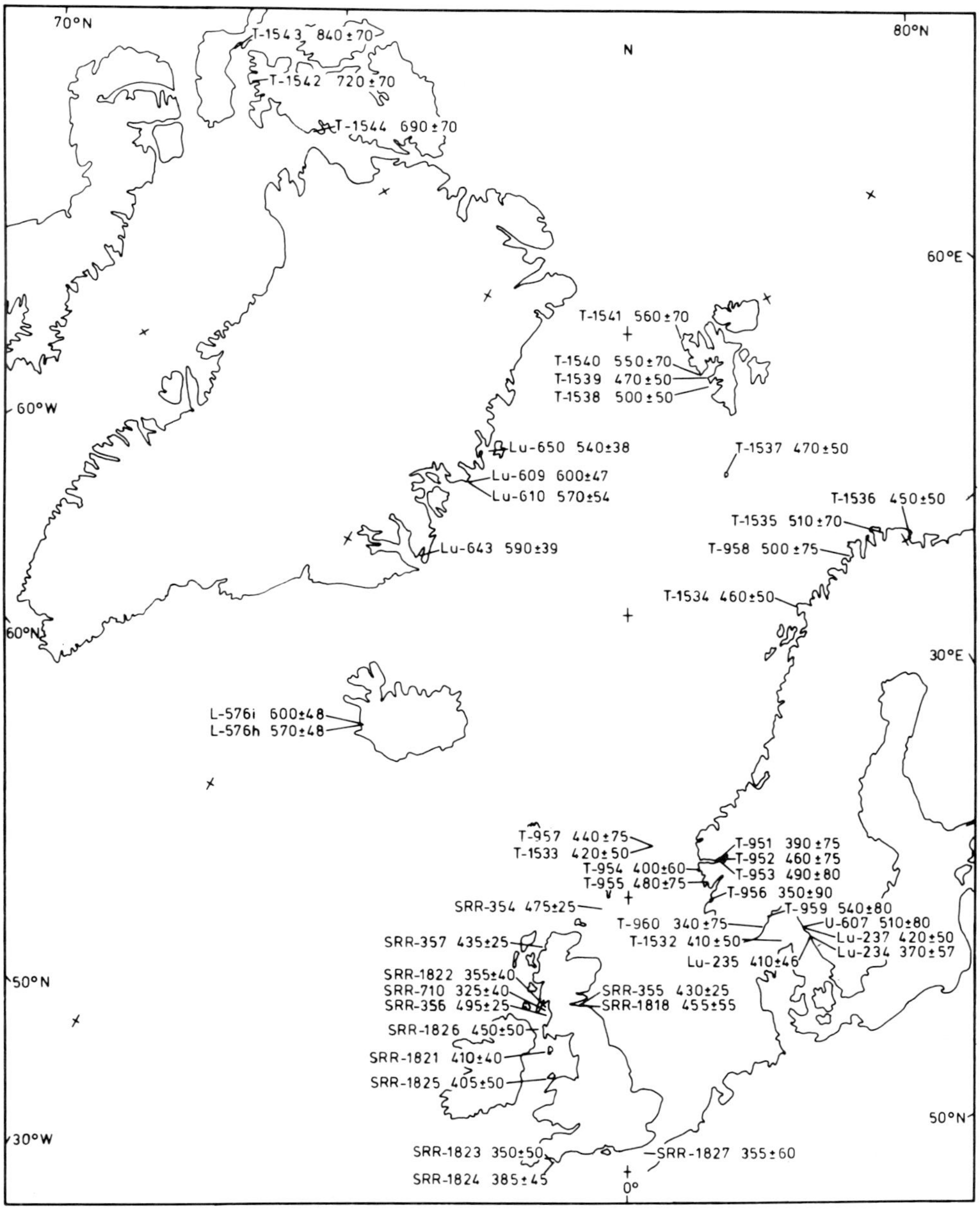

Figure 1. The apparent age of recent marine shells around the North Atlantic and adjacent seas. Based on Mangerud and Gulliksen (1975) and Harkness (in press).

that of computing a ^{14}C age; secondly, the range of natural isotopic fractionation in marine carbonates is such that age normalization may be as little as 320 years or as much as 480 years; and thirdly, isotopic fractionation may be induced in the laboratory during pretreatment (Harkness, 1979) and this should be measured and allowed for. Rather, it is to be recommended that marine shell ^{14}C dates should be reported as having been normalized of isotopic fractionation (Stuiver and Polach, 1977) and the subsequent age of sea water correction applied by the user in the light of the geological situation of the sample (Polach and Golson, 1966; Olsson, 1972; Sutherland, in press).

A further matter of potential concern in the laboratory treatment of marine carbonates is that radon gas (derived from the U-series elements present in variable quantities in marine carbonates; see pp. 145–148) may be released together with CO_2 during preparation for proportional gas counting (Salvigsen and Nydal, 1981). The radon decay can lead to anomalously high sample counts with an apparent younging effect that may be severe with samples that have relatively high uranium concentrations and low ^{14}C activity. The short half-life of radon (*c.* 3·8 days; Broecker and Bender, 1972) means that this problem can be overcome by storage for up to several months prior to counting (Salvigsen and Nydal, 1981).

(b) The relationship of a dated sample to its contemporaneous shoreline is frequently difficult to establish. Marine shells or corals are found either *in situ* or as detrital accumulations and as each of these occurrences has a different possible relationship to contemporaneous sea level it is necessary to distinguish between them in the field.

A number of criteria can be applied to establish whether the samples being studied are *in situ*. For bivalves both valves should occur together and molluscs may have the periostracum attached. Infaunal species should be in the position of growth. Gastropods may be found attached to rock surfaces though large boulders may have been transported by sea ice or icebergs. If a number of species can be identified, particularly of associated microfauna or flora, then a coherent population should be present. In addition, the associated sediment must be that in which the species is normally found.

Detrital accumulations are probably found more frequently and can result from normal current and wave activity in the littoral or nearshore zone as well as from reworking of older shell beds or from glacier transport. The various criteria used to identify *in situ* samples generally do not apply to detrital accumulations. Bivalves rarely have valves together, the periostracum is usually missing, broken shell fragments occur frequently and there may be signs of abrasion. The fauna does not form a life population, certain species frequently being over-represented and out of their sedimentary context.

Once a sample is established to be *in situ*, there are further problems in relating the sample to a particular shoreline. Littoral shells are only rarely found *in situ* because of the active nature of the marine processes in this zone. In certain restricted environments littoral samples may be preserved *in situ*, such as the stromatolitic algal mats recorded in the hypersaline lagoons along the Mauritanian coast (Einsele *et al.*, 1974). A rather different example comes from Australia where Hopley (1974) has indicated that beach rock in the form of flat horizontal benches is precipitated in relation to the local water table just below mean high water mark of Spring tides (M.H.W.S.T.).

More typically, however, the *in situ* samples that have been dated are species of mollusc that have a considerable range of depth of occurrence. The actual depth of the sea at the time can frequently be approximated by study of the associated microfauna but ranges of up to or above 10 m are often reported even in detailed studies (e.g. Peacock *et al.*, 1977, 1978). Close observation of the ecological succession in the near-shore zone can allow the establishment of a succession of faunal zones that can then be applied to similar relict successions to establish a direct relationship between sample and shoreline (e.g. Chappell, 1974; Einsele *et al.*, 1974).

Detrital accumulations are often more easily associated with particular shorelines as they may occur in shingle ridges, beach gravels, delta foreset beds or in channels and depressions on low gradient sand or mud flats. The main problem with detrital accumulations is whether the shells relate to the shoreline with which they are associated or whether they are derived from older sediments. In certain circumstances, such as when the shoreline being studied is transgressive across previously deposited marine sediment or whether a shelly glacial deposit is being eroded, there is a considerable possibility for reworking of old shells. In general, however, it is frequently tacitly assumed that detrital accumulations are overwhelmingly derived from the contemporaneous sea and this assumption is only questioned when an "anomalous" date is recorded. The use of the relatively cheap and quick amino acid technique (see pp. 148–150) to assay a number of shells from the same deposit can help circumvent this problem and avoid expensive ^{14}C dating (cf. Blake, 1980).

Deposits of marine origin are abundantly preserved in the sedimentary record and are often the prime evidence from which sea-level changes can be inferred and dated. Care is necessary in assessing ^{14}C ages of this type of material but providing such care is exercised reliable dates can be achieved. The ability to date certain marine carbonates by U-series or amino acid methods (see pp. 145–150) also means that cross-checks on the same or associated samples are possible.

Deposits of terrestrial or freshwater origins

Perhaps the majority of ^{14}C dates relating to sea-level changes have been related to sediments of non-marine origin. A variety of materials have been dated, such as driftwood, salt marsh or fen peats, soil horizons, freshwater carbonates or limnic sediments in isolation basins. In general, most of these materials cause fewer technical problems for ^{14}C dating than marine carbonates and such questions as have been raised are mentioned below. The interpretation, however, of the relationship between dated samples and the corresponding shorelines or sea levels is widely debated and is now considered for various types of material.

Driftwood

Particularly in Arctic areas driftwood has been found buried by or on the surface of raised shoreline deposits (e.g. Blake, 1961, 1970; Olsson and Blake, 1962; Salvigsen, 1981). In more temperate latitudes driftwood has been discovered preserved at the base of blanket peat covering shoreline deposits (e.g. Marthinussen, 1962; Sutherland, 1981).

The contemporaneity of the driftwood with the beach on which it is found must be assumed rather than established. Generally, with the exception of the possibility of human transport, the driftwood will be older than the shoreline on which it was found. This age difference may be no more than the age of the wood (up to several hundred years if the central part of a large tree was dated) plus the length of time adrift (probably less than ten years at maximum; cf. Binns (1972) on pumice). Natural agencies, however, may result in reworking with consequent mis-associations to be measured in thousands of years. Thus rivers eroding previously deposited material as the sea regresses may have carried wood down to be redeposited along a lower shoreline, whilst in areas of marine transgression driftwood could be refloated and carried upwards to the maximum of the transgression. The interpretation of dates on driftwood from areas of marine transgression should take account of these possibilities (Marthinussen, 1962).

^{14}C dates on wood are widely considered reliable and of the various fractions that may be dated, cellulose is the most stable (Olsen and Broecker, 1958; Olsson, 1972).

Salt marsh and fen peats

Along many coasts of the world complicated interdigitations of marine and terrestrial deposits record the fluctuating position of the shoreline that has resulted from the interplay of eustatic, isostatic and tectonic movements. Frequently the marine deposits can be related to specific shorelines on the basis of stratigraphical or morphological evidence, and ^{14}C dates on the upper or lower contacts of the associated terrestrial deposits provide limiting ages for the formation of the shorelines. The greatest uncertainty associated with this approach is the assurance with which the dated sample can be related to sea level.

On open coasts, particularly, transgressive contacts may have been subject to erosion with the result that the dated sample provides a maximum age for the time of the transgression. At regressive contacts a period may have elapsed, especially on well drained sediments, prior to the commencement of peat growth, thus resulting in a minimum age for the marine regression. In order to minimize such errors palynological and diatom studies have been undertaken to identify a marine-brackish-freshwater seral development, the presence of which reduces the likelihood of hiatuses at the contacts. Such studies also provide valuable additional information concerning the marine origin of certain sediments and pollen analysis can indicate the approximate age of the contacts (pp. 131–132) (e.g. Mörner, 1969; Sissons and Brooks, 1971; Tooley, 1978; Devoy, 1979; Smith *et al.*, 1980).

A distinct argument that has been advanced is that fen peats start to form on a land surface as the water table rises above the surface in response to rising sea level. Dates on the fen peats, excluding those from enclosed depressions or very flat areas with poor drainage, should therefore date the rise of sea level to the corresponding altitudes (Jelgersma, 1961; Kidson and Heyworth, 1973). The material dated should be peat and not associated tree trunks or branches as these can sink in peat bogs and hence be appreciably younger than the surrounding material (Kaye and Barghoorn, 1964; Tooley, 1978). Furthermore, detailed local stratigraphies should be established for the surrounding area for in certain circumstances peat growth can match the rate of rise of sea level (Sissons and Smith, 1965), thus apparently implying that sea level was always below the base of the peat.

Scholl (1964, 1965) has demonstrated the two lines of evidence considered above for tropical mangrove swamp environments. He showed that in Southern Florida a calcite-rich layer had developed on the land surface in response to a rising water table and that this was succeeded by seral mangrove development and then marine conditions. The transition from the freshwater carbonate-rich deposits to the mangrove peats was controlled by a rise in the water level of as little as 12 cm, thus relating the sequence very closely to tidal regime. ^{14}C dates on the freshwater carbonates had an apparent age of *c.* 400 years due to underlying limestone (Scholl and Stuiver, 1967) but when corrected for this they agreed closely with dates on the organic fraction of the same sediment.

Soil horizons buried by marine deposits have been used as evidence of transgressions and a particular type of buried organic soil termed "pitchy soil" (von Post, 1968) has been recorded in Sweden, which Mörner (1969) regards as giving particularly accurate ^{14}C dates.

The majority of studies that have employed the above types of evidence have been of large estuaries where relatively quiet water sedimentation has preserved a detailed sedimentary record. Most of the studies have emphasized the problems of relating the terrestrial-marine contacts to a particular portion of the tidal cycle, a matter reviewed in detail by Tooley (1978). This is an important problem not only because tidal variations can result from short-term coastal changes that may give rise to apparent transgressions or regressions with no true accompanying relative movement of sea level (Johnson and Winter, 1927; Heyworth and Kidson, 1982), but also because tidal characteristics in estuaries may vary through time as the change in sea level results in changes in coastal configuration. Even where a good relationship to sea level may seem established new evidence can result in a different interpretation. Thus Jelgersma (1961) originally related the formation of groundwater induced fen peats to high tide level whilst she now suggests that it is more likely that they relate to mean tide level (Jelgersma, 1980).

A further problem in relating the dated sample to sea level is the autocompaction of peat (Kaye and Barghoorn, 1964) and the compaction and consolidation of associated sediments. Where the dated sample is from the base of peat that has formed on sandy or gravelly surfaces little post-depositional change of this nature is likely, but in sequences of peats and silts or clays a variable response may be expected. Certain studies have attempted to correct for these effects (e.g. Kidson and Heyworth, 1973) whilst others, although noting the problem, have made no corrections (e.g. Tooley, 1978; Devoy, 1979). In other regions, such as the lagoonal areas of the Netherlands that possess excellent sedimentological evidence for mean sea level, these effects are considered to be so irregular that data from these deposits cannot be used for sea-level curves (Jelgersma, 1980).

The above two problems can be largely eliminated if the sedimentary sequences being dated are directly related to shorelines that are traceable along the coast. The incorporation of detailed sedimentological work into a shoreline sequence has a number of important advantages. First, the shoreline can be independently related to the tidal cycle on morphological grounds (Sissons *et al.*, 1966; Sutherland, 1981; Gray, this volume) and the shoreline trace also defines the changes in coastal configuration that may have produced significant tidal changes thus opening the way to palaeotidal mathematical modelling. Secondly, the preservation of sedimentary sequences may be best in restricted coastal embayments that are characterized by exceptional tidal regimes. Dates from these embayments, however, can be applied to the general altitude of the corresponding shoreline outside the embayment where tidal regimes were neither damped nor amplified by a restricted coastal setting (e.g. Smith *et al.*, 1980). Thirdly, it allows samples to be selected that accord to the maxima of transgressions (i.e. the corresponding shoreline) rather than having age-limiting dates during or following the transgressive sea-level movements. Fourthly, areas of faulting, subsidence, compaction or consolidation can be identified by anomalous shoreline altitudes (e.g. Sissons, 1972). And finally, it provides an independent relative chronology that places limits on the ^{14}C dates (for the use of shoreline sequences as a dating technique, see p. 151).

Unfortunately in a number of areas of detailed study shorelines appear poorly or not at all developed, whilst on open coasts exposed to extreme conditions only storm beaches may be preserved. However, shoreline studies have been complemented by detailed sedimentological, palynological and diatom studies in Scotland (Sissons, this volume) and in Scandinavia (e.g. Mörner, 1969).

^{14}C dating of the terrestrial organic deposits mentioned above has a number of potential problems associated with it. In dating soils, for example, Matthews (1980) has shown that not only do different soil fractions have different ^{14}C ages, but also there is a steep age-depth gradient such that erosion of even a few centimetres from the top of a soil profile may result in errors of several hundred years.

Peat is generally reliable as a dating material but the presence of rootlets is an inevitable problem. Scholl and Stuiver (1967) observed that roots from *Rhizophora mangle* in mangrove swamps penetrated as much as 4·5 m into the underlying sediments, whilst in more temperate latitudes Kaye and Barghoorn (1964) traced *Spartina* rootlets up to 38 cm (15 in) below the rhizome level. Tooley (1978) has recorded rootlets penetrating intercalated peat and silty clay horizons. The extent of contamination by rootlets is very variable, depending on the proportion of the peat that they make up as well as the age difference between the roots and the host material. Olsson (1979) has indicated that rootlets may contain significantly less carbon than the host organic matter. From a sample of gyttja that was 10% by weight rootlets she found that only 1/600th of the CO_2 evolved from the sample was derived from the roots. Streif (1972) adopted an empirical approach to the problem and divided 12 samples of *Phragmites* peat into two fractions, one consisting of rootlets and rhizomes, the other being the remainder of the peat. The rootlet fractions were always younger by up to a maximum of 845 ± 200 years and, whilst the "remainder" fractions of samples from corresponding horizons gave ages in close agreement, the rootlet fractions were much more variable in age. Notwithstanding Olsson's result reported above it should therefore be routine practice to remove, as far as possible, rootlets from samples that are to be dated.

For brackish sediments Berglund (1971) has argued that ^{14}C dates should have an apparent age correction similar to that of marine carbonates, mammals and plants (Olsson, 1980). On this basis Berglund altered his dates by *c.* 300 years. This is not normal practice, however, largely because the typical 2-4 cm sediment slice that is dated is mostly (or all) from freshwater peat. The ^{14}C dating of intertidal or brackish water algae and plants (Olsson, 1980) does suggest that apparent age effects may be experienced with these materials and if the organic fraction of a dated sample was derived chiefly from such sources then an apparent age correction may be necessary. More empirical data is required on this problem, however, for its accurate resolution.

Isolation basins

The relative movements of sea level that gave rise to transgressive or regressive contacts described in the previous section also produced successive layers of marine and limnic sediments in enclosed basins as the fluctuating sea level rose above or fell below lake thresholds. The study of these "isolation basins" has been carried out in most detail in the glacio-isostatically uplifted area of Scandinavia (e.g. von Post, 1968; Mörner, 1969; Persson, 1973; Berglund, 1971; Kjemperud, 1981) where separate basins with thresholds at different altitudes have been used to construct sea-level curves that, in turn, provide the ages of the associated shorelines.

A major limitation of isolation basin studies is the possibility that the present altitude of the basin threshold is different from that at the time of the event being dated. Thus Sutherland (1981) was able to demonstrate by comparison with a sea-level curve based on open coast marine-terrestrial contacts in a neighbouring area that the threshold of Loch Lomond in western Scotland had been eroded by *c.* 4 m since the time of the marine transgression recorded there by Dickson *et al.* (1978). It is therefore preferable to study basins in which the threshold is of rock and hence less likely to have been eroded.

The identification of a marine event may be based on a number of independent criteria. Frequently a distinct lithostratigraphic unit accords to the event and this may be characterized by marine macrofauna. Diatom identification can be a very accurate method of tracing varying salinity conditions. Dynoflagellate cysts, foraminifera and, frequently, increased numbers of derived spores and pollen are typically recorded in marine layers along with increases in the proportions of, for example, Chenopodiaceae and *Artemisia* on pollen diagrams. The use of palaeomagnetism for identification of marine events is discussed on pp. 133–134. The clarity with which marine strata may be identified is a function of the size of the basin, the amount of freshwater input and the rate of sea-level change. Small basins with low freshwater inputs in areas of rapid sea-level change are likely to give the sharpest marine-limnic sedimentary contacts.

^{14}C dating of limnic sediments can produce erratic results largely because of the diverse sources for the organic fraction of the sediment. Inwashed old organic material due to soil erosion by either natural or human agencies (cf. Dickson *et al.*, 1978) can produce erroneously old dates as can hard-water error in areas of carbon-rich bedrock (Olsson, 1972). Dating of the limnic sediments above or below marine strata does appear to give more reliable results than dating the marine sediments themselves as these latter have additional problems to those mentioned above. There is a possibility of an unknown apparent age correction for marine carbonates or plants that may be contributing to the organic component of the sediment and with such diverse organic fractions it may be difficult to adopt an appropriate laboratory pretreatment to minimize potential contamination. In such circumstances independent dating methods such as palynology (pp. 131–132) or palaeomagnetism (pp. 133–134) are essential whilst sequences of ^{14}C dates stratigraphically related to each other provide valuable checks on any one dated sample.

Conclusion on radiocarbon dating

^{14}C dating has undoubtedly been the single most significant factor in the revolution which has taken place in knowledge of sea-level changes relating to the last deglacial hemicycle. Organic material capable of being ^{14}C-dated has been found in close association with shorelines throughout the world and many thousands of samples have been dated from the various geological contexts described above. The discussion of ^{14}C dating in this chapter, however, has not taken into consideration the absolute accuracy of this method—in essence the assumption as to the constancy of the $^{14}C/^{12}C$ ratio through time. Dendrochronological studies have shown that there has been a deviation of up to *c.* 10% in this ratio in the last 7000 years (see papers in Olsson (1970) for detailed discussion of this problem) whilst other less accurate methods have suggested that for earlier periods the ^{14}C time-scale may be in error rather less than this (Stuiver, 1978). Furthermore, ^{14}C dates are computed on the basis of the "Libby" half-life of 5568 ± 30 years and this introduces an error of *c.* 3% into all ^{14}C dates as more accurate recalculation has indicated a half-life of 5730 ± 30 years (Godwin, 1962). For these reasons ^{14}C dates should continue to be quoted as "years B.P." and studies that involve the computation of rates based on the ^{14}C time-scale (such as sea-level curves, isostatic uplift curves or shoreline gradient against time plots) thus are extended by *c.* 10% in one part of the time-scale and, possibly, compressed by a similar amount in another part.

URANIUM-SERIES DATING

The two naturally occurring isotopes of uranium, ^{238}U and ^{235}U, give rise, by radioactive decay, to two series of daughter elements. In an ideal natural system that contains uranium, the quantities of the daughter elements should be in direct proportion to the quantities of the parent isotopes, because the amount of a particular element being created by radioactive decay of its parent will attain an equilibrium with the amount of that element that is decaying to produce a further daughter. Leaching or preferential deposition of either parents or daughters can lead to excess or deficiency in daughter elements relative to the parent. If the substance containing the uranium and its daughters is then isolated from the process that produced the disequilibrium (by death of organisms, precipitation of carbonates, etc.) then a new equilibrium will begin to be attained, the rate of this attainment being a function of the half-lives of the various daughter elements (Broecker and Bender, 1972; Ford and Schwarcz, 1981). In principle, therefore, there are two approaches to uranium-series dating, the daughter deficiency method and the daughter excess method. In both methods the measure of time is gained from the extent to which the system has returned to the ideal from a particular degree of disequilibrium. Radioactive disintegration is by emission of beta and alpha particles, these latter being helium nucleii. This helium is stable and hence accumulates throughout the decay history and thus, if it is assumed that there is no initial helium in the substance, the quantity of helium relative to the quantity of uranium is a measure of the time-lapse since the decay process began (Fanale and Schaeffer, 1965; Broecker and Bender, 1972; Bender, 1973).

There are three main parent-daughter relationships that have been used for dating purposes: ^{230}Th:^{234}U; ^{231}Pa:^{235}U; and ^{4}He:U. Two other such relationships that in principle may be used to estimate ages are ^{234}U:^{238}U and ^{226}Ra:^{230}Th, but in practice these are normally used as checks on the validity of the ages determined on the basis of the other ratios (most commonly ^{230}Th:^{234}U). One exception to this has been at the Trondheim laboratory where ^{226}Ra:^{238}U ages have been calculated on the assumption of specific values for the ^{226}Ra:^{230}Th and ^{234}U:^{238}U ratios in the samples (Salvigsen and Nydal, 1981; Andersen *et al.*, 1981). Thorium has a half-life of *c.* 75 000 years and protactinium a half-life of *c.* 32 000 years (Broecker and Bender, 1972) and these substances are thus suitable for age estimation in the *c.* 5000-350 000 and *c.* 5000–300 000 year ranges respectively (Ford and Schwarcz, 1981). Helium continues to accumulate through time and may have a theoretical upper limit of the order of 10^8 years, although in practice helium leakage or recrystallization of samples (Bender, 1973; see below) makes the upper limit of this technique 10^6-10^7 years. At the lower end of its time-scale uncertainties as to initial concentrations indicate a lower practical limit of around 200 000 years for the helium method.

In the context of sea-level change studies U-series dating has been most successfully applied to corals. Reliable dates have also been recorded for speleothems which, in certain circumstances, can be used to place age limits on high or low sea-level stands (e.g. Harmon *et al.*, 1978). Considerably more controversial, however, is the application to marine molluscs, this approach having produced rather variable results. Certain publications maintain that such dating of molluscs is unreliable (e.g. Kaufman *et al.*, 1971; Broecker and Bender, 1972) whilst others claim that the majority of samples dated are of use in solving dating problems (e.g. Stearns and Thurber, 1965; Butzer, 1975; Andrews and Miller, 1980). Bernat *et al.* (1978) have shown that ^{230}Th:^{234}U dates on molluscs can give consistent results and Szabo (1979) has reported a mollusc sample that gave a ^{230}Th:^{234}U age that is in agreement with similarly derived ages on corals and in accord, within the limits of analytic precision, with ^{231}Pa:^{235}U ages. Szabo *et al.* (1981), however, have indicated that ^{231}Pa:^{235}U ages of molluscs are less reliable than ^{230}Th:^{234}U ages.

The U-series dating of molluscs is being increasingly widely applied (e.g. Miller *et al.*, 1977; Hjort, 1981; Salvigsen and Nydal, 1981; Szabo *et al.*, 1981) particularly in association with amino acid dating (see pp. 148–150) to provide a relative age framework in which the reliability of the U-series dates may be assessed. Work in southern Norway (Andersen *et al.*, 1981) and in Spitsbergen (Salvigsen and Nydal, 1981) in which successive fractions of the molluscs leached from the outer to the inner parts were dated (by ^{226}Ra:^{238}U), showed a general tendency for the outer layers to be older than the inner ones due either to radium loss or uranium gain in the outer layers. These studies suggested that the U-series dates tended to overestimate the ages of molluscs, a relationship that is counter to that of Andrews and Miller (1979) and Szabo *et al.* (1981) who considered that U-series (^{230}Th:^{234}U) dates tended to underestimate the true age of molluscs.

The variable results that have been achieved for U-series dating methods when applied to molluscs seem largely due to the irregular uptake of uranium by molluscs shortly after death (Kaufman *et al.*, 1971). The general reliability, however, of U-series (and ^{14}C) dating of marine carbonates is dependent upon the validity of the closed-system assumption. The extent of recrystallization is normally taken as a guide to the validity of the closed-system assumption and dating of samples is best restricted to those that are unrecrystallized or only slightly so. Although Ragland *et al.* (1979) have suggested that certain diagenetic changes do occur in unrecrystallized shells, it remains to be demonstrated whether these minor effects can influence the dating results. Recrystallization, however, does so, and numerous field and laboratory methods have been applied to check on the presence and extent of it. A further possible influence on the dates is the presence of contamination by extraneous material that has become lodged in the carbonate matrix. Various field and laboratory criteria for selection of more reliable samples are now discussed.

Field criteria for sample selection

In describing the field criteria for selection of corals from the Huon Peninsula, Bloom *et al.* (1974) suggested that massive corals were more suitable than finely branching ones. As the corals have an aragonitic mineralogy extensive recrystallization (to calcite) could be identified in the field by examination of freshly broken corals for calcitic cleavage.

The mineralogy of molluscs is more variable than that of corals, certain species being composed of aragonite, others of calcite and others a combination of the two, the proportions being a function of the temperature and salinity of the surrounding sea (Milliman, 1974). In the field massive shells should be collected in preference to those with high surface to volume ratios that increase the possibility of recrystallization. As Mangerud (1972) suggested, preference should be given to those molluscs that primarily consist of only aragonite or high-magnesium calcite so that alteration can be detected in laboratory studies. The state of preservation of the surface is also important. Those shells with the periostracum still attached are to be preferred and subsequent degrees of alteration are detectable along the scale of slight pitting, loss of detail such as growth lines, to a cover of dusty calcite that may be removed by touch. Thick, mature shells should also be preferentially selected for dating purposes as these allow the outer portion to be discarded and, if desired, the subsequent division of the sample into successive fractions that can be analysed separately as checks on the degree of contamination of a sample.

Laboratory criteria for sample selection

As an initial laboratory procedure Bloom *et al.* (1974) examined corals under a binocular microscope for the presence of void-filling detrital material or cementing products such as aragonite needles or micrite crusts. Those samples that passed this test were submitted to X-ray diffraction analysis to determine the mineralogy. If high-Mg calcite or more than 3% low-Mg calcite was detected the sample was rejected. Further checks were run by atomic absorption analysis for magnesium and stronium, high magnesium values suggesting the presence of high-Mg calcite and low strontium values suggesting recrystallization to calcite. Of the 38 samples that had passed initial visual inspection and were thus considered for laboratory analysis by Bloom *et al.* (1974) only 24 were then dated.

In a study of corals and molluscs using X-ray diffraction and thin section examination, Chappell and Polach (1972) demonstrated two forms of recrystallization, a "sparry calcite" type that was interpreted as evidence of open-system conditions and a "subtle coarsening" type that took place within a closed system. Those samples that had only undergone a subtle coarsening type of recrystallization often yielded reliable ages.

The application of X-ray diffraction to mollusc samples from middle and high latitudes can be complicated by the variable mineralogy of many of the shell species in these regions. It is a valid test for recrystallization, none the less, for those samples whose original shell matrix is entirely aragonitic and for which any recrystallization to calcite could be detected (e.g. Mangerud *et al.*, 1981).

Measurements made during the dating of the samples can also be used to check on the closed-system assumption. This is particularly true of U-series ages. Thus, for example, a sample that gives a reliable ^{230}Th:^{234}U age should also:

(i) have a uranium concentration equivalent to that of living species. This is usually true of corals (Thurber *et al.*, 1965; Broecker and Bender, 1972) but frequently untrue of molluscs (Kaufman *et al.*, 1971; Salvigsen and Nydal, 1981);

(ii) the ^{234}U:^{238}U ratio, corrected for age (as calculated by ^{230}Th:^{234}U) should be 1·15 within the limits of analytical precision. This is often not observed for molluscs (Kaufman *et al.*, 1971);

(iii) the ^{226}Ra:^{230}Th ratio should be consistent with the age of the sample; and

(iv) there should have been no contaminating ^{232}Th as revealed by a ^{230}Th:^{232}Th ratio in excess of 20 (Thurber *et al.*, 1965).

In ^{14}C dating of marine carbonates the $\delta^{13}C$ value (measured as an essential part of age calculation, see pp. 137–138) can in certain circumstances indicate recrystallization because groundwater has a much lower $\delta^{13}C$ value than marine carbonate (Chappell and Polach, 1972), but this seems rarely to apply in practice probably because the extent of recrystallization necessary to alter materially the $\delta^{13}C$ value would be readily visible and the sample rejected at an early stage.

The independent dating of successive inner fractions has been widely applied in ^{14}C dating and to a lesser extent in U-series dating. This practice provides direct evidence, within the limits set by the analytical procedures, for the reliability of shell dates and is to be recommended (Olsson and Blake, 1962; Mangerud, 1972; Andersen *et al.*, 1981; Sutherland, in press).

Uranium-series methods are of great value in dating shorelines whose age is beyond the reliable limits of the ^{14}C method. Greatest success has been achieved with corals (and, of indirect relevance, with speleothems), whilst the ubiquitous molluscs, of greater utility in dating shorelines, have given variable results. All dated samples must be accurately related to particular sea levels and the problems associated with this are discussed on pp. 138–139.

AMINO ACID DATING

Diagenesis of proteinaceous material proceeds by a number of complex reactions such as racemization, hydrolysis and leaching. Amino acid racemization is the process whereby the amino acids of proteins undergo conversion (after death) from L-configured isomers (typical of living organisms) to D-configured isomers. Hydrolysis is the breaking of the peptide bonds that link the amino acids in the protein thus producing "free" amino acids. Time and temperature are fundamental controls on the extent and rate of these reactions and hence for use as an accurate dating method a full knowledge of both the temperature history and reaction kinetics must be available. Given the magnitude of Quaternary climatic changes, however, and the uncertainty in estimates of temperatures during such changes, the temperature control is clearly a major constraint on the derivation of accurate dates by this method. Different approaches to this problem are discussed below.

In addition to the time and temperature variables it is known that different amino acids have different reaction kinetics and it has also been shown that results of analyses on particular amino acids are genus-specific (Lajoie *et al.*, 1980). The method is further dependent upon the amino acids being protected from outside contamination which in this instance also includes anomalous temperature effects such as may result from surface or near-surface heating. Thus only particular amino acids from certain genera that provide adequate protection to the proteinaceous material can be utilized to achieve consistent results. Fortunately for the purposes of dating shorelines certain marine molluscs and foraminifera have been found to be excellent material for this method (Andrews and Miller, 1980; Miller and Hare, 1980). Depending therefore on the temperature control amino acid dating can be applied to marine deposits from less than 20 000 years old to those in excess of one million years old. The problems of relating specific samples to their contemporaneous sea level were discussed on pp. 138–139. Contamination of samples is a subject of current investigation, with certain studies indicating that recrystallization of mollusc shell (see pp. 146–148) has not affected the racemization reaction (e.g. Wehmiller *et al.*, 1977). Preferential loss of amino acids by certain organisms has been noted (Schroeder and Bada, 1976) although Miller and Hare (1980) suggest that after an initial decrease the amino acid compositions remain stable for periods of the order of $> 10^6$ years. The major problem, however, appears to be the temperature dependence of the reactions and hence near-surface (i.e. < 1 m depth) samples should be avoided because of direct insolation-induced anomalously high local temperatures.

In using enantiomeric (D:L) ratios or the ratio of "free" to total amino acids as measures of the time elapsed since death the problem of temperature control has been approached by the following ways. First, it has been argued that if samples are recovered from a relatively small geographical area then contemporaneous deposits will have experienced similar temperature regimes whilst older deposits will have experienced that thermal history plus an earlier one. The enantiomeric ratios of the amino acids should therefore provide a relative chronology for that area irrespective of the absolute temperatures experienced. This approach has been widely applied (e.g. Miller *et al.*, 1977; Andrews *et al.*, 1979; Mangerud *et al.*, 1981). Of particular utility in distinguishing sedimentary units of different ages are plots of enantiomeric ratios from one amino acid against enantiomeric ratios of another amino acid from corresponding samples, thus providing graphs of relative rates of racemization (Lajoie *et al.*, 1980). An analogous type of graph plots the enantiomeric ratios for a given amino acid in the "free" against the ratios in the "total" fractions. Such graphs are often capable of clearer discrimination between deposits of different ages (without themselves giving a direct estimation of age) than are the calibration graphs discussed below (e.g. Mangerud *et al.*, 1981).

A second approach has been through the application of calibration graphs in which the amino acid ratios of certain samples (again from a relatively small geographical area) of known age are plotted against those ages. The amino acid ratios of samples of unknown age can then be plotted on the graph and their "absolute" ages interpolated or extrapolated. This method assumes that the racemization reaction proceeds according to a linear model. This assumption, together with laboratory information on the temperature sensitivity of the reaction and the present-day temperature of the area studied, has been elaborated to indicate average temperatures during the time that has elapsed since the deaths of the relevant organisms (Mitterer, 1975). Linear kinetic models, however, have not been found appropriate to the racemization reactions of foraminifera from deep-sea cores (Bada and Schroeder, 1975) and non-linear models have been applied to molluscs from various shoreline deposits (Wehmiller and Belknap, 1978). This problem of the appropriate kinetic model is of increasing importance with age of sample. Calibration graphs are also entirely dependent upon the age estimation of the "known" points which will have the uncertainties associated with the other dating methods discussed in this chapter. So far, only rather wide estimates of ages have been achieved by the calibration method.

Amino acid dating has demonstrated its utility as a relative dating technique within limited areas where the samples have experienced similar thermal histories. The technique is applicable to the entire Quaternary using a wide range of naturally occurring and readily recoverable organisms. Other advantages are that only small samples are needed for single determinations and there is a possibility of field equipment being developed that will permit selective and hence more efficient sampling for further detailed laboratory work. The disadvantages currently associated with the method relate first to its temperature dependence. Thus it is not a "universal" technique since a given enantiomeric ratio (for a particular amino acid from a certain genus) implies different ages for samples depending on their location (and hence temperature history). Its application as an absolute dating method remains to be convincingly demonstrated for a variety of environments. A second concern is the extent to which results are reproducible between laboratories (Kvenvolden, 1980; Mangerud *et al.*, 1981), a matter that should be resolved by publication of cross-analysed samples from the various laboratories performing this work.

VARIOUS RELATIVE DATING METHODS

A wide variety of relative dating methods have been devised in the past to allow correlation of shoreline fragments from place to place around a coastline or to permit the extrapolation of the ages of shorelines within a shoreline sequence where only a limited number of shorelines have been dated.

Typical of the first line of approach are those methods that use marker horizons such as specific faunal elements, the occurrence of pumice or tephra horizons, unique geomorphic development or the relationship to specific glacial stages, as means to correlate shorelines within a given region. To be of greatest utility a marker horizon should be the result of a discrete event and should be capable of unambiguous association with a particular shoreline. Thus, although pumice produced by specific volcanic eruptions has been widely used as a marker in the arctic regions (Donner and West, 1957; Marthinussen, 1960; Blake, 1961, 1970; Boulton and Rhodes, 1974), the possibility of separate eruptions producing similar pumice (cf. Binns, 1972) or of old pumice being reworked into younger deposits, particularly during transgressions (Marthinussen, 1960), suggests that errors could result from this method. The introduction or disappearance of particular faunal elements may be diachronous, an example being the transition from the Ancylus Lake to the Littorina Sea in the area of the present-day Baltic that may have taken 200–300 years to be effective throughout the basin (Tauber, 1970).

Presumed systematic relationships between shoreline altitudes or gradients from tectonic or isostatic movements have been utilized in simple mathematical models to permit the extrapolation of ages within particular shoreline sequences. In tectonically active areas this approach has been used to most effect in Barbados (Matthews, 1973) and New Guinea (Chappell, 1974; Bloom *et al.*, 1974) both to estimate the age of shorelines and to infer the "eustatic" sea level at the time of their formation. This approach has been based on a combination of simple tectonic models and assumptions as to "eustatic" sea levels at certain times. Stearns (1976), however, has indicated that such assumptions should be subject to further independent verification and not uncritically accepted for other studies.

In glacio-isostatically affected areas simple models relating shoreline gradient to time have been developed, perhaps first by Gronlie (1941) in Norway and in more recent years particularly in North America (e.g. Andrews, 1970). These models depend upon the accurate dating of certain shorelines and the assumption that isostatic uplift is a smoothly decelerating function of time. Locally, however, isostatic uplift has been demonstrated to be non-uniform (Sissons, 1972; Sissons and Cornish, 1982), whilst a major readvance that interrupted general ice decay would slow, halt or even reverse isostatic uplift (e.g. Sutherland, 1981) such that a simple mathematical model would be inappropriate. The use of such models is therefore limited by local circumstances, the degree of accuracy desired and the length of time over which the extrapolations of age are made.

CONCLUSION

The dating techniques outlined above have a variety of ranges and accuracies. Figure 2 summarizes the time-ranges of the various methods from which it may be seen that present techniques allow dating throughout the whole of the Quaternary. The degree of accuracy that is acceptable varies depending on the period being dated: $\pm 10\%$ may be a sufficiently accurate figure for Middle or Early Quaternary events but for Late Quaternary time $\pm 1\%$ is likely to be necessary in order to unravel the wealth of detail that the geological and geomorphological record holds for this period.

Achievement of these levels of accuracy has largely been the result of the development and refinement of laboratory techniques during the last three decades.

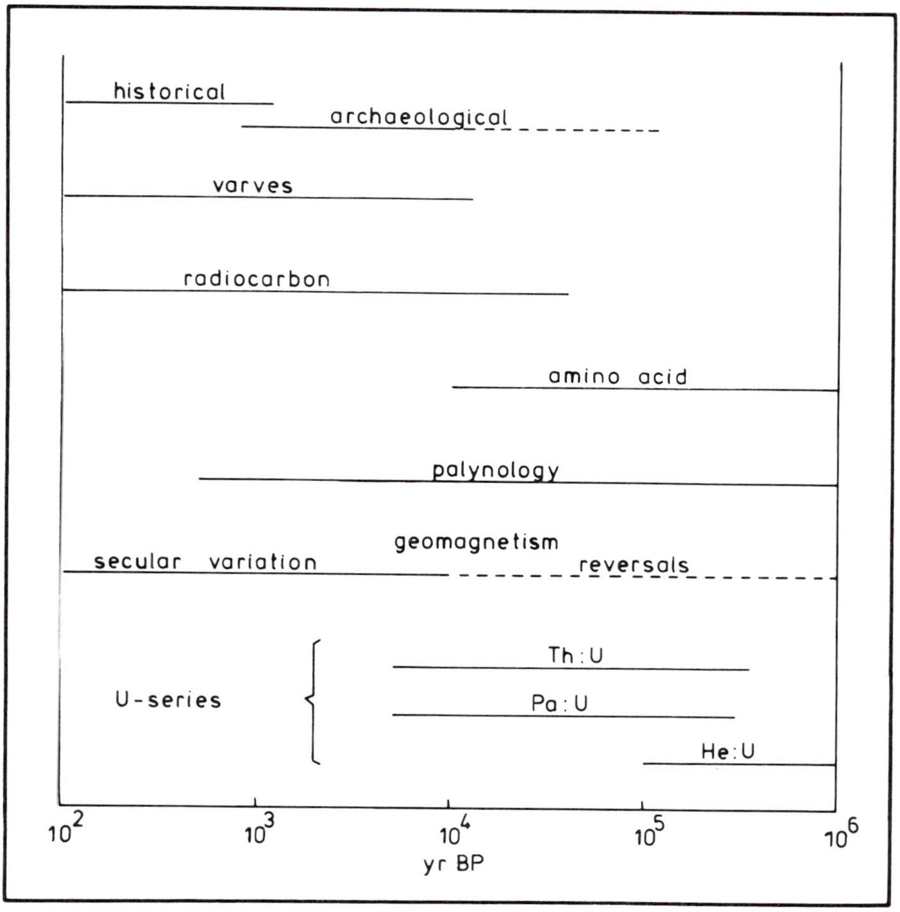

Figure 2. The ranges of the various dating techniques during the last one million years. Logarithmic time-scale.

Much of the dating research activity has seemed to have moved from the field to the laboratory during this period and it has become too easy for field scientists to recover material for dating and base their interpretation on the laboratory result. Whilst there is clearly a role for dated samples in any coherent interpretation of geological events, this chapter has been written from the point of view that field relationships are paramount and that all samples must be placed in a rigorous stratigraphical and geomorphological context. Such a context is a prime form of independent dating control and, where possible, subsequent laboratory dating methods should be subject to cross-checks as to the geochemical integrity of the samples and the validity of the dates.

Research during the last two decades has been oriented principally towards shoreline changes during the last *c.* 20 000 years and has been most dependent upon radiocarbon dating. Whilst this period will continue to be a major focus of research, in the future it seems likely that the wider availability and increasing accuracy of the longer-range dating techniques will result in more attention being paid to sea-level changes in the earlier parts of the Quaternary.

REFERENCES

Akeroyd, A. V. (1972). *Phil. Trans. R. Soc.*, **A272**, 151-169.

Andersen, B. G., Nydal, R., Wangen, O. P. and Østmo, S. R. (1981). *Boreas* **10**, 297-314.

Andrews, J. T. (1970). "A geomorphological study of post-glacial uplift with particular reference to Arctic Canada". *Inst. Brit. Geogr. Spec. Publ.* No. 2, 156 pp.

Andrews, J. T. and Miller, G. H. (1980). *In* "Timescales in Geomorphology" (R. A. Cullingford, D. A. Davidson and J. Lewin, eds), pp. 263-287. John Wiley, Chichester.

Andrews, J. T., Bowen, D. Q. and Kidson, C. (1979). *Nature* **281**, 556-558.

Antevs, E. (1925). *Mem. Geol. Surv. Canada* **146**, 142 pp.

Bada, J. L. and Schroeder, R. A. (1975). *Naturwissenschaften* **62**, 71-79.

Barbetti, M. and McElhinny, M. (1972). *Nature* **239**, 327-330.

Bender, M. L. (1973). *Geochim. cosmochim. Acta* **37**, 1229-1247.

Berglund, B. E. (1971). *Geol. För. Stockh. Förh.* **93**, 625-652.

Bergsten, F. (1954). *Geogr. Annlr.* **36**, 81-111.

Bernat, M., Bousquet, J. C. and Dars, R. (1978). *Nature* **275**, 302-303.

Beschel, R. (1961). *In* "Geology of the Arctic", Vol. 2 (G. Raasch, ed.), pp. 1044-1062. Toronto.

Binns, R. E. (1972). *Bull. geol. Soc. Am.* **83**, 2303-2324.

Björck, S., Dearing, J. A. and Jonsson, A. (1982). *Boreas* **11**, 99-111.

Blake, W. (1961). *In* "Geology of the Arctic", Vol. 1 (G. Raasch, ed.), pp. 133-145. Toronto.

Blake, W. (1970). *Canadian Journal of Earth Sciences* **7**, 634-664.

Blake, W. (1980). *In* "Biogeochemistry of Amino Acids" (P. E. Hare, T. C. Hoering and K. King, eds), pp. 453-461. John Wiley and Sons, New York.

Bloom, A. L., Broecker, W. S., Chappell, J. M. A., Matthews, R. K. and Mesolella, K. J. (1974). *Quaternary Research* **4**, 185-205.

Bonhommet, N. and Zahringer, J. (1969). *Earth Planet. Sci. Lett.* **6**, 43-46.

Boulton, G. S. and Rhodes, M. (1974). *Geol. Mag.* **111**, 481-500.

Bowen, D. Q. (1978). "Quaternary Geology: A Stratigraphic Framework for Multidisciplinary Work". Pergamon Press, Oxford.

Bradshaw, R. (in press). *In* "Geomagnetism of Baked Clays and Recent Sediments" (K. M. Creer, P. Tucholka and C. E. Barton, eds). Elsevier, Amsterdam.

Broecker, W. S. and Bender, M. L. (1972). *In* "Calibration of Hominoid Evolution" (W. W. Bishop and J. A. Miller, eds), pp. 19-38. Edinburgh.

Broecker, W. S., Gerard, R., Ewing, M. and Heezen, B. C. (1960). *J. geophys. Res.* **65**, 2903-2931.

Butzer, K. W. (1975). *In* "After the Australopithecines" (K. W. Butzer and G. L. Isaac, eds), pp. 25-71. Beresford Book Services, The Hague.

Chappell, J. (1974). *Bull. geol. Soc. Am.* **85**, 553-570.

Chappell, J. and Polach, H. A. (1972). *Quaternary Research* **2**, 244-252.
Craig, H. (1953). *Geochim. cosmochim. Acta* **3**, 53-92.
Creer, K. M. (1974). *Earth Planet. Sci. Lett.* **23**, 34-42.
de Geer, G. (1912). *Comptes Rendues 11th Int. geol. Congr. Stockholm 1910* **1**, 241-258.
de Geer, G. (1940). *K. svenska Vetensk-Akad. Handl. Ser. 3* **18:6**, 1-360.
Devoy, R. J. N. (1979). *Phil. Trans. R. Soc.* **B285**, 355-407.
Dickson, J. H., Stewart, D. A., Thompson, R., Turner, G., Baxter, M. S., Drndarsky, N. D. and Rose, J. (1978). *Nature* **274**, 548-553.
Donner, J. J. (1964). *Soc. Sci. Fenn. Comm. Phys. Math.* **30**, 47 pp.
Donner, J. J. (1978). *Soc. Sci. Fenn. Comm. Phys. Math.* **48**, 11-38.
Donner, J. J. and West, R. G. (1957). *Skr. norsk Polarinst.* **109**, 29 pp.
Donner, J. J., Eronen, M. and Jungner, H. (1977). *Norsk geogr. Tidsskr.* **31**, 103-128.
Einsele, G., Herm, D. and Schwarz, H. U. (1974). *'Meteor' Forsch. Ergebnisse* **18C**, 43-62.
Fanale, F. P. and Schaeffer, O. A. (1965). *Science* **149**, 312-317.
Flemming, N. C. (1969). *Geol. Soc. Am. Spec. Paper* **109**, 125 pp.
Flemming, N. C. (1972). *In* "The Mediterranean Sea: A Nature Sedimentation Laboratory" (D. J. Stanley, ed.), pp. 189-201. Stroudberg.
Ford, D. C. and Schwarcz, H. P. (1981). *In* "Geomorphological Techniques" (A. Goudie, ed.), pp. 284-287. Allen and Unwin, London.
Fromm, E. (1970). *In* "Radiocarbon Variations and Absolute Chronology" (I. U. Olsson, ed.), pp. 163-172. *Twelfth Nobel Symposium Uppsala, 1969*. New York.
Gaunt, G. D., Bartley, D. D. and Harland, R. (1974). *Bull. geol. Surv. Gt. Br.* **48**, 1-23.
Godwin, H. (1943). *J. Ecol.* **31**, 199-247.
Godwin, H. (1962). *Nature* **195**, 984.
Goudie, A. (ed.) (1981). "Geomorphological Techniques". London.
Grant, D. R. (1970). *Canadian Journal of Earth Sciences* **7**, 676-689.
Gronlie, A. (1941). *K. norske Vidensk. Selsk. Forh.* **14**, 12.
Harkness, D. D. (1979). *Br. Antarct. Surv. Bull.* **47**, 43-59.
Harkness, D. D. (in press). *Proc. 1st Inter. Symp.* ^{14}*C and Archaeology, Groningen, 1981.*
Harmon, R. S., Schwarcz, H. P. and Ford, D. C. (1978). *Quaternary Research* **9**, 205-218.
Heyworth, A. and Kidson, C. (1982). *Proc. Geol. Ass.* **93**, 91-111.
Hibbert, F. A., Switsur, V. R. and West, R. G. (1971). *Proc. R. Soc.* **B177**, 161-176.
Hjort, C. (1973). *Geol. För. Stockh. Forh.* **95**, 132-134.
Hjort, C. (1981). *Boreas* **10**, 259-274.
Hopley, D. (1974). *Proc. 2nd Inter. Coral Reef Symp.* **2**, 551-562.
Huntley, B. and Birks, H. J. B. (1982). "An Atlas of Past and Present Pollen Maps for Europe: 0-13 000 Years Ago". Cambridge.
Jelgersma, S. (1961). *Meded. geol. Sticht. C.* **Vi.7**, 1-100.
Jelgersma, S. (1980). *In* "Earth Rheology, Isostasy and Eustasy" (N. A. Morner, ed.), pp. 435-447. John Wiley, Chichester.
Jelgersma, S., Oele, E. and Wiggers, A. J. (1979). *In* "The Quaternary History of the North Sea" (E. Oele, R. T. E. Schuttenhelm and A. J. Wiggers, eds), pp. 115-142. *Acta Univ. Ups. Symp. Univ. Ups. Annum Quingentesimum Celebrantis* **2**. Uppsala.
Johnson, D. and Winter, E. (1927). *Proc. Am. phil. Soc.* **66**, 465-496.
Kaufman, A., Broecker, W. S., Ku, T.-L. and Thurber, D. L. (1971). *Geochim. cosmochim. Acta* **35**, 1155-1183.
Kaye, C. A. and Barghoorn, E. S. (1964). *Bull. geol. Soc. Am.* **75**, 63-80.
Kidson, C. and Heyworth, A. (1973). *Proc. Ussher Soc.* **2**, 565-584.

Kjemperud, A. (1981). *Boreas* **10**, 27-38.
Krog, H. and Tauber, H. (1974). *Danm. geol. Unders. Arbog 1973*, 93-105.
Kvenvolden, K. A. (1980). *In* "Biogeochemistry of Amino Acids" (P. E. Hare, T. C. Hoering and K. King, eds), pp. 223-232. John Wiley and Sons, New York.
Lajoie, K. R., Wehmiller, J. F. and Kennedy, G. L. (1980). *In* "Biogeochemistry of Amino Acids" (P. E. Hare, T. C. Hoering and K. King, eds), pp. 305-340. John Wiley and Sons, New York.
Libby, W. F. (1955). "Radiocarbon dating". Chicago.
Liden, R. (1938). *Geol. För. Stockh. Förh.* **98**, 45-51.
Lundqvist, J. (1975). *Boreas* **4**, 47-54.
Mackereth, F. J. H. (1971). *Earth Planet. Sci. Lett.* **12**, 332-338.
Mangerud, J. (1972). *Boreas* **1**, 143-172.
Mangerud, J. and Gulliksen, S. (1975). *Quaternary Research* **5**, 263-273.
Mangerud, J., Gulliksen, S., Larsen, E., Longva, O., Miller, G. H., Sejrup, H.-P. and Sonstegaard, E. (1981). *Boreas* **10**, 447-462.
Marthinussen, M. (1960). *Norges geol. Unders.* **208**, 416-432.
Marthinussen, M. (1962). *Norges geol. Unders.* **215**, 37-67.
Matthews, J. (1980). *Geogr. Annlr.* **A62**, 185-208.
Matthews, R. K. (1973). *Quaternary Research* **3**, 147-153.
Miller, G. H. and Hare, P. E. (1980). *In* "Biogeochemistry of Amino Acids" (P. E. Hare, T. C. Hoering and K. King, eds), pp. 415-443. John Wiley and Sons, New York.
Miller, G. H., Andrews, J. T. and Short, S. K. (1977). *Canadian Journal of Earth Sciences* **14**, 2824-2857.
Miller, G. H., Hollin, J. T. and Andrews, J. T. (1979). *Nature* **281**, 539-543.
Milliman, J. D. (1974). "Marine Carbonates". Berlin.
Mitterer, R. M. (1975). *Earth Planet. Sci. Lett.* **28**, 275-282.
Mörner, N.-A. (1969). *Sver. geol. Unders. Ser.* **C640**, 487 pp.
Mörner, N.-A. (1980). *In* "Earth Rheology, Isostasy and Eustasy" (N.-A. Mörner, ed.), pp. 251-284. John Wiley and Sons, Chichester.
Mörner, N.-A., Lanser, J. P. and Hospers, J. (1971). *Nature Phys. Sci.* **234**, 173-174.
Movius, H. L. (1953). *Am. J. Sci.* **251**, 697-740.
Nilsson, E. (1968). *K. svenska Vetensk-Akad. Handl. Ser. 4*, **12**, 117 pp.
Noel, M. (1975). *Geol. För. Stockh. Förh.* **97**, 357-367.
Nydal, R., Gulliksen, S. and Lovseth, K. (1972). *Radiocarbon* **14**, 418-451.
Oldfield, F. (1981). *In* "Geomorphological Techniques" (A. Goudie, ed.), pp. 306-326. Allen and Unwin, London.
Olsen, E. A. and Broecker, W. S. (1958). *Trans. N. Y. Acad. Sci. Ser. II* **20**, 593-604.
Olsson, I. U. (ed.) (1970). "Radiocarbon Variations and Absolute Chronology". *Twelfth Nobel Symposium, Uppsala 1969*. New York.
Olsson, I. U. (1972). *Acta Univ. Oul. A 3 Geol.* **1**, 9-37.
Olsson, I. U. (1979). *Boreas* **8**, 203-207.
Olsson, I. U. (1980). *Radiocarbon* **22**, 662-675.
Olsson, I. U. and Blake, W. (1962). *Norsk geogr. Tidsskr.* **18**, 47-64.
Peacock, J. D., Graham, D. K., Robinson, J. E. and Wilkinson, I. (1977). *In* "Studies in the Scottish Lateglacial Environment" (J. M. Gray and J. J. Lowe, eds), pp. 89-100. Pergamon Press, Oxford.
Peacock, J. D., Graham, D. K. and Wilkinson, I. P. (1978). *Rep. Inst. Geol. Sci.* **78/17**.
Persson, G. (1973). *Sver. geol. Unders. Ser.* **C684**, 47 pp.

Polach, H. A. and Golson, J. (1966). *Austr. Inst. Aboriginal Stud. Manual* **2**, 42 pp.
Rafter, T. A. and O'Brien, B. J. (1970). *In* "Radiocarbon Variations and Absolute Chronology" (I. U. Olsson, ed.), pp. 355-378. *Twelfth Nobel Symposium Uppsala, 1969*. New York.
Ragland, P. C., Pilkey, O. H. and Blackwelder, B. W. (1979). *Chem. Geol.* **25**, 123-134.
Rose, J. (1981). *In* "Geomorphological Techniques" (A. Goudie, ed.), pp. 327-341. Allen and Unwin, London.
Salvigsen, O. (1981). *Geogr. Annlr.* **A63**, 283-291.
Salvigsen, O. and Nydal, R. (1981). *Boreas* **10**, 433-446.
Sauramo, M. (1923). *Bull. Commn. geol. Finl.* **60**, 164 pp.
Scholl, D. W. (1964). *Marine Geol.* **1**, 344-366.
Scholl, D. W. (1965). *Marine Geol.* **2**, 343-364.
Scholl, D. W. and Stuiver, M. (1967). *Bull. Geol. Soc. Am.* **78**, 437-454.
Schroeder, R. A. and Bada, J. L. (1976). *Earth Science Reviews* **12**, 347-391.
Sissons, J. B. (1972). *Trans. Inst. Br. Geogr.* **55**, 145-159.
Sissons, J. B. and Brooks, C. L. (1971). *Nature Phys. Sci.* **234**, 124-127.
Sissons, J. B. and Cornish, R. (1982). *Nature* **297**, 213-214.
Sissons, J. B. and Smith, D. E. (1965). *Scott. J. Geol.* **1**, 247-255.
Sissons, J. B., Smith, D. E. and Cullingford, R. A. (1966). *Trans. Inst. Br. Geogr.* **39**, 9-18.
Smith, A. G. and Pilcher, J. R. (1973). *New Phytol.* **72**, 903-914.
Smith, D. E., Morrison, J., Jones, R. L. and Cullingford, R. A. (1980). *In* "Time scales in Geomorphology" (R. A. Cullingford, D. A. Davidson and J. Lewin, eds), pp. 225-245. John Wiley, Chichester.
Sparks, B. W. and West, R. G. (1970). *Phil. Trans. R. Soc.* **B285**, 1-30.
Stearns, C. E. (1976). *Quaternary Research* **6**, 445-449.
Stearns, C. E. and Thurber, D. L. (1965). *Quaternaria* **7**, 29-42.
Streif, H. (1972). *Geol. För. Stockh. Förh.* **94**, 281-299.
Stuiver, M. (1978). *Nature* **273**, 271-274.
Stuiver, M. and Polach, H. A. (1977). *Radiocarbon* **19**, 355-363.
Sutherland, D. G. (1981). "The raised shorelines and deglaciation of the Loch Long/Loch Fyne area, western Scotland". Unpublished Ph.D. Thesis, University of Edinburgh.
Sutherland, D. G. (in press). "A review of Scottish marine shell radiocarbon dates, their standardization and interpretation". *Scott. J. Geol.*
Szabo, B. J. (1979). *Marine Geol.* **29**, M11-M15.
Szabo, B. J., Miller, G. H., Andrews, J. T. and Stuiver, M. (1981). *Geology* **9**, 451-457.
Tauber, H. (1970). *In* "Radiocarbon Variations and Absolute Chronology" (I. U. Olsson, ed.), pp. 173-196. *Twelfth Nobel Symposium, Uppsala, 1969*. New York.
Thom, B. G. (1973). *Progress in Geography* **5**, 167-246.
Thompson, R. (1977). *J. geol. Soc. Lond.* **133**, 51-59.
Thompson, R. and Turner, G. M. (1979). *Geophys. Res. Lett.* **6**, 249-252.
Thurber, D. L., Broecker, W. S., Blanchard, R. L. and Potratz, H. A. (1965). *Science* **149**, 55-58.
Tooley, M. J. (1978). "Sea-level Changes in North-West England during the Flandrian Stage". Clarendon Press, Oxford.
Turner, G. M. and Thompson, R. (1981). *Geophys. J.R. astr. Soc.* **65**, 703-725.
von Post, L. (1968). *Geol. För. Stockh. Förh.* **90**, 37-110.
Wehmiller, J. F. and Belknap, D. F. (1978). *Quaternary Research* **9**, 330-348.

Wehmiller, J. F., Lajoie, K. R., Kvenvolden, K. A., Peterson, E., Belknap, D. F., Kennedy, G. L., Addicott, W. O., Vedder, J. G. and Wright, R. W. (1977). *U.S. Geol. Surv. Open File Rept.* 77-680.

Wenner, C.-G. (1968). *Stockholm Contributions in Geology* **18**, 75-95.

West, R. G. (1972). *Phil. Trans. R. Soc.* **A272**, 87-98.

Zagwijn, W. H. (1979). *In* "The Quaternary History of the North Sea" (E. Oele, R. T. E. Schuttenhelm and A. J. Wiggers, eds), pp. 31-42. *Acta Univ. Ups. Symp. Univ. Ups. Annum Quingentesimum Celebrantis* **2**, Uppsala.

PART FOUR

Shorelines and glacio-isostasy

SEVEN

Biostratigraphical evidence for Late Weichselian and Holocene sea-level changes in southern Norway

U. HAFSTEN

INTRODUCTION

The long coastline of southern Norway bordering the North Sea and the North Atlantic (Fig. 1) contains widespread evidence for relative sea-level changes. This results largely from the complex interaction between eustatic sea-level oscillations and isostatic crustal movements during the Late Weichselian and Holocene. Evidence of relative sea levels before the Late Weichselian is rare, exceptions being the marine sediments deposited during the Late Saalian, Eemian and Early Weichselian at Fjøsanger in Bergen (Mangerud *et al.*, 1981); during the Eemian to Middle Weichselian at Jæren and Karmøy (Andersen *et al.*, 1981); and during the Middle Weichselian at Sunnmøre (the "Ålesund interstadial", cf. Mangerud, 1981).

The frequent occurrence in southern Norway of subfossil mollusc beds, glaciofluvial terraces, raised beaches and strandlines at considerable altitudes above present sea level, has attracted the attention of many research workers (e.g. Keilhau, 1838; Sars, 1865; Kjerulf, 1879; Brøgger, 1900–01; Holtedahl, 1924, 1953, 1960).

METHODS AND CHRONOLOGY

Information on sea-level changes, obtained from biostratigraphical investigations, was first outlined by Axel Blytt in his pioneering essay of 1876. Blytt observed that both the thickness and the stratigraphical complexity of peat bogs increases from low levels towards the highest marine level in accordance with the time that has passed since the various levels have been isolated from the sea. He also realized that isostatic recovery and related sea-level changes have been continual processes, although the rates have varied during the Late Weichselian and Holocene.

The introduction of pollen analysis (von Post, 1916) and also the increasing knowledge of various phytoplankton groups, particularly diatoms and their

SHORELINES AND ISOSTASY
ISBN 0 12 652960 4

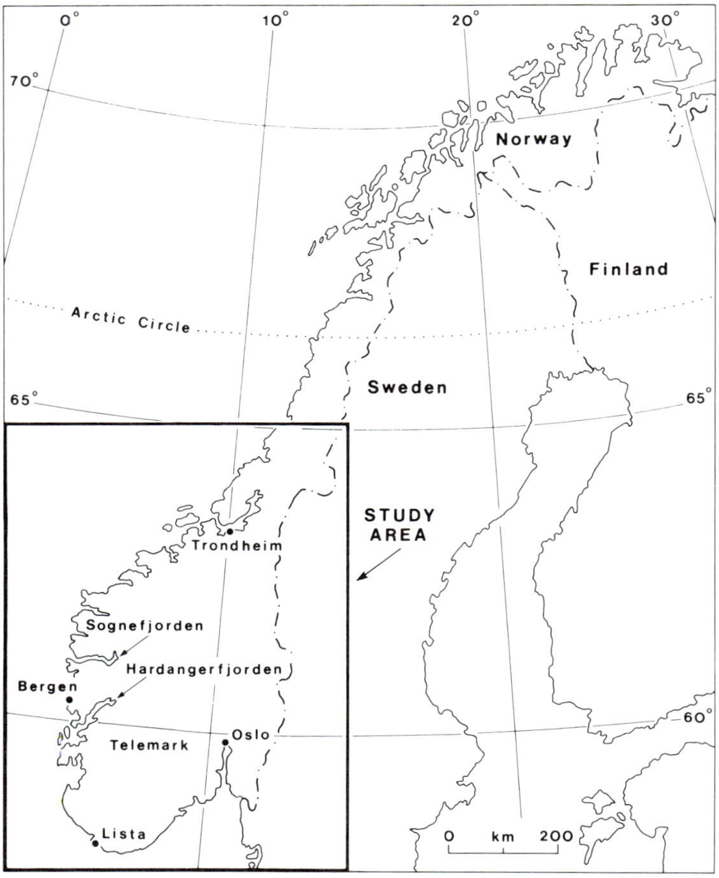

Figure 1. Location map, with main area mentioned in the text.

ecological requirements (Kolbe, 1927), provided possibilities for obtaining very precise information about former sea-level changes through analyses of deposits in lakes and bogs. It should be realized, however, that most studies of this kind do not provide any more than a general picture of sea-level changes. The reason is that the need to study relatively small areas, to avoid errors caused by differential land uplift, means that in any one area there are seldom a sufficient number of suitable basins between the earliest marine limit and present sea level to reconstruct all the smaller sea-level oscillations that may have taken place. One should not ignore either that the methods employed might not have a sufficient "power of resolution" for a more detailed reconstruction, or that some important method of discrimination has been utilized in an inadequate way or has not been applied at all.

The present contribution is restricted to results obtained primarily from lithostratigraphical and biostratigraphical investigations of selected bog and lake deposits, normally in circumstances outlined in Fig. 2. From such studies, 15 shore displacement curves have been compiled (Fig. 3). Most of the curves are based upon ^{14}C dates, but some are based only upon pollen-chronological data. In the case of the

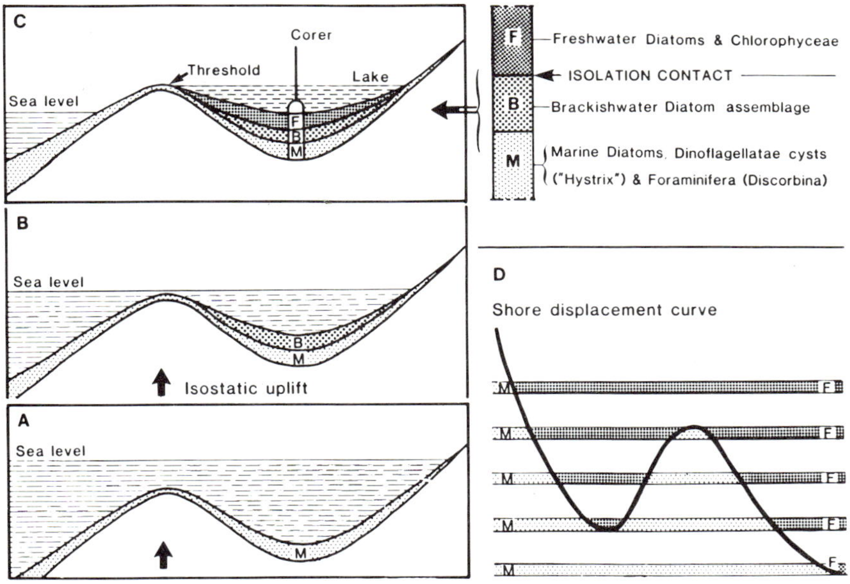

Figure 2. Depositional conditions (a) when a basin or trough on the sea bottom is isolated from the sea during isostatic uplift (A, B, C), and (b) when eustatic rise in sea level exceeds isostatic uplift, causing a marine transgression (D). A, B and C adapted after Holtedahl (1953).

latter, however, because most Holocene pollen zone boundaries have been shown to be diachronous, or time-transgressive, from south to north in southern Norway (Hafsten, 1970); adjustments to zone boundaries based upon recent ^{14}C dates have been made in redrawing these curves. In the case of the shore displacement curve for the Oslo area, ^{14}C dates of most lake sediments are affected by the presence of inactive carbon derived from Cambro-Silurian sedimentary rocks in the area, and therefore is less reliable. In the shore displacement diagrams, full curves are based upon calibrated ^{14}C ages (MASCA) and broken curves are not. The scales of the curves are the same for comparison, with Jessen's pollen zones and also archaeological subdivisions indicated. The disputed zone boundary between the Preboreal and Boreal, which in the south-eastern and south-western parts of southern Norway is defined pollen-analytically at the early Holocene birch/pine transition and coincident establishment of hazel, is drawn at 9400 radiocarbon years before present (Hafsten, 1970). The Boreal/Atlantic zone boundary, which is determined pollen-analytically primarily by the establishment of alder, has been fixed at 8000 B.P., in agreement with the proposal of Mangerud *et al.* (1974).

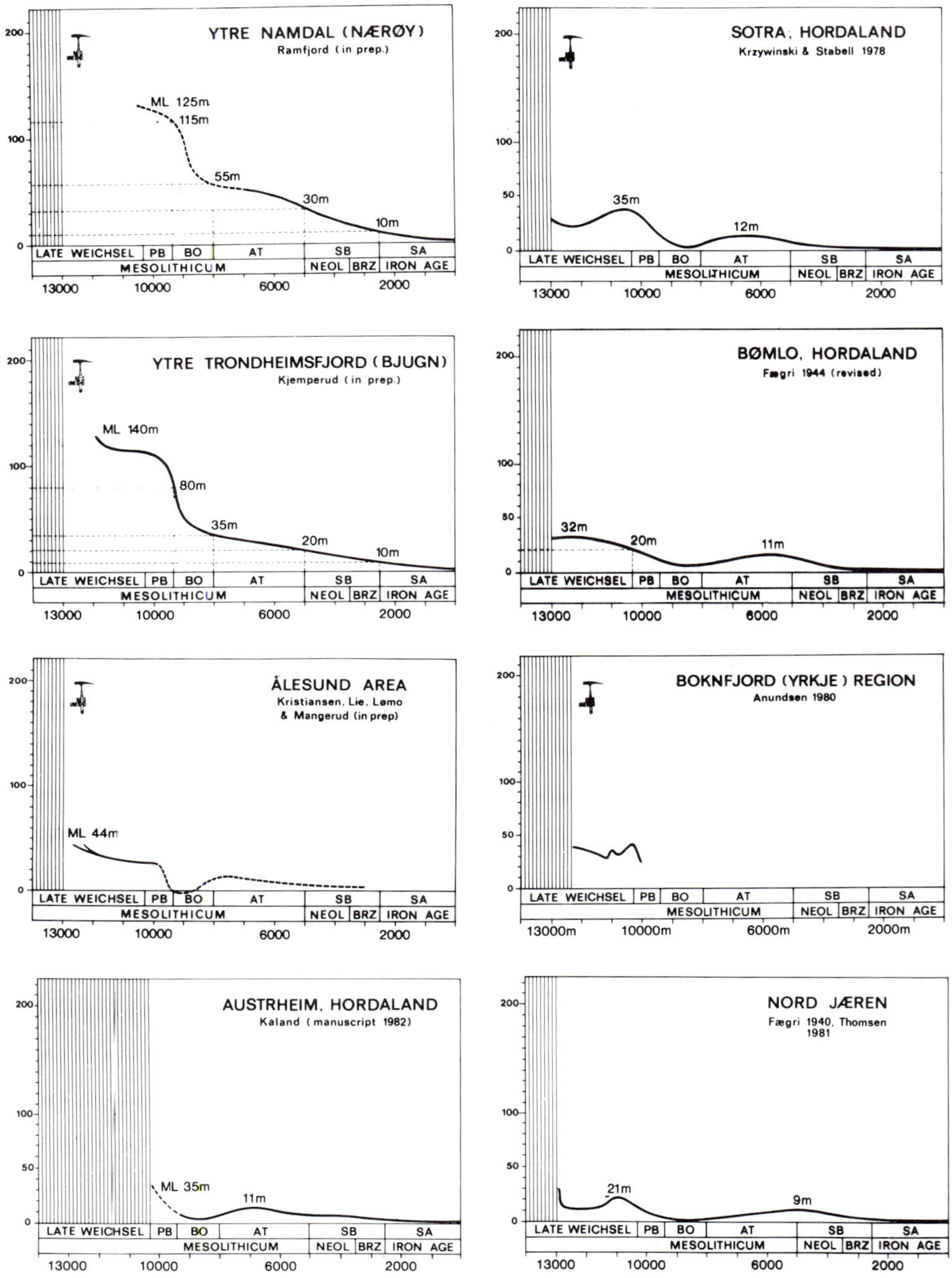

Figure 3. Late Weichselian and post Weichselian (Holocene) shore displacement curves from southern Norway. On the key map, the dotted line is the Younger Dryas glacial limit, full circles identify the place for which curves are drawn and broken circles other locations discussed. On the graphs, ML

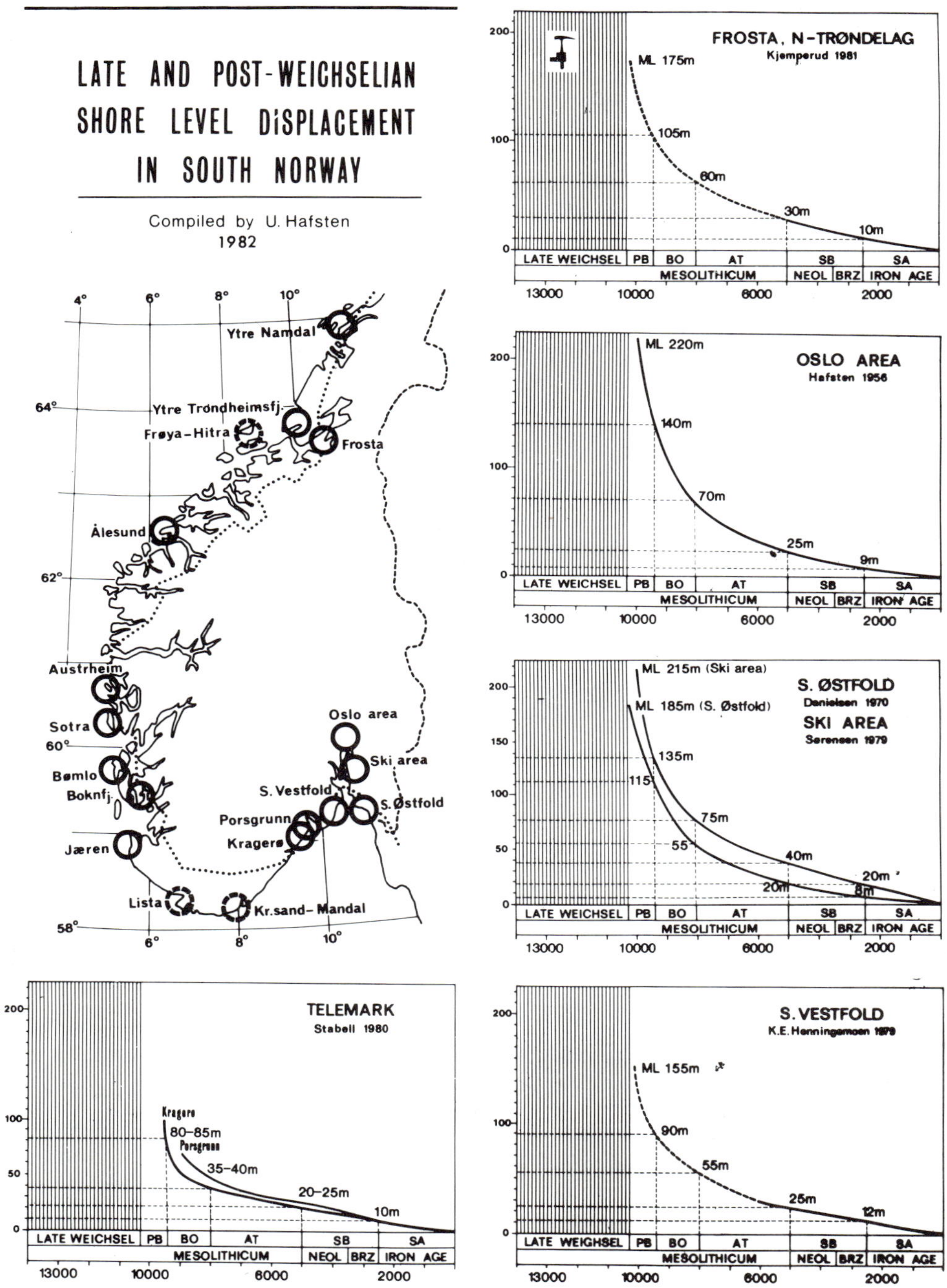

is the marine limit, and periods of ice cover are shaded. Information partly from the Norwegian contribution to IGCP projects 24 and 61, supported by the Norwegian Research Council for Science and Humanities.

DIFFERENTIAL CRUSTAL MOVEMENTS AND VARIATIONS IN THE HIGHEST MARINE LIMIT

As a result of differential uplift, the Late Weichselian and/or Holocene highest marine limit varies considerably from one region to another along the coast of southern Norway (Fig. 3). One obvious difference is that the highest marine limit (i.e. the highest recorded relative sea level) is located at much greater altitudes in the eastern areas (Fig. 3, right-hand page) than in the western and extreme southern areas (Fig. 3, left-hand page).

The highest marine limit in Norway occurs in the Oslo area, at lat. 60 °N, where the early Preboreal sea level reached an altitude in excess of 220 m above present sea level (Hafsten, 1956). Farther south, the early Holocene marine limit drops to about 185 m in Østfold (Danielsen, 1970) and to about 155 m in South Vestfold (Henningsmoen, 1979). Farther south, on the coast of Telemark, it falls to about 140 and 120 m in the Langesunds- and Kragerøfjord, respectively (Stabell, 1980). These altitudes imply a regional decline in early Preboreal marine limit from north to south in the Oslofjord region by *c.* 0.8 m km^{-1}.

In the Tronheimsfjord region, located approximately at the same longitude as the Oslofjord region, but 3.5° latitude farther north, the early Holocene shoreline can be traced at about 175 m in the eastern part (Frosta peninsula—Kjemperud, 1981) and at about 140 m in the western, coastal part (Ytre Trondheimsfjord—Kjemperud, 1982). Farther north in the Nærøy region, at about 65° N, the highest shoreline occurs at *c.* 125 m (Ramfjord, 1982).

On the farthest south-west and west coasts which were deglaciated during the early Late Weichselian substage, the highest shoreline nowhere exceeds an altitude of 45 m above present sea level. The highest shoreline rises in altitude from south to north; e.g. it lies at *c.* 8 m in Lista (Lat. 58° N), *c.* 10 m in the southern part of Jæren and 25 m in the northern part of Jæren (Fægri, 1940; Thomsen, 1981), *c.* 32 m in the Bømlo archipelago in the outer part of the Hardangerfjord region (Fægri, 1944), *c.* 35 m in Sotra (Krzywinski and Stabell, 1978) and Austrheim, near Bergen (Kaland, in press), and *c.* 44 m at Ålesund in the north-west (Kristiansen *et al.*, in preparation). In all extreme west coast areas referred to above (except the Austrheim area which was occupied by readvancing Younger Dryas ice) the highest shoreline was produced during the Late Weichselian. In the southern areas this maximum sea level probably occurred during the marked transgression phase of the Allerød (see p. 176). In Lista, an almost complete Late Weichselian sequence, deposited entirely above sea level, was found in a basin at 7 m altitude, and it seems that the highest sea level during the Late Weichselian is unlikely to have greatly exceeded this figure. It is believed that in this area (and nowhere else in Norway) the highest marine deposits were in fact produced during the Tapes or Littorina transgression (Hafsten, 1963, 1979).

Studies on the deglaciation history of the fjord district in the west, based primarily on the observed altitudes of glaciofluvial terraces related to moraines and on

^{14}C dating of shell material embedded in till, have shown that the isobases of shorelines produced during ice-sheet deglaciation decline in altitude from 120–130 m within the inland part of the fjord systems (eastern tributary branches of the Sognefjord and Hardangerfjord) to 20 m or less in the outer coast regions (Holtedahl, 1960; Aarseth and Mangerud, 1974) (see Fig. 4).

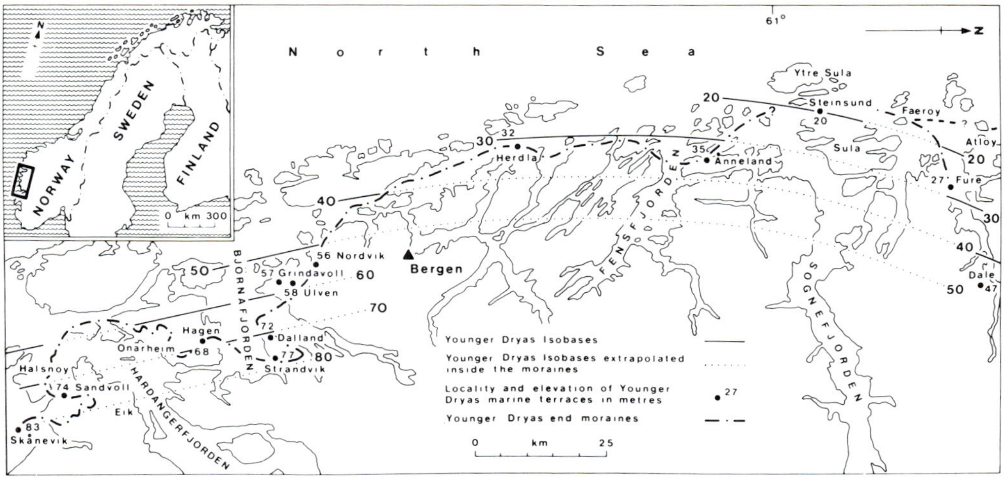

Figure 4. Palaeogeographical map of part of the Norwegian west coast at Younger Dryas time, showing the position of Younger Dryas end moraines and corresponding shoreline isobases (in m). Locations and elevations of Younger Dryas marine terraces are marked with dots. After Aarseth and Mangerud (1974).

LATE WEICHSELIAN SEA-LEVEL CHANGES (13 000–10 000 B.P.)

Stratigraphical studies providing information on Late Weichselian shoreline changes in southern Norway are available from six selected areas along the extreme west coast, between the Lista peninsula in the south and the Trondheimsfjord in the north. These are the very exposed Jæren area (Fægri, 1940), the inner Yrkjefjord region at the north side of Boknfjord (Anundsen, 1978, 1980), the Bømlo archipelago (Fægri, 1944), Sotra island (Krzywinski and Stabell, 1978; Stabell, 1982), the Ålesund region in Sunnmøre (Kristiansen *et al.*, in preparation), and finally, the region at the mouth of the Trondheimsfjord (Kjemperud, 1982). Their locations are shown in Fig. 3.

Jæren

The occurrence of Late Weichselian marine transgressions in Jæren was proposed by Fægri (1940), primarily on the basis of morphological studies of the old beach ridges which are so well preserved in that area. Fægri suggested that two Late

Weichselian transgressions had occurred in this area: (a) the Alvevatn transgression of pre-Allerød (Bølling?) age that reached an altitude of 19 m in the northern area, and (b) a later Allerød transgression that culminated at an altitude of *c.* 24 m above present sea level. In northern Jæren, Fægri also described a system of beach ridges and abrasion terraces located at a level somewhat lower (*c.* 16 m) than the ridges belonging to the Alvevatn transgression. A comparison of the altitudes of the early Alvevatn transgression observed for the southern, central and northern part of Jæren indicates a regional tilting (or a difference in submergence) of as much as $0{\cdot}5\ \mathrm{m\,km^{-1}}$ from south to north.

Biostratigraphical investigations undertaken in north Jæren (Thomsen, 1981) have partly corroborated and completed Fægri's previous results, identifying a regression from above 25 m altitude and dated at more than 13 000 B.P., to a level between 11 and 14 m during the early Bølling, followed by a rather stable sea level persisting until the Allerød when the sea level rose rapidly by 6–9 m to a transgression maximum at about 21 m above present sea level at the transition between the Allerød and Younger Dryas. During the Younger Dryas a new regression took place and was associated with a fall in sea level from 21 to 5 m above present sea level.

Boknfjord region

In the inner Yrkjefjord area, to the north of Boknfjord, between Jæren and Bømlo but farther inland (i.e. in an area of higher shoreline isobases), stratigraphical and radiometric analyses of five basins situated at altitudes of 25, 30, 34, 38 and 45 m respectively, have provided evidence for a marine transgression during the Younger Dryas that reached a maximum altitude as late as *c.* 10 200 B.P. (Anundsen, 1978, 1980). The transgression is stratigraphically well-documented in basins located up to 38 m altitude, but is not observed in the 45 m basin. This indicates that the maximum altitude of the transgression lies between 38 and 45 m. Anundsen inferred (1) that the maximum level of this late "Yrkje transgression" represents virtually the highest marine level in that area, and (2) that an earlier (Allerød) transgression occurred in the area but did not exceed 38 m. However, an earlier transgression phase, dated at *c.* 11 400 B.P. (i.e. Allerød), is also documented in his 34 m basin. The regression minimum that preceded the Younger Dryas transgression lies at between 25 and 30 m altitude and is dated between 11 200 and 10 500 B.P. The major reason for a comprehensive marine transgression during Younger Dryas—a period characterized by climatic deterioration and increased glacial activity—is considered to be a substantial isostatic subsidence caused by renewed ice accumulation during this period.

Bømlo

Farther north in Bømlo, there are no raised beach ridges or other morphological features on the basis of which Late Weichselian sea levels can be determined. In spite of the fact that the Late Weichselian stratigraphy is not well known, Fægri (1944) reported stratigraphical evidence for an early transgression from below 27 m up to 32 m, which he considered synchronous with Alvevatn transgression in Jæren. He emphasized, however, that at that time he was unable to demonstrate synchroneity.

Sotra

At Sotra, where detailed biostratigraphical investigations involving both pollen and diatom analysis and ^{14}C dating have been undertaken, a very distinct marine transgression has been interpreted as having commenced during the late Bølling at *c.* 25 m and to have culminated in the late Allerød at *c.* 40 m (Krzywinski and Stabell, 1978; Stabell, 1982). When the ice disappeared from Sotra about 13 000 years ago, sea level appears to have been at *c.* 30 m above present.

Ålesund region

As part of the deglaciation studies in Sunnmøre in the north-west corner of southern Norway, investigations on sea-level changes have also been undertaken, though so far only at Sula island just south of Ålesund (Lie and Lømo, 1981). The Late Weichselian part of the shore displacement curve constructed for this area is based on stratigraphical investigations as well as radiometric datings from four basins.

From a highest marine level of 44 m above present sea level during probably the late Bølling, the local shore displacement curve suggests a continuous regression during the entire Late Weichselian period to *c.* 26 m at the end of Younger Dryas. The fact that no marine transgression phases have been detected in the Late Weichselian deposits of this area contradicts the results obtained for the south-west corner of southern Norway. The explanation may be that the Ålesund area is situated somewhat farther east and thus subject to more rapid glacio-isostatic uplift than those areas further south. Compared with the patterns of Late Weichselian shore displacement for the coastal areas situated farther north (viz. Ytre Trondheimsfjord), the Ålesund curve represents a rather logical transition.

Ytre Trondheimsfjord

Biostratigraphical studies in the outer part of the Trondheimsfjord region have also unveiled deposits reaching well into the Late Weichselian (Kjemperud, 1982). Pollen and diatom analyses as well as radiocarbon measurements of deposits from four high-lying lakes permit the construction of a Late Weichselian shore displacement curve back to the transition between the Older Dryas and the Allerød. In this area the highest marine deposits are at 130–140 m. Thereafter, a gradual decline in sea level is indicated for the Allerød and Younger Dryas to *c.* 110 m at the beginning of the Preboreal.

HOLOCENE SEA-LEVEL CHANGES (AFTER 10 000 B.P.)

Regional patterns

From a comparison of the sea-level curves in Fig. 3, it will be seen that there are distinct differences between the extreme western coast areas and eastern fjord areas.

Eastern areas

In the Oslofjord and inner Trondheimsfjord regions, which were deeply submerged during deglaciation, post-glacial isostatic uplift resulted in an uninterrupted recession of relative sea level that reached a maximum rate during the Preboreal and Boreal when as much as two-thirds of the total post-glacial shoreline recession took place (Table 1). The Preboreal and Boreal periods are here treated together in order to avoid the controversial definition of the Preboreal/Boreal zone transition mentioned above.

Even during the Atlantic, when a considerable rise in sea level took place in the extreme western and southern areas, a progressive fall in sea level occurred in most eastern areas although at rates that were only one-fifth or less of those calculated for the two preceding periods. The shore displacement curves from Porsgrunn and Kragerø on the coast of Telemark do not show an Atlantic submergence, although one here may speak of a "syngression", where the rate of eustatic rise balanced that of isostatic uplift (Stabell, 1980, p. 80). This indicates that the area where the Tapes (Littorina) transgression first manifests itself stratigraphically by marine sequences along the south-east coast of southern Norway, is to be found not far south of the Telemark coast. On the eastern Skagerak coast marine transgression sequences have not been observed farther north than in the Kolbengtserød valley just south of Uddevalla in central Bohuslän (Sweden), at a distance of about 200 km south of Oslo (Persson, 1973).

During the Subboreal and Subatlantic shoreline recession in the Oslofjord and inner Trondheimsfjord regions slowed down further, to an average rate for the Subatlantic period of 0.3–0.4 m per century (Oslo, S. Østfold, Frosta). This value corresponds closely with the rates being measured by the Geographical Survey of Norway for modern sea-level recession within those regions (Fig. 5).

Holocene marine transgressions have been traced only in the most southern and western coast regions (see below). The absence of transgressions in the eastern areas (the Oslofjord and Trondheimsfjord—Namdal regions), is attributed to the rapid post-glacial isostatic recovery that prevented a eustatic submergence. Only a reduced rate of sea-level change can be observed in these regions during the post-glacial transgression phases (note the bend in the displacement curves at the Boreal/Atlantic zone boundary). In the Oslofjord diagrams this bend may be traced at 70–75 m in the Oslo and Ski area, at *c.* 55 m in S. Østfold and S. Vestfold, and at *c.* 50 m on the coast of Telemark where this "transgression marker" becomes more

TABLE 1

Shore displacement data for five eastern areas in Fig. 3

		Preboreal and Boreal 10 000–8000 B.P.			Atlantic 8000–5000 B.P.			Subboreal and Subatlantic 5000 B.P.–present		
Region	Highest marine limit (m)	Total shore displacement (m)	Mean shore displacement per 100 years (m)	Total Holocene shore displ. (%)	Total shore displacement (m)	Mean shore displacement per 100 years (m)	Total Holocene shore displ. (%)	Total shore displacement (m)	Mean shore displacement per 100 years (m)	Total Holocene shore displ. (%)
Oslo area[b]	220	150	7·5	68	45	1·5	20	25	0·5	11
Ski area[a]	215	140	7·0	65	38	1·3	18	38	0·8	18
S.Østfold[b]	185	130	6·5	70	35	1·2	19	20	0·4	11
Frosta[a]	175	115	5·8	66	30	1·0	17	28	0·5	16
S. Vestfold[a]	155	105	5·3	68	25	0·8	16	25	0·5	16

[a]Based upon ^{14}C measurement (Ski area data based in part upon shell dates).
[b]Based upon pollen analysis.

Figures are approximate only.

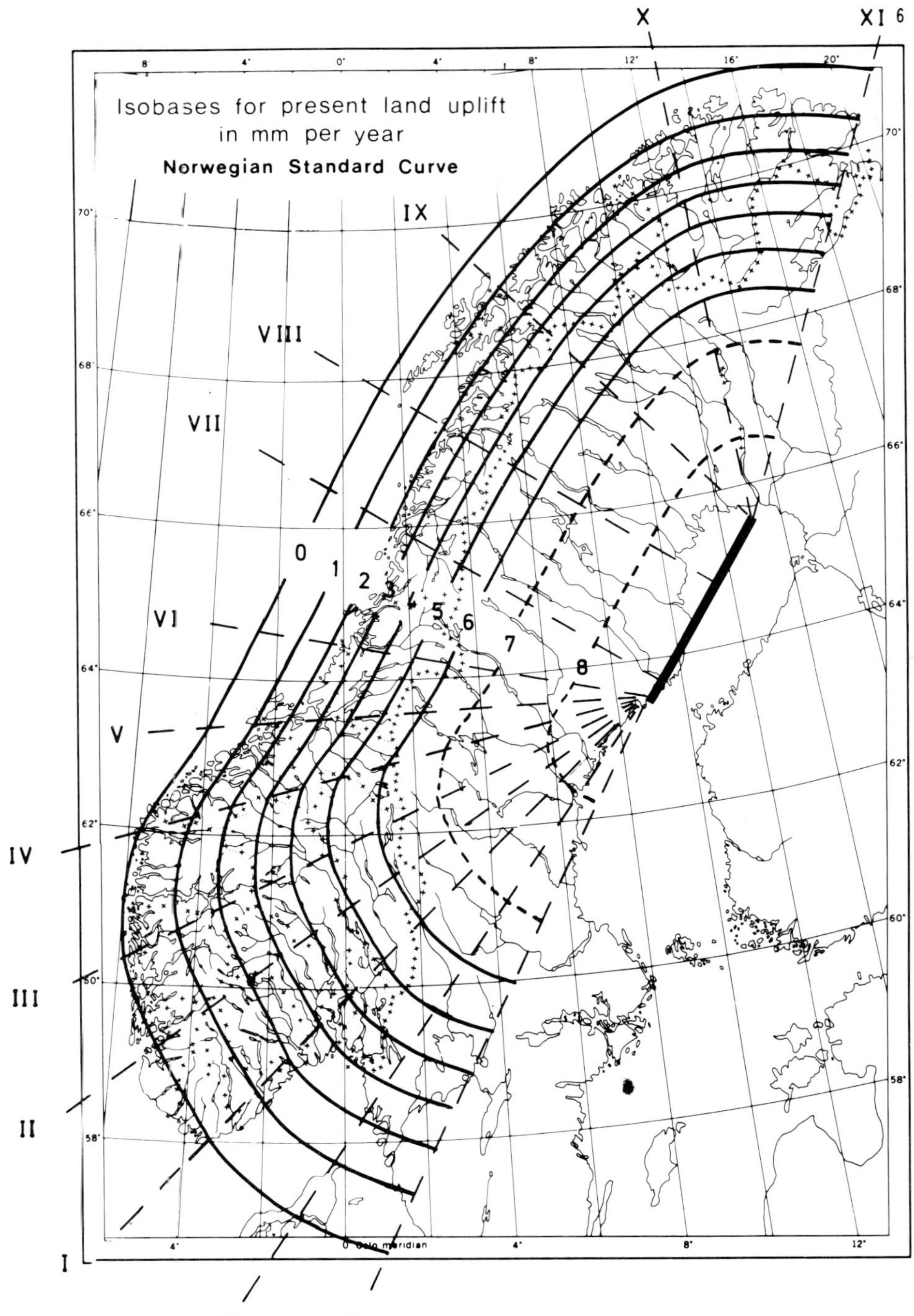

Figure 5. Isobases for the present land uplift in West Scandinavia, in mm per year. Norwegian standard curve after S. Bakkelid at the Geographical Survey of Norway.

pronounced. This feature is even more pronounced in the diagrams from the two northernmost areas at the mouth of Trondheimsfjord and, particularly, in the Nærøy area of outer Namdal. Apparently, the latter two areas represent a transitional type of displacement curve between the eastern, exponential type and the more complex, extreme south and west coast type.

Western areas

In the western coastal area of southern Norway the glacial depression was moderate and the subsequent isostatic uplift relatively small. In these areas, deposits at low altitudes provide evidence of a eustatic rise in sea level which for a long time exceeded the rate of isostatic recovery. These marine deposits rest on terrestrial peat or detritus mud of freshwater origin and are overlain by deposits of lacustrine or

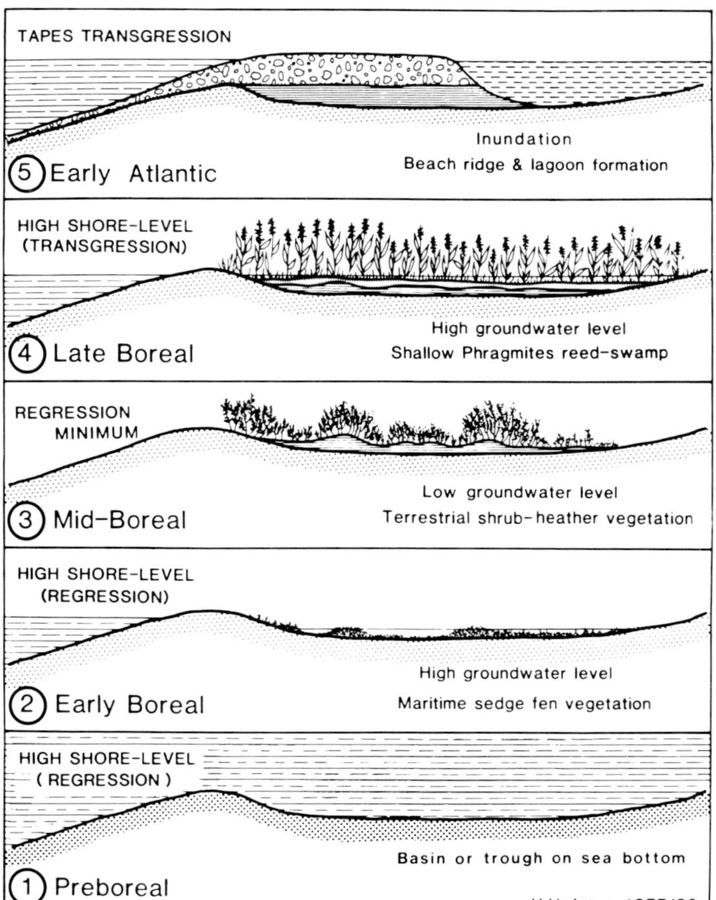

Figure 6. Sketch illustrating fluctuations in shore level, groundwater level and local vegetation from Preboreal time (1) up to the Tapes transgression in the early Atlantic (5). Such beach ridges occur frequently along the Norwegian west coast. After Hafsten and Tallantire, 1978.

terrestrial origin. In many places beach ridges, consisting of thick layers of pebbles resting on terrestrial peat, extend for many kilometres along the coast (Hafsten and Tallantire, 1978). In certain areas the beach ridges have impounded bogs and lakes (e.g. in Lista and Jæren) and elsewhere are succeeded landwards by dried or artificially drained depressions that represent earlier lagoonal locations, e.g. at the coast of Møre and Romsdal. Figure 6 illustrates the sequence of events often found in these western areas. A basin or trough below sea level during Preboreal time (1) is exposed during marine regression (2) as isostatic uplift develops. A peat bog develops (3), but as relative sea level rises becomes an area of shallow reed-swamp (4). Eventually the bog is inundated during the ensuing transgression, and a beach ridge forms on its seaward side (5). Subsequent regression may leave a peat-moss landward of the ridge, which may become artificially drained.

The Holocene transgression phase can be correlated with the younger beds in a system of mollusc beds (e.g. the *Tapes*, *Trivia* and part of the *Ostrea* beds) of the Oslofjord and Trondheimsfjord regions, which in the Oslofjord region occur up to 75 m (Hafsten, 1958a). For this reason the Holocene transgression(s) in the Skagerrak-Kattegat region is widely known as the Tapes transgression, which is regarded as equivalent to the Baltic Littorina transgression.

In addition to the transgression areas shown in Fig. 3, a number of other localities are known for which continuous displacement curves have not yet been constructed but where typical Tapes transgression sequences are present. Many of these localities are on the extreme south coast, particularly in the Kristiansand-Lista region (see pp. 177–178), whilst on the extreme west coast, beach ridges, abrasion platforms and sedimentary sequences with evidence of the Tapes transgression may be found up to 10–15 m above present sea level.

The Boreal regression minimum

The moment when the eustatic rise in sea level equalled and thereafter began to exceed the rate of isostatic recovery took place during the Boreal period and is represented by a glacio-isostatically tilted shoreline. In Lista, where peat and lacustrine gyttja of Boreal age have been found offshore at 4 m below present sea level, the shoreline at that time must have been distinctly lower than present sea level (Holmboe, 1909; Sollesnes and Fægri, 1951). This may be true also for Jæren, although it seems more likely that the most probable Boreal shoreline here lies at about present sea level (Fægri, 1940).

Further north, at Bømlo, Sotra and Austrheim, the Boreal regression minimum occurs above present sea level, *c.* 5 m in Bømlo (Fægri, 1944) and close to 4 m in Sotra and Austrheim (Krzywinski and Stabell, 1978; Kaland, in press). A stratigraphical section described from Borgund (Hafsten, 1979) indicates that the Boreal regression minimum in the Ålesund area was at least 0.7 m lower than present sea level (see p. 176). Evidence from a transgression sequence described from the Kristiansand-Mandal region on the extreme south-east coast (Dostad in Harkmark, Gabrielsen, 1959), indicates a mid-Boreal sea level at 2 m above present sea level.

The marine transgression(s)

In South Scandinavia the Tapes, or Littorina, transgression does not represent only one, prolonged phase of flooding of previously isolated ground, but is found to include intervals of temporary marine recession (e.g. the four transgression phases described from Denmark (Iversen, 1937; Jessen, 1937; Krog, 1968)) and the many Holocene shoreline oscillations reported from southern Sweden (Florin, 1944; Berglund, 1964; Mörner, 1969).

On the south-west coast of Norway, long before the introduction of ^{14}C dating, oscillations in the Tapes transgression were also proposed. Here evidence was presented of at least one interval of temporary late Atlantic recession of relative sea level that separated an "older Tapes transgression" from a "younger" one (Fægri, 1940, 1944). The two transgression phases were dated pollen-analytically to the mid-Atlantic and early Subboreal respectively. The younger one was reported to represent the higher level (i.e. the Holocene transgression maximum). In Jæren this maximum was reported to reach 6 m in the south (Ogna) and 8 m in the north (Randaberg), whereas the older, mid-Atlantic phase was reported to have reached an altitude of only 3 m above present sea level in the central Figgjo valley. The corresponding altitudes for Bømlo, which were based primarily on the altitudes of raised pumice layers, are at 11 and 8 m respectively, whereas the altitude of the intermediate regression minimum was determined at between 5 and 6·5 m, indicating a temporary 2–3 m lowering of relative sea level during the late Atlantic.

There is, however, some reason to doubt these interpretations. A re-examination of the stratigraphical section on which the evidence of the two transgression phases in Bømlo was based, indicates that the postulation of a transient, late Atlantic lowering of sea level is the result of a stratigraphical misinterpretation (Kaland, personal communication). This may be true also for the two transgression phases postulated during the pioneering work in Jæren, because the existence of these phases has not been verified by the recent biostratigraphical investigations referred to above. The revised Bømlo curve is, in fact, very similar to the one presented recently for southern Sotra, in showing only one, very protracted, transgression peak approaching the present 12 m level (Krzywinski and Stabell, 1978; Stabell, 1982). The transgression maximum of 11–12 m calculated for Bømlo and Sotra thus represents a total rise in sea level of between 6 and 8 m since the Boreal period.

The shore displacement curve constructed for Austrheim, in northern Hordaland, 130 km north of Bømlo, differs from those described for Jæren and Bømlo, both with regard to the timing of the transgression maximum and to the general pattern of the displacement curve. Detailed stratigraphical studies combined with many ^{14}C datings suggest only one Atlantic transgression phase, followed by a prolonged and slow regression during late Atlantic and Subboreal time. The altitude of the transgression maximum in this area of *c.* 11 m and of the Boreal regression minimum of 4 m, gives a maximum post-Weichselian rise of sea level in Austrheim of *c.* 7 m, which is in good accordance with the results obtained for Bømlo and Sotra.

In contrast to the Late Weichselian Alvevatn transgression, which indicates a tilting of Jæren at that time of as much as 0·5 m km^{-1} from south to north, the Tapes shorelines reflect practically no tilting at all. From measurements of the altitudes of beach ridges formed during the transgression maximum, Fægri (1940) estimates the south–north gradient of the highest shoreline (the "Tapes line") to be *c.* 0·1 m km^{-1}, and the west–east gradient, at right-angles to the isobases, to be 0·3 m km^{-1}. A gradient similar to the latter has also been estimated along a west–east line in the Hardangerfjord and Sognefjord areas (Undås, 1955, 1964).

The north-west coast

North of Austrheim, at the north-west corner of southern Norway, present knowledge of Holocene shore displacement is based mostly on casual investigations as far as biostratigraphical studies are concerned. The Holocene part of the shore displacement curve presented here for the Ålesund area (Fig. 3) is based on investigations of three basins only and, therefore, may be regarded merely as tentative (indicated by pecked curve).

From Borgund, just east of Ålesund, and from Haramsøy, an outlying island some 25 km to the north of Ålesund, stratigraphical sequences have been studied, which according to pollen analysis and ^{14}C dating provide information about the Tapes transgression in that region (Hafsten, 1979; Hafsten and Tallantire, 1978). In Haramsøy a 0·95 m thick layer of compact, terrestrial peat or drift-mud, probably formed in a shallow lagoon, is overlain by a 1·3 m thick beach ridge (similar to the situation in Fig. 6). The ridge has a level top at a height of 10 m above present sea-level and extends continuously along the southern coastal plain of the island for more than 5 km. Pollen analysis and six ^{14}C datings indicate that the peat layer was formed mainly during the Boreal (between *c.* 9500 B.P. (bottom) and 7300 B.P. (top)) and thus dates beach-ridge formation to the Tapes (Atlantic) period. The altitude of the beach-ridge crest at 10 m (9·6 m in the north-west and 10·3 m in the south-east) corresponds with the maximum level of the Tapes transgression observed at many places within this region (see the 10 m Tapes isobase, drawn close to the south-eastern point of Haramsøy, in Holtedahl, 1953, Plate 18).

The sequence from Borgund (Fig. 7), collected from the deposits underlying the site of the medieval "*Borgund-kaupangen*" trade centre, comprises a very sharp boundary between basal terrestrial peat or lacustrine gyttja and overlying marine clay-mud, which was dated both pollen-analytically and radiometrically to the late Boreal (8250 ± 250 B.P.). The difference in age of the flooding at Haramsøy and at Borgund (approximately 1000 years) may be attributed to the difference in altitude above sea level of the two locations; the top of the peat layer at Haramsøy reaching an altitude of 2·3 m above present sea level, and the boundary between terrestrial peat and marine clay-mud at Borgund being 0·7 m below present sea level. The stratigraphical evidence of the extremely low Boreal sea level at Borgund is hardly to be questioned, because the vegetational development reflected in the pollen diagram reveals no signs of redeposition or other stratigraphical irregularities. The

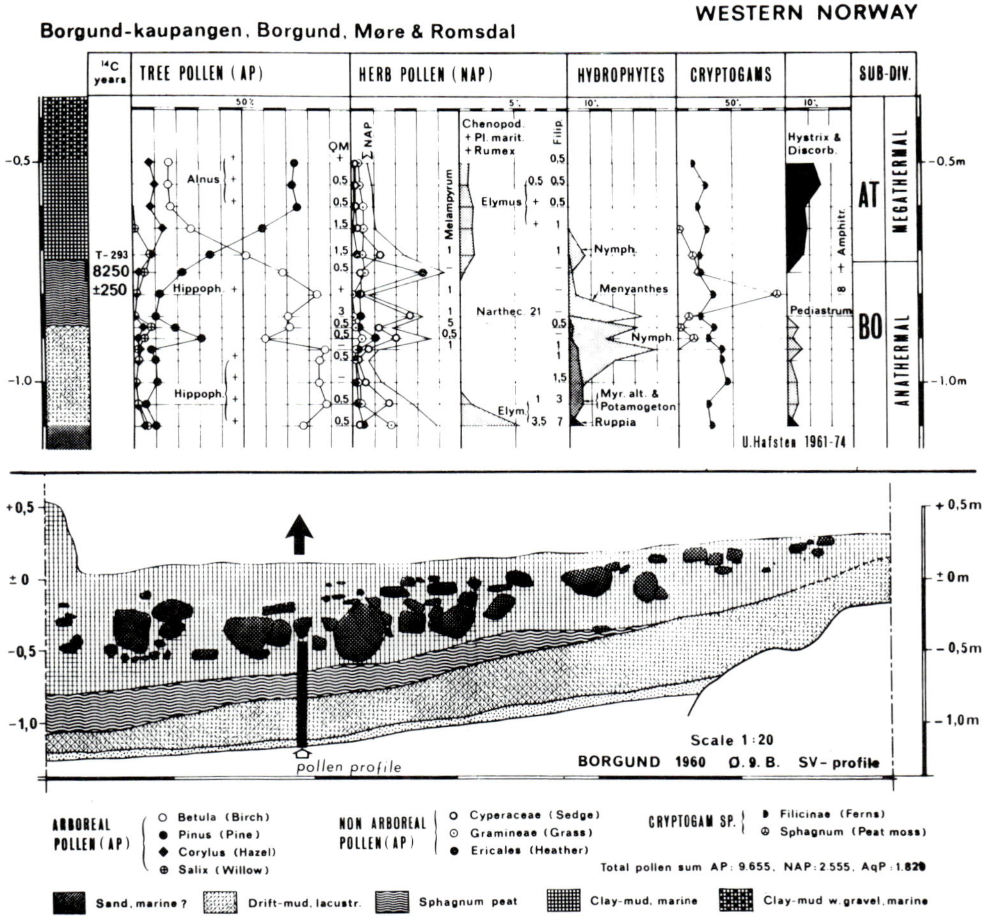

Figure 7. Stratigraphical section and pollen diagram from the Tapes transgression sequence underlying the medieval trade centre at Borgund on the north-west coast of southern Norway. The figure illustrates the method of investigation which forms the basis for the shore displacement curves presented.

incidence of such a low shoreline at that time is sustained also archaeologically, by evidence of human settlements older than 8000 B.P. at present sea level.

The south coast

The extreme south coast, particularly the Lista area, is of considerable interest in sea-level research because it corresponds more closely to areas farther south in Scandinavia than the remaining areas of the Norwegian coast. This concerns particularly the marine limit, which here coincides with the Tapes transgression maximum and not with the highest Late Weichselian shorelines as in Jæren and the coastal regions farther north. An accurate altitude for this maximum has not yet been determined by stratigraphical studies and ^{14}C dating. However, the presence of distinct marine sections in various basins located up to 7 m above present sea level (Hafsten, unpublished), sustains the assumption maintained by Holtedahl (1953, 1960) and Andersen (1960) that the top level of the Holocene beach-ridge system in Lista at 8 m represents the highest raised shoreline produced during the last 13 000 years. Thus it would appear that the Lista peninsula represents the most stable area in Norway.

Radiometric dating of terrestrial peat or lacustrine gyttja at the ingression contact in basins situated at low altitudes in the coastal area eastwards of Lista, has provided dates similar in age to the Boreal regression minimum. From a basin at 0·5 m altitude some 25 km south-east of Lista (at Vage close to Lindesnes, the southernmost point of Norway) peat just below the ingression contact has been dated at 7950 ± 170 B.P. (T-208, Gabrielsen, in Nydal *et al.*, 1962, p. 171). In a basin situated 2 m above sea level at Dostad, about midway between Lindesnes and Kristiansand, a subfossil peat layer, 1·5–1·7 m thick, lying between basal marine clay and overlying "marine diatomaceous earth" was formed during the first part of the Boreal, between 9410 ± 220 (T-91, bottom) and 8770 ± 100 B.P. (T-80, Gabrielsen, 1959; Nydal, 1959, p. 78). The latter date, which refers to the top of the peat, indicates that the marine flooding of the 2 m level had already taken place by the mid-Boreal. In another basin within the same area at 8 m (Launesmyr), Gabrielsen collected shell fragments mostly of *Ostrea edulis*, beneath a 5·5 m thick peat deposit and dated "the age of the maximum transgression" at 6800 ± 170 B.P. (T-292, Nydal, 1962, p. 170). From Søgne, midway to Kristiansand, Gabrielsen studied basins at 4 m (Ravnemyr) and 5 m altitudes (Oftenes), with ingression contacts dated at 7400 ± 250 (T-209) and 8300 ± 250 B.P. (T-273), respectively (Nydal, 1962, p. 170). Finally, at Oddernes in Kristiansand, at *c.* 10 m, marine clay-mud more than 2 m in thickness has been found, overlying a 1·2 m thick peat deposit containing well preserved pine trunks (Hafsten, 1958b). The pollen diagram provided an accurate dating of the ingression contact at exactly the Boreal/Atlantic zone boundary, i.e. about 8000 B.P.

The studies carried out so far within the southernmost coastal region have not confirmed more than one Holocene transgression phase. However, until more systematic investigations have been carried out, one cannot exclude the possibility that transient regression phases, probably of less extent than those described from areas farther south in Scandinavia, may be discovered. The fact that the south coast, particularly Lista, represents the most stable part of southern Norway as far as Late Weichselian and Holocene sea-level changes are concerned, makes such studies difficult, since the total amplitude within which sea-level oscillations have taken place is here very narrow. In addition, in Lista particularly, the investigations are hampered also by the fact that Boreal regression minimum and, probably also, some of the critical Late Weichselian shoreline regressions, are situated below present sea level and, on this exposed coastal strip, buried by thick layers of sand.

SUMMARY

Relative sea-level changes during the Late Weichselian and Holocene in southern Norway show important regional differences. In the east of the area, both around Trondheimsfjord and in the Oslofjord regions, isostatic depression was considerable, and thus during both Late Weichselian and Holocene times, uplift was rapid. In these areas, relative sea level was subject to a comprehensive, almost exponential, fall. In the west of the area, land uplift was slower and here marine transgressions can be identified for both the Late Weichselian and Holocene.

As might therefore be expected, the highest marine limits in southern Norway occur in the Trondheimsfjord and Oslofjord areas at *c.* 175 m and *c.* 220 m respectively. Recession of the sea in these areas was uninterrupted, and maximum rates were achieved in the early Holocene, when during the Preboreal and Boreal as much as two-thirds of the total Holocene shoreline recession took place. In the west and south-west of southern Norway, by contrast, the highest marine limit is considerably lower, lying at *c.* 8 m in Lista and rising northwards and westwards, for example attaining 35 m in Sotra and 44 m in Ålesund. Over most of the area, the highest shorelines are of Late Weichselian age, but around Lista they are of Holocene age, belonging to the Tapes (Littorina) transgression.

Evidence of marine transgressions is found only in the west and south-west. In Jæren and Sotra, a widespread marine transgression commencing in the late Bølling and culminating in the late Allerød has been identified. This event was followed by a rapid regression which reached a minimum during the Boreal, perhaps as much as 4 m below present sea level in the Lista area. Following this, the Tapes or Littorina transgression reaches a maximum altitude of 11-12 m on the extreme west coast, 6-8 m above the Boreal regression minimum.

ACKNOWLEDGEMENTS

I am thankful to my colleagues and collaborators who most willingly have put unpublished or only internally published data at my disposal. My thanks also go to Mrs Ruth Waadeland who carried out most of the work on the illustrations and to Dr A. G. Dawson and Dr D. E. Smith who revised the English manuscript.

REFERENCES

Aarseth, I. and Mangerud, J. (1974). *Boreas* **3**, 3-22.

Andersen, B. G. (1960). *Nor. geol. Unders.* **210**, 142 pp.

Andersen, B. G., Nydal, R., Wangen, O. P. and Østmo, S. R. (1981). *Boreas* **10**, 297-314.

Anundsen, K. (1978). *Boreas* **7**, 49-60.

Anundsen, K. (1980). Quaternary geological studies in Sunnhordland and Nord-Rogaland, Southwest Norway. Unpublished Dr. philos. thesis, Bergen University.

Berglund, B. E. (1964). *Sver. geol. Unders.* **C599**, 47 pp.
Blytt, A. (1876). "Essay on the Immigration of the Norwegian Flora During Alternating Rainy and Dry Periods". Kammermeyer, Kristiana (Oslo).
Brøgger, W. C. (1900-01). *Nor. geol. Unders.* **31**, 731 pp.
Danielsen, A. (1970). *Univ. Bergen Årb. Mat. naturvitensk. Ser. 1969* **14**, 146 pp.
Fægri, K. (1940). *Bergens Mus. Årb. 1939-40, Naturvitensk.* **R7**, 201 pp.
Fægri, K. (1944). *Bergens Mus. Årb. 1943, Naturvitensk.* **R8**, 100 pp.
Florin, S. (1944). *Geol. För. Stockh. Förh.* **66**, 551-634.
Gabrielsen, G. (1959). *Nature* **183**, 1616-1617.
Hafsten, U. (1956). *Univ. Bergen Årb. 1956, Naturvitensk.* **R8**, 161 pp.
Hafsten, U. (1958a). *Nor. geogr. Tidsskr.* **16**, 74-99.
Hafsten, U. (1958b). *Nor. geol. Tidsskr.* **38**, 313-325.
Hafsten, U. (1963). *Grana palynol.* **4**, 326-337.
Hafsten, U. (1970). *Palaeogeogr. Palaeoclim. Palaeoecol.* **7**, 279-296.
Hafsten, U. (1979). *In* "The Quaternary History of the North Sea" (E. Oele, R. T. E. Schuttenhelm and A. J. Wiggers, eds), pp. 45-59. *Acta Univ. Ups. Symp. Univ. Ups. Annum Quingentesimum Celebrantis* **2**, Uppsala.
Hafsten, U. and Tallantire, P. A. (1978). *Boreas* **7**, 109-122.
Henningsmoen, K. E. (1979). *In* "Fortiden i søkelyset" (R. Hydal, S. Westin, U. Hafsten and S. Gulliksen, eds.), pp. 239-247.
Holmboe, J. (1909). *Naturen* **33**, 235-243.
Holtedahl, O. (1924). *Skr. Vidensk. Selsk. Kristiania, Mat.—naturv K1.* No. 14.
Holtedahl, O. (1953). *Nor. geol. Unders.* **164**, 1118 pp.
Holtedahl, O. (1960). *Nor. geol. Unders.* **208**, 540 pp.
Iversen, J. (1937). *Medd. Dansk Geol. Foren.* **9**, 223-232.
Jessen, K. (1937). *Medd. Dansk Geol. Foren.* **9**, 232-236.
Kaland, P. E. (in press). "Holocene shore displacement and shorelines in Hordaland, western Norway". *Boreas.*
Keilhau, B. M. (1938). *Nyt mag. Naturvid* **1**, 105-254.
Kjemperud, A. (1981). *Nor. geol. Tidsskr.* **61**, 1-15.
Kjemperud, A. (1982). Late Weichselian and Holocene shoreline displacement in parts of Trøndelag, Central Norway. Unpublished Dr. scient. thesis, Oslo University.
Kjerulf, Th. (1971). *Univ. Progr. 1870, 1ste Sem.* 101 pp. Kristiania.
Kjerulf, Th. (1879). "Udsigt over det sydlige Norges Geologi". Christiania (Oslo).
Kolbe, R. W. (1927). *Pflanzenforschung* **7**, 138 pp.
Krog, H. (1968). *In* "Means of correlation of Quaternary succession, 8", pp. 421-435. *Proc. VII, Congr. Int. Assoc. Quaternary Res.* University of Utah Press.
Krzywinski, K. and Stabell, B. (1978). *Arkeo* **1**, 27-31.
Lie, S. E. and Lømo, L. (1981). En litho- og biostratigrafisk undersøkelse av marine og limniske sedimenter i Ålesund-området. Unpublished Cand. real. thesis, Bergen University.
Mangerud, J. (1981). *Boreas* **10**, 381-393.
Mangerud, J., Andersen, S. T., Berglund, B. E. and Donner, J. J. (1974). *Boreas* **3**, 109-127.
Mangerud, J., Sønstegaard, E., Sejrup, H. P. and Haldorsen, S. (1981). *Boreas* **10**, 137-208.
Mörner, N.-A. (1969). *Sver. geol. Unders.* **C640**, 487 pp.
Nydal, R. (1959). *Radiocarbon* **1**, 76-80.
Nydal, R. (1962). *Radiocarbon* **4**, 160-181.
Nydal, R., Løvseth, K. and Syrstad, O. (1970). *Radiocarbon* **12**, 205-237.

Persson, G. (1973). *Sver. geol. Unders.* **C684**, 47 pp.
Ramfjord, H. (1982). *Nor. geol. Tidsskr.* **62**, 192-205.
Sars, M. (1865). "Om de i Norge forekommende fossile Dyrelevninger fra Quartærtiden". Christiania (Oslo).
Sollesnes, A. F. and Fægri, K. (1951). *Blyttia* **9**, 41-58.
Stabell, B. (1980). *Nor. geol. Tidsskr.* **60**, 71-81.
Stabell, B. (1982). "The response of diatom floras during late Quaternary shore line displacement in southern and western Norway". Unpublished. Dr. scient. thesis, Oslo University.
Sørensen, R. (1979). *Boreas* **8**, 241-246.
Thomsen, H. (1981). *Geol. För. Stockh. Förh.* **103**, 447-468.
Undås, I. (1955). *Nor. geor. Tidsskr.* **35**, 169-178.
Undås, I. (1964). *Nor. geor. Tidsskr.* **19**, 291-296.
von Post, L. (1916). *Geol. För. Stockh. Förh.* **38**, 384.

EIGHT

Late Weichselian and Holocene shore displacement in Finland

M. ERONEN

INTRODUCTION

The Nordic countries have a long tradition in the study of shoreline displacement, with important scientific observations dating back well into the last century. The latter half of that century was signified by some excellent investigations by Swedish scientists into events connected with vertical movements of the earth's crust and variations in sea level. This was followed in the present century by a great expansion of research in all Nordic countries, so that the history of the Baltic has now been substantially worked out in all the countries that border onto it (Fredén, 1979; Hyvärinen and Eronen, 1979; Gudelis, 1979; Ludwig, 1979).

A wide variety of views on the events connected with shoreline displacement have thus been put forward and major controversies have not been unusual. Nowadays specialists in the field are generally agreed on the main outlines, but there remain many details upon which opinions still differ, and in spite of the long traditions of investigation, many points regarding the history of the Baltic remain unresolved.

One common source of differences in interpretation is methodological inaccuracy, and the study of shoreline displacement has been perpetually dogged by a lack of reliable dating techniques or excessive inaccuracies in the dates obtained. The morphologically distinct ancient shorelines serve as indicators of past water levels, but they do not usually contain material of use in dating these levels. Pollen and diatom analyses have played major rôles in shoreline displacement studies in Finland, both being very useful when working with lake or mire basin sediments. The diatom flora show a sharp change in species composition at the point where a basin rose above the water level in the Baltic as a consequence of land uplift and became an independent lake or mire, but even this does not provide a direct date for such an event. For this purpose pollen analysis was for a long time in a key position, but this proved a somewhat inaccurate method. A substantial improvement in dating methods was achieved by the discovery of the ^{14}C method of age determination, and it is this method that has come to occupy a crucial position in shoreline displacement research over the last 20 years or so. This has helped to solve

SHORELINES AND ISOSTASY
ISBN 0 12 652960 4

many problems, but has still not carried research forward as rapidly as had been initially expected, and many points have arisen which complicate the interpretation of ^{14}C dates in embarrassingly many instances.

THE ESTIMATION OF LAND UPLIFT

The connection between the lowering of shorelines and their seaward retreat and the process of land uplift was obvious to Scandinavian scientists back in the nineteenth century (Fredén, 1979). The first accurate measurements of contemporary land uplift in Finland, however, were obtained only in the 1950s, when it became possible to correlate the results of two successive precise levellings (Kääriäinen, 1953). Since then precise levellings have continued, and estimates of land uplift are now available for the whole country (Kiviniemi, 1980). Confirmation of these results has been obtained from mareograph readings taken on the coast.

Present land uplift around the Baltic Sea is depicted in Fig. 1, in which isobases denote the extent of uplift in relation to sea level. It should be noted, incidentally, that maps published for land uplift in Finland (Kääriäinen, 1953, 1966; Kiviniemi, 1980) also make allowance for a theoretical annual eustatic rise in sea level of 0·8 mm, a value which dates back to the 1950s, when such a rise was assumed to be taking place. This cannot now be regarded as reliable, however, as research into variations in sea level suggests that areal differences occur in the extent of these variations, and may be related to changes in the surface of the geoid. We still have no idea of the extent of these changes in the surface of the geoid within Finland, and thus the best means of indicating present-day land uplift is in relation to sea level,* because calculation in absolute terms, with reference to the centre of the earth, remains impossible.

The long-term course of land uplift, covering many thousands of years, has been determined principally from the results of research into ancient shorelines and past water levels. During certain periods of time, however, the Baltic waters have been at levels higher than that of the ocean outside. To all this we must then add the possibility of changes in the form of the geoid, the order of magnitude of which remains equally uncertain in the long-term as in the short-term.

Dating accuracy is a further factor involved in the estimation of past rates of land uplift. If there are errors in the dating of ancient shore levels, these will immediately affect the calculated land uplift values. Nevertheless, even though ^{14}C dates may often be erroneous, so many dates have now accumulated from Finland that it is possible to formulate at least a general picture of the course of shoreline displacement during the Holocene. This suggests that land uplift, and thus the lowering of relative sea level, was extremely rapid immediately following deglaciation and slowed considerably thereafter.

*The lack of appreciable tides in the Baltic Sea means that sea level in a locality is relatively close to mean sea level. In this account, altitudes are referred to in terms of sea level in this context.

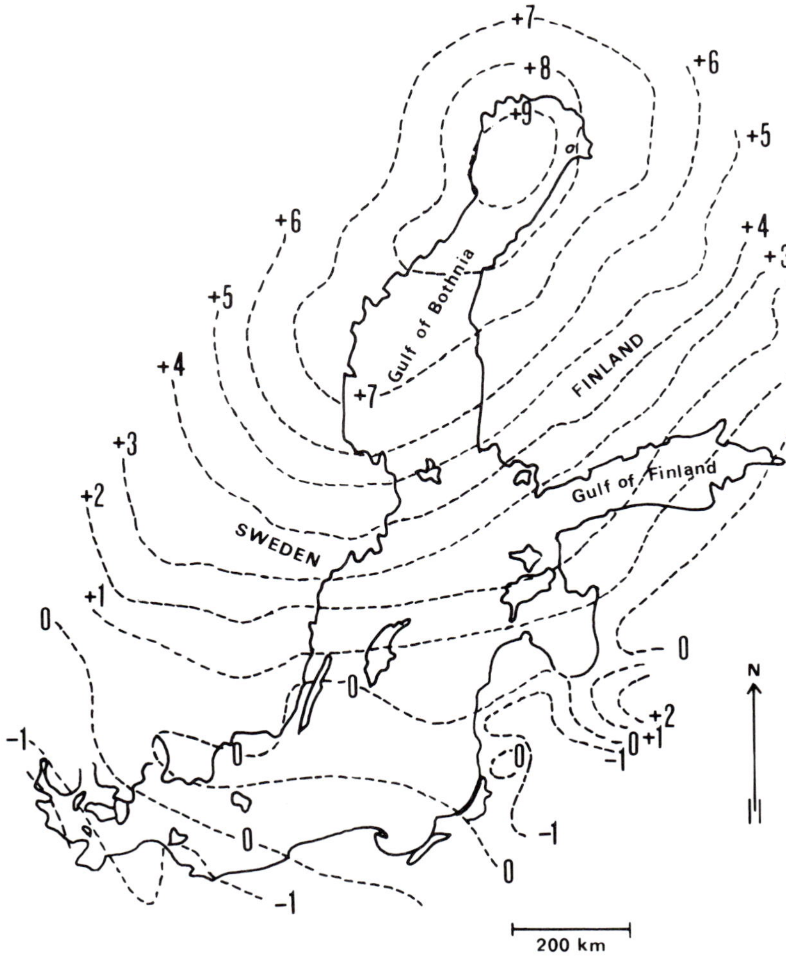

Figure 1. Present-day relative crustal movements in the Baltic area. Land uplift and submergence values (in mm/yr) are given in relation to sea level (cf. Fig. 14). The isobases are derived from precise levellings and tide-gauge data (after Flodén and Winterhalter, 1981).

Although it may be difficult to determine precise values for land uplift rates, it has been possible to achieve some measurements within certain limits, and even when the various error factors inherent in the calculation of "absolute" values are taken into account, it remains clear that land uplift in Fennoscandia is glacio-isostatic in nature, and that the inclusion of any other components in the calculations would, at least at this stage, seem to be no more than speculation.

DEGLACIATION AND THE POSITION OF THE HIGHEST SHORELINE

The southernmost areas of Finland were initially deglaciated during the Late Weichselian and then were submerged by waters of the Baltic. Thereafter, a pause in ice retreat took place during the Younger Dryas with the ice margin located in the area where the two arcs of end moraines known as the Salpausselkä ridges are now found (Fig. 2). These two ridges, the outer Salpausselkä I and the inner Salpausselkä II formations, are generally 10–20 km apart along their lengths and this represents the distance of ice retreat during the period *c.* 11 200–10 200 B.P., including phases during which the ice margin was stationary at the points represented by the end moraines (see Donner, 1978).

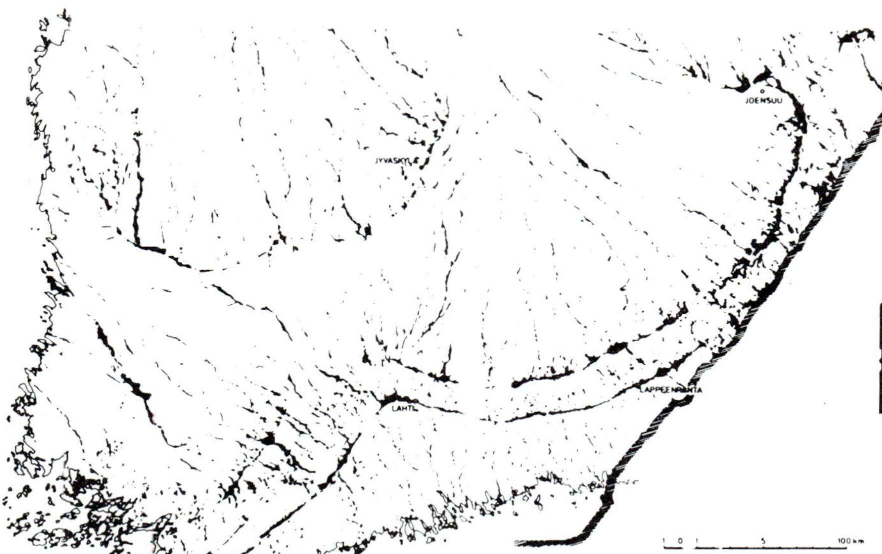

Figure 2. This map of gravel and sand resources in southern Finland shows the arc-shaped form of the ice-marginal features. The Salpausselkä I, II and III moraines can be seen in the south-west, whilst Salpausselkä I and II form long arcs that extend into eastern Finland (after Niemelä, 1979; Ignatius *et al.*, 1980).

At the time when the bulk of the Salpausselkä moraines were being deposited the southern parts of the Baltic basin were covered by the waters of the Baltic Ice Lake, which were at a level higher than that of the ocean outside. These waters were then released as the ice retreated from Salpausselkä II and the water level fell. Thus the highest water level in the Salpausselkä zone is attributable to the various phases of this Baltic Ice Lake (including the so-called g-phase). These are discussed in more detail below.

The highest shoreline in the area inside the Salpausselkä zone was produced during the Yoldia stage, which immediately followed drainage of the Baltic Ice Lake. This highest level becomes progressively younger in age north-westwards from the Salpausselkä zone, since deglaciation proceeded in this direction. The whole area of Finland was free of ice by around 9000 B.P. (see Ignatius *et al.*, 1980).

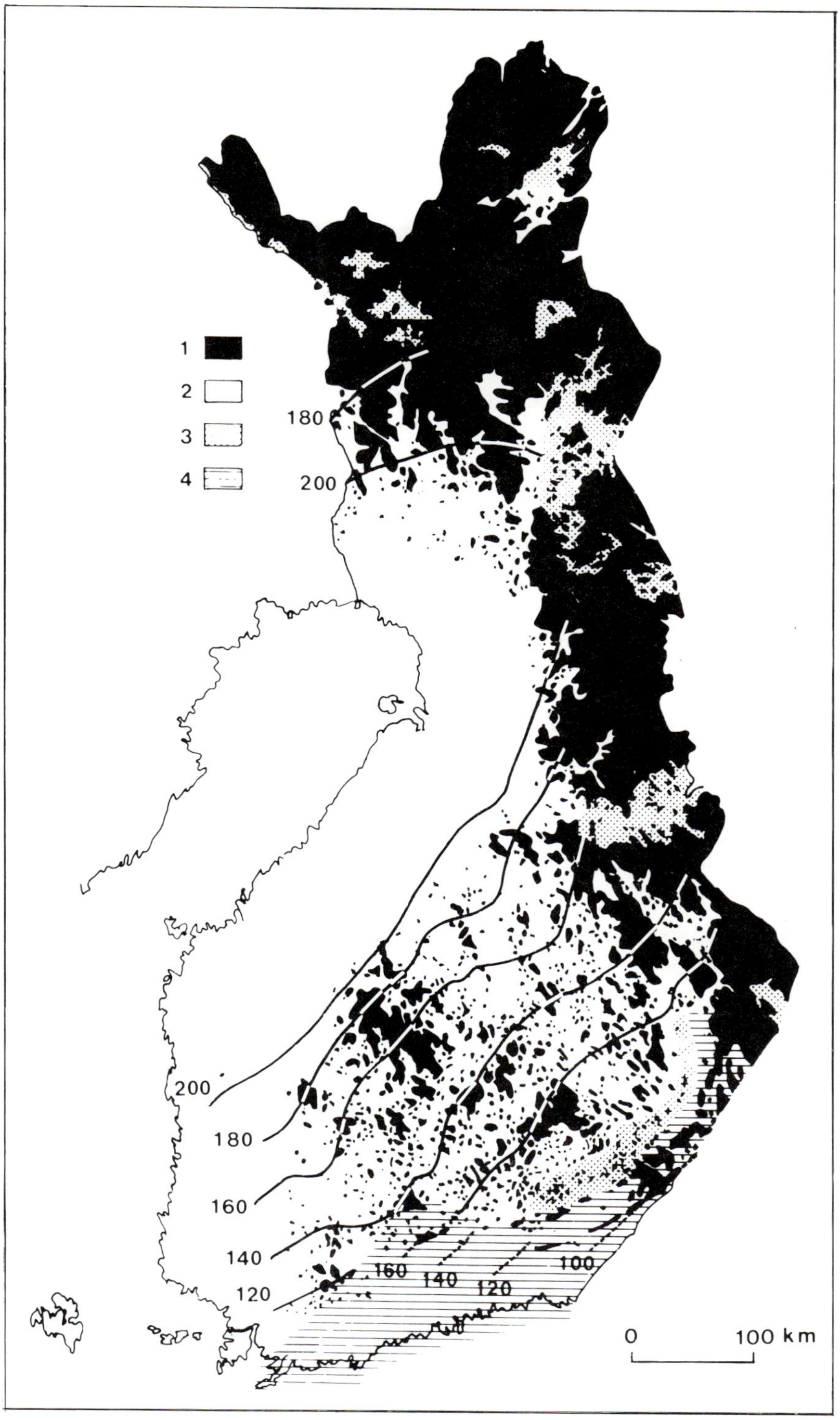

Figure 3. Highest shoreline of the Baltic in Finland. 1, supra-aquatic areas; 2, subaquatic areas; 3, areas covered by ice-dammed lakes; 4, areas covered by the Baltic Ice Lake. Solid lines are isobases for the metachronous highest shoreline features formed during the Yoldia and Ancylus Lake stages, and dotted lines those for the highest shore level of the Baltic Ice Lake on Salpausselkä I, i.e. the B I Level (after Eronen and Haila, 1981). Map compiled for the Atlas of Finland and published by permission of the Editorial Board.

Around 9500 B.P. the history of the Baltic entered a new stage, that of the Ancylus Lake, and consequently by the time the northernmost parts of the basin were emerging from the ice, the low-salinity Yoldia stage had been replaced by the freshwater Ancylus Lake as shown by diatom analyses (Eronen, 1976). Thus the highest shoreline on the coastal fringe of the northern Gulf of Bothnia was produced along the margin of the Ancylus Lake.

As can be appreciated from the pattern of deglaciation, the highest shoreline differs markedly in age from one part of Finland to another. Moreover, each raised shoreline is glacio-isostatically tilted. Because the latter phenomenon was most pronounced in the region of the Gulf of Bothnia, the uppermost shoreline in this area is also the highest in Finland, in spite of the fact that deglaciation took place later here than elsewhere in Finland. The highest marine limit in Finland occurs at *c.* 220 m above sea level (as may be inferred from the isobases in Fig. 3) and can be compared with a level of 290 m in Sweden (Lundqvist, 1961). It might in theory be possible to find ancient shorelines at this level or even slightly above it in Finland, but the terrain on the present coast of the Gulf of Bothnia is so flat that only young, low-level shores are represented there.

ANCIENT LATE WEICHSELIAN SHORE LEVELS

All the Late Weichselian shorelines in Finland are located in the Salpausselkä zone, the area to the south of this being so low-lying that it was entirely submerged during this period, while the country to the north-west and north of these end moraines was not deglaciated until the Holocene.

Salpausselkä ridges I and II form a huge arc across south-eastern Finland, with their westernmost parts curving to form relatively straight ridges oriented NE–SW. An additional ridge, Salpausselkä III, is found on their proximal side, in south-western Finland. The Salpausselkä III moraine was probably deposited at the very beginning of the Holocene (Aartolahti, 1972; Donner, 1978), and cannot be linked with a clearly demonstrable cold climatic phase in the same way as the other Salpausselkä ridges.

The Salpausselkä moraines are composed principally of glaciofluvial gravels and sands although locally their proximal slopes are composed of till. The gravels and sands represent former proglacial sediments and form parts of extensive deltas and sandur-delta formations at altitudes corresponding to the water levels that prevailed during the Salpausselkä stages (Fig. 4). It is these primary delta surfaces occurring at various levels in the Salpausselkä zone which have enabled fluctuations in water level in the Baltic basin during the Late Weichselian to be determined. The altitudes (above sea level) of the proximal parts of the deltas are believed to correspond to the ancient water levels at the ice margin.

The ancient shorelines of the Salpausselkä zone as a whole form a highly complex system, since land uplift has differed in magnitude in different parts of the moraines, and thus contemporaneous shoreline fragments now exist at different

Figure 4. Several airports have been constructed in southern Finland on the extensive flat glaciofluvial delta plateaux of the Salpausselkä formations. One of them is that of Nummela, shown here, built on the "g plateau", at 112 m a.s.l. Photograph by M. Eronen, 1981.

altitudes (Fig. 5). Thus it is extremely difficult to reconstruct the various phases involved, especially since it is hard to date the shoreline surfaces accurately. The main methods used for dating are varve chronology and pollen analysis (Donner, 1978), and radiocarbon dates have been obtained only in certain exceptional cases (e.g. Hyvärinen, 1973), since one does not usually find any organic deposits which are contemporaneous in origin with the deltas. Thus the determination of the fluctuations in water level that took place at the time of deposition of the Salpausselkä moraines has been based principally on deductions made from the relative positions and altitudes (above sea level) of the glaciofluvial deltas, with the inevitable outcome that a wide variety of interpretations have been reached. Divergent opinions on these fluctuations are constantly being put forward, backed up by evidence which is to a greater or lesser degree convincing.

At the time when the Salpausselkä formations were created, the Baltic basin consisted of a vast lake, the Baltic Ice Lake (Fig. 6), which bordered onto the continental ice sheet to the north. Numerous works by Sauramo (e.g. Sauramo, 1958) were devoted to the various phases in the history of this lake. Later, Okko (1962) proposed a model which departed markedly from that of Sauramo, and this was then altered somewhat by Donner (1969, 1978) in order to explain the altitudes of the various delta surfaces. Recently, however, new models have been put forward (Synge, 1980, 1982; Saarnisto, 1982) which deviate from the ideas of Donner, and investigations into the same phases in Sweden (Björck, 1979) have similarly led to the proposal of certain revisions to the model presented by Donner (1978). It is not possible to go into the details of all these shore-level systems here, but it is worth considering briefly Donner's model (largely as presented in Donner, 1978), which is simple, but adequate for explaining the field data.

Two primary delta levels are recognized on Salpausselkä I. These are normally referred to by the abbreviations g and B I (see Fig. 5), where the former is derived

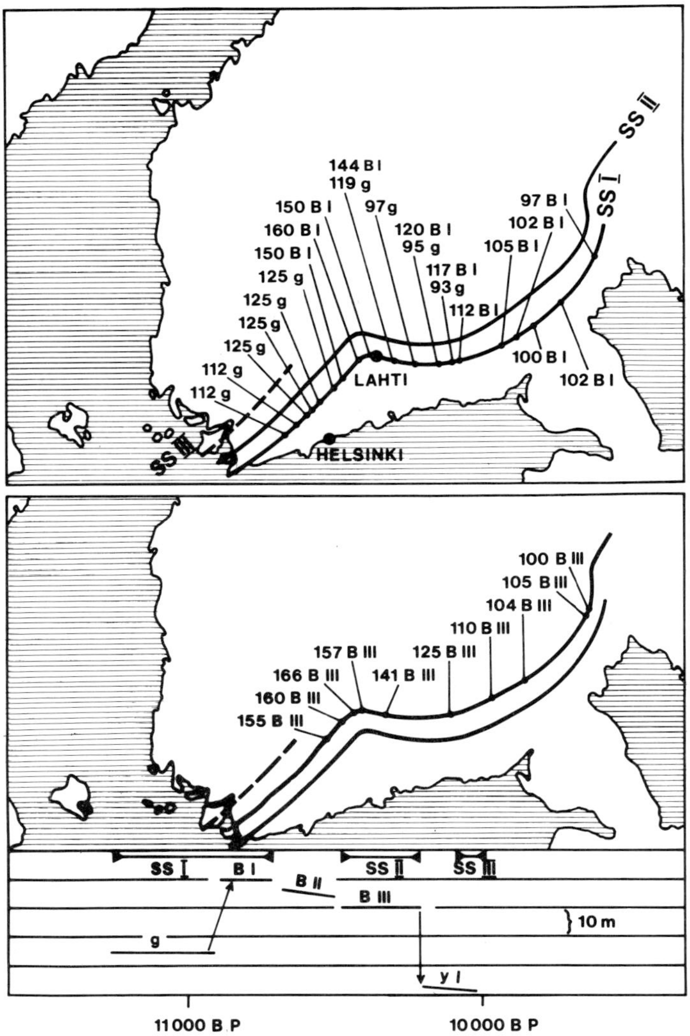

Figure 5. Altitudes (m above sea level) of the major glaciofluvial delta plateaux in the Salpausselkä zone. g, B I and B III are abbreviations for Late Weichselian levels of the Baltic and yI refers to the highest Yoldia stage shoreline (see text). The lower part of the figure gives the relative changes in level during the formation of the Salpausselkä ridges. Redrawn from Donner (1969, 1976, 1978).

from the shore-line nomenclature once employed by V. Tanner on the Arctic Ocean coast, and the latter alludes to the first phase of the Baltic Ice Lake, which is to be seen in the form of a delta level high up on Salpausselkä I. The g level is assumed to have been produced at a time when the Baltic basin was connected with the ocean by some route, which it has nevertheless not proved possible to map, and it has not yet been possible to correlate this level with any particular ocean level. The g level is a low delta level, lying 25 m below the B I deltas on Salpausselkä I, and according to Donner it is the oldest of the Salpausselkä delta surfaces. The water surface is

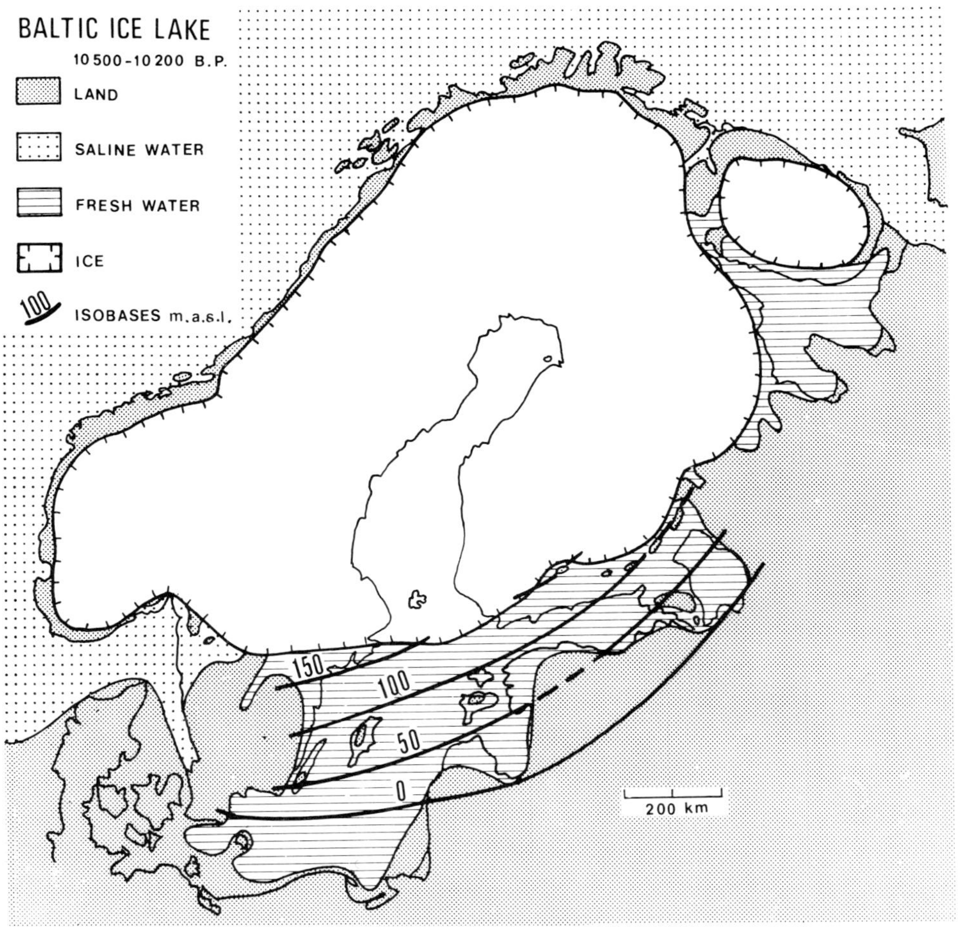

Figure 6. The Baltic Ice Lake. Map compiled by M. Eronen and H. Haila for the Atlas of Finland and published by permission of the Editorial Board.

assumed to have then risen by 25 m before the deposition of the B I deltas. The generally accepted view is that this latter level represents the surface of an ice lake whose waters have been held back at a level above that of the outside ocean. According to this scheme the rise in water level did not destroy the morphological features of the g level. At some places the surfaces of the g deltas are deformed, fossil ice-wedges being located beneath a sand layer (Aartolahti, 1970). B I surfaces are found only on Salpausselkä I, since Salpausselkä II did not yet exist owing to the presence of glacier ice.

The Ice Lake phase lasted uninterrupted until the point at which the Salpausselkä II moraines were deposited. In the terminology of Donner (1969, 1978), phase B II consists of those levels which were established in the intervening area between the Salpausselkäs as the ice was retreating from Salpausselkä I to Salpausselkä II, while level B III (see Fig. 3) corresponds to the phase at which it was stationary at Salpausselkä II. This is seen on the latter in the form of numerous extensive delta plateaux laid down at this level, and is also a clearly distinguishable ancient shoreline on Salpausselkä I, represented by a terrace eroded into the slope and located 10 m below the B I level (e.g. at Lahti).

The altitudes of the delta surfaces on the Salpausselkäs and a scheme for their arrangement according to altitude and the sequence of the events is given in Fig. 5. In summary, it may be said that many open questions still remain concerning the fluctuations in water level in the Baltic Ice Lake. In particular, it is not known how great a part was played in these fluctuations by land uplift (or possibly even changes in the form of the geoid) and what part is represented by variations in the water level of the lake, caused either by changes in meltwater influx or by changes occurring at the outflow threshold.

In all events, the Baltic Ice Lake stage came to an end with a drop in water level, which took place just as the ice was beginning to retreat from Salpausselkä II. The key area for the interpretation of this last phase is the hill of Billingen in Central Sweden, since it is thought that the waters of the lake were able to escape into the ocean as the ice margin retreated north of this hill. The only problem is that there are no obvious signs in the area of such a mighty outflow having taken place (Strömberg, 1977). In other words, the question of exactly how and why this drop in water level took place is something of an enigma. Nevertheless, there is clear evidence of such a drop in the level of the Baltic Ice Lake in Finland, with a whole set of double deltas visible on the proximal side of Salpausslkä II, the upper delta being deposited at the B III level and the lower one on the proximal side of this and *c.* 28 m lower (Donner, 1969, 1978; Glückert, 1977). There is thus strong evidence in favour of such a drop, but as yet no agreement as to how this happened, whether as a sudden catastrophic outpouring or as a gradual process spread over several decades or perhaps centuries.

It has been convincingly demonstrated, however, that the water level in the Baltic basin fell to the level of the ocean outside as a consquence of this event, marking the beginning of the Yoldia stage, as represented by the lower delta surface encountered on Salpausselkä II, which forms the uppermost Yoldia level, Y I.

THE YOLDIA STAGE AND ANCYLUS LAKE

The fall in the water level of the Baltic Ice Lake has been dated at *c.* 10 200 B.P. (Donner, 1978). Although an apparently exact date of 8213 B.C. exists for this event (Nilsson, 1968), based on varve chronology, the radiocarbon evidence suggests that it happened about 200 years before the date agreed for the beginning of the Holocene (Donner, 1978). Thus the onset of the Yoldia stage occurred at the very end of the Late Weichselian period.

Once we turn to examine shoreline displacement during the Holocene we find that there are rather more ^{14}C dates from which to work, since such dates can be obtained from mire and lake sediments in basins which have become separated from the Baltic itself at various stages. Even so, it is not possible yet to formulate a precise, detailed account of the events connected with shoreline displacement, since it appears that such dates frequently involve errors arising from a variety of sources. This can easily be appreciated when comparing the dating results with the pollen chronology or when investigating cases of conflicting dates. A general impression of the trend in shoreline displacement can nevertheless be formed under such circumstances by obtaining additional dates and by trying to eliminate the sources of error in various ways.

Eventually a more or less consistent picture has emerged for shoreline displacement events during the Holocene on the basis of stratigraphical research, in which the Yoldia stage is one of rapid regression, while the subsequent Ancylus Lake stage features a transgression that led to the renewed inundation of the land laid bare by marine regression in southern Finland.

The connection between the Baltic basin and the ocean through Central Sweden which marked the beginning of the Yoldia stage (Fig. 7) naturally also constituted

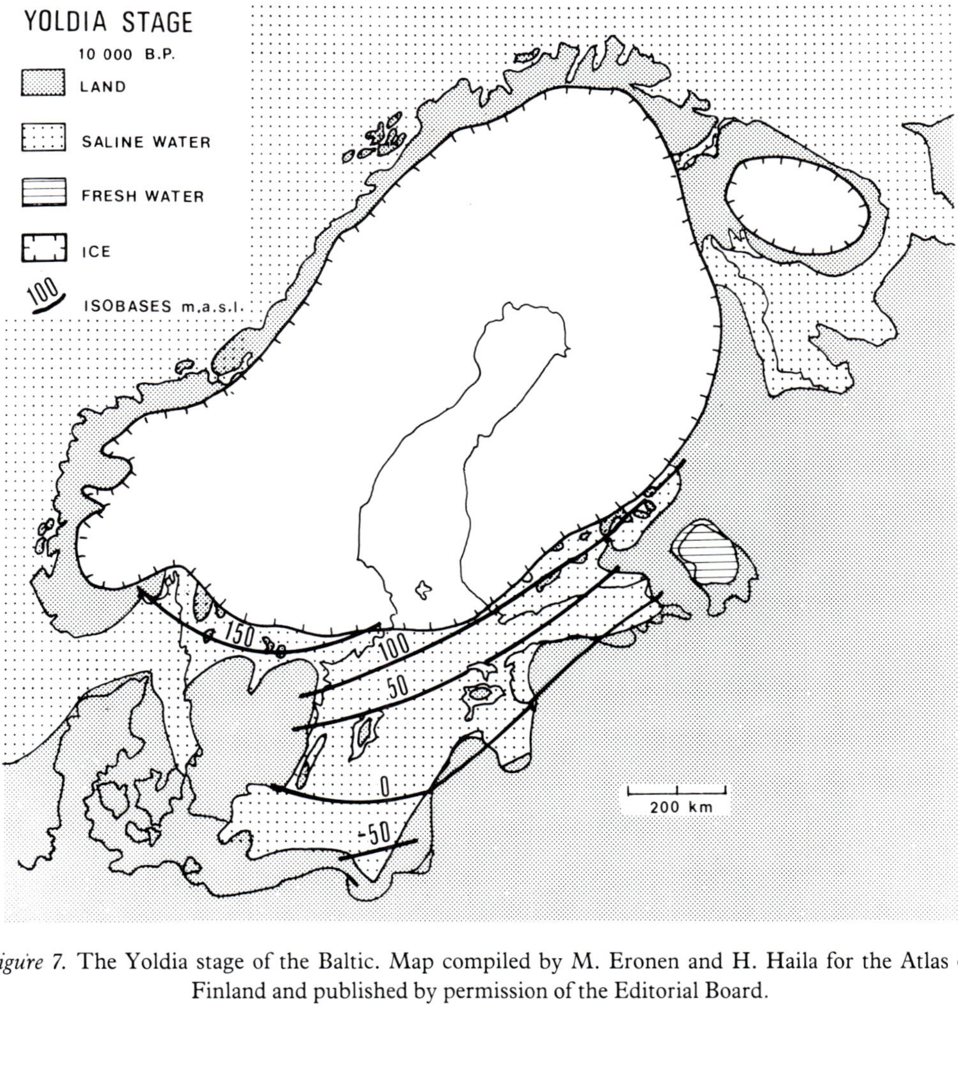

Figure 7. The Yoldia stage of the Baltic. Map compiled by M. Eronen and H. Haila for the Atlas of Finland and published by permission of the Editorial Board.

a channel through which salt water was able to pass into the Baltic (see Agrell, 1979). The fact that this salt water extended as far as the Finnish coast is indicated by diatom studies of Yoldia stage sediments which show taxa indicative of brackish water conditions (see Eronen and Haila, 1982). It is none the less probable that this "saline" zone was restricted to the south of the country, and that it would scarcely have extended far beyond the Salpausselkä moraines. This is hard to verify, however, on account of the presence of redeposited diatoms, quite probably in relatively large quantities, in the clays which accumulated close to the ice margin, and especially in Ostrobothnia, since many of these are evidently derived from interstadial or interglacial sediments laid down in saline water (see Niemelä and Tynni, 1979). The redeposited diatoms, originating from old clays and having been mixed with fine sediments during deglaciation, greatly complicate the drawing of any conclusions from clays dating from the beginning of the Holocene (see Eronen, 1974).

It should be noted that "Yoldia clays" containing salt-water diatoms have been found in the northern parts of the Baltic basin (Fromm, 1965; Eronen, 1974), even though this area emerged from the ice only some time after 9500 B.P., by which stage the connection with the ocean through Central Sweden could no longer be utilized owing to glacio-isostatic land uplift (Agrell, 1979; Fredén, 1979). Thus the diatoms attributed to the early Holocene in the northern Baltic would not seem to reflect true brackish water conditions.

It is difficult to draw a precise boundary between the Yoldia stage and the Ancylus Lake stage, since although it is customary in diatom analysis to regard as "Ancylus sediments" any clays in which one finds large numbers of freshwater large-lake diatoms and no brackish water ones, even the "Yoldia clays" themselves only contain a very small percentage of brackish water taxa (Eronen, 1974; Alhonen, 1979). This boundary is an indeterminate one not only in the south of the country, but also in the north, where the change described above implies nothing more than the almost complete disappearance of redeposited salt-water diatoms from the flora. Thus the "Yoldia diatoms" disappear and only freshwater "Ancylus diatoms" are left. This change is due to the retreat of the ice margin, since with the ice front more distant, redeposition disappears.

By far the most reliable way of defining the beginning of the Ancylus Lake stage (Figs 8, 9) is by reference to the onset of the so-called Ancylus transgression, during which water levels rose over extensive areas of southern Finland. Sediments associated with this event may be dated by pollen analysis and ^{14}C methods.

The ^{14}C dates obtained for the indicators of the Ancylus transgression in southern Finland, i.e. peat and lake sediment horizons buried beneath clays, have been very variable (Donner and Eronen, 1981; Eronen and Haila, 1982; Glückert and Ristaniemi, 1982), but in general terms the transgression may be said to have begun *c.* 9500 B.P. and to have culminated in the south of the country around 9000 B.P. It was most probably of greater amplitude in the south-east than farther west, since land uplift was less pronounced there, and correspondingly, the present altitudes of its shoreline fragments are greater towards the west (see Fig. 8). Thus transgression

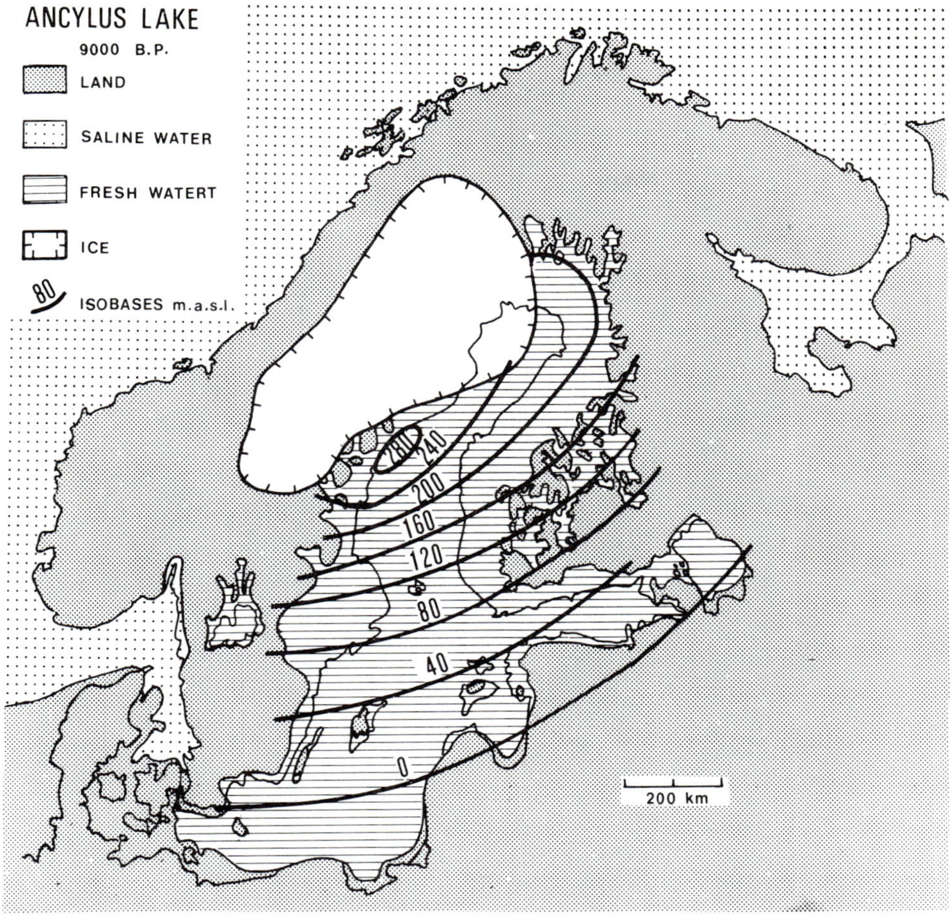

Figure 8. The Ancylus Lake. Map compiled by M. Eronen and H. Haila for the Atlas of Finland and published by permission of the Editorial Board.

Figure 9. Sandy beach ridges from the Ancylus Lake located one below the other on a slope in the Rokuanvaara hills in Ostrobothnia. The ridges are covered by forest, while the depressions between them are occupied by shallow mires. Photograph by M. Punkari, 1980.

horizons are still encountered at levels above 70 m above sea level in western Uusimaa, where they indicate an amplitude of a few metres, whilst farther east the amplitude may well have been as much as 10 m (Fig. 10), but this has not yet been studied in detail. The amplitude in the southern Baltic was probably in excess of 40 m in places (Agrell, 1979; see also Eronen, 1976; Kolp, 1979).

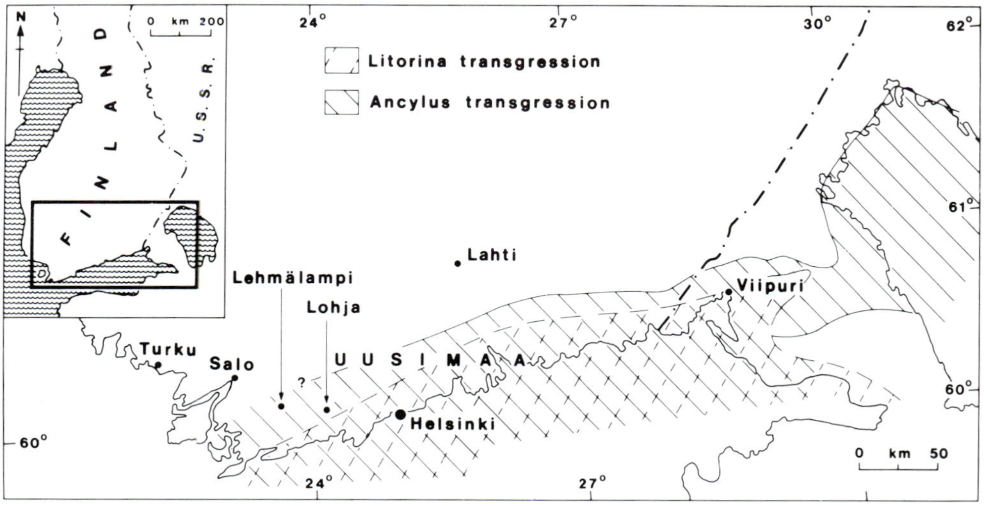

Figure 10. Approximate extent of the Ancylus and Litorina transgressions in southern Finland. Indicators of the Ancylus transgression have been found in Lake Lehmälampi (71 m above sea level) (after Glückert and Ristaniemi, 1980).

A pronounced regression commenced after 9000 B.P., causing the water level in the Helsinki area, for example, to drop some 30 m in the interval 9000–8500 B.P. Calculated on the 34 m isobase for the Litorina Sea, this amounted to a drop from a level of 64 to 34 m (Eronen and Haila, 1982). However, the causes of the Ancylus transgression and the subsequent major regression remain unresolved.

One putative explanation for the Ancylus transgression is that the outflow channel in Central Sweden ceased to function. The early theory was that a great waterfall had developed at that point, allowing the waters from the great lake of the Baltic to pass into the sea through the area of the present Lake Vänern. Fredén (1967, 1979) nevertheless showed that the only feasible site for this outlet channel could never have seen any very rapid discharge of water since the till cover was not eroded. Thus the channel must have been abandoned so quickly that no proper river bed was able to form. What, then, is the explanation for the transgression seen even at isobases of over 70 m? It is possible, of course, that the whole of the Baltic remained without any outflow channel for a short time until a new route, probably through Öresund, opened up in the south-west once the water level had risen sufficiently as a result of the pronounced transgression (Eronen, 1976). There are still many open questions concerning the history of the Ancylus Lake, however, and these can only be solved as a result of detailed research. A great deal of work has been undertaken recently in Finland, for example, to determine the extent of the Ancylus transgression and the present altitudes above sea level of the resulting deposits (Eronen and Haila, 1982; Glückert and Ristaniemi, 1980, 1982).

The cause of the Ancylus regression may be assumed to be the opening up of a new outflow channel, although the rôle of the deglaciation which was still going on in the northern Baltic at this time should not be forgotten. The Gulf of Bothnia region was still covered by ice around 9500 B.P., for instance (Hyvärinen, 1973), and there was still a large ice mass in northern Sweden in 9000 B.P. Thus the last of the ice had apparently melted by about 8500 B.P. (Lundqvist, 1961). The Ancylus regression therefore coincides with the final disappearance of the ice mass, an event which must have caused major alterations in the hydrological balance of the Baltic basin with the cessation in the influx of meltwater.

As may be seen from the above, shoreline displacement during the Ancylus Lake period is awkward to explain and there are still several critical pieces of evidence missing from the puzzle. One might propose that changes in the form of the geoid could have played some part in events at this stage, for example, this being all the more likely on account of the rapid rate of land uplift at the time (or at least a rapid drop in water levels in the northern Baltic), which may have been associated with movements in the earth's crust. This is all very much a matter for speculation as yet, however.

THE MASTOGLOIA PHASE AND EARLY LITORINA SEA STAGE

Following the rapid progress of the Ancylus regression, the general lowering of the shoreline slowed down during the interval 8500–8000 B.P., presumably as a result of a decline in the rate of land uplift. This period also saw the opening up of a new outflow channel into the ocean, allowing the eustatic rise of ocean levels occasioned by world-wide melting of ice to reach the level of the Baltic basin and slow down the process of shoreline retreat even in those areas with a high rate of land uplift.

This connection was established, therefore, as a result of the ocean level rising above the threshold points in the Straits of Denmark. Relatively little detailed research has been published on the question of exactly when this took place (cf. Krog, 1979), but it is probable that the connection was created around 8500–8000 B.P. This now allowed salt water to penetrate into the Baltic basin, as reflected in diatom analyses by the fact that some brackish water species begin to appear alongside the freshwater large-lake "Ancylus" flora. The first of these indicators to appear are members of the genus *Mastogloia*, and it is from these that the transition phase between the Ancylus Lake and the Litorina Sea* comes to be known as the Mastogloia phase (Eronen, 1974; Donner and Eronen, 1981).

Not all investigators have been prepared to recognize the Mastogloia phase as a separate entity in the history of the Baltic, however, and it has been variously

*In this chapter, Litorina is spelt in accordance with practice in Finland.

included within either the Ancylus Lake or Litorina Sea stages. It is nevertheless an important phase in that it is indicative of the existence of a connection with the sea, and because the beginning of the Litorina Sea is an extremely clearly-defined boundary, the distinguishing of a separate Mastogloia phase is in many respects essential to maintain the clarity of the system. On the other hand, it is true that very similar clays were deposited in the Baltic basin at this time to those of the Ancylus Lake, so that it is not even possible to distinguish this phase sedimentologically, nor is it apparent in every biostratigraphical profile. It can be detected repeatedly, however, in studies where the material is sufficiently broad-based (Eronen, 1974).

The beginning of the Litorina Sea stage (Fig. 11) is marked by the most clearly distinguishable of all the stratigraphical boundaries in the history of the Baltic. Similarly, the highest Litorina shoreline is clearly defined, being found at the highest level at which obvious brackish water sediments are encountered. In general two very clear limits related to the shoreline displacement history may be distinguished

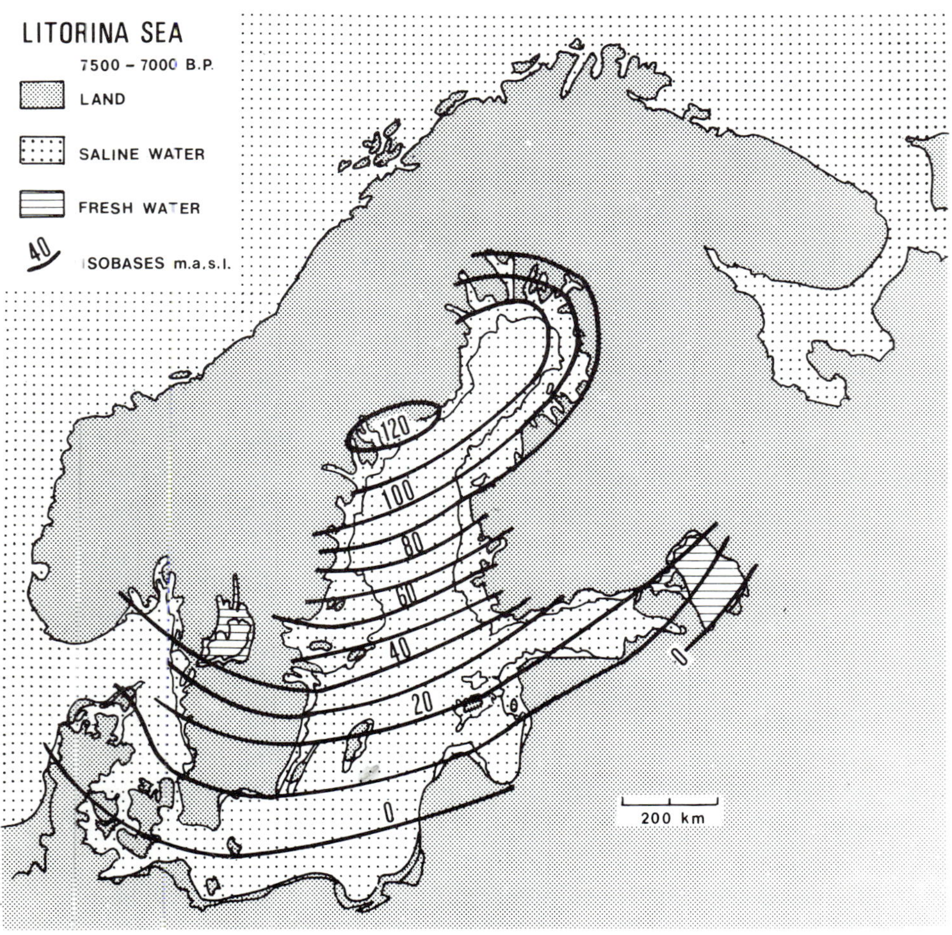

Figure 11. The Litorina Sea. Map compiled by M. Eronen and H. Haila for the Atlas of Finland and published by permission of the Editorial Board.

in Finland. The highest shoreline is represented by a variety of raised shoreline landforms, and below this one finds a range of ancient shorelines at various altitudes, bearing witness to a continuous regressive trend in areas of rapid land uplift.

The beginning of the Litorina Sea period may be seen in littoral sediments in the form of the "Clypeus" limit, defined from diatom analyses. The name is derived from the species *Campylodiscus clypeus*, which is an extremely common one in brackish-water sediments, especially where these have accumulated in shallow bays. The term "Clypeus flora" is then used to describe diatom assemblages of the kind that accumulate under such conditions. This Clypeus limit can also be interpreted as the upper limit of the Litorina Sea when examined in the field, denoting in effect the uppermost point at which a Clypeus flora is to be found. This limit lies at approximately 30 m above sea level in the Helsinki district and at over 100 m above sea level in the area north and north-east of the Gulf of Bothnia.

Quite a number of ^{14}C dates are available from Finland for the beginning of the Litorina Sea stage (Eronen, 1974; Alhonen *et al.*, 1978; Hyvärinen, 1980, 1982; Eronen *et al.*, 1982), dates of around 7500-7300 B.P. having been obtained in the south and just one date so far, of approximately 7000 B.P., for the Gulf of Bothnia region in the north. This would suggest that the brackish water spread to the northernmost parts of the Baltic basin somewhat later than to the coast of southern Finland (Eronen, 1974).

The highest shoreline of the Litorina Sea varies slightly in age from one part of the country to another in any case, being possibly a little younger in the north than in the south because of the age difference mentioned above, and somewhat younger in the south-east than in the south-west. This latter difference is due to the fact that the highest Litorina shoreline in the south-west is that which represents the upper limit of the Litorina transgression, which postdates the actual beginning of the Litorina stage. The difference is nevertheless only a matter of a few hundred years, and certainly not more than 1000 years, since the transgression had given way to a regression even on the south-east coast of Finland by 6500 B.P.

THE LITORINA TRANSGRESSION AND SUBSEQUENT SHORELINE DISPLACEMENT

Numerous papers have been published on the Litorina transgression in Finland from the early part of this century onwards (see Eronen, 1974; Hyvärinen and Eronen, 1979), and detailed stratigraphical research has been carried out in recent years by Hyvärinen (1980, 1982) to determine its extent. In his interpretation, the 34 m isobase for the Litorina Sea denotes the limit below which the sea was still transgressive, while at this point the rise in water level meant only a pause in the progress of shoreline retreat. Since the 34 m isobase is located close to Helsinki in southern Finland, the Litorina transgression is limited to the south-east coast, where evidence for it has been found in many mires and lakes (Eronen, 1974).

No very exact estimates have been obtained as yet for the amplitude of this transgression, although within Finland it would seem to have amounted to a rise in water level of a few metres. It also remains somewhat unclear as to when this rise began. One may assume that it began during the Mastogloia phase, at the time when the ocean connection was established, but there is no proof for this. This is partly on account of the fact that the Mastogloia diatom flora does not differ so very markedly from a freshwater flora, whereas the Clypeus flora of the Litorina Sea is clearly distinct from a lake flora. Thus it is easy to see from diatom analyses when a basin which had become isolated as a separate lake due to land uplift had been temporarily united with the waters of the Baltic again on account of the Litorina transgression.

As noted above, the transgression which began at the commencement of the Litorina Sea stage, or possibly during the Mastogloia phase, came to an end on the coast of Finland by 6500 B.P., possibly doing so a few hundred years earlier in the areas of more rapid land uplift than it did farther towards the south-east (Eronen, 1974). The water level in the Helsinki area remained virtually stationary throughout the period 8500–6500 B.P. (Hyvärinen, 1980, 1982).

A general regressive trend commenced on the coast of Finland after 6500 B.P., and with the cessation of the eustatic rise in ocean levels by 6000–5000 B.P. (Clark and Lingle, 1979), shoreline displacement even in the south of the country became largely governed by land uplift. In the north, of course, land uplift had been pronounced throughout, and no transgression occurred. A great deal of discussion has centred around the question of whether the Litorina Sea stage includes any other transgressions. Although the existing evidence was interpreted in Eronen (1974) as implying that no transgression has occurred since the major one at the beginning of the Litorina Sea period, a number of Swedish scholars believe that a succession of small transgressions can be detected (Berglund, 1964, 1971; Digerfeldt, 1975; Liljegren, 1982; Miller, 1982; see also Agrell, 1979). Similarly the conclusion has been reached on the southern shore of the Baltic that a number of fluctuations in water level must have taken place (Klug, 1980). At least some of these deposits interpreted as indicators of a transgression may be regarded as sediment changes or biostratigraphical variations attributable to other causes, and there may also have been local irregularities in land uplift which have given rise to restricted "transgressions", but it is also possible that there may have been some temporary fluctuations in water level of small amplitude which affected the whole Baltic basin. Further research would be needed to establish these with any certainty, however. It should be noted, incidentally, that attention has very rarely been paid to shoreline development in the later part of the Holocene (i.e. in the last 5000 years) in stratigraphical research carried out in Finland.

As far as shoreline development in the later part of the Holocene is concerned, it may be generally stated that a steady regressive trend has continued throughout, slowing down gradually towards the present day, although small fluctuations may have occurred within this trend. This decline in the regressive trend is obviously closely connected with the gradual retardation in land uplift. Comparison between the present rates of uplift and the water-level data for the early Litorina Sea stage, for instance, indicates that the rate of uplift has decreased considerably (Eronen *et al.*, 1982). The same can also be seen from shoreline displacement studies. Shoreline retreat as a consequence of land uplift has also brought about many changes on the coasts of Finland in historical times. Harbours have had to be abandoned, and whole towns have had to be moved closer to the receding shore (see Jones, 1977).

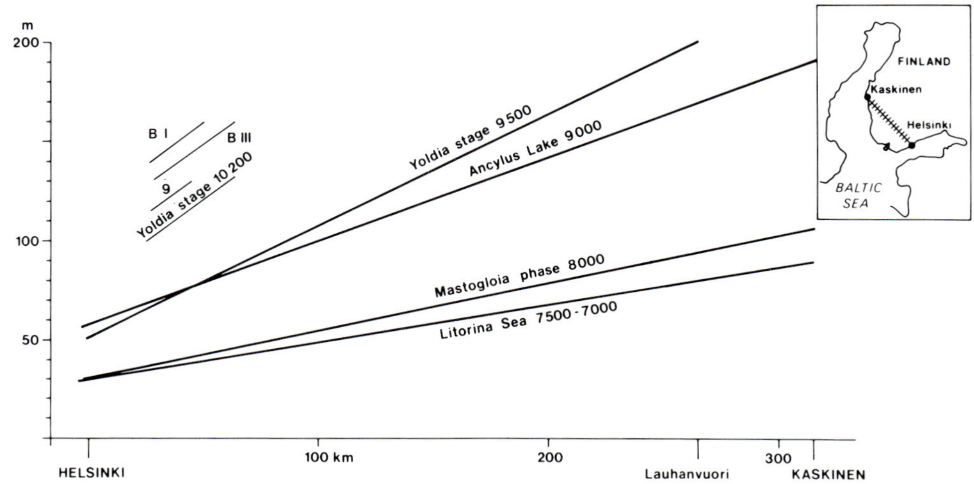

Figure 12. Tilted ancient shorelines along a profile crossing southern and western Finland.

Present day altitudinal relationships of the main shorelines of the events described above are shown in Fig. 12, where their altitudes are plotted along a line approximately normal to the isobases. This line is practically the same as used by Donner (1969, 1980). Curves for shoreline displacement in five areas are shown in Figs 13 and 14. Figure 13 shows shoreline displacement in the Porvoo-Askola area for 10 000–5000 B.P. The ^{14}C dates marked on the diagram refer to sites of biostratigraphical studies (Eronen, 1974; Eronen and Haila, and Eronen and Matiskainen, in preparation), detailed descriptions of which will be published separately. The map included in the picture shows that all the sites used for the diagram are not situated on exactly the same isobase of land uplift. Figure 14 shows shoreline displacement for four other areas of Finland, though, like in Fig. 13, the ^{14}C dates used are not all from exactly the same isobases as the curves. In this latter diagram, the land uplift values are not the same as in Fig. 1, since those used here allow for a constant rate of eustatic rise in sea level of 0·8 mm per year (Kääriäinen, 1953, 1966; Kiviniemi, 1980).

ARCHAEOLOGICAL EVIDENCE—COASTAL DWELLING SITES

Archaeological research in Finland has been traditionally very closely concerned with the question of shoreline displacement, insofar as the early prehistoric dwelling

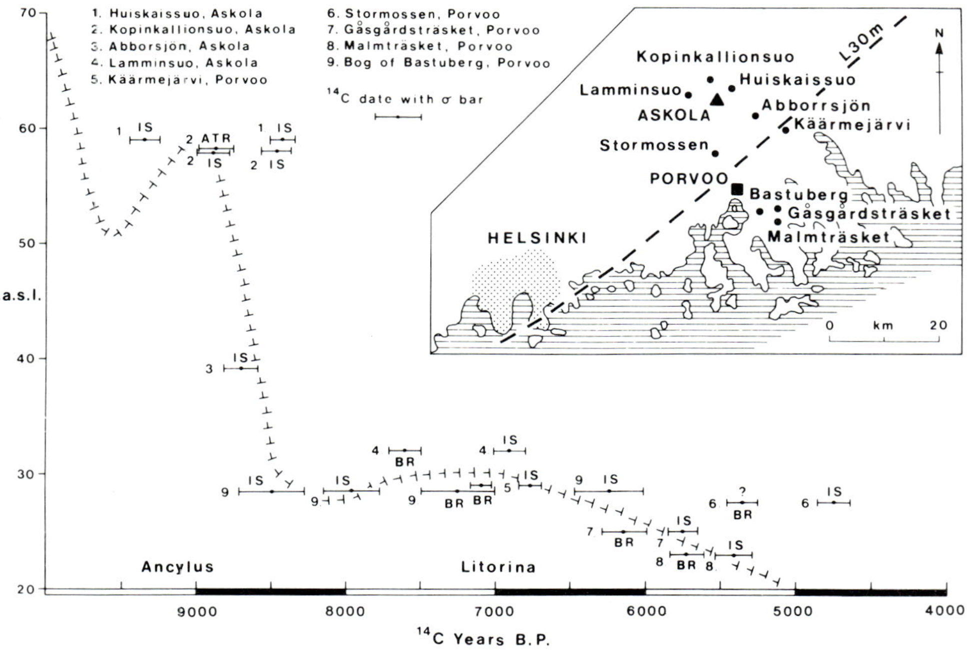

Figure 13. Curve for shoreline displacement in the Porvoo-Askola area around 10 000–5000 B.P. IS, isolation point; ATR, Ancylus transgression; BR, layer deposited in brackish water; L 30 m refers to the 30 m isobase of the Litorina Sea.

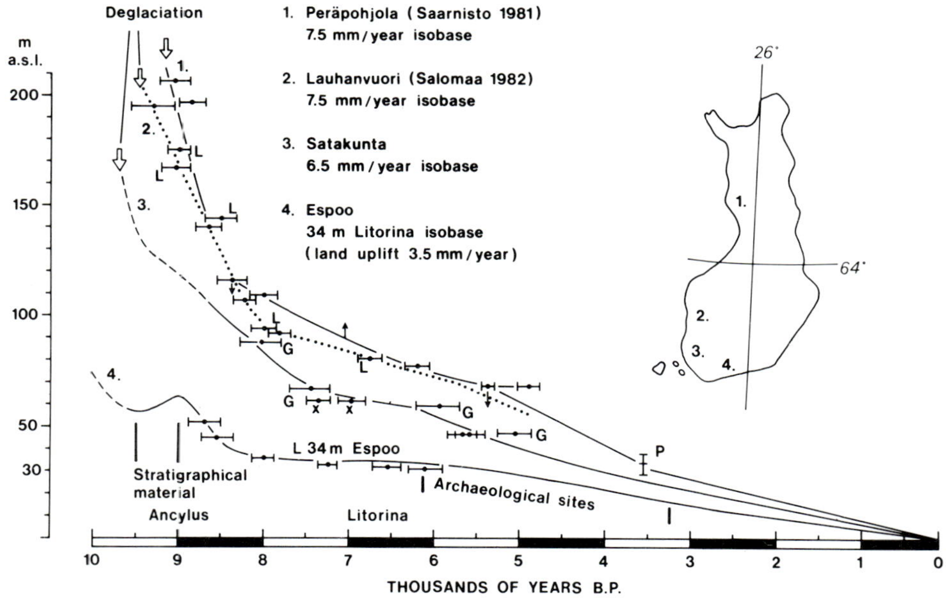

Figure 14. Shoreline displacement curves for four areas in Finland. The ^{14}C dates for curve 3 are from Eronen (1974) (X), Eronen *et al.* (in press) and Glückert (1976) (G). The older part of curve 4 is from Eronen and Haila (1982) and the younger part from Hyvärinen (1980). L refers to Lauhanvuori.

sites are intimately connected with ancient shorelines. The Stone Age population depended to a great extent on fishing for their livelihood, and therefore their dwelling sites were normally located close to the shore, and as the shoreline retreated over the centuries and millennia, the dwelling sites also moved to progressively lower levels. Thus it is relatively easy to place archaeological finds in order of age, simply according to the relation which their sites bear to the altitudes of the various phases of the Baltic in the locality in question. Just as the study of archaeological finds is related to the history of the Baltic in this way, so the archaeological material available has come to serve as a valuable tool in the history of the Baltic.

A population representing the Mesolithic Suomusjärvi culture inhabited Finland about 9000 B.P., and this was followed by the Neolithic Combed-Ware culture. Since these people were producers of clay vessels, different phases in their culture can be distinguished on the basis of the style of decoration used on these vessels. The boundary between the Suomusjärvi and Combed-Ware cultures has been dated to around 6200 B.P. (Siiriäinen, 1970, 1973).

The oldest known Suomusjärvi dwelling sites are associated with the history of the Ancylus Lake, while the later sites and those representing the various phases of the Combed-Ware culture are linked with the development of the Litorina Sea (see Siiriäinen, 1969, 1973, 1982; Eronen, 1974; Nunez, 1978). It has also been possible to study the history of the major lakes in the Lake Region of Finland using the determination of Stone Age dwelling sites as one of the principal methods (Saarnisto, 1970, 1971; Siiriäinen, 1970).

MORPHOLOGICAL OBSERVATIONS OF ANCIENT SHORELINES

The morphological identification of ancient shorelines occupied an important position in the early research into the history of the Baltic (see Hyvärinen and Eronen, 1979), but was later superseded in importance by stratigraphical investigations, since it was these that could be used for dating the events. Morphological shorelines have nevertheless continued to be used as one method for determining the phases of shoreline displacement.

Ancient shorelines are common features on the slopes of eskers, where the sandy soil has been most susceptible to the action of littoral forces. The extensive sand fields on the Salpausselkä moraines naturally form a subject of their own, but the Salpausselkä zone also contains large numbers of later shorelines in addition to the primary delta plains. The ancient shorelines are often found on eskers in the form of banks, or "shore niches", located one below the other, each having been formed as the water level descended.

Although attempts have been made in some works to combine raised shoreline fragments into more extensive reconstructed levels (Glückert, 1976), it has been pointed out by Donner (1964, 1980) that the majority are scattered irregularly and, owing to glacio-isostatic uplift, occur at different altitudes. Thus, individual shorelines are difficult to correlate. There are nevertheless certain altitude zones at which more shoreline fragments than usual are to be found, and it is these in southern Finland which represent the transgressions. Thus a clear concentration of ancient shoreline fragments occurs around the upper Litorina limit in south-western Finland, due to the retardation in shoreline displacement during the Litorina transgression (Pyökäri, 1976; see also Donner, 1980). Extremely well developed, raised shorelines are also observed close to the upper limit of the Ancylus transgression (Glückert and Ristaniemi, 1980, 1982; Eronen and Haila, 1982).

Raised shorelines are also present on till soils, where they are usually represented by boulder barricades on the slopes of rising ground. These shoreline landforms also represent clearly detectable leaching limits, enabling one to distinguish the boundaries between subaquatic and supra-aquatic areas (Eronen and Haila, 1981).

CONCLUSIONS

Although the study of shoreline displacement has a long tradition in Finland and the broad outlines of the history of the Baltic basin have been determined fairly reliably, there remains a great deal for modern research to achieve in filling in the details and settling the many dating problems. The phases of the Baltic Ice Lake during the Late Weichselian are indeterminate at many points, whereas very much more reliable information is available on events during the Holocene. The early part of this period was characterized by very rapid regression, which was then interrupted in southern Finland by the Ancylus transgression. The subsequent Ancylus regression was again a period of rapidly receding shorelines, until this in turn was interrupted for some time in southern Finland by the Litorina transgression. Thereafter shoreline displacement was distinctly slower than during the Early Holocene, and a gradual retardation is noted towards the present day as a result of the progressive decline in land uplift.

The beginning of the Litorina Sea stage forms by far the most easily distinguishable stratigraphical boundary in the history of the Baltic, for it was at this point that the freshwater lake basin altered fairly abruptly into a nutrient-rich brackish-water sea. This change can be detected best in diatom analysis, but the Litorina Sea sediments also regularly contain more organic material than those originating from early phases of the Baltic (Eronen, 1974; Ignatius *et al.*, 1981).

No dramatic changes have taken place in the Baltic basin since the beginning of the Litorina Sea, although its salinity has declined somewhat with the narrowing of the connecting channel through the Straits of Denmark as a consequence of crustal movements in the area (Eronen, 1974; Eronen *et al.*, 1979). Moreover, the shoreline has continued to recede as a result of land uplift. It is possible to speak of the Baltic Sea in the modern sense from about 2500 B.P. onwards (i.e. from the end of the post-glacial climatic optimum) but no distinct stratigraphical limit exists by which it can be separated from the Litorina Sea.

REFERENCES

Aartolahti, T. (1970). *Ann. Acad. Sci. Fennicae A III* **107**, 26 pp.
Aartolahti, T. (1972). *Fennia* **114**, 84 pp.
Agrell, H. (1979). *In* "The Quaternary History of the Baltic" (V. Gudelis and L.-K. Königsson, eds), pp. 219-239. *Acta Univ. Upsaliensis, Symp. Univ. Ups. Annum Quingentesimum Celebrantis* **1**.
Alhonen, P. (1979). *In* "The Quaternary History of the Baltic" (V. Gudelis and L.-K. Königsson, eds), pp. 101-113. *Acta Univ. Upsaliensis, Symp. Univ. Ups. Annum Quingentesimum Celebrantis* **1**.
Alhonen, P., Eronen, M., Nunez, M., Salomaa, R. and Uusinoka, R. (1978). *Bull. geol. Soc. Finland* **50**, 69-79.
Berglund, B. (1964). *Sveriges geol. Unders.* **C599**, 47 pp.
Berglund, B. (1971). *Geol. För. Stockh. Förh.* **93**, 625-652.
Björck, S. (1979). "Late Weichselian stratigraphy of Blekinge, SE Sweden, and water level changes in the Baltic Ice Lake." University of Lund, Department of Quaternary Geology, Thesis 7, 248 pp.
Clark, J. A. and Lingle, G. S. (1979). *Quaternary Research* **11**, 279-298.
Digerfeldt, G. (1975). *Boreas* **4**, 125-142.
Donner, J. (1964). *Soc. Sci. Fennica, Comment. Phys.-Math.* **30** (5), 47 pp.
Donner, J. (1969). *Bull. geol. Soc. Finland* **41**, 135-150.
Donner, J. (1976). *Univ. Helsinki, Department of Geology, Stencil* **1**, 264 pp.
Donner, J. (1978). *Soc. Sci. Fennica, Comment. Phys.-Math.* **48**, 11-38.
Donner, J. (1980). *In* "Earth Rheology, Isostasy, and Eustasy" (N.-A. Mörner, ed.), pp. 285-293. John Wiley and Sons, Chichester.
Donner, J. (1982). *Ann. Acad. Sci. Fennicae A III* **134**, 13-28.
Donner, J. and Eronen, M. (1981). "Excursion Guide". *Univ. Helsinki, Department of Geology, Stencil* **5**, 53 pp.
Eronen, M. (1974). *Soc. Sci. Fennica, Comment. Phys.-Math.* **44**, 79-195.
Eronen, M. (1976). *Boreas* **5**, 65-76.
Eronen, M. and Haila, H. (1981). *In* "Florilegium Florinis Dedicatum" (L.-K. Königsson and K. Paabo, eds), pp. 157-158. *Striae* **14**.
Eronen, M. and Haila, H. (1982). *Ann. Acad. Sci. Fennicae A III* **134**, 111-129.
Eronen, M., Jungner, H. and Uusinoka, R. (1979). *Terra* **91**, 209-218.
Eronen, M., Heikkinen, O. and Tikkanen, M. (1982). *Fennia* **160** (2), 195-223.
Flodên, T. and Winterhalter, B. (1981). *In* "The Baltic Sea" (A. Voipio, ed.), pp. 1-54. Elsevier Oceanography Series 30. Elsevier, Amsterdam.
Fredên, C. (1967). *Geol. För. Stockh. Förh.* **89**, 239-267.
Fredên, C. (1979). *In* "The Quaternary History of the Baltic" (V. Gudelis and L.-K. Königsson, eds), pp. 59-74. *Acta Univ. Upsaliensis, Symp. Univ. Ups. Annum Quingentesimum Celebrantis* **1**.
Fromm, E. (1965). *Sveriges geol. Unders. Ca* **39**, 236 pp.
Glückert, G. (1976). *Ann. Acad. Sci. Fennicae A III* **118**, 92 pp.
Glückert, G. (1977). *Publications, Department of Quaternary Geology, University of Turku* **32**, 22 pp.
Glückert, G. and Ristaniemi, O. (1980). *Publications, Department of Quaternary Geology, University of Turku* **41**, 22 pp.

Glückert, G. and Ristaniemi, O. (1982). *Ann. Acad. Sci. Fennicae A III* **134**, 99–110.

Gudelis, V. (1979). *In* "The Quaternary History of the Baltic" (V. Gudelis and L.-K. Königsson, eds), pp. 29-40. *Acta Univ. Upsaliensis, Symp. Univ. Ups. Annum Quingentesimum Celebrantis* **1**.

Hyvärinen, H. (1973). *Boreas* **2**, 85-102.

Hyvärinen, H. (1980). *Bull. geol. Soc. Finland* **52**, 207–219.

Hyvärinen, H. (1982). *Ann. Acad. Sci. Fennicae A III* **134**, 139–149.

Hyvärinen, H. and Eronen, M. (1979). *In* "The Quaternary History of the Baltic" (V. Gudelis and L.-K. Königsson, eds), pp. 7-27. *Acta Univ. Upsaliensis, Symp. Univ. Ups. Annum Quingentesimum Celebrantis* **1**.

Ignatius, H., Axberg, S., Niemistö, L. and Winterhalter, B. (1981). *In* "The Baltic Sea" (A. Voipio, ed.), pp. 54-104. Elsevier Oceanography Series 30. Elsevier, Amsterdam.

Ignatius, H., Korpela, K. and Kujansuu, R. (1980). *Boreas* **9**, 217-228.

Jones, M. (1977). "Finland, Daughter of the Sea". Studies in Historical Geography. Archon Books, Dawson.

Kääriäinen, E. (1953). *Fennia* **77** (2), 106 pp.

Kääriäinen, E. (1966). *Ann. Acad. Sci. Fennicae A III* **90**, 187-190.

Kiviniemi, A. (1980). *Tectonophysics* **71**, 65-71.

Klug, H. (1980). *Eiszeitalter Gegenw.* **30**, 237-252.

Kolp, O. (1979). *In* "The Quaternary History of the Baltic" (V. Gudelis and L.-K. Königsson, eds), pp. 261-274. *Acta Univ. Upsaliensis, Symp. Univ. Ups. Annum Quingentesimum Celebrantis* **1**.

Krog, H. (1979). *In* "The Quaternary History of the Baltic" (V. Gudelis and L.-K. Königsson, eds), pp. 207-217. *Acta Univ. Upsaliensis, Symp. Univ. Ups. Annum Quingentesimum Celebrantis* **1**.

Liljegren, R. (1982). Paleoekologie och strandförskjutning i en Littorinavik vid Spjälkö i mellersta Blekinge. University of Lund, Department of Quaternary Geology, Thesis 11, 95 pp.

Ludwig, A. O. (1979). *In* "The Quaternary History of the Baltic" (V. Gudelis and L.-K. Königsson, eds), pp. 41-58. *Acta Univ. Upsaliensis, Symp. Univ. Ups. Annum Quingentesimum Celebrantis* **1**.

Lundqvist, G. (1961). *Sveriges geol. Unders. Ba* **18**, 148 pp.

Miller, U. (1982). *Ann. Acad. Sci. Fennicae A III* **134**, 185–211.

Niemelä, J. (1979). *Rep. geol. Surv. Finland* **42**, 119 pp.

Niemelä, J. and Tynni, R. (1979). *Bull. geol. Surv. Finland* **253**, 79 pp.

Nilsson, E. (1968). *K. svenska Vetensk. Handl. 4. Ser.* **12** (1), 117 pp.

Nunez, M. (1978). *Iskos* **2**, 25-51.

Okko, M. (1962) *Bull. Comm. géol. Finlande* **202**, 162 pp.

Pyökäri, M. (1976). *Publ. Inst. Geogr. Univ. Turkuensis* **73**, 40 pp.

Saarnisto, M. (1970). *Soc. Sci. Fennica, Comment. Phys.-Math.* **37**, 107 pp.

Saarnisto, M. (1971). *Soc. Sci. Fennica, Comment. Phys.-Math.* **41**, 149-170.

Saarnisto, M. (1981). *Ann. Acad. Sci. Fennicae A III* **130**, 24 pp.

Saarnisto, M. (1982). *Ann. Acad. Sci. Fennicae A III* **134**, 61–79.

Salomaa, R. (1982). *Ann. Acad. Sci. Fennicae A III* **134**, 81–97.

Sauramo, M. (1958). *Ann. Acad. Sci. Fennicae A III* **51**, 522 pp.

Siiriäinen, A. (1969). *Suomen Museo* 1969, 40-73.

Siiriäinen, A. (1970). *Bull. geol. Soc. Finland* **42**, 119-127.

Siiriäinen, A. (1973). *Finskt Museum* 1973, 5–22.
Siiriäinen, A. (1982). *Ann. Acad. Sci. Fennicae A III* **134**, 173–184.
Strömberg, B. (1977). *Z. Geomorph. N.F. Suppl. Bd.* **27**, 89–111.
Synge, F. M. (1980). *Geol. För. Stockh. Förh.* **102**, 235–249.
Synge, F. M. (1982). *Ann. Acad. Sci. Fennicae A III* **134**, 29–60.

NINE

Shorelines and isostasy in Scotland

J. B. SISSONS

The foundations of present interpretations of relative changes of sea level in Scotland were laid by Jamieson in his classic paper in 1865. He identified a period of high sea level, this being followed by a major regression that allowed peat to accumulate down to about present sea level. Since the peat is covered by estuarine silts and clays Jamieson inferred a later rise in relative sea level, which was followed by a lowering to the present level. He was also the first person to recognize that the prime cause of land uplift was recovery from glacier loading of the earth's crust.

By the end of last century Jamieson's higher feature had come to be known as the "100-foot (30 m) beach" and his lower feature as the "25-foot (7-8 m) beach", terms that, despite Jamieson's (1906) objections, continued to be linked to a concept of essentially horizontal raised shorelines until rejected in the 1960s (cf. Sissons, 1962). During the last 20 years studies of former sea-level changes have provided detailed evidence for various parts of Scotland: these studies have elaborated Jamieson's basic interpretations and have also resulted in new concepts.

GLACIATION

Before considering sea-level changes further it is necessary to outline possible recent glacial events in Scotland. "Possible" must be emphasized since there is little firm evidence from Scotland for events during the Devensian (last) glaciation: the limit of the last ice sheet in Scotland or the surrounding sea areas is nowhere established and ^{14}C dates associated with the glaciation are sometimes contradictory and often of uncertain geological significance (Sissons, 1981).

That the last brief full interglacial conditions were centred on 120 000–125 000 B.P. is indicated, for example, by evidence from deep-sea cores from the north-east Atlantic (Ruddiman *et al.*, 1980) and by speleothem dates from northern England (Gascoyne, 1981). A speleothem date of 122 000 ± 12 000 B.P. from north-west Scotland is in accord (Lawson, 1981). Ruddiman *et al.* proposed major build-up of the last Scandinavian ice sheet at *c.* 75 000 B.P., the necessary moisture being supplied by an ocean that, until that time, was warm to high latitudes (60°N in the north-east Atlantic). Certain Scottish evidence (p. 220) is consistent with a major

SHORELINES AND ISOSTASY
ISBN 0 12 652960 4

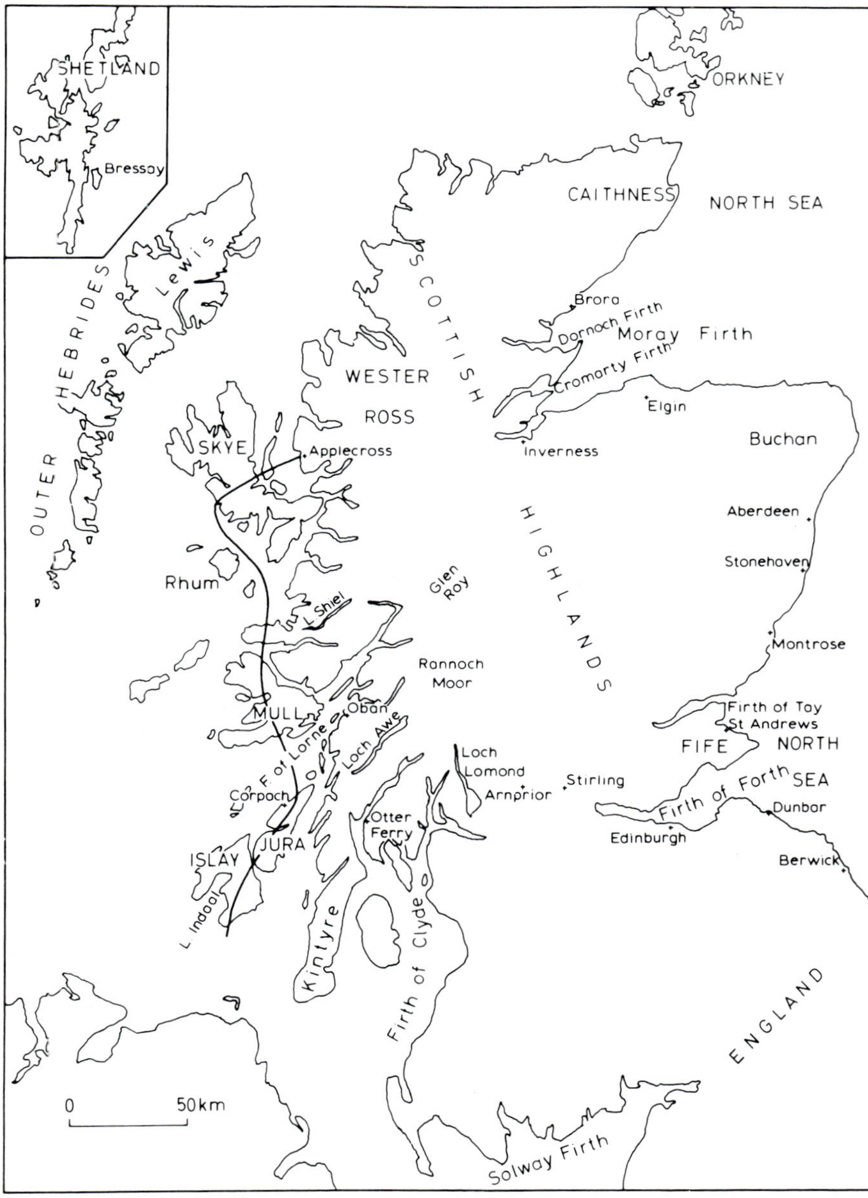

Figure 1. Places mentioned in the text. The continuous line is the inner limit of high rock platforms in western Scotland.

build-up of the Scottish ice sheet at the same time (Sutherland, 1981a). ^{14}C dates for basal organic sediments from pollen sites at several points deep within the Scottish Highlands indicate extensive ice-sheet decay by 13 000–12 500 B.P.

Evidence for stillstands or, more probably, readvances of the last ice sheet has been found in three coastal areas. By far the most extensive is in part of north-west Scotland, where a former ice limit that is often defined by a clear end moraine has been traced for some tens of kilometres; the term Wester Ross Readvance has been proposed (Robinson and Ballantyne, 1979). The other two areas are mentioned later.

The Lateglacial Interstadial lasted from 13 000 to 11 000 B.P. and was succeeded by the Loch Lomond (Younger Dryas) Stadial, which may have lasted for *c.* 750 years. During the stadial glaciers descended to present sea level at many places in the west and south-west Highlands and in Mull.

SOUTH-EAST SCOTLAND

This area comprises the coasts of the Firths of Forth and Tay, the lowlands that extend inland from the firths, and the coastline between the firths. Basic fieldwork in the area involved mapping, at a scale of 1:10 560, of all identifiable marine, estuarine, fluvial, fluvioglacial and glacial landforms below an altitude of *c.* 100 m. This was followed by accurate instrumental levelling, tied to Ordnance Survey bench marks, of all fossil shorelines and outwash deposits at intervals of *c.* 50 m: this provided some 12 000 altitudes. Over 2500 boreholes have been made with hand-operated equipment and a similar number of commercial borehole records has been obtained. About 70 ^{14}C dates related to sea-level changes are available, although most lie between 10 000 and 6000 B.P. Pollen and/or diatom analyses have been made for numerous ^{14}C-dated sites.

Detailed studies have also been made for eastern Mull and the adjacent mainland coast (Gray, 1972), for western Jura and northern Islay (Dawson, 1979), for the coasts of the sea lochs leading into the upper Firth of Clyde (Sutherland, 1981b) and for the area between the Firth of Tay and Stonehaven (Cullingford and Smith, 1980). The evidence from south-east Scotland is the most complete, however, largely owing to its extensive preservation in lowlands extending far inland from the open sea. Hence this area will be considered in most detail.

The oldest and most steeply inclined shorelines of south-east Scotland are those in East Fife (Fig. 2). Their gradients range between 0·94 and 0·54 m km^{-1} as determined by linear trend surface analyses that include data from farther north (Cullingford and Smith, 1980). Shoreline formation contemporaneously with westward retreat of the ice-sheet margin is demonstrated by the tendency of younger shorelines to extend farther west, ice-margin positions in some instances being recorded by outwash spreads that begin in kame-and-kettle topography. Continued ice-margin retreat towards the west is indicated by other shorelines that merge westwards into outwash, the most extensive being the Main Perth Shoreline (Figs 2 and 3). Trend-surface analysis of over 500 shoreline altitudes gave a gradient of 0·43 m km^{-1} (Fig. 4). The raised beach merges westwards into outwash and is succeeded in the Tay area by lower raised beaches whose shoreline gradients successively diminish to 0·21–0·24 m km^{-1}, this slight variation being related to the plane of projection used (Sissons and Smith, 1965a; Smith *et al.*, 1969; Cullingford, 1977).

Continued emergence eventually led to a low sea level that is recorded by evidence for extensive marine erosion. At the head of the Firth of Forth, far from the open sea, an area of at least 28 km^2 was planed, this being demonstrated by *c.* 1000 boreholes that reveal a buried gravel layer that rises landwards from the firth (Fig. 3). The gravel normally overlies soft glacial marine deposits but till and bedrock

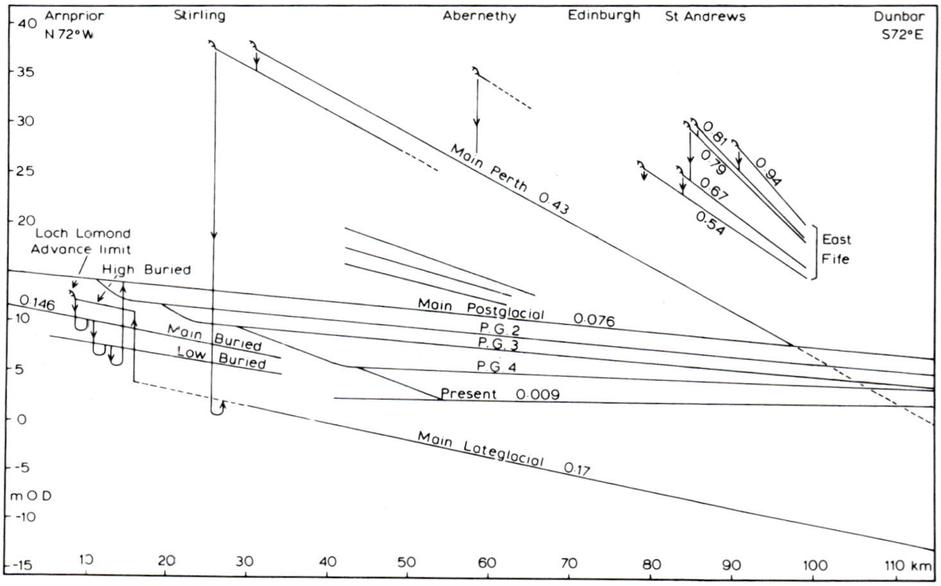

Figure 2. Height-distance diagram of shorelines in south-east Scotland. Gradients are in m km^{-1}. A small part of the diagram is shown in more detail in Fig. 6.

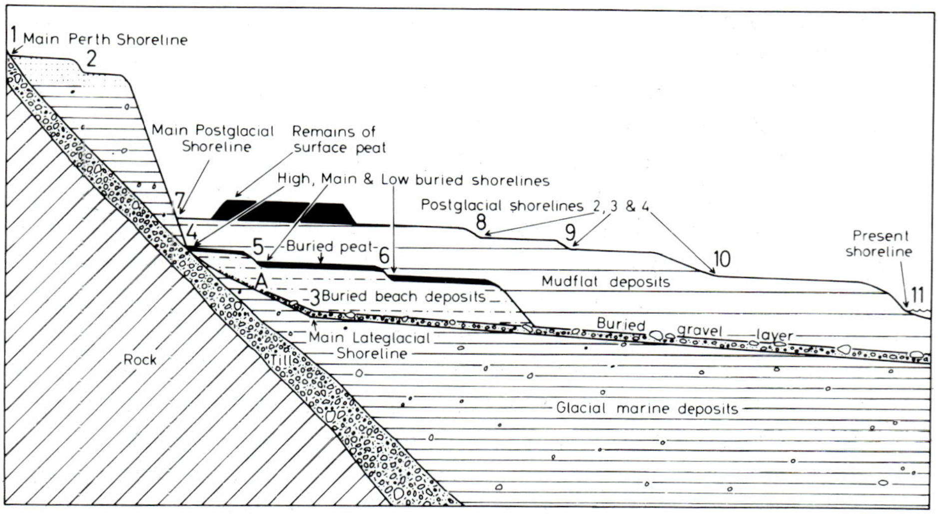

Figure 3. Diagrammatic representation of morphology, buried morphology and stratigraphy at the head of the Firth of Forth. Numbers indicate the shoreline sequence.

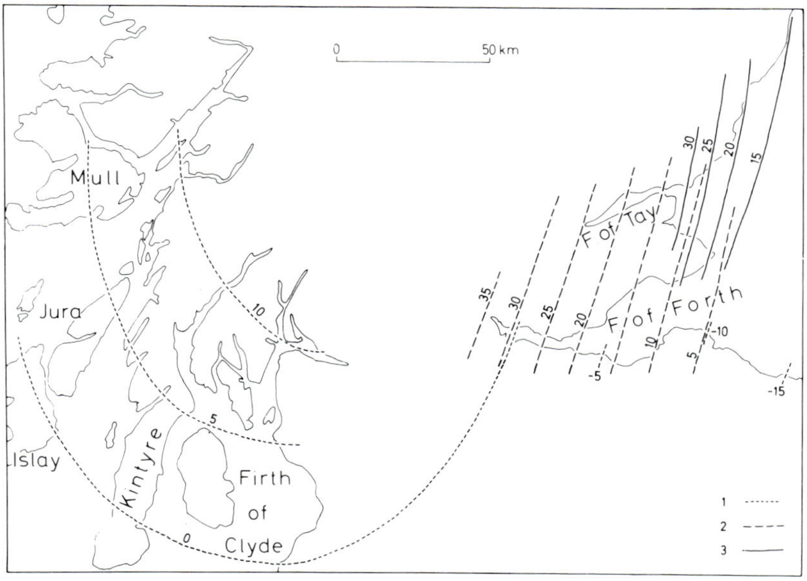

Figure 4. Isobases at 5 m intervals for certain shorelines. (1) Main Lateglacial Shoreline, isobases not calculated. (2) Main Perth Shoreline, quadratic trend surface. (3) An early shoreline in eastern Scotland, quadratic trend surface. Based mainly on Smith *et al.* (1969), Gray (1974b, 1978), Sissons (1976), Dawson (1980a), Cullingford and Smith (1980) and Sutherland (1981).

were truncated locally (Sissons, 1969). The main area of planation, which rises very gently landwards through 6 m and probably reflects slow marine transgression, terminates at the Main Lateglacial Shoreline (Fig. 3). This shoreline slopes down the Firth of Forth at *c.* 0·17 m km^{-1} (Fig. 2): at Edinburgh it backs a buried platform up to 1 km wide that is cut mainly in till and locally in bedrock (Sissons, 1976) and on the coast 9 km north of Berwick, where the shoreline is at −18 m, is associated with a submerged rock platform probably about 600 m broad (Eden *et al.*, 1969). The buried channel of the River Tweed apparently correlates with this low sea level because, at Berwick, its floor is at *c.* −20 m and, significantly, does not contain till (Rhind, 1972). The Main Lateglacial Shoreline may be at −8 m at the entrance to the Firth of Tay (Paterson, 1981).

The pronounced erosion is attributed mainly to the very cold climate of the Loch Lomond Stadial, although erosion may have begun with climatic deterioration before the end of the preceding interstadial. In some localities the buried gravel layer and associated planation surface continue landwards from the Main Lateglacial Shoreline, rising *c.* 7 m at a steeper gradient than seawards from this shoreline (A in Fig. 3): this indicates continued marine erosion accompanying transgression in such localities and suggests that the rate of transgression had increased. Erosion ended when a beach, the highest of three beaches that are now buried by later deposits, began to be deposited. This High Buried Beach (Figs 2 and 3) occurs outside the major end moraine that marks the limit of the Loch Lomond Advance at the head of the Forth lowlands but is absent within the morainic arc (Sissons, 1966).

The beach locally overlies outwash associated with meltwater channels cut in the moraine. Hence the beach was being formed when ice still stood at the moraine and its age is likely to be 10 300 ± 200 B.P.

Subsequent marine regression was probably interrupted by minor transgressions accompanying deposition of the Main and Low buried beaches. The average gradient of the Main Buried Shoreline in the Forth valley is 0·146 m km^{-1}, but this figure conceals significant variations (p. 221). Pollen analyses indicate that marine regression from the gently-sloping surfaces of the Main and Low buried beaches was accompanied by peat accumulation (Newey, 1966; Brooks, 1972): ^{14}C dates for the basal peat indicate that the ages of the shorelines in the western Forth valley are respectively 9600 and 8700 B.P. (Sissons, 1966, unpublished data; Sissons and Brooks, 1971). In part of the Tay area the two shorelines have been dated as 9600 and 8500–8600 B.P. (Cullingford *et al.*, 1980).

Marine regression in the western Forth valley probably terminated at *c.* 8600 B.P. (but at least 200 years later in the Tay area) and was followed by the Main Postglacial Transgression during the most rapid part of which relative sea level, despite isostatic uplift, rose as much as 5 m in 700 years. Westwards from the present head of the Firth of Forth and in the Firth of Tay of that time extensive deposition of mudflats occurred, with consequent concealment of the now-buried beaches and their overlying peat. (The thin peat on the High Buried Beach accumulated during the transgression.) In two areas, however, each 1–2 km broad, peat growth was sufficiently rapid to prevent mudflat deposition (Sissons and Smith, 1965b).

The culmination of the transgression is marked by the Main Postglacial Shoreline, whose average gradient is 0·076 m km^{-1}. The shoreline was formed at 6800 B.P. at the western end of the Forth lowland, which is where this shoreline was most affected by isostatic uplift in south-east Scotland. Shortly before the culmination (at *c.* 6850 B.P.) a sand layer a few centimetres thick, whose marine origin is demonstrated by diatoms and derived spores, was deposited. Since a similar sand layer has been similarly dated at several sites along the east coast of Scotland as far as a point 20 km north of Aberdeen, the layer probably resulted from a major storm surge. Intermittent regression from the Main Postglacial Shoreline to present sea level is indicated by three shorelines identified in the Forth valley (Smith, 1968), where extensive mudflat deposition continued in the more sheltered areas.

Two modifications of the pattern of events described above have been proposed. First, Paterson *et al.* (1981) argued that, in the Tay area, the Main Lateglacial Shoreline was formed before the Loch Lomond Stadial and that during the stadial deep (now buried) channels were cut consequent on a "drastic fall of sea-level" (p. 18). As Paterson *et al.* recognized, their interpretation is contradicted by the evidence of Peacock *et al.* (1978) in the west of Scotland. It is also contradicted by the evidence around the head of the Firth of Forth, which is far more detailed than that in the Tay area (Sissons, 1969, e.g. Figs 5 and 6).

Secondly, it has been claimed that laminated clays that contain foraminifera and, occasionally, mollusca (normally very fragmented) demonstrate marine transgression far to the west of, and to higher altitudes than, the shorelines associated with ice-sheet retreat shown in Fig. 2 (Browne, 1980; Browne *et al.*, 1981a). Smith and Cullingford (1981) have pointed out that this interpretation is inconsistent with the westward termination of some of these shorelines in outwash deposits, a point conceded by Browne *et al.* (1981b). This dispute is currently unresolved but the possibility that at least some of the marine fossils have been derived from earlier deposits has not been considered in the exchange of views.

PERIPHERAL AREAS

Projected eastwards the shorelines of south-east Scotland that were associated with ice-sheet retreat intersect the bed of the North Sea (Sissons, 1981). Even when subsequent sedimentation and probable eastward decrease in shoreline gradients are taken into account (and hydro-isostasy would have operated in the opposite sense) it appears that a large area of the present North Sea bed was then land as far north as the latitude of Buchan, the sea itself then being confirmed to a relatively narrow area along the east coast of Scotland that terminated southwards off north-east England. An extensive buried erosion surface beneath the central North Sea at *c.* −60 to *c.* −100 m has been shown to be partly Devensian in age (Holmes, 1977; Thomson and Eden, 1977): in the places where it is of this age its final fashioning may be partly attributed to eastward marine transgression as the relatively narrow North Sea (mentioned above) increased in width owing to eustatic rise of sea level and glacio-isostatic recovery.

Minimal sea level in the North Sea during the last glaciation appears to have been *c.* −110 m (Jansen, 1976). On the opposite side of Scotland by the island of St. Kilda, 60–70 km west of the Outer Hebrides, there is an extensive rock platform at *c.* −120 m, which probably relates to the minimal sea level of the last and earlier glaciations. Another extensive platform that occurs between *c.* −80 and *c.* −45 m is probably composite (D. G. Sutherland, personal communication). On St. Kilda and other peripheral islands such as Orkney and Shetland raised beaches are absent; they also appear to be absent from the Outer Hebrides except in a small area in Lewis (McCann, 1968) where raised beach deposits are considered to predate the last glaciation (Peacock, 1981). In these peripheral areas submergence has prevailed, the later part of the submergence being demonstrated by peat below sea level at innumerable points and by archaeological evidence.

RAISED SHORELINES ASSOCIATED WITH ICE-SHEET GLACIATION

Raised beaches and associated shorelines

Inwards from the peripheral areas fossil shorelines associated with ice-sheet glaciation rise above present sea level and their upper limit initially increases in altitude fairly rapidly towards the centre of isostatic uplift in the Rannoch Moor area. It may be inferred that individual shorelines have relatively steep gradients (many have not yet been investigated in detail), probably comparable with those of the East Fife shorelines, and that they terminated at contemporary ice-sheet margins. Such shorelines are apparently absent from extreme north-eastern England but they attain 28 m in East Fife. They are at low altitudes in extreme north-eastern Scotland but attain 35 m at Stonehaven (Cullingford and Smith, 1980) and *c.* 25 m around Elgin (Ogilvie, 1923). Shorelines associated with the last ice sheet are absent or at a very low altitude in northern Caithness but may reach *c.* 25 m around Brora. They are similarly absent or low in the far north-west of the mainland but reach 29-30 m near Applecross (Robinson, 1977). Lateglacial raised marine features are absent from the southern islands of the Outer Hebrides but shingle ridges occur up to an altitude of 30 m in Rhum (McCann and Richards, 1969). The marine limit (theoretical) also rises rapidly inwards across the Solway Firth.

Inwards from the zone where the marine limit rises rather rapidly there is a zone in which most of the terminations of successive shorelines at positions of the ice-sheet margin occupy a limited range of altitude. The failure to recognize a succession of shorelines was largely responsible for the long-held view that there is an essentially horizontal shoreline at around 30 m. In the Forth area and on the open coast of East Fife the western ends of successive shorelines normally lie between 25 and 38 m (Fig. 2). On the coasts of the Moray Firth inwards from the vicinities of Elgin and Brora the comparable range appears to be roughly 25-30 m. On the other side of the country along the coasts of Mull (outside the limit of the Loch Lomond Advance), Jura and the adjacent mainland, as well as around the head of the Firth of Clyde, some shorelines attain, or the highest reported shoreline fragments are at, *c.* 30-41 m (Synge and Stephens, 1966; Gray, 1974a; Gray and Sutherland, 1977; Sutherland, 1981b; Dawson, 1982). This area is probably part of a much larger western area with most ice-sheet-associated shorelines terminating in this altitude range.

For some areas information is more detailed. For example, the most widespread lateglacial raised beach in Jura is absent north-east of Corpach, suggesting proximity of the ice-sheet margin. The associated shoreline falls south-westward from 40 m at Corpach to 15 m in extreme western Islay at a gradient of 0·59 m km^{-1}. This shoreline is succeeded by a slightly lower one with a gradient of 0·56 m km^{-1}. Visible shorelines associated with ice-sheet decay terminate eastwards in central Islay, one of them merging (at 21·5 m) into outwash by Loch Indaal. The outwash is associated with an end moraine at which the lateglacial marine limit drops at least 12 m (Dawson, 1982).

Shorelines associated with westward ice-sheet retreat between Stonehaven and the entrance to the Firth of Tay have been correlated with the East Fife shorelines, which are similarly related to ice retreat. Five shorelines common to the two areas have successive gradients, as calculated by linear trend surface analysis, of 0·94, 0·81, 0·79, 0·67 and 0·54 m km^{-1} (Cullingford and Smith, 1966, 1980).

The zone in which most shorelines associated with ice-sheet decay terminate inland at *c.* 25–41 m ends abruptly in some places and in others is succeeded by evidence for a considerable fall of sea level while the ice-sheet margin retreated a limited distance. The altitudes of raised shorelines and outwash spreads by the Cromarty and Dornoch firths indicate a considerable fall of sea level during ice retreat here (Ogilvie, 1923). Immediately outside the area covered by the Wester Ross Readvance the lateglacial marine limit represents an approximately synchronous shoreline corresponding with the culmination of the readvance, the shoreline gradient being between 0·33 and 0·39 m km^{-1}. During subsequent ice retreat of 17 km relative sea level fell by 11 m (Sissons and Dawson, 1981).

Outwash terraces that begin in kame-and-kettle topography at the south-west end of Loch Awe slope down towards the coast. The altitudes of the terrace termini and of adjacent raised shorelines indicate relative sea-level fall of at least 26 m during terrace formation (Gray and Sutherland, 1977). At Otter Ferry the marine limit abruptly drops 20 m (where there is also an end moraine) and it drops 22 m (or more) and 26 m at localities situated respectively 6 and 27 km to the east (Sutherland, 1981b). Using faunal evidence and ^{14}C dates from two sites in the same general area, Peacock *et al.* (1977, 1978) plotted this relative change of sea level against time. Using additional evidence Sutherland (1981b) produced a revised graph, which indicates rapid sea-level fall from *c.* 13 000 to *c.* 12 000 B.P., this being followed by a decelerating fall to *c.* 11 000 B.P., the total relative movement of sea level being *c.* 30 m. In the period 13 000–12 000 B.P. seven, possibly eight, raised shorelines were considered to have been formed, the best defined one, with a gradient of 0·33 m km^{-1}, being correlated with the Main Perth Shoreline of south-east Scotland.

Rock shorelines

Rock platform fragments at *c.* 20–25 m occur at a few points on the east coast as, for example, between Dunbar and Berwick. In at least some instances the platform fragments pre-date the last ice sheet. Raised platforms are far better developed in much of the Inner Hebrides and on limited parts of the mainland coast, the shorelines normally having been recorded as lying between 18 and 44 m, although a shoreline at 51 m has been reported (Wright, 1911; Bailey *et al.*, 1924; McCann, 1968; McCann and Richards, 1969; Dawson, 1979). The platforms are sometimes hundreds of metres wide, the extreme reported width exceeding a kilometre, and the backing cliffs range in height up to nearly 90 m. Some platform fragments are striated, ice-moulded and/or overlain by till and hence all the features have normally been interpreted as interglacial in age. Yet firm evidence that some platform fragments in Skye have been glaciated is lacking (Richard, 1969) and in western Mull, where platform fragments are very extensive, evidence that they have been glaciated is restricted to a few points. It has been recently proposed that the features were produced by severe frost action and wave action, the former being dominant in sheltered areas and the latter important in areas exposed to open Atlantic fetch. Clear platforms and cliffs apparently terminate eastwards along the line shown in Fig. 1, which is interpreted as marking a relatively stable position of the margin of the last ice sheet (and, probably, earlier ice sheets) that was occupied for long periods. An implications of this interpretation is that unglaciated rock shorelines should correlate with depositional shorelines (Sissons, 1982). In Wester Ross the shoreline mentioned above that corresponds with the lateglacial marine limit demonstrated this relationship; furthermore, rock platform fragments above this limit have been glaciated (Robinson, 1977; Sissons and Dawson, 1981).

THE MAIN LATEGLACIAL SHORELINE

A rock platform and cliff at low or relatively low altitudes occur along part of the west coast of Scotland, being especially prominent around the Firth of Lorne and in western Jura. The platform is typically 20-30 m broad, but occasionally exceeds 150 m, and the cliff is in places over 15 m high. The features were originally regarded as Holocene in age (e.g. Bailey *et al.*, 1924) and were subsequently considered to be older than the last glaciation (McCallien, 1937; McCann, 1968; Gray, 1974b), but a Lateglacial age was later proposed (Sissons, 1974). On this view the rock shoreline correlates with the Main Lateglacial Shoreline of south-east Scotland and was formed during the Loch Lomond Stadial and, possibly, during part of the Lateglacial Interstadial. Outside the limit of the Loch Lomond Advance the platform and cliff are unglaciated but within the limit only the cliff, or part of the cliff, is normally visible. The feature is often very well developed in very sheltered areas and its formation is attributed to severe frost action assisted by wave action. The shoreline is at *c.* 11 m slightly north of Oban, whence it slopes down westwards at an average gradient of $0{\cdot}16\ \mathrm{m\,km^{-1}}$, while its south-westward gradient in Jura is $0{\cdot}13\ \mathrm{m\,km^{-1}}$ (Gray, 1974b, 1978; Dawson, 1980a) (Fig. 4). Independent evidence for a low or relatively low sea level at the end of the Loch Lomond Stadial is provided by the altitudes to which certain outwash deposits descend: such deposits associated with the limits of Loch Lomond Advance glaciers or with slightly later ice-margin positions occur at the western end of Loch Shiel, at points in Mull and by the Firth of Lorne and at the southern end of Loch Lomond (Bailey *et al.*, 1924; McCann, 1966; Gray, 1975; Dickson *et al.*, 1978).

As implied by the isobases in Fig. 4, the Main Lateglacial Shoreline is below sea level in south-west Scotland. Evidence for a low Lateglacial sea level is also provided by radiocarbon dates ranging between *c.* 12 300 and *c.* 10 300 B.P. for peat close to present sea level near the head of the Solway Firth (Bishop and Coope, 1977).

HOLOCENE SEA-LEVEL CHANGES

Intermittent early Holocene marine regression recorded by buried beaches in south-east Scotland has not been specifically identified elsewhere in Scotland, although ^{14}C dates on peat or wood indicate low sea level in south-west Scotland in the early

Holocene (Jardine, 1975; Bishop and Coope, 1977). At almost all these sites peat is covered by marine or estuarine deposits laid down during the Main Postglacial Transgression. The youngest date in this series is *c.* 6800 B.P., while the oldest date for the base of peat resting on the marine or estuarine deposits is *c.* 6650 B.P. It thus appears that the transgression culminated between these dates. This interpretation was rejected by Jardine (1975) but is in accord with the evidence he presented. Walker (1966) concluded that the maximum of the transgression was between *c.* 6950 and *c.* 6600 B.P. on the south side of the Solway Firth and it has been placed between *c.* 7050 and *c.* 6700 B.P. near Montrose in eastern Scotland. On the other hand, some evidence from the Firth of Tay area suggests a slightly later date (Smith *et al.*, 1980; Morrison *et al.*, 1981).

The altitude of the Main Postglacial Shoreline, which marks the culmination of the transgression, is shown by isobases in Fig. 5. These indicate only the generalized form of subsequent uplift since there are no accurate data for long stretches of the shoreline and, in some places where such data are available, there are hints that the pattern of uplift is not so simple. ^{14}C-dated peat from Bressay in Shetland (Hoppe, 1965) indicates that the Main Postglacial Shoreline there is at least 9 m below sea level. The transgression converted lochs Lomond and Shiel into arms of the sea,

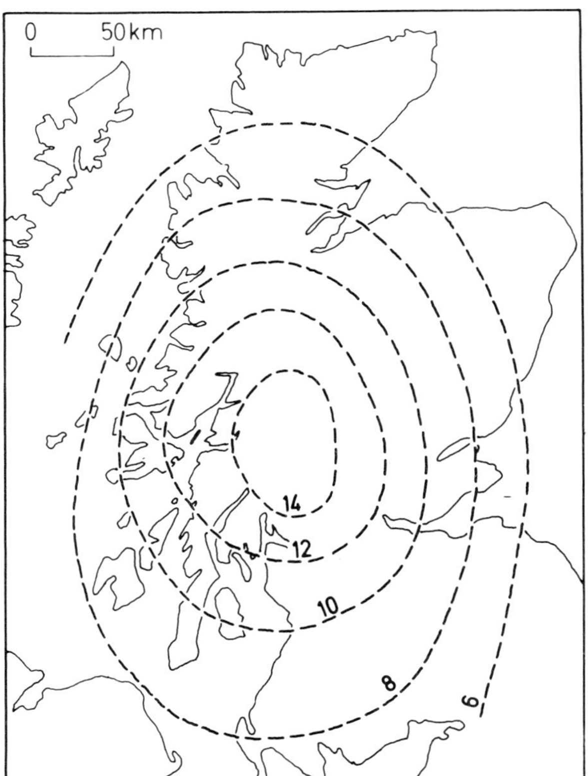

Figure 5. Isobases at 2 m intervals for the Main Postglacial Shoreline. Based on numerous sources.

subsequent marine regression from the former (present surface altitude *c.* 8 m) being placed at roughly 5500 B.P. and from the latter (*c.* 3 m) at roughly 4200 B.P. (Dickson *et al.*, 1978; Thompson and Wain-Hobson, 1979). Shorelines formed during marine regression to present sea level, and perhaps representing temporary reversals of the dominant trend, have been identified; for example, at *c.* 8 and *c.* 4 m in eastern Mull, where the Main Postglacial Shoreline is at *c.* 12 m (Gray, 1974a). A shoreline at 5-6 m by the Solway Firth, but only 1-2 m above present high-water mark, has been dated as *c.* 2000 B.P. (Jardine, 1975).

HIGH-LEVEL MARINE SHELL BEDS

Marine shell beds covered by till are known from a number of localities within a few kilometres of the Scottish coastline. Their altitudes range between *c.* 34 and *c.* 90 m except for one bed near Inverness, which is at 145-150 m. Much discussion has centred on whether the shell beds have been transported by glacier ice or are *in situ* (e.g. Horne *et al.*, 1893, 1896), but Sutherland (1981a) has given strong reasons for accepting the latter interpretation. Since radiocarbon dates associated with two of the shell beds and dates for shells believed to have been transported by the last ice sheet are all infinite, he has argued that it is improbable that this ice sheet built up after *c.* 27 000 B.P. as has been generally accepted. Instead Sutherland proposed major build-up of the Scottish ice sheet at *c.* 75 000 B.P., in accord with the evidence from deep-sea cores (e.g. Ruddiman *et al.*, 1980). He argued that the Scottish ice sheet, owing to maritime location and its relatively small size, would probably have built up, in relative terms, more rapidly than the Scandinavian and North American ice sheets. Thus early advance of the Scottish ice sheet occurred when only limited lowering of eustatic sea level had taken place. Since the Scandinavian and North American ice sheets were still extensive when the Scottish ice sheet finally retreated, relative sea level was much higher during advance than during retreat, thus explaining the comparatively high altitudes of the till-covered shell beds, although the site near Inverness appears anomalous.

INTERGLACIAL SEA LEVELS

Platform fragments up to hundreds of metres wide occur between tide marks on many parts of the Scottish coastline. Although being fashioned at present, their extent renders wholly recent formation improbable and in many instances this is eliminated by their being striated, ice-moulded and/or passing beneath till. The platform fragments in some areas are scattered and vary greatly in width, features that Dawson (1980b) attributed to glacial erosion. He also proposed that intertidal fragments over a large area constitute a single horizontal platform and that this platform was cut during a prolonged period of interglacial sea-level stability. While interglacial formation appears consistent with the available evidence, the proposed long period of sea-level stability is not in accord with the sea-level curves derived from deep-sea cores. Neither is it certain that intertidal platform fragments can be correlated with each other over a large area: it appears more likely that, owing to removal of overlying drift, the intertidal fragments are the most obvious members of a series of features at various altitudes. Such platform fragments have been reported, for example, from Caithness, Wester Ross, Skye and Kintyre (Crampton and Carruthers, 1914; Richards, 1969; Robinson, 1977; Gray, 1978). In Kintyre associated features include raised stacks, caves and geos: the highest feature recorded here is a platform fragment at 13 m that passes beneath till. Since shallow-water submarine features have not been investigated, a zone of interglacial marine erosion extending above and below present sea level may exist. Individual features are likely to have been produced during more than one interglacial, in the same way that the present intertidal platform is now being re-shaped.

IRREGULAR GLACIO-ISOSTATIC UPLIFT

In describing and illustrating shorelines affected by glacio-isostasy in the preceding pages a simple pattern of uplift and tilting has normally been implied. A more complicated pattern cannot usually be identified owing to the fragmentary nature of the evidence. However, in certain instances where individual shorelines are well preserved for considerable distances, complexity has been demonstrated: thus Gray (1974b, 1978) has shown that the Main Lateglacial Shoreline has been irregularly uplifted in the Firth of Lorne area and in part of Kintyre.

In the western Forth valley the Main Postglacial Shoreline (formed at 6800 B.P.) and the Main Buried Shoreline (9600 B.P.) are extensively preserved. Except at A in Fig. 6, where the upper shoreline has probably been slightly lowered owing to compaction of thick underlying peat, the gradients of the two shorelines are almost identical. However, the lower shoreline is dislocated at two places: one dislocation, of 1 m, occurs at the end moraine marking the limit of the Loch Lomond Advance and the other, of nearly 1·5 m, corresponds with a major fault. Figure 6 shows that two parts of the western Forth valley have been uplifted without measurable tilting in the plane of projection during the last 9600 years. During the first 2800 years of

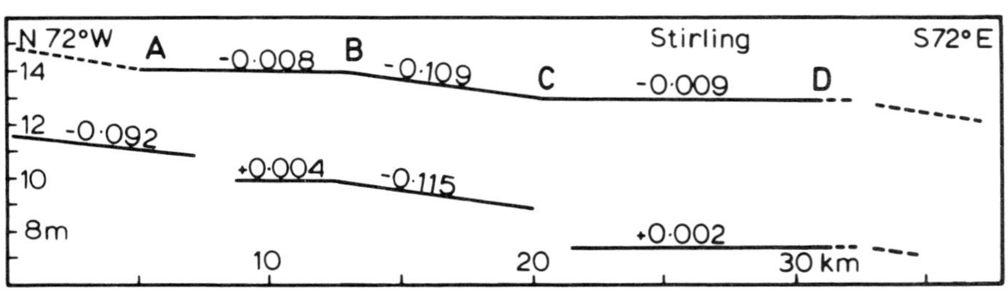

Figure 6. Deformation of the Main Postglacial Shoreline (upper line) and Main Buried Shoreline in part of the Forth valley. Letters are referred to in the text.

this period there was no tilting of stretch AD: block uplift was accompanied by production of the two dislocations. During the last 6800 years, however, part BC has been tilted and there has been no detectable dislocation (Sissons, 1972).

The correspondence between the more westerly dislocation in Fig. 6 and the Loch Lomond Advance limit implies that the advance affected crustal recovery from previous ice-sheet loading. Such a relationship has also been demonstrated in Glen Roy and vicinity, where the shorelines of former ice-dammed lakes have been levelled at 1760 points (Sissons and Cornish, 1982). In part of Glen Roy block uplift without detectable tilting has occurred. In another part eight crustal blocks were differentially tilted in a distance of 7 km: such complexity probably reflects the structural complexity of the Caledonian orogenic belt and contrasts with the differential movement of three crustal blocks in 24 km in the Forth Valley, which is underlain by the simpler structures of the Hercynian orogenic belt. Several short lengths of shoreline in the Glen Roy area have gradients exceeding considerably the highest average gradients calculated for fossil marine shorelines in Scotland, the highest value obtained being 4·6 m km^{-1} (which is minimal since the direction of maximal tilt is unknown). The most pronounced displacement in Glen Roy, which produced a fault scarp, involves a rise of *c.* 3 m in shoreline altitude followed by a descent of *c.* 3 m in a distance of 2 km. Immediately adjacent to this locality a river terrace formed very soon after lake drainage has suffered no dislocation, implying that the pronounced shoreline displacement occurred earlier. This suggests very rapid crustal response to unloading caused by glacier retreat and/or by sudden lake drainage consequent on glacier retreat (Sissons, 1979).

Total glacio-isostatic uplift in the Glen Roy and western Forth areas since the maximum of the Loch Lomond Advance has been *c.* 40-50 m. Hence the dislocations and block tilting represent a minor set of movements superimposed on a major movement. That the latter was mainly or entirely crustal recovery from ice-sheet loading is demonstrated by tilted Holocene marine shorelines extending far beyond the limits of the Loch Lomond Advance: hence it probably represents long-term sub-crustal flow. The minor set of movements appears to represent the immediate response of the earth's crust to unloading as glaciers retreated from their Loch Lomond Advance maximum.

The complications introduced in this final section raise the possibility that some miscorrelation of marine shoreline fragments has occurred in the parts of Scotland where fossil shorelines are very fragmented. However, there is no reason to suppose that the broad pattern of events outlined in the preceding sections is significantly in error owing to these complications.

ACKNOWLEDGEMENT

The writer wishes to thank Dr D. G. Sutherland for his comments on the draft of this paper.

REFERENCES

Bailey, E. B., Clough, C. T., Wright, W. B., Richey, J. E. and Wilson, G. V. (1924). *Mem. geol. Surv. Scot.*

Bishop, W. W. and Coope, G. R. (1977). *In* "Studies in the Scottish Lateglacial Environment" (J. M. Gray and J. J. Lowe, eds), pp. 61-88. Pergamon Press, Oxford.

Brooks, C. L. (1972). *Trans. Inst. Br. Geogr.* **55**, 161-170.

Browne, M. A. E. (1980). *Scott. J. Geol.* **16**, 221-230.

Browne, M. A. E., Armstrong, M., Paterson, I. B. and Aitken, A. M. (1981a). *Quaternary Newsletter* **34**, 8-15.

Browne, M. A. E., Armstrong, M., Paterson, I. B. and Aitken, A. M. (1981b). *Quaternary Newsletter* **35**, 14-17.

Crampton, C. B. and Carruthers, R. G. (1914). *Mem. geol. Surv. Scot.*

Cullingford, R. A. (1977). *In* "Studies in the Scottish Lateglacial Environment" (J. M. Gray and J. J. Lowe, eds), pp. 15-32. Pergamon Press, Oxford.

Cullingford, R. A. and Smith, D. E. (1966). *Trans. Inst. Br. Geogr.* **39**, 31-51.

Cullingford, R. A. and Smith, D. E. (1980). *Boreas* **9**, 21-38.

Cullingford, R. A., Caseldine, C. J. and Gotts, P. E. (1980). *Nature* **284**, 159-161.

Dawson, A. G. (1979). "Raised shorelines of Jura, Scarba and NE Islay". Unpublished Ph.D. thesis, University of Edinburgh.

Dawson, A. G. (1980a). *In* "Studies in the Lateglacial of North-West Europe" (J. M. Gray and J. J. Lowe, eds), pp. 45-53. Pergamon Press, Oxford.

Dawson, A. G. (1980b). *Proc. Geol. Ass.* **91**, 339-344.

Dawson, A. G. (1982). *Scott. J. Geol.* **18**, 253-265.

Dickson, J. H., Stewart, D. A., Thompson, R., Turner, G., Baxter, M. S., Drndarsky, N. D. and Rose, J. (1978). *Nature* **274**, 548-553.

Eden, R. A., Carter, A. V. F. and McKeown, M. C. (1969). *Mar. Geol.* **7**, 235-251.

Gascoyne, M. (1981). *Quaternary Newsletter* **34**, 36-37.

Gray, J. M. (1972). "The inter-, late- and post-glacial shorelines, and ice limits of Lorn and eastern Mull". Unpublished Ph.D. thesis, University of Edinburgh.

Gray, J. M. (1974a). *Boreas* **3**, 129-138.

Gray, J. M. (1974b). *Trans. Inst. Br. Geogr.* **61**, 81-99.

Gray, J. M. (1975). *Proc. Geol. Ass.* **86**, 227-238.

Gray, J. M. (1978). *Trans. Inst. Br. Geogr.* **3** , 151-164.

Gray, J. M. and Sutherland, D. G. (1977). *In* "Studies in the Scottish Lateglacial Environment" (J. M. Gray and J. J. Lowe, eds), pp. 163-181. Pergamon Press, Oxford.

Holmes, R. (1977). *Rep. Inst. geol. Sci. Lond.* 77/14.

Hoppe, G. (1965). *Geogr. Annlr.* **A47**, 195-203.

Horne, J., Robertson, D., Jamieson, T. F., Fraser, J., Kendall, P. F. and Bell, D. (1893). *Rep. Br. Ass.* 438-514.

Horne, J., Robertson, D., Jamieson, T. F., Fraser, J., Kendall, P. F. and Bell, D. (1896). *Rep. Br. Ass.* 378-399.

Jamieson, T. F. (1865). *Q. Jl. geol. Soc. Lond.* **19**, 235-259.

Jamieson, T. F. (1906). *Geol. Mag.* **53**, 22-35.

Jansen, J. H. F. (1976). *Neth. J. Sea Res.* **10**, 1-43.

Jardine, W. G. (1975). *Boreas* **4**, 173-196.

Lawson, T. J. (1981). *Scott. J. Geol.* **17**, 301-303.
McCallien, W. J. (1937). *Proc. Soc. Antiq. Scot.* **71**, 174-206.
McCann, S. B. (1966). *Scott. J. Geol.* **2**, 84-95.
McCann, S. B. (1968). *In* "Geography at Aberystwyth" (E. G. Bowen, H. Carter and J. A. Taylor, eds), pp. 22-34. University of Wales Press, Cardiff.
McCann, S. B. and Richards, A. (1969). *Scott. J. Geol.* **5**, 15-25.
Morrison, J., Smith, D. E., Cullingford, R. A. and Jones, R. L. (1981). *Proc. Geol. Ass.* **91**, 197-209.
Newey, W. W. (1966). *Trans. Inst. Br. Geogr.* **39**, 53-59.
Ogilvie, A. G. (1923). *Trans. R. Soc. Edinb.* **53**, 377-404.
Paterson, I. B. (1981). *Rep. Inst. geol. Sci. Lond.* 81/1.
Paterson, I. B., Armstrong, M. and Browne, M. A. E. (1981). *Rep. Inst. geol. Sci. Lond.* 81/7.
Peacock, J. D. (1981). *Quaternary Newsletter* **35**, 45-54.
Peacock, J. D., Graham, D. K., Robinson, J. E. and Wilkinson, I. P. (1977). *In* "Studies in the Scottish Lateglacial Environment" (J. M. Gray and J. J. Lowe, eds), pp. 89-100. Pergamon Press, Oxford.
Peacock, J. D., Graham, D. K. and Wilkinson, I. P. (1978). *Rep. Inst. geol. Sci. Lond.* 78/17.
Rhind, D. W. (1972). *Trans. Nat. Hist. Soc. Northumb.* **4**, 159-164.
Richards, A. (1969). *Scott. geogr. Mag.* **85**, 122-131.
Robinson, M. (1977). "Glacial limits, sea-level changes and vegetational development in part of Wester Ross". Unpublished Ph.D. thesis, University of Edinburgh.
Robinson, M. and Ballantyne, C. K. (1979). *Scott. J. Geol.* **15**, 271-277.
Ruddiman, W. F., McIntyre, A., Niebler-Hunt, V. and Durazzi, J. T. (1980). *Quaternary Research.* **13**, 33-64.
Sissons, J. B. (1962). *Trans. geol. Soc. Edinb.* **19**, 83-99.
Sissons, J. B. (1966). *Trans. Inst. Br. Geogr.* **39**, 19-29.
Sissons, J. B. (1969). *Trans. Inst. Br. Geogr.* **48**, 19-50.
Sissons, J. B. (1972). *Trans. Inst. Br. Geogr.* **55**, 145-159.
Sissons, J. B. (1974). *Boreas* **3**, 41-48.
Sissons, J. B. (1976). *Scott. geogr. Mag.* **92**, 17-29.
Sissons, J. B. (1979). *J. geol. Soc.* **136**, 215-224.
Sissons, J. B. (1981). *Boreas* **10**, 1-17.
Sissons, J. B. (1982). *Trans. Inst. Br. Geogr.* N.S. **7**, 205-216.
Sissons, J. B. and Brooks, C. L. (1971). *Nature phys. Sci.* **234**, 124-127.
Sissons, J. B. and Cornish, R. (1982). *Quaternary Research.* **18**, 268-288.
Sissons, J. B. and Dawson, A. G. (1981). *Proc. Geol. Ass.* **92**, 115-124.
Sissons, J. B. and Smith, D. E. (1965a). *Trans. R. Soc. Edinb.* **66**, 143-168.
Sissons, J. B. and Smith, D. E. (1965b). *Scott. J. Geol.* **1**, 247-255.
Sissons, J. B., Smith, D. E. and Cullingford, R. A. (1966). *Trans. Inst. Br. Geogr.* **39**, 9-18.
Smith, D. E. (1968). *Z. Geomorph.* **12**, 388-408.
Smith, D. E. and Cullingford, R. A. (1981). *Quaternary Newsletter* **35**, 12-14.
Smith, D. E., Sissons, J. B. and Cullingford, R. A. (1969). *Trans. Inst. Br. Geogr.* **46**, 45-52.
Smith, D. E., Morrison, J., Jones, R. L. and Cullingford, R. A. (1980). *In* "Timescales in Geomorphology" (R. A. Cullingford, D. A. Davidson and J. Lewin, eds), pp. 225-245. John Wiley, Chichester.
Sutherland, D. G. (1981a). *Boreas* **10**, 247-254.
Sutherland, D. G. (1981b). "The raised shorelines and deglaciation of the Loch Long/Loch Fyne area, western Scotland". Unpublished Ph.D. thesis, University of Edinburgh.

Synge, F. M. and Stephens, N. (1966). *Trans. Inst. Br. Geogr.* **39**, 101-125.
Thompson, R. and Wain-Hobson, T. (1979). *J. geol. Soc.* **136**, 383-388.
Thomson, M. E. and Eden, R. A. (1977). *Rep. Inst. geol. Sci. Lond.* 77/17.
Walker, D. (1966). *Phil. Trans. R. Soc. Lond.* **B251**, 1-210.
Wright, W. B. (1911). *Geol. Mag.* **48**, 97-109.

TEN

Late Quaternary shorelines in Ireland: an assessment of their implications for isostatic land movement and relative sea-level changes

R. J. DEVOY

TECTONIC AND STRUCTURAL FRAMEWORK

The Pliocene-Pleistocene in Ireland forms a time of relative tectonic stability recording perhaps only small-scale earth movements. In long-term perspective this apparent stability has not been maintained. Ireland's location on the periphery of north-west Europe belies its former more focal role in the region's structural development (Naylor and Mounteney, 1975). For much of geological time it formed part of a tectonically dynamic zone of the earth's crust. The interaction particularly of the North American-Greenland and north-west European-Fennoscandian crustal plates since mid-Palaeozoic time, represented initially in the Caledonian orogeny, has resulted in major phases of rock folding, faulting and crustal subsidence (Kent, 1975; Laughton, 1975; Holland, 1981) (Fig. 1). From the Carboniferous Period onward the Irish Sea and Celtic Sea areas have been affected by a complex series of subsiding grabens recording progressive sediment infill, interspersed with phases of uplift, faulting and erosion.

These grabens formed part of the extensive northern North Atlantic-Eurasian Arctic rift systems, the opening up of which led to the separation of the crustal plates during the early Palaeogene and onset of the present phase of ocean spreading (Le Pichon *et al.*, 1977). Faulting and associated igneous activity identified in the North of Ireland resulted from this process, together with the development of northerly aligned structures and faults in the southern Irish Sea (Dobson, 1977a,b; Dobson and Whittington, 1973). Instability at this time, in combination with the later more minor Neogene earth movements, may have led to the widespread differential uplift of the coastal margins and relative downwarping of the central lowlands. Whatever the exact geological evolution of these areas, whether influenced by predominantly structural or geommorphological/erosional factors (Davies and Stephens, 1978), a distinctive upland rim has been produced around Ireland composed of lithologically varied but relatively resistant rocks (Fig. 2). The

SHORELINES AND ISOSTASY
ISBN 0 12 652960 4

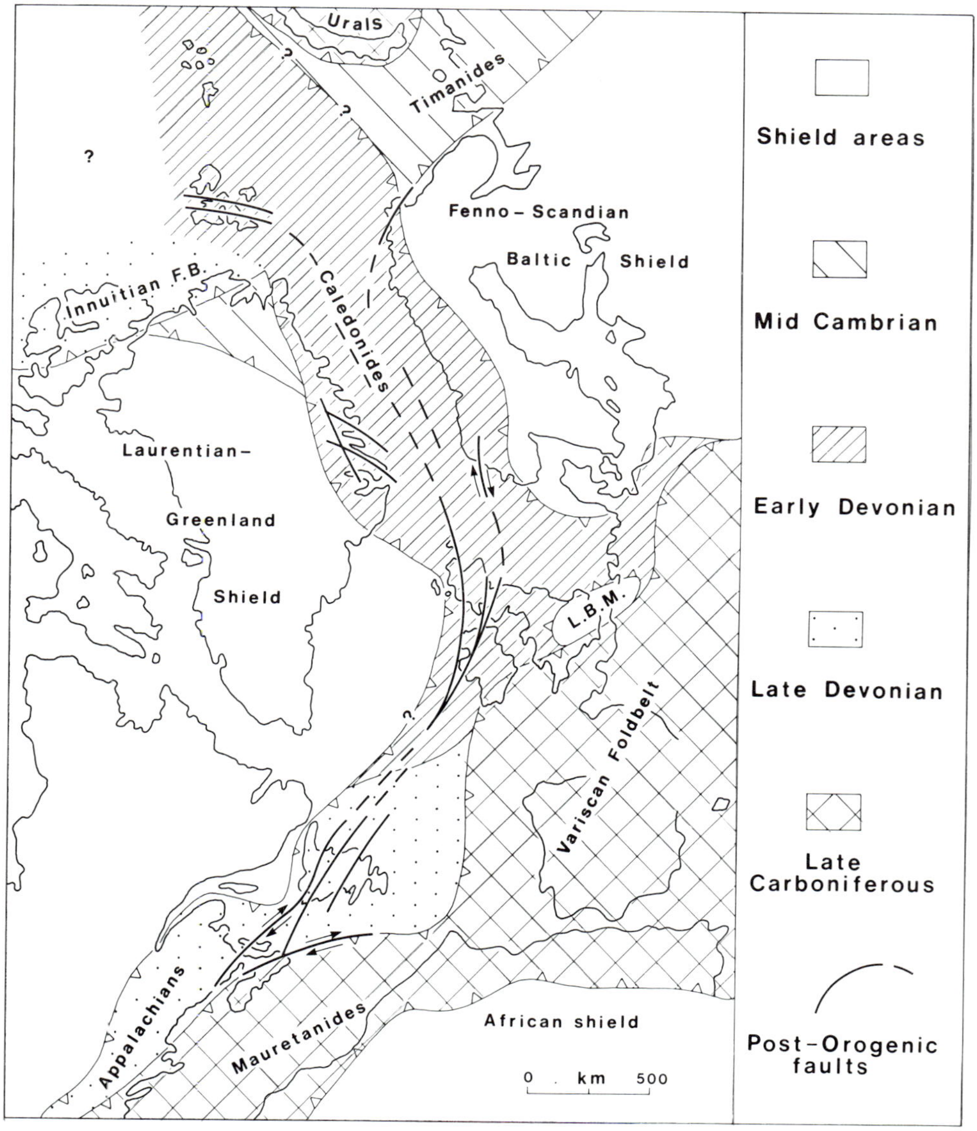

Figure 1. Structural and tectonic setting of Ireland during the Palaeozoic Era, showing the development of the main orogenic belts (after Ziegler and Louwerens, 1979).

presence of this feature in the coastal zone combined with strong erosional shore processes characteristic particularly of areas exposed to westerly Atlantic storms, have produced long stretches of spectacularly cliffed coastline. It is these zones that have recorded through the preservation of tilted raised shorelines, the recent dominance of isostasy, as opposed to tectonic activity, in movements of the land surface.

In terms of shoreline analysis the pattern of tectonic and structural development has perhaps only direct relevance for the form of Neogene sea-level changes.

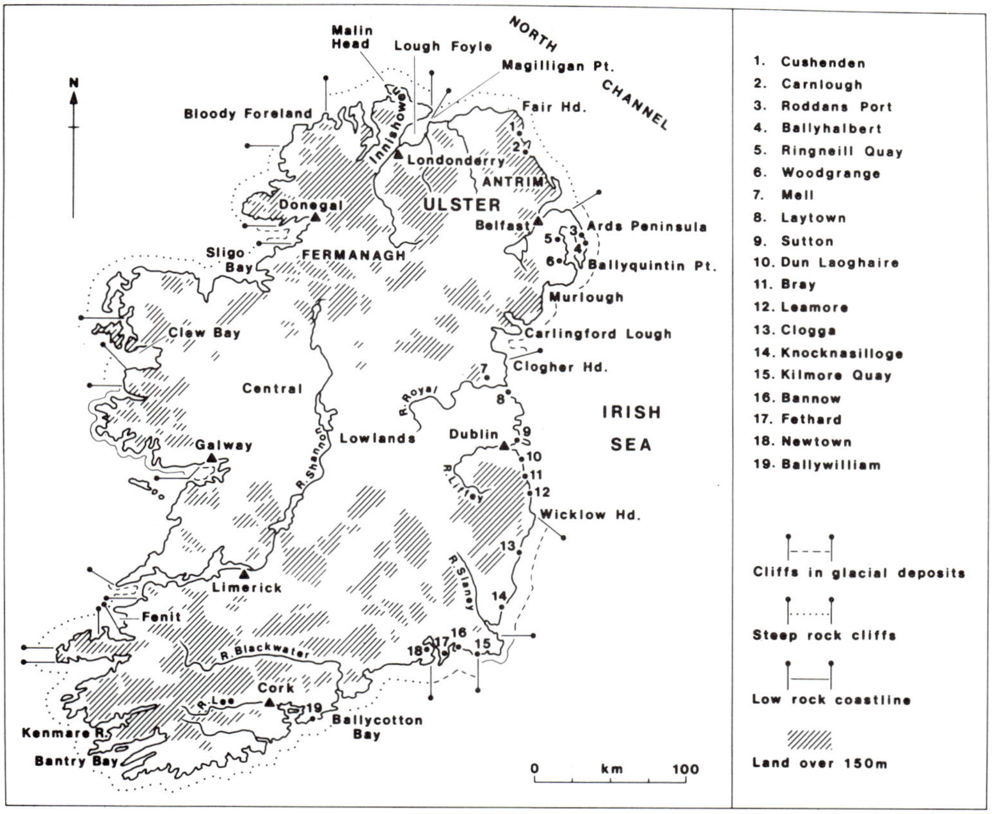

Figure 2. Coastal morphology and site location.

Distinct crustal movements have, however, been identified during the Pleistocene (Synge and Stephens, 1966; Dobson, 1977b) and show the likelihood of movement along major basin faults in the southern Irish Sea during this time. Such evidence, particularly in the context of the established long-term record of crustal instability, suggests that the possible influence of tectonic activity should not be ignored in the reconstruction of more recent shoreline features.

THE LITERATURE AND DATA BASE

The collection and analysis of shoreline/sea-level information has been undertaken by comparatively few people, though there has been a surge of work particularly upon northern areas in the past 30 years (Synge, 1970; Stephens *et al.*, 1975). Most writers on related Quaternary geological themes in Ireland have, at some stage, worked directly upon or reviewed sea-level problems (Praeger, 1893, 1896; Farrington, 1933, 1965, 1966; Wright, 1937; Charlesworth, 1957; Stephens, 1968, 1970; Mitchell, 1972, 1977; Synge, 1977a). Much information has been obtained indirectly through primarily archaeological studies as at Cushendun (Movius, 1940, 1942, 1953) and at other sites (Knowles, 1889; May and Batty, 1948; Mitchell,

1956). Excepting the early work of Wright and Muff (1904), few detailed litho- and biostratigraphical studies of sea-level indicators, accompanied by area-based levelling programmes, have been attempted on any scale until those of more recent times (Morrison and Stephens, 1965; Stephens and Synge, 1965; Singh and Smith, 1966, 1973). As a result there is a comparative lack of accurate shoreline/sea-level data that would conform to the specifications established by the IGCP Project 61 (van de Plassche, 1977).

Existing studies have been based upon a wide variety of sea-level indicators, with reconstruction of even local patterns of sea surface movement derived from a composite of evidence types. Marine inorganic deposits interleaved with freshwater organic sediments have been found in river estuaries and other low-lying zones throughout the coastline. Inside the main areas of isostatic uplift (Fig. 3) much of the data is of a morphological type, represented by raised beaches, wave-cut notches, shore platforms/planation surfaces and marine washing limits. Along the northern and eastern coasts many of the shorelines are represented by sediment accumulations. The higher lateglacial features, well displayed on the Inishowen peninsula and Londonderry coasts (Stephens and Synge, 1965), take the form of massive, rounded shingle and coarse-sized gravel storm beaches. The altitudinally lower beaches of presumed Holocene age found northward from Clew Bay to Dublin Bay are frequently composed of finer sand and gravel material. It would be incorrect, however, to use the criterion of particle size as confirmation of the age separation of these shorelines and a variety of lithologies and material sizes occur in all beaches. A more useful guide perhaps in distinguishing older from younger features is the presence or absence of periglacial structures and shell material. The former commonly show evidence of cryoturbation with frost shattering of *in situ* beach material and contain few shell remains. In contrast younger shorelines have large quantities of shells, particularly of *Littorina* spp. and *Patella* spp. and appear unaffected by periglacial activity.

The appearance of raised caves and wave-cut notches, as in Antrim, clearly expresses former phases of coastal erosion. Although taken to indicate marine activity in an upper intertidal environment, care is needed in interpreting them as evidence of former sea-level positions (Stephens, 1963). In areas of sub-horizontally bedded rock, as in the Carboniferous limestone of the Burren (Sweeting, 1955) or the chalk and basalt of north-east Ulster, difficulty may arise in distinguishing between true marine notches and structural or sub-aerially weathered benches. Shore platforms are frequently associated with these notches. Often traceable beneath neighbouring glacial sediments, these features may show evidence of ice striation and periglacial activity, with modern wave, stream and weathering processes retrimming their form (Stephens and Synge, 1965). In glacial sediments the preservation of shore platforms is more rare, though an extensive feature (*c.* 1 km wide) has been recognized between Laytown and Clogher Head on the east coast (Synge, 1977b). Marine washing limits defining the upper surface of wave erosion have been recognized at numerous sites (Synge and Stephens, 1966). As detailed by Synge (1969) and by Synge and Stephens (1966) the morphological

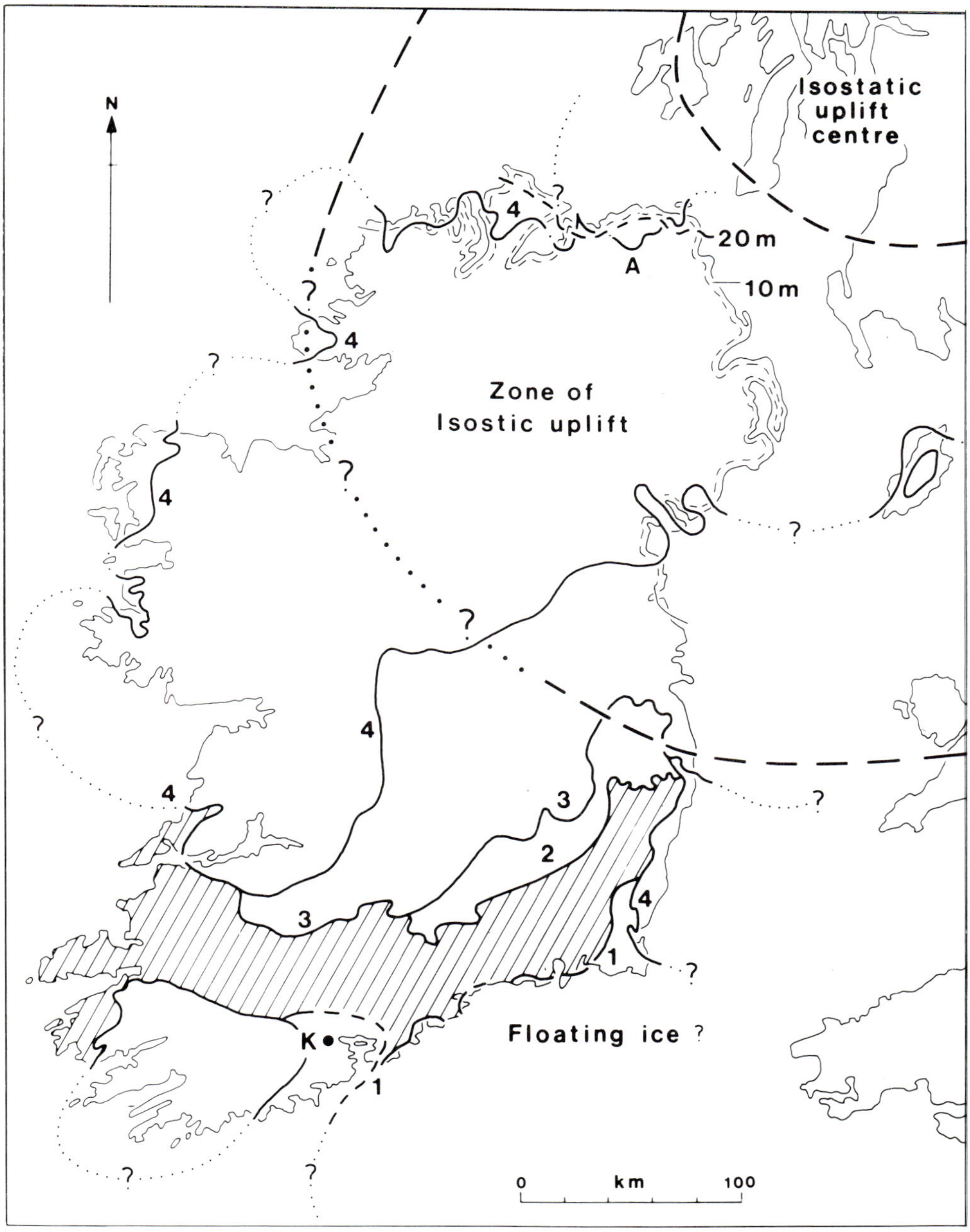

Figure 3. Zones of glacio-isostatic uplift and Midlandian stage ice limits (derived from sources cited in the text). Glacial sediments lying beyond the maximum known limits of Midlandian ice are shown as a shaded area. These deposits have usually been ascribed to the Munsterian stage. Continuous lines represent the position of major moraine limits. 1, Ballycroneen/Garryvoe substage; 2, Hacketstown substage; 3, Ballylanders and Blessington substage (K = Killumney moraine); 4, Drumlin Readvance substage (A = Armoy moraine). _ _ 20 m, isolines on marine limits of presumed lateglacial and Holocene age.

expression of this feature is varied. Most commonly, as at Clogher Head, the marine limit is marked by the upper level of bare rock though occasionally it may be recognized on a coastline by the lowest elevation of "perched" blocks.

DIFFICULTIES IN THE INTERPRETATION OF SHORELINE DATA

Analysis of factors affecting the development and subsequent interpretation of shorelines is presented elsewhere in this volume. Brief reference here is made to some of the problems particularly relevant to shoreline information in Ireland.

Sea surface levels

Much of the Irish coastline forms part of the Atlantic swell wave, meso-tidal environment characteristic of the British Isles (Kidson, 1977a). Maximum Spring Tidal ranges around the coastline vary between *c.* 1.0-4.5 m, with larger values recorded in restricted estuary and bay locations (Fig. 4) (Admiralty Tide Tables, 1981). These alterations in height for a particular tidal position result primarily from the influences of coastal shape, changing positions of amphidromic points, storm activity and degree of fetch. In lateglacial time modification of the coastal environment by ice may also have been of significance (Clarke *et al.*, 1978). The seasonal addition of meltwater from downwasting glaciers may have further locally raised water levels in restricted ice-bay environments. Although arguably of minimal importance, the operation of this factor may have affected the altitude of marine deltas and washing limits, such as those found in the Boyne Valley (Synge, 1977a,b). Palaeotidal work to assess the importance of such spatial and temporal changes in tidal height upon the recording of land/sea surface movements has not been undertaken in Ireland.

A related problem to that of tidal variation lies in determining the height relationship of a sea-level indicator to the sea surface that formed it. The vertical and lateral height ranges of features along a coastline, particularly of shore platforms, notches and beach ridges, frequently vary in relation to a defined sea surface position (Kidson, 1971; Jardine, 1977a,b; Carter and Orford, 1981). Recognition of these vertical changes in elevation (Stephens and Synge, 1965; Synge and Stephens, 1966) has rarely led to the especial treatment of such features or their avoidance in shoreline reconstruction. An exception is the possibly more selective use of data from shore platforms, and perhaps stems from a long-held understanding of their possible complex origin and often indeterminate age. Generally only platforms and notches cut in glacial deposits of apparently known age have been considered in shoreline studies.

The relationship of river and fluvioglacial outwash terraces to former sea-level positions are similarly problematical (Rose, 1978) and in this regard the many delta and terrace sequences associated with Irish rivers (Charlesworth, 1924; Farrington,

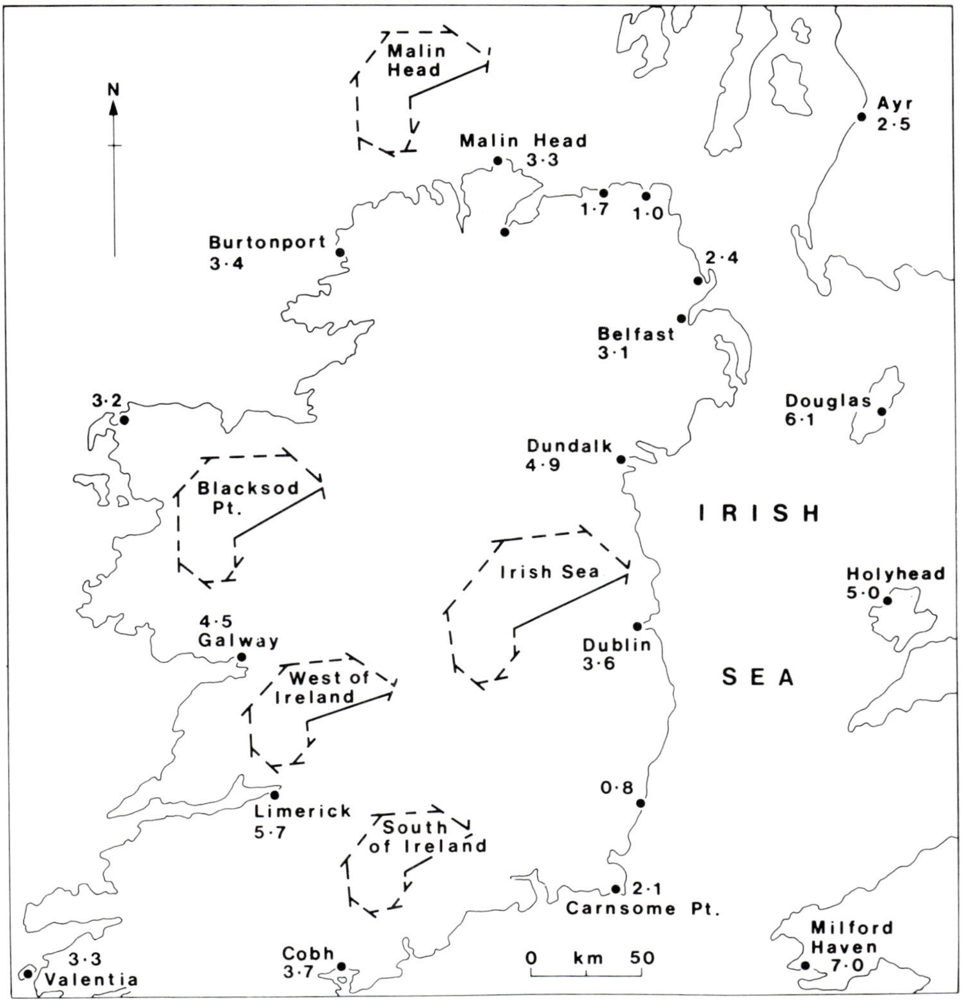

Figure 4. Spring tidal ranges (in metres) and resultant wind direction on Irish coasts (tidal and wind data based respectively on Admiralty Tide Tables, 1981 and Davies and Stephens, 1978; Atlas of Ireland, R.I.A., Dublin, 1979).

1954, 1959; Stephens, 1963), are probably not exceptional. Surface gradients may be related more to sediment supply and changes in water discharge than to sea-level movements and the features are of limited value in accurately fixing sea surface positions. In a discussion on the use of data from marine limits as a whole, Synge and Stephens (1966) indicate the difficulties in field identification of this type of feature. Possibility of the simulation of marine limits by mass movement phenomena or processes of sub-aerial weathering and erosion, however, was not mentioned.

Height measurements

Confusion exists in the literature on selection of a common levelling datum. Shoreline data have been referred variably to Ordnance Datum at Newlyn, Cornwall (O.D.), Dublin (D) and to Belfast (O.D.(B)). Alternatively Synge (1977a) computes his own value for mean sea level (m.s.l.). In other studies (Wright and Muff, 1904; Farrington, 1966; Synge and Stephens, 1966) high water mark (H.W.M.) has been used, although indication is rarely given as to which H.W.M. was employed or how it was established. The use of different levelling systems has resulted in a lack of clarity as to the wider inter-regional significance of the specific level used and potentially introduces unnecessary error in converting between the different data. The problem of the Dublin datum is that it is an arbitrary level, only indirectly related to sea surface positions represented by indicator points found on the coast (Admiralty Tide Tables, 1981). It further suffers from a lack of ready comprehension by those used to levelling systems based upon m.s.l. Although approximate inter-regional conversion between levels may be possible (Dixon, 1949; Jardine, 1976), it is suggested that for ease of comparability with other European datums O.D.(B) is used in future and all heights given here refer to this level. Based upon m.s.l. at Belfast, it has not been geodetically levelled to O.D. (Newlyn), but has the advantage of being an internationally defined datum. Selection of this in preference to the proposed new Irish datum at Malin Head results from the possible continued influence at the latter of isostatic land warping (Todd, 1981).

Dating

Most of the range of dating methods available in shoreline studies (Bowen, 1978; Rose, 1981) have been used in Ireland. Shoreline chronology, however, is based largely upon inference of age from morpho- and lithostratigraphic sources. Dating here is dependent upon the validity of the regional/national Quaternary geological framework. The sequence and areal pattern of lateglacial ice movements in Ireland, let alone those of earlier glacial and interglacial phases, is still open to debate (McCabe and Hoare, 1978) (see pp. 238–242), and reliance upon a chronology derived from such sources would be unwise (Sissons, 1967). As illustrated at many sites (Mitchell, 1976), archaeology and palynology have long been applied to the dating of coastal sequences. Limitations resulting from the relative and time transgressive nature of their dating framework in space, has led to their replacement by, or use only in conjunction with direct measures. Alternatively, both archaeological remains and the examination of pollen from interleaved marine and terrestrial sediments have been valuable in demonstrating the common occurrence of unconformities in shoreline development, as illustrated at Ballyhalbert, Co. Down (Morrison and Stephens, 1960). Here, as at numerous similar sites at which organic sediments now outcrop in the coastal zone (Godwin, 1975; Mitchell, 1976), a date for the marine inundation cannot be inferred from the age of the biogenic material.

Use of radiometric dating methods has been limited to some extent by the absence of suitable material, although generally little work has been attempted. The abundant shell remains in Holocene beaches have not been studied in detail, although a programme to investigate these is now underway (Synge, personal communication). The reliability of the radiocarbon dates that do exist, derived largely from organic muds and plant material, may now be in question (Pearson, 1980) and results should be used cautiously. Recent developments in dating by amino acid diagenesis has re-stimulated efforts by Bowen and co-workers (Department of Geography, University College of Wales) to estimate the age of the otherwise chronologically enigmatic "pre-glacial" shore features around the coastline.

Shoreline correlation diagrams

Shoreline height/distance diagrams and related isobase maps have been produced in a number of papers (e.g. Synge and Stephens, 1966; Synge, 1977b). Both techniques of showing level change with distance assume that the shoreline features correlated are of the same age. Few direct dates for these exist and studies use the properties of height and gradient as a means of inter-areal correlation between shoreline fragments.

The absence of precise shoreline dating, accurate stratigraphic correlation and the influence of spatially variable patterns of deglaciation on the timing of shore features (Hillaire-Marcel, 1979; Gray *et al.*, 1980), makes the value of such height/distance diagrams for land-/sea-level reconstruction questionable. Studies often use a variety of sea-level indicators in diagram construction (Synge and Stephens, 1966) and the resultant shorelines rarely represent a single sea surface position. This practice has been criticized in land-/sea-level studies as a whole (Sissons, 1967; Tooley, 1978; Devoy, 1979). Unless a precise height reconstruction of the relationship of an indicator to a common sea surface is available errors may result in determining the trend and gradient of shore planes. Differences in sea surface position represented by the data have at best only been estimated. No account has been taken in the "smoothed" shoreline gradients of the probiable inter-site variations in tidal amplitude (Fig. 4). The insertion by hand of shoreline planes lacks objectivity and unlike studies from elsewhere (Gray, 1975; Andrews, 1967), regression analysis has not been applied to the data. The value of this latter technique, however, will also be limited in determining gradients unless applied to a uniform data set.

The possible metachroneity of some shorelines in Ireland (Prior, 1968; Stephens and McCabe, 1977) undermines the validity of establishing isobase maps for the country and only isolines for the height of the marine limit have been drawn (Fig. 3). The employment of the shoreline relation diagram in regional correlation of data also represents problems (Andrews, 1970; Marthinussen, 1960). Difficulties in its use centre in Ireland, as elsewhere, upon accuracy in dating and doubts as to the uniformity of the pattern of lateglacial ice loading and rates of land uplift.

RELATIVE HIGH-LEVEL PLEISTOCENE SHORELINES

The "pre-glacial" raised beach and rock platform

Definition of the "pre-glacial" beach and associated shore platform has been made by Wright and Muff (1904). Their significance for Quaternary geology in Ireland has been much discussed (Mitchell, 1962, 1972), with the shorelines seen variably as developing during phases of interglacial or even interstadial high sea levels (Synge, 1977a). The platform occurs commonly along the south and east coasts (Fig. 5B) though its appearance is less frequent in northern and western areas (Davies and Stephens, 1978). Preserved almost exclusively where developed in rock, the platform appears as a composite feature (Fig. 6). In the south three levels have been observed (Farrington, 1966), occupying a vertical height range between *c.* − 1.0 m to *c.* +7.6 ± 0.5 m O.D.(B). The surface of the "Main Platform" (Wright and Muff, 1904) maintains a level of *c.* +5.2 ± 1.0 m O.D.(B) for approximately 350 km west of Carnsore Point. On the east coast the platform and associated notch rise northward in height from *c.* +9·0 to +20·0 m O.D.(B). Synge (1981) associates the high levels in the Dublin Bay area with tectonic uplift of this zone, northward of which the elevation of the platform falls again toward Belfast Lough (Stephens, 1957). At Malin Head an inclined rock platform truncated seaward at *c.* +9·5 m O.D.(B) may also be correlated with this feature.

Between headlands exposed to high wave energy levels and the lower energy bay environments platform altitudes may differ by *c.* 2–3 m. Apart from possible isostatic and tectonic influences, these changes in bay–headland height, consequent upon shoreward changes in wave energy dispersion, may, together with spatial alterations in tidal amplitude, explain much of the recorded vertical range in platform elevation. Further, the three-dimensional problems of measuring sections influenced by subsequent changes in coastal shape may also complicate the accurate height correlation of levels. Where the platform is fragmented and no upper limit or cliff-notch is visible, great difficulty arises in fixing the position of the exposed platform surface in the feature's original cross-sectional profile. Comparison of heights between coastal sections may thus be misleading with incorrect correlations established between unrelated surfaces. This may account for the differences in platform level noted at Ballywilliam, Co. Cork and at other locations (Farrington, 1966).

A composite section showing the stratigraphy overlying the platform is given in Fig. 5A. Descriptions of individual coastal sections are detailed in Wright and Muff (1904) and Farrington (1966). Evidence for a raised beach above the platform comes almost exclusively from southern and western coasts. Beach heights have been given as *c.* +3 to +5 m O.D.(B) (Mitchell, 1962; Synge, 1981). Disparities with levels given by Farrington (1966) may result from the latter's use of H.W.M. as a datum. Identification of the raised beach has depended much upon subjective interpretations. Quantitative analysis of beach material, in terms of particle shape, size and

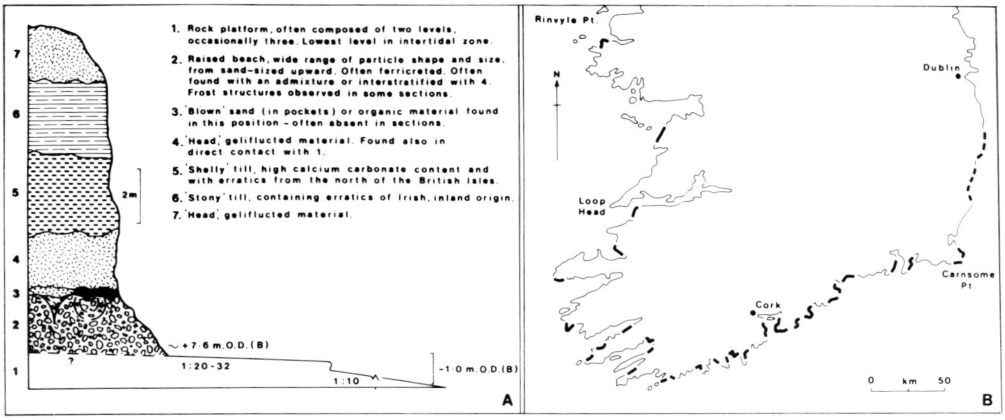

Figure 5. (A) Composite stratigraphy of sediment sequences found overlying the "preglacial" rock platform; (B) generalized distribution of "pre-glacial" rock platform exposures.

Figure 6. "Pre-glacial" rock platform at Howe's strand, Co. Cork, showing the two platform levels commonly found.

lithological composition or the comparison of sections with studies from modern beaches has rarely, if ever, been undertaken. Distinction between freshwater, terrestrial and marine sediments, all of which occur commonly together in coastal exposures may be difficult on visual grounds alone. Despite differences in structural forms, the common denominator of water flow produces strong similarities between deposits of both marine and fluvial/fluvioglacial origin. At Fethard a ferricreted beach sand and gravel has been identified as overlying the rock platform (Huddart, 1977). Recent examination of the section has aroused doubts as to its marine origin (Carter and Orford, 1981). The deposit appears as an angular, stratified grit resembling more the overlying head than material found in either modern beach environments or analogous coastal exposures. Possible incorrect identification of such sediments as representing the raised beach further highlights the areas of probable error in height measurement and inter-site correlation of the feature as a whole.

Beach/platform age and their stratigraphical implications

Establishment of the Munsterian and Midlandian glacial chronology, in which the interpretation of these shorelines has become an integral part, has been criticized as unnecessarily complex (Synge, 1977a, 1981; Huddart, 1977; Culleton, 1978; Warren, 1979). Details of alternative Quaternary successions are given in Table 1. To a degree shoreline dating has been achieved through a circular argument, whereby the lithostratigraphy dates the platform/beach in one locality, whilst at another the presence of these features is used as a stratigraphic marker in interpreting the age of associated Quaternary deposits. Such correlations and the assumption that the shorelines may form isochronous surfaces (Warren, 1979; Synge, 1981) could be looked at more closely in the light of stratigraphic principle and the lessons of marine process studies. In many localities the identification of the glacial stages rests almost entirely upon differences in the derivation of tills and related glacigenic sediments. Rarely is there an objective measure of the time-span represented either by the glacial deposits themselves or by the boundary zones between them. As present day studies of glacial physics and ice-depositional behaviour suggest (Paterson, 1981), there is no reason why the different till units identified should not all belong to the same glacial stage (McCabe and Hoare, 1978). On this basis the timing for the earliest formation of the raised beach would thus move forward from the Gortian (Hoxnian/Holsteinian) to the last interglacial stage (Ipswichian/Eemian).* In other situations changes in the spatial location of glacial deposits, drift morphology (Charlesworth, 1928; Wright, 1937; Farrington, 1954) and the identification of soil weathering horizons (Farrington, 1944; Warren, 1977) have been used to distinguish the two cold-stage sequences. Each, however, suffers from the same inability to measure a finite time-scale for events and their chronological significance must be treated with caution. The influence of subjective assessment is again important in the stratigraphic interpretation of these and other deposits. At Kilmore Quay, the Irish Sea till overlies a periglacial material which seals the underlying raised beach and shore platform. Close by at Nemestown

*The use of these terms in the text, based on the current stratigraphic scheme for regions within the British Isles and their correlation with those in Europe (Mitchell *et al.*, 1973) is included only in an attempt to aid clarity. It is not intended to imply necessarily, by their inclusion here, their time equivalence.

TABLE 1

Selected regional stratigraphic models proposed for the Irish Quaternary

Stages	Warren (1979) (general)	Mitchell *et al.* (1973) (south-east + south)	Synge (1981) (south)
Littletonian			Raised beaches north of Wicklow
Midlandian (last cold stage)	Nahanagan moraine Woodgrange and Derryvree muds "Midlandian" tills Ballyvoyle till Garryvoe till Ballycroneen till Ballycroneen till	Nahanagan cold sub-stage Shortalstown Upper silt Shortalstown Lower silt Blessington moraine Shortalstown Upper shelly till Irish Sea formations including the Screen Hills formation, Rosslare formation	W6 Nahanagan sub-stage W5 ? cirque glaciers/Bantry sub-stage W4^{b} Wicklow/Colbinstown sub-stage W4^{a} Blessington/Athdown/Glenealy/ Killumney sub-stage W3 Hacketstown/Brittas/Screen Hills sub-stage W2 Kilmore Quay stage W1 Ballycroneen/Garryvoe sub-stage
Interglacial (Ipswichian)	Courtmacsherry raised beach Newtown and Ballymakegoge peats	Shortalstown Estuarine sands	Interglacial plant beds (Kilbeg and assoc. sites) Courtmacsherry raised beach
Munsterian (penultimate cold stage?)	Baggotstown till	Garryvoe + Ballycroneen shelly tills Shortalstown Lower shelly till Bannow till Brittas. moraine Kilmore Quay Upper till Kilmore Quay shelly till Kilmore Quay Head	? Aughrim sub-stage/ice from E. in N. Co. Kerry (Connachtian?)
Interglacial (Gortian?)		Kilmore Quay and Courtmacsherry raised beaches Ballykeerogemore, Ballymakegoge, Newtown and Kilbeg peats and muds	Marine rock platform
Cold stage			Bannow/Clogga tills/ice from N. in E. Co. Cork
Culleton (1978) (south-east)	Huddart (1981a,b) (south-east)	Synge (1977b) and McCabe and Hoare (1978) (Model A) (east)	McCabe and Hoare (1978) (Model B) (east)
		Sutton, Port and assoc. raised beaches ca. +2 to +6 m O.D.(B)	Sutton, Port and assoc. raised beaches
Forth Mountain Gravels Blackwater formation Ballinaclash member Screen member Knocknasilloge member Macamore member Greenore member Bannow formation includes Blackhall member Old Ross member	Screen hills complex	Lateglacial shorelines rising northwards, +7 m O.D.(B) Loughshinny to +21 m O.D.(B) Port Dunany formation Togher formation (Proudfootstown and association gravels) Gilltown formation Gormanstown formation Mell formation Drogheda formation	Lateglacial shorelines Dunnany formation Togher formation Gilltown formation
	Knocknasilloge member Shortalstown Estuarine sands Castleannesley till/Kilmore Quay shell till Head Courtmacsherry raised beach and platform Ballytrent and Nemestown tills		Gormanstown formation Mell formation Drogheda formation Slievethoul formation

a distinctive sandy deposit underlies the Irish Sea till and has been considered variably as a gumbotil (Farrington, 1939), a gleyed periglacial deposit (Mitchell, 1962) and a weathered till (Huddart, 1977). If the latter view together with the re-interpretation of the overlying till as Midlandian are correct, then the Nemestown beds may provide some evidence for deposits of the earlier (Munsterian) cold stage, although detailed analytical work is lacking.

Recognition (Synge, 1979, 1981) of possibly another cold stage (Connachtian), with its relevance for the age of the rock platform, may equally be seen as the product of stratigraphic speculation (Table 1). Argument that the Clogga, Bannow and associated glacial deposits may represent a third cold stage (Elsterian) in Ireland is based upon speculative and probably impossible stratigraphic correlation with European Pleistocene sequences. Development of the argument to a search for the theoretical existence of deposits of the intervening second cold stage (Connachtian/Saalian), encounters the same problem of unnecessary stratigraphic complexity mentioned previously (Synge, 1977a, 1981; Warren, 1979). At Clogga (Fig. 2) an exposure of the rock platform rises inland from the present intertidal zone to a cliff notch at + 12 to + 14 m O.D.(B). This surface merges laterally with an adjoining planed level truncating the basal Clogga till. Both surfaces are overlain by Irish Sea till. Assuming the uniformity in age and origin of the two surfaces, the importance of this drift-cut section of the shore platform is unique, as a timing for the Clogga till would allow, at least in this locality, a definition of the platform's age. Interpretation of the Irish Sea till as Midlandian has led formerly to identification of the Clogga till, on position and lithological criteria, as Munsterian in age (Synge, 1977a). Formation of the shore platform in this locality would on this basis therefore have occurred probably during the last (Ipswichian) interglacial. There is no stratigraphic reason, however, why either the Clogga or Bannow till in Wexford with which it is correlated should be of Munsterian age. Their formation in the Midlandian, nevertheless, would make an interglacial origin for both the platform and overlying raised beach impossible. Re-interpretation of the stratigraphy and construction of the Connachtian again conveniently places development of both these features in interglacial stages.

Doubts about the reliability of the biostratigraphic underpinning by Gortian interglacial sediments of the two-stage glacial stratigraphy and therefore interglacial (Gortian/Hoxnian) age of the raised beach have been discussed by Warren (1979) and Synge (1981). Assuming the glacial sedimentary record to represent only a single cold stage, it is argued the Gortian sites on stratigraphic grounds alone must be of last interglacial (Ipswichian) age. The fragmentary nature of the vegetation record of many "Gortian" deposits makes inter-site correlation difficult and the possibility exists that the sites represent more than one interglacial sequence (Mitchell, 1976; Warren, 1979). This coupled with the common existence of an unconformity with the suprajacent glacial sediments and doubts as to whether such deposits are of glacial origin at all (Synge, 1981), reduces the need to question the floristic correlation of the Gortian with the Hoxnian/Holsteinian. At both Newtown and Fenit the organic deposits overlying the raised beach could equally be of interstadial origin (Watts, 1959). The low pollen percentages at the latter of *Rhododendron*, *Taxus*, *Abies*, *Ilex* and other species taken as indicative of Gortian age, may be the product of inwashing and reworking and neither site can be used in dating the shoreline sequence.

Further rejuvenation of the one or two cold-stage argument and thus dating of the shoreline features has been aroused by Huddart (1981a,b). Identification of a shallow-water (< 20 m), brackish interglacial foraminiferal fauna at Knocknasilloge and Clogga re-establishes the possibility that the underlying Irish Sea tills belong to a previous (Munsterian) cold stage. The correlation of the assemblages with the last (Ipswichian) interglacial is based upon arguments of stratigraphic probability. The results at Clogga appear to be at variance with earlier analyses showing the sediments to contain an arctic marine fauna (Huddart, 1977). The possibility is acknowledged that the foraminifera found at both sites may derive from glacial and fluvioglacial reworking of earlier marine deposits. The *in situ* nature of the Shortalstown estuarine sands and gravels (Colhoun and Mitchell, 1971) with which these sediments are correlated come under similar suspicion, showing evidence of distortion and lateral movement by ice thrusting. In the southern Irish Sea the existence of marine clastic sediments containing ostracoda and foraminifera inter-leaved between tills of Irish Sea origin (Garrard, 1977), are seen as forming the offshore facies of the Knocknasilloge member and to be of comparable Ipswichian age. The degraded and fragmentary nature of the macrofauna combined with the presence of a small arctic microfaunal element suggests that the sediments here may also be derived, having been deposited under high energy conditions in the area of the St. George's Channel. The lack of reworking exhibited by the dominant marine boreal microfaunal assemblage would not support this view. There seems little reason, however, why the deposits should not be of interstadial rather than interglacial origin. Criticism of Huddart's views based on sedimentological and stratigraphic criteria have been levelled by Thomas and Summers (1981). They consider the Knocknasilloge member to represent deposition in a glacio-lacustrine and not marine environment. The high elevation above the rock platform of the presumed marine Askintinny member at Clogga poses further problems of interpretation. Unless the area has undergone either some isostatic or tectonic displacement for which there is little supporting evidence, difficulties may arise in explaining the presence of such sediments, representative of an estuarine or off-shore marine environment, in a position occupied elsewhere along the south coast by near-shore beach material.

The rock platform-raised beach show similarities in position and morphology with features found along many European coastlines at *c.* +1 to +15 m m.s.l. and dated in part to the last interglacial (Guilcher, 1969; Bowen, 1977; Kidson, 1977b; Mottershead, 1977; Dawson, 1980). These criteria coupled with stratigraphic data have led Bowen (1973) to propose a similar age for the Irish feature. Subsequent amino acid diagenesis of shells from the similarly positioned raised beach in South Wales broadly confirms the earlier dating conclusions (Andrews *et al.*, 1979). Studies place the stratigraphically younger ("*Patella*") beach recognized here in amino acid Group 1 and the older ("*Inner*") beach in Group 2. Conversely recent interpretation of mammalian fauna from these and related sites has strengthened doubts as to its single interglacial age (Sutcliffe, 1981). To a degree this view is aided by the time gap in beach formation implicit in the amino acid stratigraphy. A perhaps puzzling aspect of the Irish raised beach is its apparent lack of shells, contrasting with its presumed counterparts elsewhere. If these beaches are contemporaneous there seems little reason in terms of ecology or subsequent environmental evolution for this difference, although some of the South Wales beaches are found in protected cave sites. Alternatively, the absence of a shell fauna may suggest its formation under quite different climatic and energy conditions. Given the wide variety of datum levels used, differences in marine and tidal environments and the lack of reliable dating evidence, the correlation of this shoreline in Ireland with those elsewhere may only be misleading. The possibility remains that the Irish raised beach need not represent a time equivalent feature.

Interpretation of the shorelines

The view that the shore platform is an inherited feature of at the earliest Gortian age, or alternatively is the product of persistent reoccupation of the shore zone during successive interglacial high sea levels, is based only upon inference. The idea conforms with a belief that the length of time needed for platform development could not be accounted for by short-lived or Holocene marine planation processes (Steers, 1952; Stephens, 1957; Wright, 1967; King, 1972). Modern studies on the efficiency of the mechanisms and rates of platform development in varying environments would disagree (Dawson, 1979). Evidence suggests that rock platforms may reach an equilibrium profile after *c.* 2500–5000 years of exposure to water-wave activity (Trenhaile, 1980). Contemporary changes in the distribution of wave energy across the littoral zone resulting from the interaction of rock structure, fetch and tidal amplitude may, along some stretches of Irish coastline, be responsible for the tiered structure characteristic of this feature (Carter, personal communication). Although the upper part of many platform sections are now covered by halophytic vegetation, giving the impression of inheritance and old age to the land form, their surfaces are still reached by storm waves and lower levels are visibly under marine attack. The probable complex process origin of the feature is further evidenced by possible indications that parts were developed during phases of cold climate. Some sites show the platform truncating both the main body and individual fragments of gelifluccted material (Orme, 1966). Elsewhere, descriptions of narrow, sub-horizontal sections of the platform with sharp, angular rock surfaces bear resemblance to profiles from polar shore platforms (Dawson, 1979). These sections contrast markedly with the more common inclined "ramp-abrasion" profile of the platform characteristic of development under humid conditions. At many sites the rock platform appears to be quickly destroyed once its protecting drift cover has been removed by wave or sub-aerial action (Farrington, 1966). This, coupled with its extensive preservation and conformity with the shape and height of the modern shoreline, may indicate that neither this feature nor the suprajacent raised beach have been exposed to prolonged marine action in the Quaternary. The suggestion, however, that the platform and raised beach were formed in the same interglacial and that the hiatus between them is of short duration is not new (Orme, 1966; Synge, 1977a). Numerous possibilities for fitting the coastal stratigraphy to contrasting patterns of shoreline age and environmental change have been produced. Further such constructions here could only add to the confusion. The widely held assumption that the platform is older than the beach and has undergone a uniform evolution must, in the absence of accurate dating and the evidence of modern shore processes, be much more rigorously tested. Work to survey the feature in detail along the southern and western coasts is at least now underway (Warren, personal communication).

As with the shore platform the beach may have a complex history being composed of sections formed at widely different times and under changed energy conditions. Exposures dominated by gravel-sized material probably represent upper tidal zone sedimentation and storm activity. The occurrences of thick sand sequences may reflect times of high sea surface and deposition at lower intertidal levels under reduced energy conditions. Alternatively, the ranges in sediment types found could be approximately contemporary in origin, resulting from the landward progradation of the shore zone and from variations in energy levels between coastal sections. The intricate association of periglacial and beach deposits indicates that parts of the raised beach were formed in a cold environment. Supraposition of till directly upon the head (Fig. 5A) argues for their formation during the same cold stage and this highlights the importance of dating the tills for beach chronology. An early Midlandian age for the basal Irish Sea till, as supported by an objective interpretation of lithostratigraphic data, would place beach formation in the last interglacial stage. Some elements of the shoreline may have formed at the end of this stage, early Midlandian time, in association with the head.

Indication of sea surface position at the opening of the last cold stage

The presence of the basal Irish Sea till along the south coast probably evidences the persistence of a relatively high sea level after the close of the last interglacial. Implicit in the current lack of any indication of isostatic uplift in the south of Ireland is that this till is unlikely to have been deposited by a land-based glacier and most probably derived from the action of marine shelf-ice. The occurrence of such floating ice, forming from an early rapid glacial advance, would explain adequately the restriction of till here to the coastal zone, together with the near constant elevation for *c.* 90 km of correlatable deposits south of Wicklow Head. The reduced energy conditions associated with such ice would be consistent with the near absence of striations on the rock platform east of Ballycotton Bay. In contrast, surfaces west of this point, formerly affected by stronger inland ice movement, exhibit pronounced ice scratching. Preservation of the raised beach almost exclusively in southern and western coastal areas, suggests that its absence from the east coast may be the result of its exposure here to the subsequent higher energy movements of land-based ice. The apparent anomaly of a high relative sea-level position at a time of presumed continental glaciation in north-west Europe, may be set in the context of the probable occurrence of independent, regional patterns of glaciation and climatic development (Andrews, 1973) and is now less difficult to accept. The possibility of a relatively high global interstadial sea level during Midlandian time (Milliman and Emery, 1968; Thom, 1973) has led Synge (1977a) to propose tentatively this event for beach formation and emplacement of the basal till. Data from the marine $^{18}O/^{16}O$ record would, however, dispute this view (Shackleton and Opdyke, 1973; Shackleton, 1977). Ocean levels probably lay below those of the present during any such rise and if so, would have prevented the shelf-ice deposits from reaching their current location. Further, if the glacial sediments are in a primary position and have not slumped over the raised beach as suggested by Synge (1977a), such a late date for the beach and overlying till would imply that most Irish Quaternary deposits result from approximately the last 35 000 years!

The Mell formation and isostatic uplift

On the east coast the occurrence of clays, sands and gravels of the Mell formation at *c.* +29 to +40 m O.D.(B) present something of an enigma for the reconstruction of sea surface movements. The deposits contain arctic to boreal-arctic fauna indicative of accumulation at depths of *c.* −90 to −110 m O.D.(B) in cold saline water probably close to an ice front (Colhoun and McCabe, 1973). In conjunction, these figures suggest that at the time of marine inundation at least the present Irish Sea coastal margin in the vicinity was isostatically depressed by *c.* 120–150 m. If the sediments are *in situ*, then marine inundation to a position close to that of the present day took place at a time of probable glaciation in Ireland and occupation of part of the Irish Sea by ice. The age of this event remains uncertain and depends upon stratigraphic evaluation of the timing for implacement of the overlying Irish Sea (Tullyallen) till. Suggested dates for the Mell formation and associated tills range from that of the Munsterian stage to late Midlandian times (Doncarney phase) (McCabe and Hoare, 1978). In the absence of firm radiometric dating or stratigraphic evidence the correlation (Synge, 1977a) of the deposits with the mid-Midlandian interstadial deposits of Co. Fermanagh remains speculative. An anomaly arises from the high values assumed for isostatic uplift in this area, which exceed the possible *c.* 80 m emergence recorded for the centre of the late Midlandian land uplift in north Donegal (Stephens *et al.*, 1975). If this figure is correct then disparity with that from Mell may indicate a need to re-examine the timing and stratigraphic position of the site, as well as giving much closer consideration to former estimates of isostatic recovery in Ireland.

LOW STANDS OF SEA LEVEL

Evidence for low relative sea-level positions during the Pleistocene is derived primarily from the incised buried channels of the river estuaries. In the Liffey a bedrock channel at *c.* 35 m below present sea level can be traced to > −80 m O.D.(B) near Dun Laoghaire (Whittington, 1977). Similar rock-cut features found in the outer estuaries of the Lagan, Shannon and rivers along the Cork/Wexford coastline attain levels of *c.* −40 to −58 m O.D.(B) (Lamplugh *et al.*, 1904; Farrington, 1959; Orme, 1964). Evidence of downcutting to a lower sea surface is also preserved in rock-cut benches at *c.* −15·6 to −18·6 m O.D.(B) in the Blackwater and Slaney rivers. The bedrock base of many of the channels is overlain directly by till, formerly dated in the case of the latter rivers to the Munsterian glaciation. The glacial deposits in the Liffey, however, are correlated with tills of mid- or late Midlandian age, found elsewhere in the Irish Sea beneath Holocene marine sediments (Pantin, 1977). In the light of the reassessment of Irish glacial stratigraphy the tills of the south coast rivers may also now be ascribed to the Midlandian stage. These glacial deposits only allow a *terminus ante quem* to be set upon the development of the channels. Observation that the shore platform conforms to the outer-estuary shape of a number of rivers (Orme, 1966) suggests that at least some stages in their development may pre-date this coastal feature. It is probable the channels represent phases of sub-aerial fluvial erosion during times of low Pleistocene sea level and that the courses have been inherited, as in the case of the Liffey, from late Cainozoic river systems. Height variations in some channel profiles may be the product of rock structure and shape modification during subsequent glaciation. The final erosional phase has been attributed to fluvioglacial meltwater action at the close of the Midlandian, with a channel cutting down through the till to a base level possibly 130 m below present sea level (Whittington, 1977). Offshore a series of sinuous north-south aligned pits and troughs at maximum depths >97 m represent the southward extension of this channel system. Dates for deglaciation of the Irish Sea places development of the drainage network to *c.* 14 000–16 000 B.P. Renewed downcutting through the till in the Blackwater and Slaney rivers to *c.* 26·3 m O.D.(B) may similarly be correlated with this time.

In Cork Harbour brackish-marine sediments, containing a possible interglacial stage II–III "Gortian" pollen assemblage, occur between *c.* −13 to −30 m O.D.(B) (Devoy, in preparation). The deposits form part of the buried channel infill of the river Lee, and if *in situ* appear to evidence a time of low interglacial sea level. Whether the organic rich clay-silt sequence here represents sedimentation in an estuarine or deeper offshore environment remains to be established, as does its time relationship with the rock platform and raised beach shorelines.

LATEGLACIAL AND HOLOCENE LAND/SEA LEVEL MOVEMENTS

Many synoptic papers of shoreline and related sea-level data are now available (Stephens, 1968; Mitchell, 1972, 1977; Mitchell and Stephens, 1974; Stephens and McCabe, 1977; Synge, 1977a,b, 1980a,b, 1981; Davies and Stephens, 1978; Carter, 1982). Reference must be made to these or to the original sources cited for detailed discussion of site information and environmental reconstruction. In view of the common occurrence of raised lateglacial shorelines in north and north-east Ulster, their apparent absence on coastlines west of Bloody Foreland, combined with the resultant anomalous distortion here of the presumed isobase lines, is puzzling. Shoreline evidence between Bantry Bay and Donegal Bay is restricted to the recognition of sand and shingle ridges. These occur in a uniform height range of *c.* +1 to +3 m O.D.(B) and are of presumed late Holocene age. Explanation for the paucity of data indicative of isostatic rebound close to the centre of ice loading in Donegal probably lies in a lack of research. Work concentrated on the islands off the coastline (Gaulin, personal communication) indicates the presence of numerous raised shoreline features and suggests that a fundamental re-appraisal of the accepted pattern of isostatic recovery is now needed. Support for this view comes from the analysis of modern tide-gauge records (Todd, 1981). These show the continued and relatively rapid rate of land uplift in the Malin Head area at values of *c.* 2·3 mm yr^{-1}, contradicting the former belief that isostatic emergence in Ireland is complete (Mitchell, 1977; Carter, 1982). The first lateglacial marine inundation of the land surface is recorded by the highest raised shorelines of the Inishowen Peninsula, reaching maximum values of *c.* +20 to +22 m O.D.(B). Although probably not synchronous, these and related shorelines within the height spectrum *c.* +6 to +22 m O.D.(B) (Fig. 7) have been broadly correlated with features at similar elevations along the east coast in Counties Down, Louth, Meath and Dublin. The oldest shoreline here occurs outside the limit of the Drumlin Readvance moraine and is represented on Clogher Head in a marine washing limit at *c.* +19 m O.D.(B). Marine limits in outwash deltas at *c.* +16 to +18 m O.D.(B) evidence inundation at this time of areas further to the south in the Boyne Valley. A subsequent removal of the marine influence is recognized here in downcutting from this level and the formation of a lower series of outwash terraces. The provenance of successively younger shorelines between Dublin and Fair Head trace the recession of lateglacial ice limits northward. Dating of shoreline L2 has been made by reference to a ^{14}C assay of 18 990 ± 150 years (Birm-270) for the minimal age of moraine at Kirkmichael, Isle of Man. Correlation of this deposit with moraine truncating shoreline L2 in the area of the Carlingford Peninsula places the feature's formation to ~ 18 000–19 000 B.P., whilst shoreline L1 is seen as a little older (Synge, 1977a,b). Wave-cut notches equated with shoreline L3/4 developed at Ballyquintin Point and other sites along the Ards Peninsula between *c.* 14·6 to 16·8 m O.D.(B), appear to be associated with the deposition of a stoneless red clay containing a cold water marine fauna. The presence of these sediments, occupying

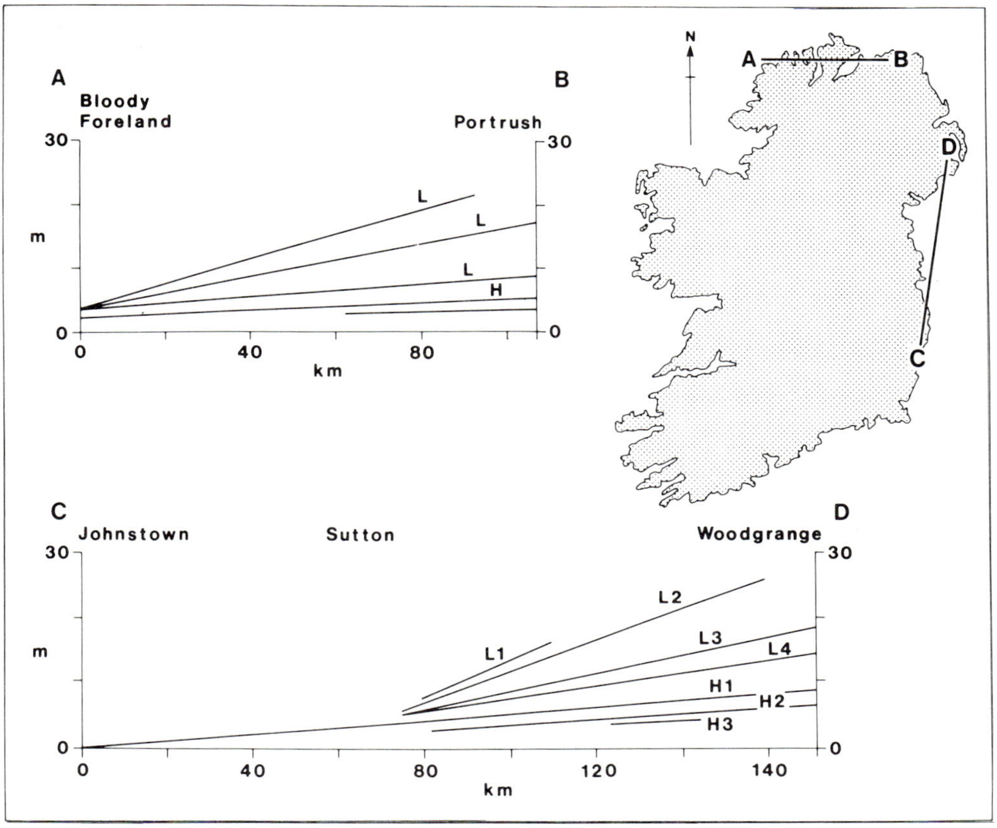

Figure 7. Shoreline height/distance diagrams for northern and eastern coasts (derived from Stephens and Synge, 1965; Synge and Stephens, 1966; Synge, 1977a,b, 1980a,b). Given the reservations as to the accuracy of such diagrams expressed on p. 235, these figures are presented only as an illustration of the generalized pattern of shoreline height change. L, lateglacial shorelines; H, Holocene shorelines.

former inter-drumlin hollow locations, indicates the existence at this time of relatively deep water conditions close to the shore zone. The removal of the marine influence from these positions to levels below the present shoreline by *c.* 12 000 B.P., is recorded in the occurrence of terrestrial and freshwater deposits above marine sediments at Roddans Port (Morrison and Stephens, 1965) and Carnlough (Prior *et al.*, 1981). Radiocarbon dating here of biogenic material overlying the marine sequences at −1·1 and −1·3 m O.D.(B) respectively, gave assays of 12 110 ± 150 B.P. (Q-360) and 11 740 ± 120 B.P. (UB-985). The extent of this relative sea-level fall is unknown, but may have reached levels below −30 m O.D.(B) on the basis of borehole data from Magilligan Point, Lough Foyle (Carter, 1982). In Bantry Bay freshwater sediments containing a zone II pollen assemblage (*c.* 11 000-12 000 B.P.) (Stillman, 1968) may indicate a sea level below *c.* −60 m O.D.(B) at this time, a picture supported by identification of a possible "recent" marine erosion surface at −90 m O.D.(B) in the North Channel off Larne (Eden *et al.*, 1973).

The possibility remains in the former case that the deposits accumulated in an initially protected inter-drumlin or back-beach bar location and have little significance for sea surface position. Timing of the onset of marine conditions at the site is prevented by the presence of an unconformity with the overlying marine sequence. Renewed inundation after *c.* 10 000 B.P. is recorded from sediments in many river estuary and other low-lying zones around the coastline. Peat in Belfast Lough at *c.* −12·0 m O.D.(B) is overlain by estuarine sediments and gives a ^{14}C assay of 9130 ± 120 B.P. (St-3006) for the onset of this sea surface rise in the north (Manning *et al.*, 1970). To the south in Cork Harbour *Phragmites* peat at similar levels of *c.* −12·0 m O.D.(B), containing a zone VI pollen assemblage and dated by radiocarbon to ~8000 B.P., passes conformably upward to brackish-marine and then into marine sediments (Devoy, in preparation). Progression of the upward trend in the sea surface rise is evidenced at the estuarine basin site of Woodgrange. Here an initial marine influence upon freshwater organic sediments at +0·5 m O.D.(B) occurs by 7650 ± 400 B.P. (LJ-904), culminating in the deposition of an estuarine clay at +0·8 m O.D.(B) by 6550 ± 300 B.P. (LJ-903) (Singh and Smith, 1966, 1973; Singh, 1970; Dresser *et al.*, 1973). The continuing importance of isostatic uplift in the north during this early phase of Holocene sea-level recovery, is shown from differential changes in the elevation of such interleaved freshwater–marine sequences. Thus peats dated to *c.* 8500 B.P. show a rise in level from *c.* −9·0 m O.D.(B) in Belfast Lough (Manning *et al.*, 1970) to 0·2 m O.D.(B) at Cushendun (Mitchell and Stephens, 1974), although differences in environmental conditions between sites make direct comparison of data difficult.

Above the present shore zone this balance between isostatic land recovery and ocean surface rise is represented along the north and east coasts in the widespread development of relict beaches and marine erosion surfaces. Maximum levels of *c.* +9·0 to +6·0 m O.D.(B) for the highest Holocene shoreline (H1) occur between Magilligan Foreland (+8·0 m) and Murlough (+9·0 m), falling to +2·0 to +3·0 m O.D.(B) at Sutton, Co. Dublin (Fig. 7). At Malin Head the highest strandline occurs at +7·5 m O.D.(B), declining in height south-westwards to *c.* +4·0 m O.D.(B) at the last recorded shoreline site in Sligo Bay (Davies and Stephens, 1978).

On the east coast Synge (1980a,b) has recognized three raised shorelines (H1-3) isostatically tilted towards the south, with the highest shoreline disappearing beneath the present intertidal zone in the area of Wicklow Head (Fig. 7). These are seen as the result of different phases of marine incursion separated by low relative sea-level positions and have been dated respectively to *c.* 5000–5500, 4000 and 2600 B.P. Analysis of the shoreline sea-level data shows, however, that the beach ridges and other shore forms used to construct these levels fall within the altitudinal range of present day stormtide activity (Carter, 1982). The apparent fossil nature of many of the beaches excludes explanation of their formation solely by contemporary processes. The features probably represent the product of a complex relationship with storm-surge phenomena during the Holocene, rather than a simple isostatically height/age controlled beach series. Comparison between these shorelines and European shoreline sea-level studies (Synge, 1977b, 1980a) as a basis for recognizing an "oscillating" Holocene sea-level curve is unsatisfactory. Difficulties arise in the correlation and interpretation of sea-level indicators of different types, formed under varied marine environmental regimes. Much greater accuracy in the dating of shorelines and a clearer understanding of the development of shore zone features in relation to sea surface behaviour is required before such comparisons can be attempted (Devoy, 1982; Shennan, 1982).

The pattern of mid- to late Holocene sea-level movement is perhaps obscure and records of accurate indicators of sea-level position are few. Inferences about the height and trend of sea surface movement have been based often upon the negative evidence of erosional surfaces, or upon material not directly recording either the height or presence of marine conditions at a site. The timing of a relative maximum elevation for sea-level rise varies around the coastline and depends upon regional differences in the interaction of eustatic and isostatic factors. As might be expected it appears earlier in areas closer to the centre of land uplift and is timed at *c.* 6000–6500 B.P. in the north (Carter, 1982), occurring possibly between *c.* 4000–5000 B.P. along the east coast north of Dublin (Mitchell and Stephens, 1974; Mitchell, 1976, 1977). At Leamore brackish-marine sediments at +3·0 to +4·0 m O.D.(B) have been dated to *c.* 5000 B.P. (Mitchell and Stephens, 1974) and post-date the development of biogenic deposits now exposed in the lower intertidal zone at Bray. A ^{14}C assay dates a phase of the peat growth here to *c.* 6750 B.P. This evidence together with marine indicators from Woodgrange (+3·2 m O.D.(B)), Ringneill Quay (+4·9 and +3·4 m O.D.(B)), Sutton (~ +4·0 m O.D.(B)) and other localities, has been used to support the view of a sea surface position higher than that of the present day, at least for eastern coastal areas. However, a common sea surface (e.g. M.H.W.S.T.) has not been recognized in the correlation of levels between sites. Confirmation of this picture, therefore, must depend upon a clear understanding of the meaning of spatial stratigraphic changes in sedimentary environments, reconstruction of the possible amplitude of storm tidal influences and upon an accurate assessment of the trends of isostatic activity in Ireland.

On southern and western coasts regional analyses of relative sea surface movements are lacking. Models of the probable pattern of isostatic behaviour in the south suggests that the land surface has remained relatively "stable" during the lateglacial and Holocene time (Valentin, 1953; Todd, 1981; Carter, 1982). If this is correct, then the problem arises in south-west Ireland of explaining the anomaly produced by the growth of an ice cap here, possibly *c.* 8500 km^2 in size (Synge, 1981), and the apparent lack of evidence for any isostatic response. Low beach ridges in Bantry Bay (Synge, personal communication), together with small beds and accumulation of oyster shells above H.W.M. along the Kenmare River (Geological Survey of Ireland, 1859) may form evidence for the existence of raised shorelines. Equally the latter may represent shell middens similar to those found in comparable situations in Cork Harbour. Possibilities in explanation of the anomaly are numerous, but in the absence of adequate field data resolution of the question is impossible.

CONCLUSION

In elucidating the form of isostatic land recovery and related relative shoreline/sea-level movements, it would appear that the techniques applied in their analysis have been too imprecise and subjective to provide other than a primary, low-level resolution interpretation of events. Some basic patterns in Quaternary environmental development have been established. Northward of an approximately elliptical line between Clogherhead and Donegal Bay a zone of isostatic land uplift may be defined. Centred upon the Inishowen area, land recovery appears to have begun during the early phases of ice retreat around 16 000–18 000 B.P. Evidence for the operation of this process through time is based largely upon the existence of a series of westerly-southerly tilted raised shorelines. Falling from a maximum altitude of *c.* +20 to +22 m O.D.(B) these disappear beneath the present shore zone on the east coast south of Wicklow Head. In terms of reconstructing sea surface movements such shorelines and other sea-level indicators record the complex interaction of isostatic and eustatic factors. An apparent removal of the marine influence is recorded in the north from levels above 0·0 m O.D.(B) by *c.* 12 000 B.P. to a position below possibly −30 to −60 m O.D.(B). Subsequent progressive inundation of the coastline during the Holocene is concomitant with a waning of the isostatic influence.

Apart from these broad statements a lack of relevant data of all types inhibits the development of a more detailed knowledge of events. A distinct imbalance in the spatial collection of information has arisen with much greater effort having been expended upon north-eastern areas. In terms of dating, the process of establishing a shoreline and related Quaternary chronology, particularly in the use of stratigraphic data, resembles more the progress of a football match or an analogous game of chance. The see-saw scoring of opposing teams, represented here by different author's viewpoints, continually alters the odds as to the outcome. Thus sedimentary sequences used in the relative dating of events continue to be interpreted in a variety of ways with resultant repercussions for both environmental and chronological reconstructions. Whilst authors cannot be blamed for speculation, it must be remembered when viewing the Quaternary in Ireland that many statements lack objective supporting data and should be treated with caution. Future clarification and extension of the patterns outlined depends upon an intensive programme of data collection, coupled with the application of modern process and modelling studies in the interpretation of palaeoenvironments. Radiocarbon and other direct dating techniques should be employed more widely in establishing a shoreline/sea-level chronology. The use of stratigraphic information where such methods are inappropriate, must pay strict regard to established international stratigraphic procedures.

REFERENCES

Admiralty (1981). "Admiralty Tide Tables: European Waters including the Mediterranean Sea", Vol. 1. Hydrographer of the Navy, Admiralty Hydrographic Department, Taunton.

Andrews, J. T. (1967). *Geogr. Bull.* **8**, 71-74.

Andrews, J. T. (1970). *Inst. Br. Geogr. Spec. Publ.* **2**, 156 pp.

Andrews, J. T. (1973). *Arctic and Alpine Research.* **5**, 185-199.

Andrews, J. T., Bowen, D. Q. and Kidson, C. (1979). *Nature* **281**, 556-558.

Bowen, D. Q. (1973). *Proc. Geol. Assoc.* **84**, 249-272.

Bowen, D. Q. (1977). *In* "The Quaternary History of the Irish Sea" (C. Kidson and M. J. Tooley, eds), pp. 223-256. Seel House Press, Liverpool.

Bowen, D. Q. (1978). "Quaternary Geology: A Stratigraphic Framework for Multi-disciplinary Work." Pergamon Press, Oxford.

Carter, R. W. G. (1982). *Proc. Geol. Ass.* **93**, 7-23.

Carter, R. W. G. and Orford, J. D. (1981). "Field Guide No. 4." Irish Quaternary Association, Dublin.

Charlesworth, J. K. (1924). *Proc. R. Ir. Acad.* **36B**, 174-314.

Charlesworth, J. K. (1928). *Q.J. Geol. Soc. Lond.* **84**, 293-344.

Charlesworth, J. K. (1957). "The Quaternary Era", 2 Vols. London.

Clark, J. A., Farrell, W. E. and Peltier, W. R. (1978). *Quaternary Research* **9**, 265-287.

Colhoun, E. A. and McCabe, A. M. (1973). *Proc. R. Ir. Acad.* **73B**, 167-206.

Colhoun, E. A. and Mitchell, G. F. (1971). *Proc. R. Ir. Acad.* **71B**, 211-245.

Culleton, E. B. (1978). *J. Earth Sci. R. Dubl. Soc.* **1**, 33-39.

Davies, G. L. A. and Stephens, N. (1978). "Ireland." Methuen, London.

Dawson, A. G. (1979). *Dept. of Geography, Bedford College, Univ. of London. Occ. Pap.* No. 6, 28 pp.

Dawson, A. G. (1980). *Proc. Geol. Ass.* **91**, 339-344.

Devoy, R. J. (1979). *Phil. Trans. R. Soc.* **B285**, 355-407.

Devoy, R. J. (1982). *Proc. Geol. Ass.* **93**, 65-90.

Dixon, F. E. (1949). *Sci. Proc.R. Dublin Soc.* **25**, 1-8.

Dobson, M. R. (1977a). *In* "The Quaternary History of the Irish Sea" (C. Kidson and M. J. Tooley, eds), pp. 13-26. Seel House Press, Liverpool.

Dobson, M. R. (1977b). *In* "The Quaternary History of the Irish Sea" (C. Kidson and M. J. Tooley, eds), pp. 93-98. Seel House Press, Liverpool.

Dobson, M. R. and Whittington, R. (1973). *Inst. Geol. Sciences Report No.* 73/11, 1-35.

Dresser, P. Q., Smith, A. G. and Pearson, G. W. (1973). *Proc. R. Ir. Acad.* **73B**, 53-56.

Eden, R. A., Deegran, C. E., Rhys, G. H., Wright, J. E. and Dobson, M. R. (1973). *Inst. of Geol. Sci. Rept. No.* 73/2. 1-27.

Farrington, A. (1933). *Ir. Nat. J.* **4**, 211-212.

Farrington, A. (1939). *Proc. Geol. Ass.* **50**, 337-344.

Farrington, A. (1944). *Proc. R. Ir. Acad.* **50B**, 133-157.

Farrington, A. (1954). *Ir. Geogr.* **3**, 47-53.

Farrington, A. (1959). *Proc. R. Ir. Acad.* **60B**, 135-166.

Farrington, A. (1965). *Ir. Nat. J.* **14**, 216-217.

Farrington, A. (1966). *Sci. Proc. R. Dublin Soc.* **2A**, 197-219.

Garrard, R. A. (1977). *In* "The Quaternary History of the Irish Sea" (C. Kidson and M. J. Tooley, eds), pp. 69-92. Seel House Press, Liverpool.

Geological Survey of Ireland (1859-1862). Geological Memoir of Co. Kerry. Dublin.

Godwin, A. (1975). "The History of the British Flora." Cambridge University Press, Cambridge.

Gray, J. M. (1975). *Dept. of Geography, Queen Mary College, Univ. of London. Occ. Pap.* No. 2, 40 pp.

Gray, J., de Boutray, B., Hillaire-Marcel, C. and Lauriol, B. (1980). *Arctic and Alpine Research* **12**, 19-30.

Guilcher, A. (1969). *Earth Science Reviews* **5**, 69-97.

Hey, R. W. (1971). *Quaternaria* **15**, 273-284.

Hillaire-Marcel, C. (1979). Les mers postglaciaires du Québec: quelque aspects. 2 Vols. Thèse D. Sc. Paris VI.

Holland, C. A. (1981). "A Geology of Ireland." Scottish Academic Press, Edinburgh.

Huddart, D. (1977). *In* "X INQUA Congress Excursion Guides" (D. Q. Bowen, ed.), pp. 1-56. Geo. Abstracts, Norwich.

Huddart, D. (1981a). *Quaternary Newsletter* **33**, 28-41.

Huddart, D. (1981b). *Quaternary Newsletter* **35**, 6-11.

Jardine, W. G. (1976). *Geol. För. Stockh. Förh.* **98**, 78-82.

Jardine, W. G. (1977a). *Nature* **267**, 138-140.

Jardine, W. G. (1977b). *In* "The Quaternary History of the Irish Sea" (C. Kidson and M. J. Tooley, eds), pp. 99-118. Seel House Press, Liverpool.

Kent, P. E. (1975). *In* "Petroleum and the Continental Shelf of North-west Europe", 1, Geology (A. W. Woodland, ed.), pp. 3-28. Applied Science Publishers, Barking.

Kidson, C. (1971). *In* "Exeter Essays in Geography" (K. J. Gregory and W. L. D. Ravenhill, eds), pp. 1-22. University of Exeter Press, Exeter.

Kidson, C. (1977a). *In* "The Quaternary History of the Irish Sea" (C. Kidson and M. J. Tooley, eds), pp. 1-12. Seel House Press, Liverpool.

Kidson, C. (1977b). *In* "The Quaternary History of the Irish Sea" (C. Kidson and M. J. Tooley, eds), pp. 257-298. Seel House Press, Liverpool.

King, C. A. M. (1972). "Beaches and Coasts" (2nd ed.). Edward Arnold, London.

Knowles, W. J. (1889). *Proc. R. Ir. Acad. Ser.* **6**, 1, 173-187.

Lamplugh, G. W. *et al.* (1904). The geology of the country around Belfast. *Mem. geol. Surv. Ireland, Dublin.*

Laughton, A. S. (1975). *Norsk. Geol. Unders. 316 Bull.* **29**, 169-194.

Le Pichon, X., Sibuet, J.-C. and Francheteau, J. (1977). *Tectonophys.* **38**, 169-209.

Manning, P. I., Robbie, J. A. and Wilson, A. E. (1970). The geology of Belfast and the Lagan valley. *Mem. Geol. Survey North Ireland.*

Marthinussen, M. (1960). *In* "Geology of Norway" (O. Holtedahl, ed.), pp. 416-429 *Norges geol. Unders.* **208**.

May, A. and Batty, J. (1948). *J.R. Soc. Antiq. Ir.* **78**, 130-156.

McCabe, A. M. and Hoare, P. G. (1978). *Geol. Mag.* **115**, 397-413.

Milliman, J. D. and Emery, K. O. (1968). *Science* **162**, 1121-1123.

Mitchell, G. F. (1956). *J.R. Soc. Antiq. Ir.* **86**, 1-26.

Mitchell, G. F. (1960). *Adv. Sci. Lond.* **17**, 313-325.

Mitchell, G. F. (1962). *Proc. Geol. Ass.* **73**, 197-213.

Mitchell, G. F. (1970). *Proc. R. Ir. Acad.* **70B**, 141-162.

Mitchell, G. F. (1972). *Sci. Proc. R. Dublin Soc.* **4A**, 181-199.

Mitchell, G. F. (1976). "The Irish Landscape." Collins, London.
Mitchell, G. F. (1977). *In* "British Quaternary Studies, Recent Advances" (F. W. Shotton, ed.), pp. 169-186. Clarendon Press, Oxford.
Mitchell, G. F. and Stephens, N. (1974). *Coll. Int. CNRS* No. 219, 115-125.
Mitchell, G. F., Penny, L. F., Shotton, F. W. and West, R. G. (1973). *Geol. Soc. Lond., Spec. Report* No. 4, 99 pp.
Morrison, M. E. S. and Stephens, N. (1960). *New Phytol.* **59**, 153-162.
Morrison, M. E. S. and Stephens, N. (1965). *Phil. Trans. R. Soc., B.* **249**, 221-255.
Mottershead, D. N. (1977). *In* "The Quaternary History of the Irish Sea" (C. Kidson and M. J. Tooley, eds), pp. 299-320. Seel House Press, Liverpool.
Movius, A. L. (1940). *Proc. R. Ir. Acad.* **46C**, 1-84.
Movius, H. L. (1942). "The Irish Stone Age; Its Chronology, Development and Relationships." Cambridge University Press, Cambridge.
Movius, H. L. (1953). *Am. J. Sci.* **251**, 697-740.
Naylor, D. and Mounteney, S. N. (1975). "Geology of the North-West European Continental Shelf", Volume 1. London.
Orme, A. R. (1964). *Ir. Geogr.* **5**, 48-72.
Orme, A. R. (1966). *Trans. Inst. Br. Geogr.* **39**, 127-140.
Pantin, H. M. (1977). *In* "The Quaternary History of the Irish Sea" (C. Kidson and M. J. Tooley, eds), pp. 27-54. Seel House Press, Liverpool.
Paterson, W. S. (1981). "The Physics of Glaciers" (2nd ed.). Pergamon Press, Oxford.
Pearson, G. W. (1980). *Radiocarbon* **22**, 337-345.
Plassche, O. van de (1977). "A manual for sample collection and evaluation of sea level data." Internal publication sponsored by UNESCO, IGCP-Project 61. Amsterdam.
Praeger, R. L. (1893). *Proc. R. Ir. Acad.* **18**, 212-289.
Praeger, R. L. (1896). *Proc. R. Ir. Acad.* **4**, 30-54.
Prior, D. B. (1968). The late-Pleistocene geomorphology of North-East Antrim. Unpublished Ph.D. thesis, Queens University, Belfast.
Prior, D. B., Holland, S. and Cruickshank, M. M. (1981). *Ir. Geogr.* **14**, 75-84.
Rose, J. (1978). *Brighton Polytechnic Geographical Society Magazine* **3**, 13-30.
Rose, J. (1981). *In* "Geomorphological Techniques" (A. Goudie, ed.), pp. 327-341. George Allen and Unwin, London.
Shackleton, N. J. (1977). *In* "British Quaternary Studies" (F. W. Shotton, ed.), pp. 1-16. Clarendon Press, Oxford.
Shackleton, N. J. and Opdybe, N. D. (1973). *Quaternary Research* **3**, 39-55.
Shennan, I. (1982). *In* "Climatic Changes in Later Prehistory" (A. F. Harding, ed.), pp. 52-67. Edinburgh University Press, Edinburgh.
Singh, G. (1970). *Proc. R. Ir. Acad.* **69B**, 189-216.
Singh, G. and Smith, A. G. (1966). *The Palaeobotanist* **15**, 230-234.
Singh, G. and Smith, A. G. (1973). *Proc. R. Ir. Acad.* **73B**, 1-51.
Sissons, J. B. (1967). *Trans Inst. Br. Geogr.* **42**, 163-173.
Steers, J. A. (1952). *Geogr. J.* **118**, 180-190.
Stephens, N. (1957). *Proc. R. Ir. Acad.* **58B**, 129-149.
Stephens, N. (1963). *Ir. Geogr.* **4**, 345-359.
Stephens, N. (1968). *In* "Means of Correlation of Quaternary Successions" (R. B. Morrison and A. E. Wright, eds). Proc. VII INQUA Conference, University of Utah.
Stephens, N. (1970). *In* "Irish Geographical Studies" (N. Stephens and R. E. Glasscock, eds), pp. 125-145. Queen's University, Belfast.

Stephens, N. and McCabe, A. M. (1977). *In* "The Quaternary History of the Irish Sea" (C. Kidson and M. J. Tooley, eds), pp. 179–198. Seel House Press, Liverpool.
Stephens, N. and Synge, F. M. (1965). *Proc. R. Ir. Acad.* **64B**, 131-153.
Stephens, N., Creighton, J. R. and Hannon, M. A. (1975). *Ir. Geogr.* **8**, 1-23.
Stillman, C. J. (1968). *Sci. Proc. R. Dublin Soc.* **3A**, 125-127.
Sutcliffe, A. J. (1981). *Quaternary Newsletter* **33**, 1-17.
Sweeting, M. M. (1955). *Trans. Inst. Br. Geogr.* **21**, 33-49.
Synge, F. M. (1969). *Geogr. Annlr.* **A51**, 193-206.
Synge, F. M. (1970). *In* "Irish Geographical Studies" (N. Stephens and R. E. Glasscock, eds), pp. 34-48. Queen's University, Belfast.
Synge, F. M. (1977a). *In* "The Quaternary History of the Irish Sea" (C. Kidson and M. J. Tooley, eds), pp. 199–222. Seel House Press, Liverpool.
Synge, F. M. (1977b). *Phil. Trans. R. Soc.* **B280**, 211-228.
Synge, F. M. (1979). *Quaternary Newsletter* **28**, 1-18.
Synge, F. M. (1980a). *Geol. För. Stockh. Förh.* **102**, 235-249.
Synge, F. M. (1980b). *Bulletin de L'Association francaise pour l'Etude du Quaternaire* **17**, 77-79.
Synge, F. M. (1981). *Geol. en Mijnb.* **60**, 305-315.
Synge, F. M. and Stephens, N. (1966). *Trans. Inst. Br. Geogr.* **39**, 101-125.
Thom, B. G. (1973). *Progress in Geography* **5**, 170-231.
Thomas, G. S. P. and Summers, A. J. (1981). *Quaternary Newsletter* **34**, 15-18.
Todd, N. W. (1981). Recent sea-level changes in North and East Ireland. Unpublished M.Sc. thesis, The New University of Ulster.
Tooley, M. J. (1978). "Sea-level Changes: North-West England during the Flandrian Stage." Clarendon Press, Oxford.
Trenhaile, A. S. (1980). *Progress in Physical Geography* **4**, 1-23.
Valentin, A. (1953). *Geogr. J.* **119**, 299-305.
Warren, W. P. (1977). *In* "South and Southwest Ireland" (C. A. Lewis, ed.), pp. 37-45. X INQUA Congress Excursion Guides, Geo. Abstracts, Norwich.
Warren, W. P. (1979). *Bull. Geol. Surv. Irel.* **2**, 315-332.
Watts, W. A. (1959). *Proc. R. Ir. Acad.* **60B**, 79-134.
Whittington, R. J. (1977). *In* "The Quaternary History of the Irish Sea" (C. Kidson and M. J. Tooley, eds), pp. 55-68. Seel House Press, Liverpool.
Wright, L. W. (1967). *Z. Geomorph.* **11**, 36-46.
Wright, W. B. (1937). "The Quaternary Ice Age" (2nd ed.). Macmillan, London.
Wright, W. B. and Muff, H. B. (1904). *Sci. Proc. R. Dubl. Soc. N.S.* **10**, 250-308.
Ziegler, P. A. and Louwerens, C. J. (1979). *In* "The Quaternary History of the North Sea" (E. Oele, R. J. J. Schuttenhelm and A. J. Wiggers, eds), pp. 7-22. *Acta. Univ. Ups. Symp. Univ. Ups. Annum Quingentisimum Celebrantis* **2**, Uppsala.

ELEVEN

Flandrian and Late Devensian sea-level changes and crustal movements in England and Wales

I. SHENNAN

Flandrian and Late Devensian sea-level and crustal movements in England and Wales have been studied generally by small groups or individual workers specializing in relatively short sections of the coastline. Comparison of local sea-level chronologies and sea-level curves for different coastal units is often reserved for a relatively brief discussion at the end of the research publication and rarely have such comparisons produced convincing results.

Churchill (1965) attempted to synthesize the radiocarbon information available to reveal crustal movements within England, Wales and southern Scotland. His paper has since been criticized (e.g. Kidson and Heyworth, 1973; Devoy, 1979) but no adequate analysis of Flandrian crustal movements in England has been forthcoming. The earlier work of Valentin (1953), using tide-gauge data of variable reliability, is still used as a model (e.g. Brown, 1979). An attempt at the regional synthesis of sea-level data has been made by Morrison (1976). This study covered a much larger area, the western seaboard of Europe, but like Churchill's suffered from an inadequate data base which was unevenly distributed in space and time. Recent reviews have been made by Jardine (1979, 1981) but no regional synthesis of the data was attempted. Flemming (1982) presents a numerical analysis of the data in a study which is more important in terms of the methodology rather than the results obtained since the data base was very limited. Thus, there is as yet no detailed pattern of isostatic movement agreed for England and Wales.

The present chapter can be divided into two main parts. The first is a review of the published sea-level data and methods employed. The second is more exploratory, looking at approaches which may be used to provide reliable correlation between areas and therefore an assessment of regionally significant factors. It must be stressed that the terminology used in this chapter has been developed in order to avoid many problems in comparing different authors' work. Terms such as "indicative meaning", "reference water level", "transgressive overlap", "regressive overlap" and "tendencies of sea-level movement" have specific unambiguous meanings. They have been explained in a number of published sources (van de Plassche and Preuss, 1978; Streif, 1979; Shennan, 1980,

SHORELINES AND ISOSTASY
ISBN 0 12 652960 4

1982a; van de Plassche, 1982; Tooley, 1982b) and reference should be made to these where details in excess of those given in the following sections are required.

REVIEW OF PUBLISHED SEA-LEVEL DATA AND METHODS OF ANALYSIS

The approach to sea-level research followed by most authors, prior to about 1980 was largely as outlined in Fig. 1. The objectives of individual research workers differed and the methods employed varied in detail. The final stage reached also varied. These local or area based studies still form the core of sea-level research in England and Wales. It should be noted that spatial units of different sizes have been given informal names using an ordinal scale (Shennan, 1982a). A "site" is the area over which a single sea-level index point is believed to be representative, thus between-site correlation leads to the study at the "area" scale, e.g. the Fenland, or the Thames Estuary; and between-area study gives a region based approach. Other terms such as "local sea-level change" should be fitted into this scheme and it is appropriate that "local" would refer to the "area" scale.

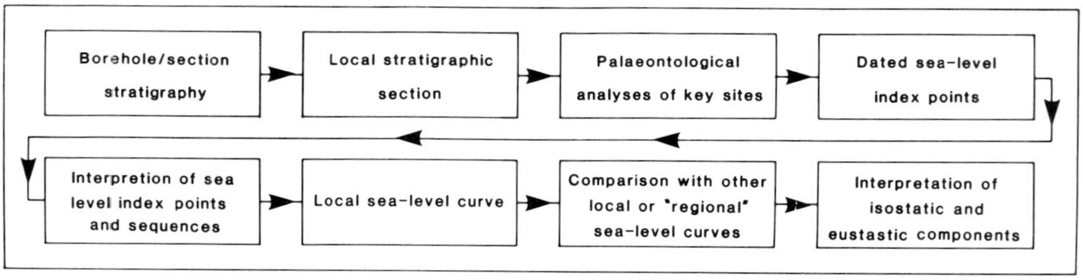

Figure 1. Model of the traditional approach to sea-level studies for coastal units of England and Wales.

Local sea-level studies

These studies are discussed using the approximate areal units studied by individual authors (Fig. 2) and described in a clockwise sequence. The reviews of these published works are brief and detailed reference should be made to the original articles. These articles, however, provide only the broadest working hypothesis for evaluating regional sea-level changes and crustal movements throughout England and Wales since the investigation of both were not always within the aims of the research. Problems of the incompatibility of definitions, interpretations and data bases effectively require a new approach specifically based on identifying regional and inter-regional phenomena. Such problems were effectively promoted by IGCP Project 61 "Sea-level movements during the last deglacial hemicycle (about 15 000 years)" (e.g. van de Plassche and Preuss, 1978), discussed further by Shennan (1980), and recently received further attention (Jardine, 1981; Heyworth and Kidson, 1982; Shennan, 1982a; Tooley, 1982a).

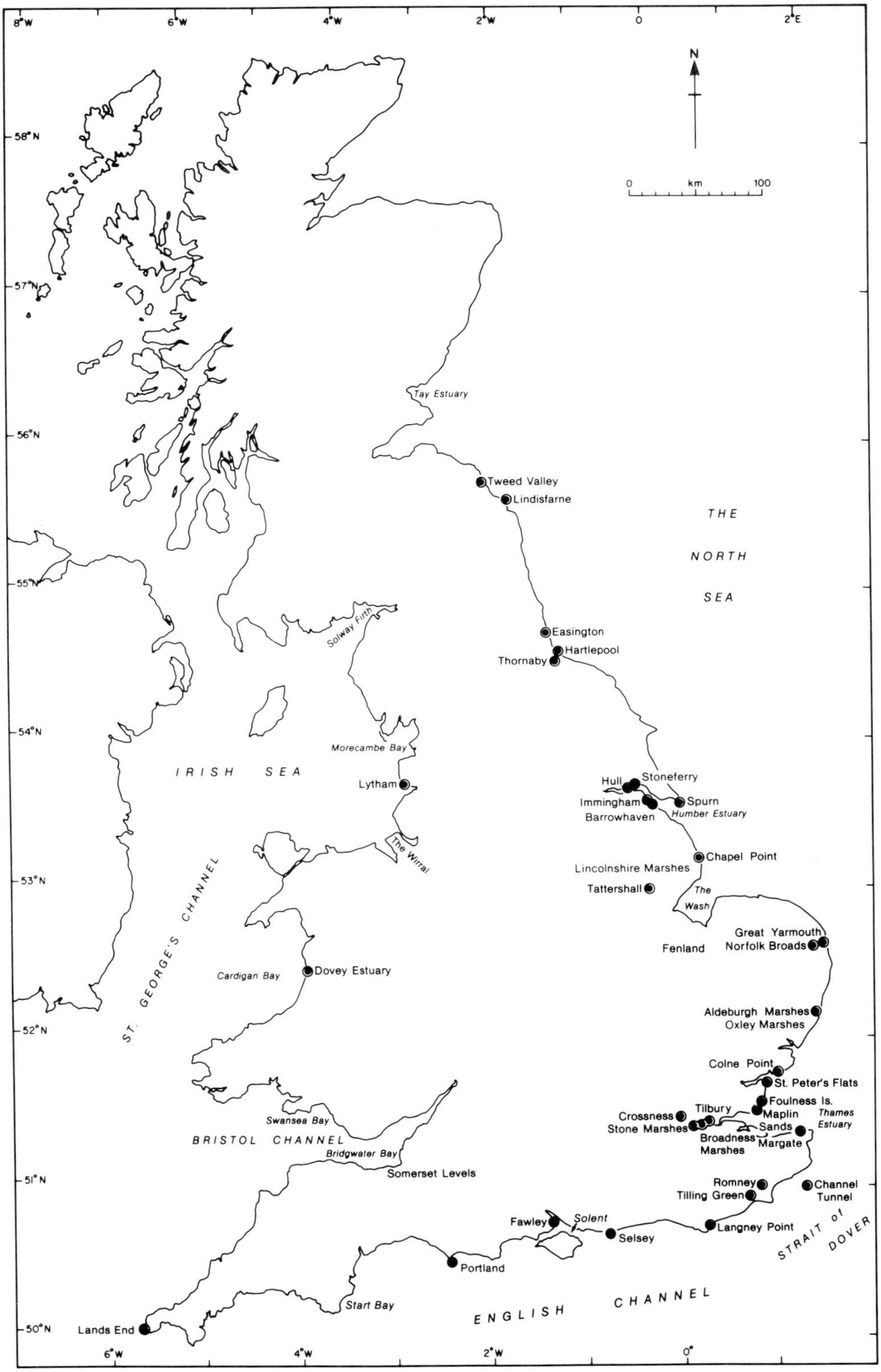

Figure 2. Location of the major areas and sites of investigation referred to in the text.

Inevitably the simple comparison of sea-level curves cannot be used to reconstruct accurately regional changes in sea level or crustal movement but it may serve as a working hypothesis.

In order to overcome some of the problems of inconsistent terminology the recommendations of Tooley (1982b) and Shennan (1982a) to adopt "transgressive overlap" and "regressive overlap" as terms descriptive of lithostratigraphic changes are followed. Furthermore, the term "shoreline" is used in a wide sense, as the line of intersection of any tide level at the given time and the terrestrial surface at the given location (cf. Jardine, 1981). The tide level will be specified according to the nature of the evidence used to identify the fossil shoreline. For comparison of different types of past sea-level indicators the relationship to the reference water level, and the conversion to a single reference water level, will be required (e.g. Shennan, 1980; van de Plassche, 1982). The terms tide level and sea level are used as synonyms unless more specific reference is made.

North-east England

This area, between the Tweed and Humber estuaries, has provided little coordinated data. Terraces and the rockhead valley of the Tweed may be related to a late-glacial sea level but the data are open to alternative explanations (Rhind, 1972). There are raised beaches of unknown age on Lindisfarne (Tooley, 1982b).

More reliable evidence has been discussed by Gaunt and Tooley (1974) and Tooley (1978a,b, 1982b). These data come from two separate areas. First, from the Hartlepool-Tees estuary area the earliest dated sea-level indicator is from Thornaby in the lower Tees valley where a 5 cm peat sample subjacent to marine deposits at −10·61 m O.D. was dated 9680 ± 100 B.P. (HAR-3711, Tooley, personal communication). Further assays from the area provided no incontrovertible chronology due to the isolated nature of the boreholes around Hartlepool Bay (see Tooley, 1978b). A transgressive overlap at −2·39 m O.D. was dated 6050 ± 90 B.P. (HAR-3714) at WH-11a and a minimal date for regressive tendencies at WH-2 is 5595 ± 135 B.P. (Hv-5216) at −0·20 m O.D. However, an assay actually on the regressive overlap at −0·34 m O.D. gave 5240 ± 70 B.P. (Hv-3459) and a preference for the latter may be suggested by the transgressive overlap at WH3 (−1·95 m O.D.) dated 5285 ± 120 B.P. (Hv-4712, Gaunt and Tooley, 1974). These data cannot be resolved into an unambiguous chronology of transgressive and regressive tendencies due to incomplete lithostratigraphic correlation. The area is important for the future investigation of sea-level and crustal movements due to its location between the subsiding zones to the south and the uplifted areas to the north and west.

The Humber estuary provides nine assays from unconsolidated Flandrian deposits in addition to dates on Bronze Age trackways and rafts (Smith, 1958; Gaunt and Tooley, 1974; McGrail and Switsur, 1975). A basal woody detrital peat started to accumulate at −11·55 m O.D. at Hull at 6970 ± 100 B.P. (IGS—C14/99) and the transgressive overlap, possibly eroded, was dated 6890 ± 100 B.P. (IGS-C14/100) at

−9·73 m O.D. *Alnus* wood at −9·14 m O.D. was dated from an intercalated peat layer at Immingham (6681 ± 130 B.P., Q-401; Gaunt and Tooley, 1974). Regressive tendencies were recorded at 6170 ± 180 B.P. (Hv3359) at Spurn, at an altitude of −2·35 m O.D.

From Stoneferry a date of 5240 ± 100 B.P. (IGS—C14/97) was obtained for the lower contact of a peat layer above an apparent freshwater organic clay at −4·0 m O.D. Intertidal shelly sand covered the peat at −3·8 m O.D. with an erosive contact. Dates on the shells were 3775 ± 100 B.P. and 3435 ± 200 B.P. (IGS—C14/98 I and II). The onset of marine conditions must have been between *c.* 5240 and 3775 B.P.

More recent changes in sea-level tendencies have been recorded from Barrowhaven. Regressive tendencies are dated 2335 ± 60 B.P. (SRR-1374) at +1·43 m O.D. and transgressive tendencies 1080 ± 40 B.P. (SRR-1373) at +2·07 m O.D. (Tooley, personal communication).

While there are reliable individual ^{14}C assays from the Tees and Humber estuaries extensive stratigraphic surveys have not been carried out and reliable chronologies of sea-level tendencies cannot be made. The data have been summarized by Tooley (1982b, Fig. 6).

The Fenland and Lincolnshire marshes

The data from this area have been studied recently by Shennan (1980, 1981, 1982a,b). The methodology developed forms the basis of the discussion in the second part of this chapter.

Lithostratigraphic evidence and 106 ^{14}C dates (Skertchly, 1877; Swinnerton, 1931; Godwin and Clifford, 1938; Godwin, 1940; Willis, 1961; Churchill, 1970; Hallam, 1970; Smith, 1970; Godwin and Vishmu-Mittre, 1975; Valentine and Dalrymple, 1975; Shennan, 1980, 1981, 1982a,b; Simmons, 1980) have been resolved into a chronology showing five periods of positive sea-level tendencies and four periods of negative tendencies prior to 2500 B.P., which are considered reliable, while further, less reliable periods of predominant tendencies have been identified for the period since that time. The within-Fenland variability of inferred past levels of M.H.W.S.T. is shown by Fig. 3. The lack of certainty with which past altitudes of sea level can be estimated must be taken into account in any comparison between areas to infer variable crustal movements.

Norfolk

Sites relating to possible changes in sea level were noted by Godwin (1943) on the Norfolk coast while further evidence was revealed from the Norfolk Broads (Jennings, 1952; Lambert *et al.*, 1960). Recent investigations of the Norfolk Broads (Coles and Funnell, 1981) have revealed a period of positive tendencies of sea-level movement from at least 7580 ± 90 B.P. (HAR-2535, at −19·3 m O.D. at Great Yarmouth) to 4-5000 B.P. (pollen dating only) at a general level of −5·5 to

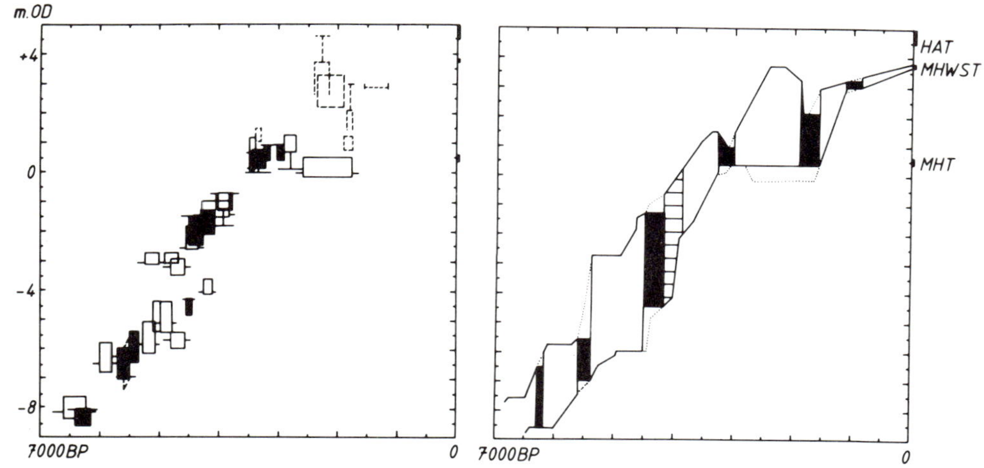

Figure 3. The rectangles on the left-hand diagram are estimates of past levels of M.H.W.S.T. in the Fenland using the most reliable indicators of altitude. The rectangle represents ± 1 standard error of the ^{14}C data (± 2 standard error is given by the horizontal line outside the box which also gives the original altitude of the sample midpoint). Shaded boxes represent dates on negative tendencies, dashed boxes represent archaeological and morphological indicators for which no ^{14}C dates are available. The right-hand diagram shows the estimated sea-level band. Shaded periods represent Fenland I,II,IV,V,VI and VII. The solid lines would assume that the tendencies reflect true fluctuations in sea level while dotted lines enclose the whole data set (after Shennan, 1982a).

−6·5 m O.D. A minimal date for the onset of a second period of positive tendencies is 1973 ± 50 B.P. (SRR-573), with marine/estuarine sediments reaching their maximum extent shortly after 1609 ± 50 B.P. (SRR575) at about −2·0 to −0·5 m O.D. Coles and Funnel infer an estimated subsidence rate of 1·5 m/1000 years from their data. This must be viewed only as an approximation since it was calculated from the *relative* sea-level curve of Tooley (1978a). Furthermore the effect of the build-up and break-down of the shingle spit at Great Yarmouth will have complicated the registration of sea-level movements.

Further investigations of the Norfolk Broads are approaching completion. The results are eagerly awaited and will show transgressive overlaps between *c.* 7500-5650 and *c.* 2400-2200 B.P. and regressive overlaps between *c.* 5000-4300 and *c.* 1800-550 B.P. (Alderton, personal communication).

The location of the coastline within the southern North Sea prior to 8000 B.P. has been estimated by various authors (Jelgersma, 1961, 1979; Kolp, 1976; Godwin, 1978; Ludwig *et al.*, 1981) but an accurate altitude of past sea level is not easy to reconstruct.

Suffolk

Borehole surveys between Aldeburgh and Oxley Marshes have revealed basal peats, intercalated peats and salt marsh clastic sediments (Carr and Baker, 1968; Carr, 1970). Limited stratigraphic details are available and the indicative meanings of the

^{14}C dates are not clear. Basal peats from Aldeburgh Marshes at approximate altitudes of − 12·7 m and − 13·7 m O.D. were dated 8640 ± 145 B.P. and 8460 ± 145 B.P. (I-2826,I-2828). Clastic sedimentation was recorded at *c.* − 12·5 m O.D. and these dates must precede such an event. A peat at − 9·25 m O.D. is overlain by pebbles and intertidal deposits. An erosive contact is likely for this indicator dated 7010 ± 130 B.P. (I-3430, Devoy, 1982, Table 3).

Clearly this area warrants detailed research to evaluate Flandrian sea-level changes.

The Essex estuaries and marshes

The Flandrian deposits between Colne Point and Maplin Sands have been studied by Greensmith and Tucker (1973, 1980) and their sea-level curve is shown in Fig. 4. Although they explicitly state that it represents a tentative assessment only it has been reproduced, without reference to the standard errors of the ^{14}C dates, by both Tooley (1978a) and Devoy (1979, 1980). Both Tooley and Devoy note the range of data used to construct the curve; peat layers, peaty silt, shell layers, overconsolidated layers and comparisons with the chronologies and altitudes suggested by Fairbridge (1961) and Mörner (1971). The curve must be viewed as tentative, even the chronology of events should not be taken as being precise.

Problems relating to the area are the paucity of boreholes, the borehole interval is sometimes over 1 km, the thickness of the Flandrian sequence, over 30 m, and the lack of continuity observed for the datable horizons. In view of these comments the Essex data must be viewed in their broadest sense.

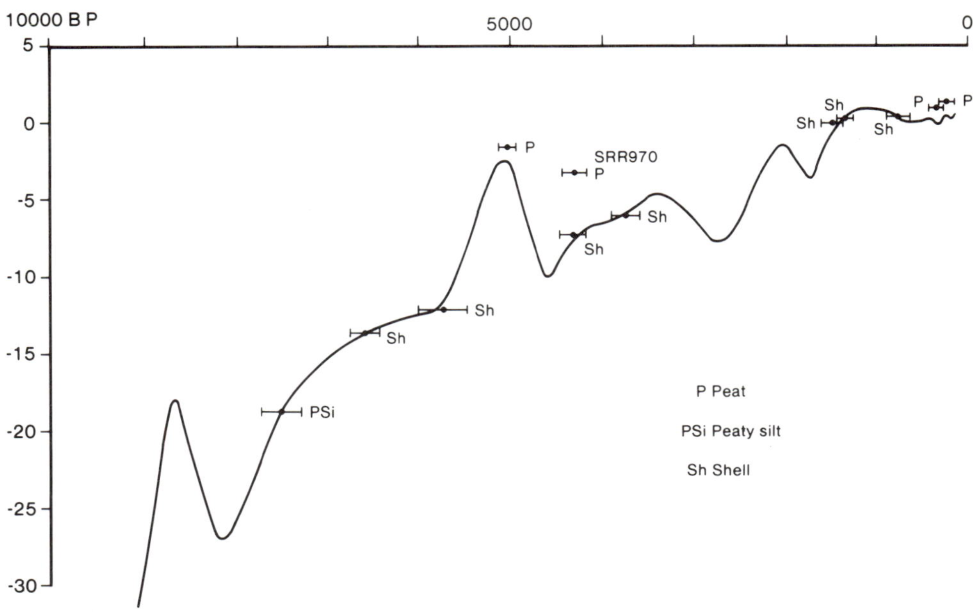

Figure 4. Changes in mean sea level at the Essex coast based on ^{14}C dates, lithological, faunal and geotechnical data and comparison with the sea-level curves of Fairbridge (1961) and Mörner (1971). Based on Greensmith and Tucker (1973, Fig. 6), with SRR-970 added.

The earliest date, on an intercalated peat layer, is from Foulness (7516 ± 250 B.P., Birm-242, −16·5 m O.D.) and the preceding fluctuation in the curve is suggested by analogy with Fairbridge and Mörner (Greensmith and Tucker, 1973). The next dated peat layer is from St. Peter's Flat (4959 ± 65 B.P., SRR-58). Greensmith and Tucker (1973) first described the peat seam as occurring at about −1·5 m O.D. interbedded with intertidal deposits. Later (Greensmith and Tucker, 1980) they give the approximate altitude as about −3·0 m O.D. but say that exact stratigraphic detail cannot be given for the sample. Therefore the nature of their curve around 5000 B.P. must be viewed with caution. However, positive tendencies were suggested by dates on *Ostrea* valves, 6620 ± 100 B.P. (IGS-C14/139) and 5650 ± 240 B.P. (IGS-C15/140). Greensmith and Tucker (1973) estimated, from various evidence, regressive tendencies from 4350 to 4260 B.P. This would be supported by an assay from an intercalated peat at Colne Point (4277 ± 95 B.P., SRR-970). The altitude of the sample was about −3·28 m O.D. and the relationship with the peat layer at St. Peter's Flat, discussed above, is not clear. Shell dates and archaeological evidence suggest renewed positive tendencies until at least 3580 ± 175 B.P. (Birm-243a) and probably continuing after that date. Subsequent changes in sea-level tendencies have not been clearly identified although there appears archaeological evidence for negative tendencies during Romano-British times (Greensmith and Tucker, 1973).

Overall, there are major problems in the interpretation of the data from Essex. The reconstruction of the past altitudes of sea level cannot be viewed as reliable due to the nature of the data base. Greensmith and Tucker (1980) have argued for differential downwarping in the area associated with a tectonic hinge structure. While the difference in levels of various sea-level indicators does not preclude the existence of such a process it does not significantly add weight to the argument due to the uncertain relationships to a common reference level.

The Thames Estuary

Ackeroyd (1972) and Devoy (1982) have discussed the evidence relating to sea-level changes and crustal movements in south-east England, from the Humber to Poole Harbour. Much of the evidence is from the Thames estuary but in general these studies differ in scale and approach from the others discussed so far.

Extremely detailed studies of sea-level change and vegetational history within the Thames estuary have been made by Devoy (1977, 1979, 1980). Much stratigraphic, micropalaeontological and ^{14}C data are available and only the general conclusions and a few specific points are discussed here.

Devoy recognizes five "transgression phases" designated Thames I-V. These are "phases of inorganic deposition" (Devoy, 1979, p. 338), but are not the same as periods of rising sea levels (Devoy, 1979, pp. 389-391) and are summarized in Table 1. These differences arise from inconsistencies in terminology. Devoy (1979, p. 362) defines Thames and Tilbury as names given to lithostratigraphic units, the inorganic and biogenic layers respectively. The names are subsequently applied as

TABLE 1

Periods of "marine transgression" and sea-level change in the Thames Estuary (Devoy, 1979, pp. 387–391) (years B.P.)

Periods of marine transgression	Periods of rises and falls in sea level	
Thames V ≈ 1750	Rise	-0
	Fall 1750	
Thames IV 2600–	Rise	-1750
	Fall 3400	
Thames III 3850–2800	Rise 4000	-3400
	Fall 5500/5000	-4000
Thames II 6575–5410	Rise 6600	-5500
	Fall 7000	-6700
Thames I 8200–6970	Rise 8500/8300	-7000

chronostratigraphic terms relating to periods of formation of the relevant deposits, and then as periods of sea-level movement and also as processes of sea-level movement. The summary schemes presently proposed for the Thames estuary are difficult to use as a result, but this does not detract from the high quality of the original data.

Interpretation of the data from the Thames estuary is not without problems. Devoy (1982) reiterates the possible effects of complex tidal gradients and flood effects, differential crustal movements, and very local factors of sediment growth and channel meandering. Models used to explain the sequences from more open coast situations, such as the Fenland or Downholland Moss, may be less applicable to the confined environment of the Thames estuary. The development of the methodology to cover such variations is discussed on pp. 275–279.

The complexities of the data are perhaps best summarized by Fig. 5 which shows the three curves used by Devoy (1979) to represent the data. First he describes two oscillating curves separated in altitude by up to 3 m, although one index point (16) lies significantly below both curves. The separation of the two curves is used as an argument for the combined effects of differential land subsidence and the effect of tidal amplitude and river discharge. However, the data are widely spaced over a relatively large area and an alternative grouping of the sites, rather than the one used for curves 1 and 2, could indicate differential downwarping over a much shorter distance (1·5 m since 7000 B.P. over *c.* 5–9 km). Alternatively, it may be better to consider the Thames data as a whole with the present index points indicating the likely range of past tide levels, representing the data as a confidence band incorporating periods dominated by rising or falling sea-level tendencies (as in Fig. 3). Devoy's third curve (3) is a mean curve calculated from all the sea-level points and designed to show the mean trend of sea level, although the variance around the mean is not indicated.

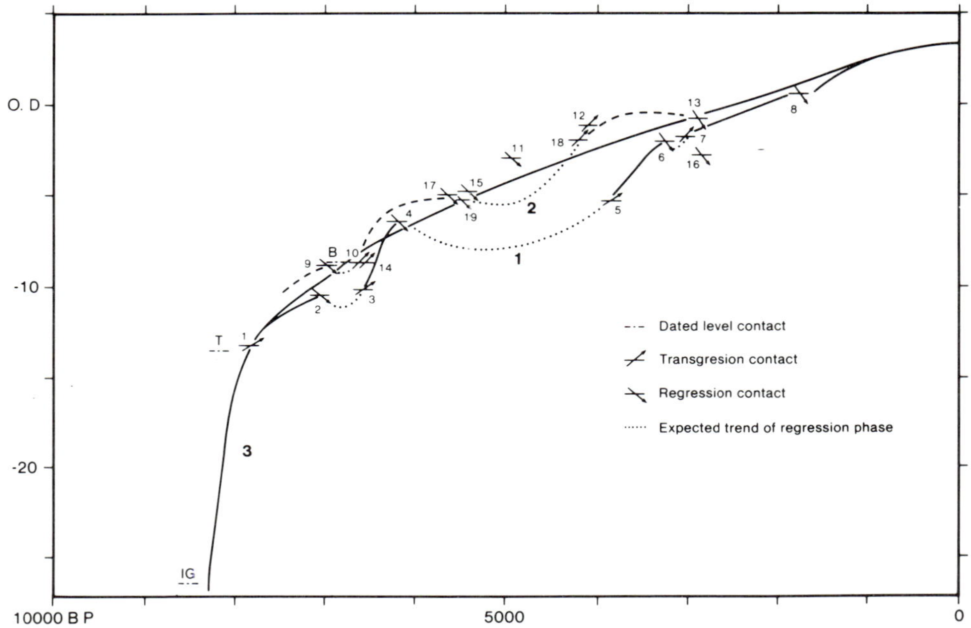

Figure 5. Relative sea-level curves for the Thames Estuary (after Devoy, 1979, Fig. 29). Curve 1 represents M.H.W.S.T. at Tilbury, curve 2 M.H.W.S.T. at Crossness, Stone Marsh, Dartford Tunnel and Broadness Marsh. Curve 3 represents the mean curve of the sampled points.

The south coast of England

The coast between Margate and Land's End has not been studied systematically with regard to sea-level changes. Numerous individual sites have provided ^{14}C-dated samples from Flandrian sediments but their relationship to contemporary tide levels is often unclear. Detailed local studies of small homogeneous areas are not available and the data are usually discussed at a general level (Hawkins, 1971; Shepherd-Thorn, 1975; Mottershead, 1977; Devoy, 1982; Heyworth and Kidson, 1982), using data from publications whose main aim was not usually the study of sea-level changes.

Shepherd-Thorn (1975) and Devoy (1982) describe the early dates from the Channel Tunnel boreholes as 10 530 ± 120 B.P. (NPL-103) at −37·0 m O.D. and 9920 ± 120 B.P. (NPL-101) at −36·5 m O.D.; from Tilling Green as 9565 ± 120 B.P. (IGS-C14/116) at −22·5 m O.D.; and from Langney Point, 8760 ± 75 B.P. (SRR-379) at −24·9 m O.D., but their relationships to sea level are unclear and so must be viewed as minimal estimates of contemporary high tide levels.

The complex changes likely to have occurred in palaeo-tidal range and coastal configuration along the south coast (e.g. Kidson and Heyworth, 1979), require that careful local palaeo-environmental conditions must be evaluated before the regional tendencies of sea-level movement may be assessed. At present, the information available (e.g. Shepherd-Thorn, 1975, Table 1; Devoy, 1982, Tables 3 and 4) cannot be reliably interpreted as showing regional events, though the data may be useful in an exploratory study to establish areas warranting intensive research.

Specific problems relate to the development of the Dungeness shingle spit and the formation of intercalated biogenic and minerogenic sediments in the vicinity, thus affecting the assays from Warehorne and Romney (Shepherd-Thorn, 1975; Cunliffe, 1980). Local phenomena may also be important in the post-glacial infilling of numerous valleys with intercalated marine/brackish and semi-terrestrial freshwater deposits.

Churchill (1965) used an assay at Fawley (6366 ± 124 B.P., Q-834) in his study of crustal movements since 6500 B.P. and it is clear that the sequence, and other deposits from the Solent area (e.g. Hodson and West, 1972; Mottershead, 1976, 1977), will provide critical information for the reconstruction of Flandrian sea-level and tidal changes and crustal movements, but only recently has the necessary detailed, coordinated research been initiated (F. M. Sutherland, personal communication).

Heyworth and Kidson (1982) list 20 ^{14}C dates from the coast between Portland and Lands End. These are, again, generally isolated sea-level index points and have yet to be adequately resolved into a scheme of regionally significant changes in sea level.

The Bristol Channel

The most important works from the Bristol Channel area are those of Kidson and Heyworth (1973, 1976, 1978, 1979), Kidson (1977), and Heyworth and Kidson (1982). Other authors have either collected numerous ^{14}C dates from the area to provide a composite sea-level curve but with no detailed research into the stratigraphic context of the samples (e.g. Hawkins, 1971; Devoy, 1979) or have provided limited sea-level information as a part of a wider study (e.g. Culver and Bonner, 1978; Evans and Thompson, 1979).

Kidson and Heyworth have presented over the past ten years, much information from a very complex area in the sense of elucidating Flandrian sea-level changes and crustal movements. The complexity is due to the presence of a wide continental shelf, an amphidromically complex situation with the largest tidal ranges around the United Kingdom, and a coastline consisting of numerous narrow valleys, entrances to which are restricted by shingle spits or barriers. The problems which these factors create, combined with the large drainage basins providing important and variable amounts of fresh water and sediments, in the interpretation of Flandrian sea-level data suggest that it will be extremely difficult to observe any regional effect due to local "noise". Kidson and Heyworth have recently stressed the problems of reconstructing past sea levels from particular indicators whose altitudinal relationships reflect storm levels, rather than say Mean High Water Mark of Spring Tides (M.H.W.S.T.). Such levels are subject to great variation in space and time and are unreliable sea-level indicators because of this. The relationship between sediments deposited in lower energy situations, as envisaged from the salt-marsh freshwater fen transition, and contemporary tide levels is also subject to error and these errors have been estimated by Shennan (1980, 1982a) and Heyworth and Kidson (1982). Similar estimates are given by both.

The early papers by Kidson and Heyworth (1973, 1976, 1978) concentrated on the Somerset Levels and Bridgwater Bay. They related 25 or 26 radiocarbon dated samples, from peats and trackways, to their contemporary drainage base-level, taken as between M.H.W.N.T. and M.H.W.S.T. An allowance was made for compaction but no recognition of sea-level tendency was made for any date (Fig. 6). Kidson and

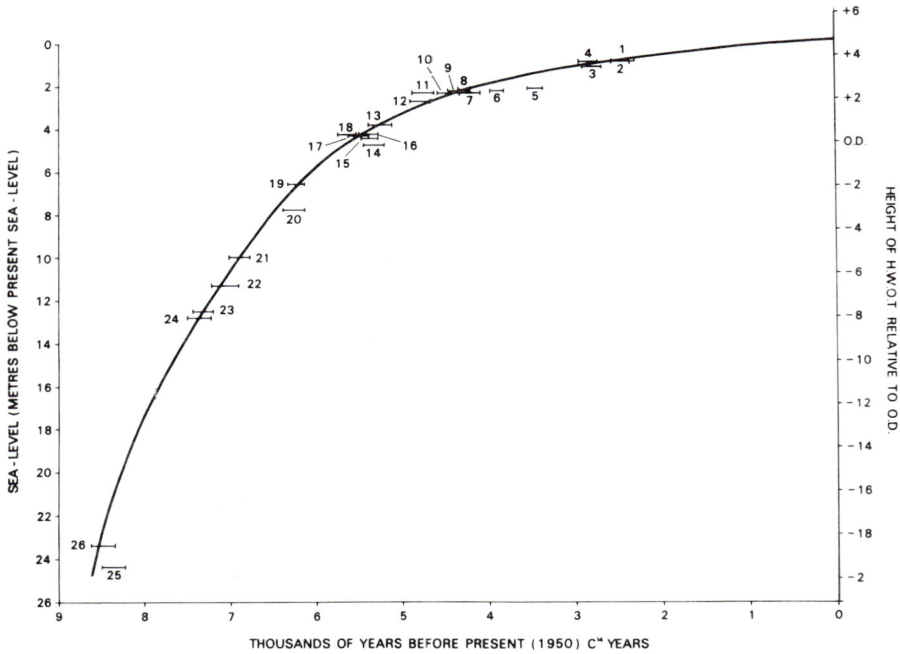

Figure 6. Sea-level rise in the Bristol Channel from Kidson and Heyworth (1976, data numbered in their Table I) assuming the dated samples formed at H.W.O.T. Note that no indication of the local sea-level tendency suggested by each sample is given.

Heyworth related changes in stratigraphy to normal local processes of coastal change (see also Kidson, 1977; Heyworth, 1978) and so proposed a smooth sea-level curve. Detailed work from the Somerset Levels by Beckett and Hibbert (1979) and Hibbert (1980) suggests that this may be too generalized a view. Similar views, based on more theoretical arguments due to their work in other areas have been put forward by Tooley (1978a, 1979), Devoy (1979) and Shennan (1980) but the controversy between the "oscillating" and "smooth schools" should not be viewed too seriously in light of the more recent papers (Devoy, 1982; Heyworth and Kidson, 1982; Shennan, 1982a; Tooley, 1982b). Little advance was made while views were polarized but the recognition of the errors affecting the reconstruction of past tide levels has largely reconciled the differences. When these errors are shown on a time-altitude diagram (e.g. Figs 3 and 7) the inadequacies of a single sea-level

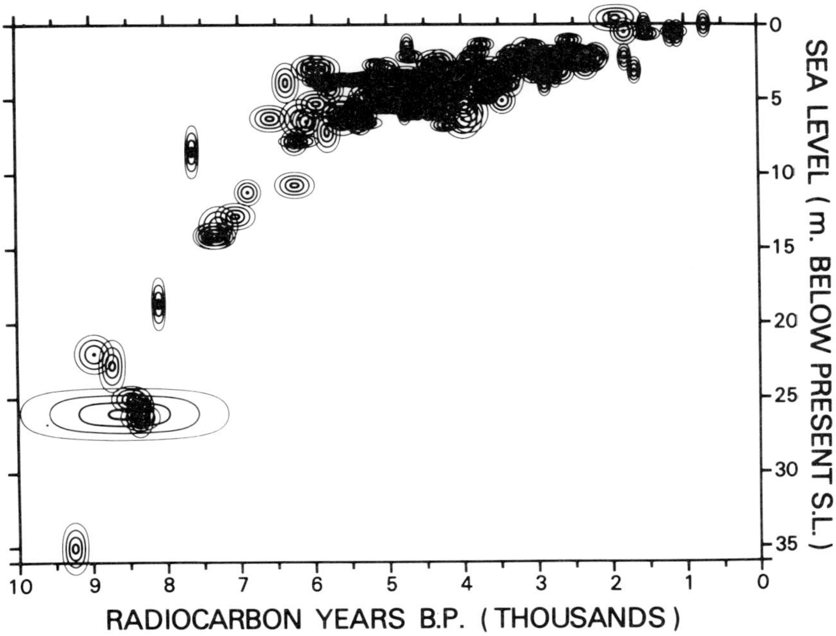

Figure 7. A time-altitude graph for 311 ^{14}C-dated sea-level index points from Heyworth and Kidson (1982). Each date is represented by four proportional ellipses to visually represent the age and altitude error estimates, but no sea-level tendencies are indicated.

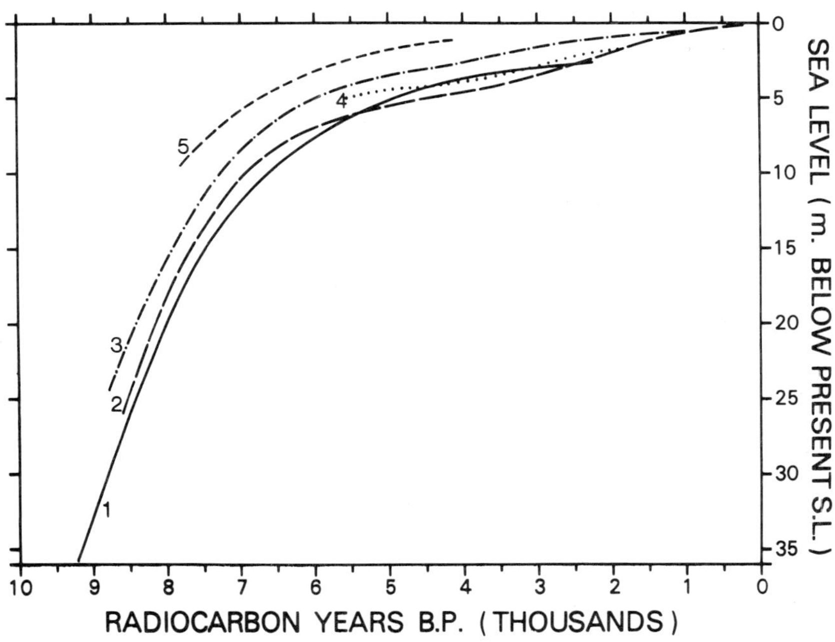

Figure 8. Sea-level curves drawn from the points in Fig. 7, separately considered as: 1, Bristol Channel; 2, English Channel; 3, Cardigan Bay; 4, Somerset Levels Trackways; 5, North Wales (Heyworth and Kidson, 1982). M.H.W.S.T. is used as the common datum (cf. Fig. 6).

curve, whether smooth or oscillatory, as a summary of the data are clear but methodological differences regarding interpretation still remain.

The evidence for crustal movement in the Bristol Channel area during the Flandrian is equivocal. Kidson and Heyworth (1973, 1976, 1978) have argued that the area has remained isostatically stable. There appear to be theoretical objections to this (e.g. Walcott, 1972) and it is presently unclear whether the evidence is reliable to prove either stability or instability. This is evident from the most recent information from Heyworth and Kidson (1982) and Fig. 7. Following a lengthy argument of the errors involved in reconstructing past altitudes for the whole section of the coasts of south-west England and Wales all the data are presented on a time-depth diagram as a series of ellipses which visually indicate the reliability of individual data points but have no strict statistical significance. This diagram is then subdivided into five curves (Fig. 8), which would appear to be at odds with their statements regarding local factors affecting the altitude of sea-level indicators, but could be justified on statistical grounds. However, their conclusions are contentious. Curves 1 and 3 appear to have a difference as great as curves 3 and 5. This may be the result of using M.H.W.S.T. as the common datum rather than M.H.W.O.T. (Kidson and Heyworth, 1978; cf. Heyworth and Kidson, 1982), but consistent differences between one series and another could well prove significant, even though they are small. Statistical analyses would be useful in the interpretation of the regional trends. At present it can be suggested that whereas there appears a subsidence of the Bristol Channel relative to North Wales (although the data from the latter areas are sparse), movement relative to Cardigan Bay is unclear. Furthermore the stability of the Bristol Channel cannot be proved, but any movement is likely to be rather small.

Shennan (1980, 1982a,b) has argued that the construction of time-depth graphs is unlikely to show unequivocal differences in crustal movement between reasonably close areas if alternating peat/clay sequences form the major data base. The results of Heyworth and Kidson clearly confer with this view. The approach of identifying tendencies of sea-level should be applied to the Bristol Channel data so that the techniques outlined by Shennan (1982a,b; Shennan *et al.*, 1983) can be tested (see pp. 275–279). Until full stratigraphic details of all the samples used by Heyworth and Kidson (1982) are available this is not possible. Devoy (1979) has abstracted 24 data points from the Bristol Channel and produced a tentative oscillating sea-level curve. The data, rather than the curve, illustrate the alternative use of the information by identifying the sea-level tendency of each sea-level index point. This is an aim for future research.

Cardigan Bay and North Wales

These areas have not been studied as intensively as the Bristol Channel. The work of Wilks (1977, 1979) effectively continues the investigations in the Dovey estuary area pioneered by Godwin (Godwin and Newton, 1983; Godwin, 1943). Further details are provided by Haynes *et al.* (1977). They obtained an assay of

8740 ± 110 B.P. (Birm-400) from salt-marsh peat overlain by intertidal sediments at −20·5 m O.D. A regressive overlap formed between *c.* 6000 B.P. (Wilks, 1979) and 5150 ± 90 B.P. (HAR-1020) at −0·25 m O.D. However, a rising water table appears to have been recorded around 5400 B.P. (Wilks, 1979; Heyworth and Kidson, 1982). Wilks explains such phenomena as being of local significance only due to the development of a coastal barrier. Further evaluation of the Cardigan Bay data is not feasible at present although Heyworth and Kidson (1982) have indicated that more data are available. These have been included in Figs 7 and 8.

Data from the North Wales coast are sparse. A regressive overlap has been dated 4725 ± 65 B.P. at +2·43 m O.D. (Tooley, 1978a) and five "transgression sequences" are suggested by Tooley (1974). Recent investigations on the Wirral (Kenna, 1978) may provide a link between the well-studied Lancashire coastline and North Wales.

North-west England

This area, between the Mersey and the Solway, has been intensively studied by Tooley (1974, 1976, 1977, 1978a,c, 1979, 1982b) and his work has been a major stimulus and controversy in sea-level research in England and Wales over the last ten years. The data are extremely well known and readily available since they are clearly presented in the numerous publications given above. The biostratigraphic, lithostratigraphic and chronostratigraphic data have been resolved into the sea-level curve shown in Fig. 9. These data are discussed individually by Tooley (1976, 1978a, 1979) and it is unnecessary to repeat them here.

Shennan (1980, 1982b) noted the inconsistencies within the formulation of the Lytham I-XI sequence proposed by Tooley and this chronostratigraphic model has now been replaced by periods of transgressive and regressive overlaps (Tooley, 1982b). The sea-level curve has been replaced by a time-depth graph (Fig. 10). The same chronostratigraphic model appears to apply to the whole of the north-west England but it remains to be tested in detail. The methods presently being developed (Shennan, 1982a,b; Tooley, 1982b; Shennan *et al.*, 1983) are discussed further on pp. 275–279.

Correlation with the sequences from the northern shore of the Solway Firth (Jardine, 1975, 1982) has not been possible and remains an aim for future research.

Temporal correlations made within areas and between areas

Tooley (1982b) and Shennan (1980, 1982a,b) have shown that the criteria used for defining changes in sea-level movement are inconsistent. Therefore the correlation schemes that have been developed, perhaps best demonstrated by Tooley (1974, Fig. 11) and Devoy (1979, Fig. 31) are invalid and misleading. Similarly, the use of such chronologies to lend support to other chronologies (e.g. Mörner, 1976; Hillaire-Marcel and Fairbridge, 1978) reflects an uncritical acceptance of the data. Effectively the process of within-area and between-area correlation must start afresh (Shennan, 1982a,b; Tooley, 1982b).

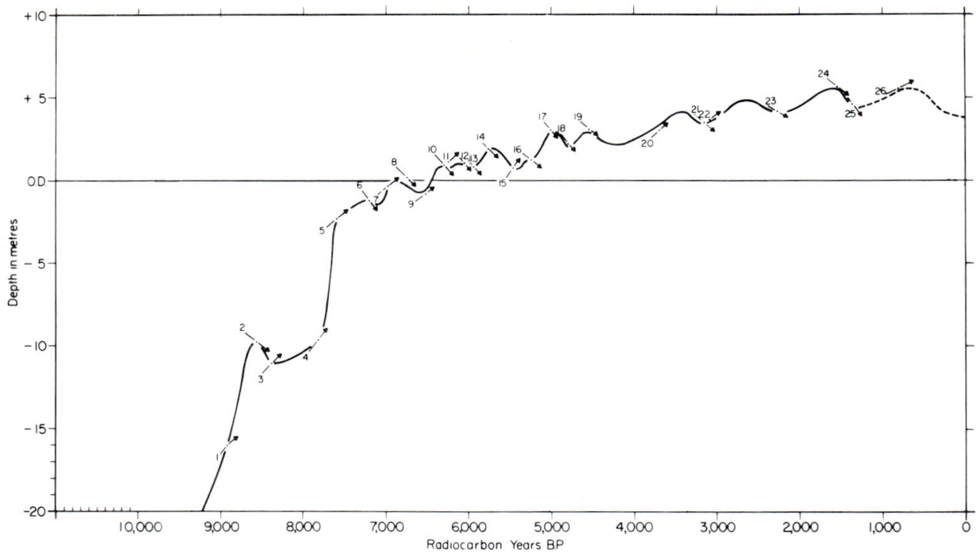

Figure 9. Relative sea-level changes in north-west England shown as a curve (from Tooley, 1979).

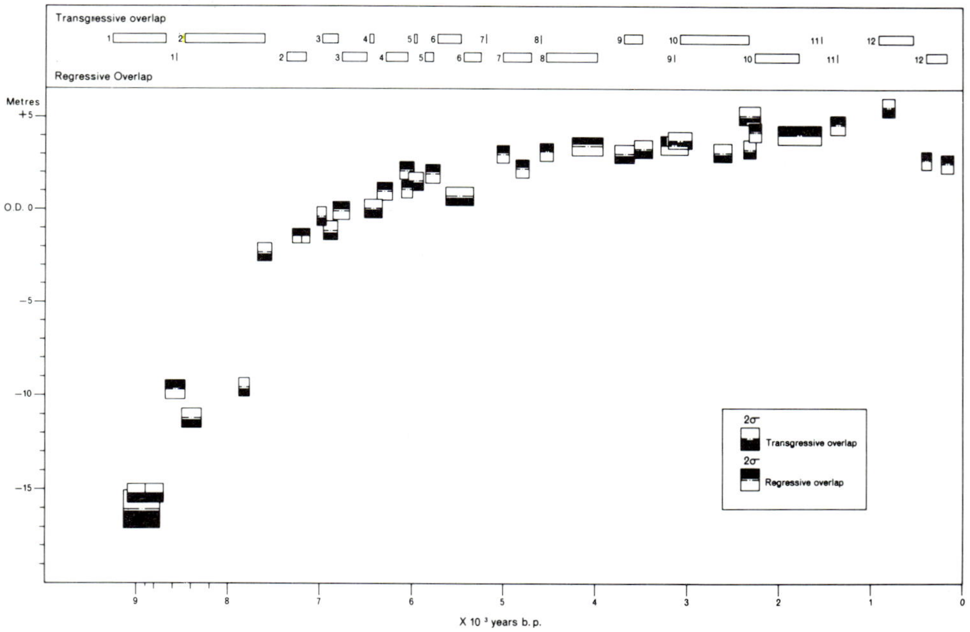

Figure 10. A time-altitude graph of sea-level index points from Liverpool Bay, south-west Lancashire and the south Fylde (Tooley, 1982b). The width of the rectangle represents $\pm 2\sigma$ of the ^{14}C date, but the height is not an estimate of the altitudinal error. Periods of transgressive and regressive overlap for north-west England are also shown.

Evidence for crustal movements within England and Wales

There is a lot of data available which reveal crustal movements within England and Wales during the Flandrian Age. However, the precision of such data is poor and it is not generally possible to calculate either exact rates of movements for individual areas or relative movements between adjacent areas. Generally, trends can be identified. The data which indicate such trends are of two types; first, those based essentially on the construction of past sea levels, and secondly those based on observations of land and sea-level changes over the past 100 years or less.

Pre-Flandrian sea-level evidence

In his discussion of the nature and mechanisms of crustal movements in north-west Europe since the Late Devensian glacial maximum, Mörner (1980b,c) has identified two components, an exponential factor and a linear factor, to uplift in Scandinavia and subsidence in the Netherlands but argues that only the exponential factor is evident in the Scottish and Cumbrian records. It is debatable whether the UK evidence allows an accurate reconstruction of crustal movements and there are some data which indicate that Flandrian crustal movements must be viewed against a background of processes operating over longer time-scales.

The long-term tilting of Britain has been discussed by Smalley (1967), Taylor and Smalley (1969) and Dunham (1972) amongst others, and attributed to convection cells within the mantle. The mechanisms relating to crustal movements are discussed further by numerous authors (in Mörner, 1980a) but no articles relate directly to England and Wales. Eden *et al.* (1978) have speculated that the North Sea may have undergone the effects of a peripheral bulge followed by possible subsidence during the Quaternary glacial-deglacial cycles (cf. Cathles, 1975; Walcott, 1972) but the data are not conclusive and whether similar phenomena should relate to England and Wales is contentious. Indeed even the on-shore data relating to pre-Flandrian sea levels and inferred crustal movements are of variable reliability and interpretation.

The most complete records of marine sediments of each interglacial are from Norfolk and Suffolk (West, 1980). West concludes that there appears no great between-site deformation of the area since Cromerian times but more evidence would be required to confirm this.

Variation in the altitudes of marine erosional and depositional features from the Ipswichian interglacial may indicate some differential movement within England and Wales although accurate dating constitutes a problem. Evidence from northern England at Easington (Smith and Francis, 1967), in the Solway Lowlands (Tooley, 1977, 1980), on the Isle of Man (Thomas, 1977) and around Morecambe Bay (Tooley, 1980) indicates relative uplift compared to the Fenland and East Anglia (West, 1972; Gaunt *et al.*, 1974; Catt, 1977; Jardine, 1979). Jelgersma (1979) has summarized the evidence for the Ipswichian (Eemian) sea-level maximum around the southern North Sea and records downwarping of the continent at least between Brugge and Schleswig-Holstein relative to the Lower Thames estuary.

Dating problems of estuarine and beach deposits along the south coast (Mottershead, 1977; Andrews *et al.*, 1979; Miller *et al.*, 1979) and Wales (Bowen, 1977; Andrews *et al.*, 1979) prevent any firm conclusions being made, but along with the more reliable evidence from the Burtle Beds in the Somerset Levels (Kidson, 1977; Kidson *et al.*, 1978) there is a strong suggestion of relative uplift of these areas compared to East Anglia and the southern North Sea basin.

While a generalized picture of relative subsidence around the southern North Sea basin emerges it would be unwise to attempt a detailed interpretation of crustal movement which may be operating over a time-scale longer than the present interglacial. The superimposition of long-term crustal movements and glacio-isostasy during the Flandrian will occur and the factors will be difficult to separate, but they should be kept in mind.

Evidence for Flandrian crustal movements

The usual methods to show crustal movement are either to compare the altitude of sea-level index points of a certain age (e.g. Churchill, 1965) or to compare sea-level curves from different areas, each curve related to its local "sea level" (Kidson and Heyworth, 1979). Conclusions made from such analyses are based on visual comparisons and compound all the errors inherent in all of the previous stages of analysis. Present evidence suggests that north-west England has been characterized by isostatic uplift until about 5000 B.P. (Tooley, 1978a; Mörner, 1976) and increased uplift to the north is suggested by the evidence from the Scottish Solway coast (Jardine, 1982). Height differences of sea-level indicators between the north-west and north-east coasts of England are explained by present tidal inequalities (Gaunt and Tooley, 1974) but there is a systematic decline in height from north to south along both the eastern and western coasts (e.g. Gaunt and Tooley, 1974; Kidson and Heyworth, 1978; Devoy, 1979; Heyworth and Kidson, 1982). Both Churchill (1965) and Hawkins (1971) suggested downwarping in excess of 6 m since 6500 B.P. for south-east England relative to the south-west. Devoy (1979) confirmed this general view except that the relative subsidence was only between 2 and 3 m.

The errors associated with the construction of sea-level curves (e.g. Figs 3 and 7; and Kidson and Heyworth, 1973; Shennan, 1982a) suggest that the measurement of accurate differential subsidence rates is not possible (see also Akeroyd, 1972). Figure 11 shows the individual data points for north-west England (Tooley, 1976), the Thames Estuary (Devoy, 1979), the Bristol Channel (Kidson and Heyworth, 1973), the maxima for each PTM shoreline for the proposed eustatic curve for north-west Europe (Mörner, 1976) and the sea-level band for the Fenland (Shennan, 1980, 1981, 1982a,b). All of the curves are reduced to "present sea level" on the assumption of constant paleo-tidal regimes. By hypothesizing Mörner's curve to be eustatic the Fenland is confirmed as subsiding, always occurring below the transgression maxima. Relative to the rest of England and Wales the situation is unclear. For apart from the relative uplift of the north-west the Fenland band occasionally overlaps the Thames and Bristol Channel index points. The significance

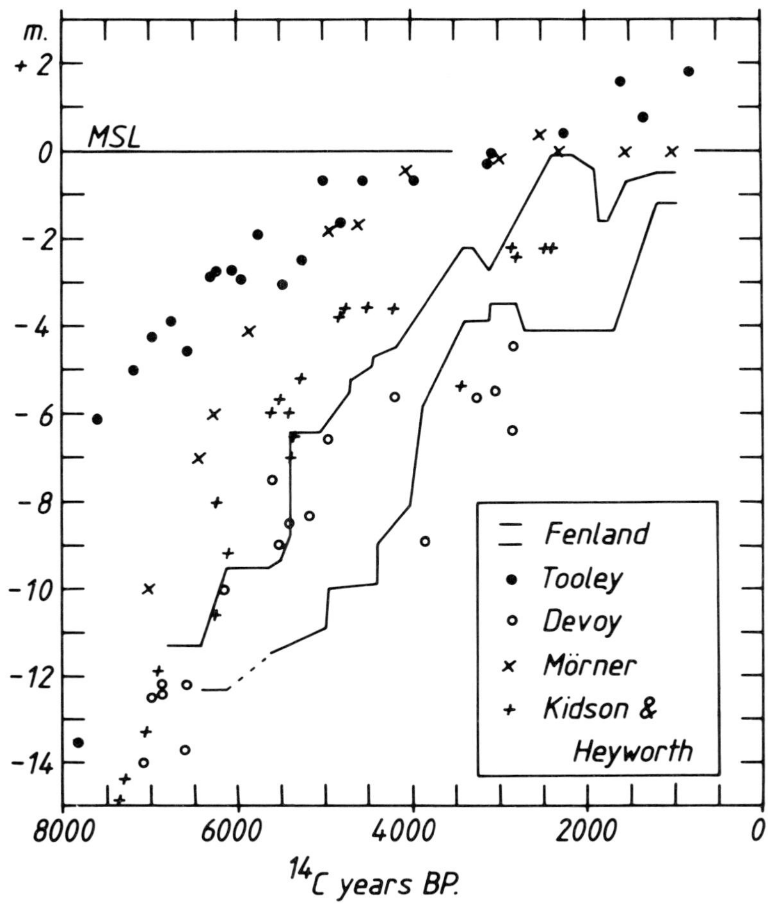

Figure 11. Sea-level band for the Fenland (Fig. 3) compared with the index points given by Tooley (1976) from Lancashire, Devoy (1979) from the Thames Estuary and Kidson and Heyworth (1973) from the Bristol Channel, and the PTM maxima of the proposed regional eustatic curve (Mörner, 1976). All are reduced to present mean sea level assuming constant palaeo-tidal ranges.

of these differences cannot be assessed when the errors for all the areas are considered (cf. Heyworth and Kidson, 1982). The degree of accuracy accepted by Mörner (1976) for example is unwarranted.

Flemming (1982) has made a statistical separation of the eustatic and geological factors of relative sea-level index points. He based his study on 143 data points, supplied by Tooley and Shennan, whose distribution was uneven both geographically and in time. The general trend of differential movement outlined above was noted by Flemming but the results must not be taken too literally. The paper represents a methodology which may be adopted and refinements to the model will be possible in the future when more data and time for analysis are available. The form of the uplift map is governed by the polynomial functions which gave the best fit results and the distribution and size of the residuals indicate the inadequacies of the data set and the present model. Flemming's study confirmed the view that the data base presently available is inadequate to produce anything other than observations of general trends, although the data base now being accumulated at Durham should improve the situation (Tooley, 1982a).

Flemming (1982) noted that there were consistent deviations from the model in the Morecambe Bay area and Tooley (1978, 1982b) has also commented on observed anomalies from the same area with respect to the regional sequences. Relative downwarping of Morecambe Bay has been suggested by Flemming (1982). Other local sea-level studies have indicated possible differential crustal movements over short distances (Devoy, 1979; Greensmith and Tucker, 1980; and pp. 262–269). In the Fenland, analysis of the altitudes attained by marine/brackish sediments of Wash III/IV age and Wash V age are seen to vary within the area (Shennan, 1980, 1981, 1982a) and it is unclear why this should be so. The problem cannot be dismissed as being due to consolidation of the sediments (Shennan, 1980) and differential crustal movements within the Fenland must be considered as a possible contributory factor.

Tide-gauge observation records

Mörner (1973) has successfully used tide-gauge data and repeated levelling information to indicate eustatic changes during the last 300 years. Analysing mean sea-level data from the British Isles using a maximum record length of 1874–1962, Rossiter (1972) indicates that Scotland is rising relative to southern England by at least $1{\cdot}5\ \mathrm{mm\ yr^{-1}}$ (using the Dunbar and Felixstowe data). Subsidence of south-east England compared to Newlyn is also suggested (Rossiter, 1972; Devoy, 1979). Valentin (1953) used the records from ten stations to produce his well-known map of the present vertical movements of the British Isles but his absolute figures are quite different to more recent calculations (Rossiter, 1972) and therefore Valentin's map should not be used as a valid model. Differences between the altitudes at fundamental bench marks observed at the second and third geodetic levellings would appear to substantiate the trends observed from the stratigraphic and tide-gauge data, but the observational accuracy of the respective geodetic surveys is inadequate to confirm this in detail (Kelsey, 1972).

Overall the numerous lines of evidence available confirm the general view of subsidence within south-east England relative to the north, and possibly to a lesser extent the west. The maps produced by Valentin (1953), Churchill (1965) and Flemming (1982) all show this but neither the absolute rates given or the patterns of isolines should be considered accurate.

General problems of sea-level research methods

There is incontrovertible evidence that microfossil analysis of intercalated peat and clastic sediments can reveal periods of rising water table and falling water table, but the relationship to sea-level movements is not necessarily so clear (e.g. Streif, 1979).

There are too many estimates of altitude to be made and the overall imprecision of the palaeo-environmental reconstruction from even the most detailed analyses prevents the elucidation of the altitudinal changes of relative sea level just by studying single sites. Although the errors involved in such reconstructions may be great they are insufficient to reject the hypothesis of a fluctuating relative sea level. The altitudinal errors hinder correlation with increasing distance between sites but do not explain the lithological changes at individual sites or the change in process associated with their formation. Therefore normal statistical smoothing of data points on a time-depth curve from different stratigraphic boundaries is invalid. Changes in the time and space environments which can be measured in both horizontal and vertical dimensions in Flandrian sequences must be assessed. The representation of the data on a time-depth graph may suppress much valuable data.

Existing research methods have polarized on the nature of the restoration of sea level during the Flandrian between various "schools of thought" (Jelgersma, 1966), but this controversy does not require consideration. The immediate problem is to review how existing research methods have suppressed relevant information and how to evaluate processes and past altitudes of relative and absolute land- and sea-levels with greater precision. Until these problems are overcome the identification of regionally significant changes in sea level, associated processes and the accurate estimation of rates of crustal movements will not be possible.

ALTERNATIVE METHODS OF SEA-LEVEL RESEARCH

The nature of sea-level research in England and Wales has changed significantly. The papers of Devoy (1982), Heyworth and Kidson (1982), Shennan (1982a,b) and Tooley (1982b) represent significant advances in the subject. As better, more detailed, and a greater number of local studies of sea-level changes are made and the number of reliable ^{14}C-dated sea-level index points increases then new methods of analysis become applicable.

Definition and interpretation of sea-level index points

In order to permit meaningful correlations to be made between the evidence collected by different authors, standardized methods of definition of sea-level index points must be used even if the interpretations of the data vary. The IGCP Project 61 guidelines (van de Plassche and Preuss, 1978; van de Plassche, 1982) developed the useful concepts of indicative meaning and reference water level of sea-level index points. The terms transgressive overlap and regressive overlap (Streif, 1979; Shennan, 1980, 1982a,b; Tooley, 1982b), and the concept of tendencies of sea-level movement, have been developed and shown to give an adequate framework for the description, interpretation and correlation of sea-level index points and sea-level chronologies (Shennan, 1982b; Shennan *et al.*, 1983). It is possible to illustrate the use of these concepts in terms of an update of Fig. 1 (Fig. 12).

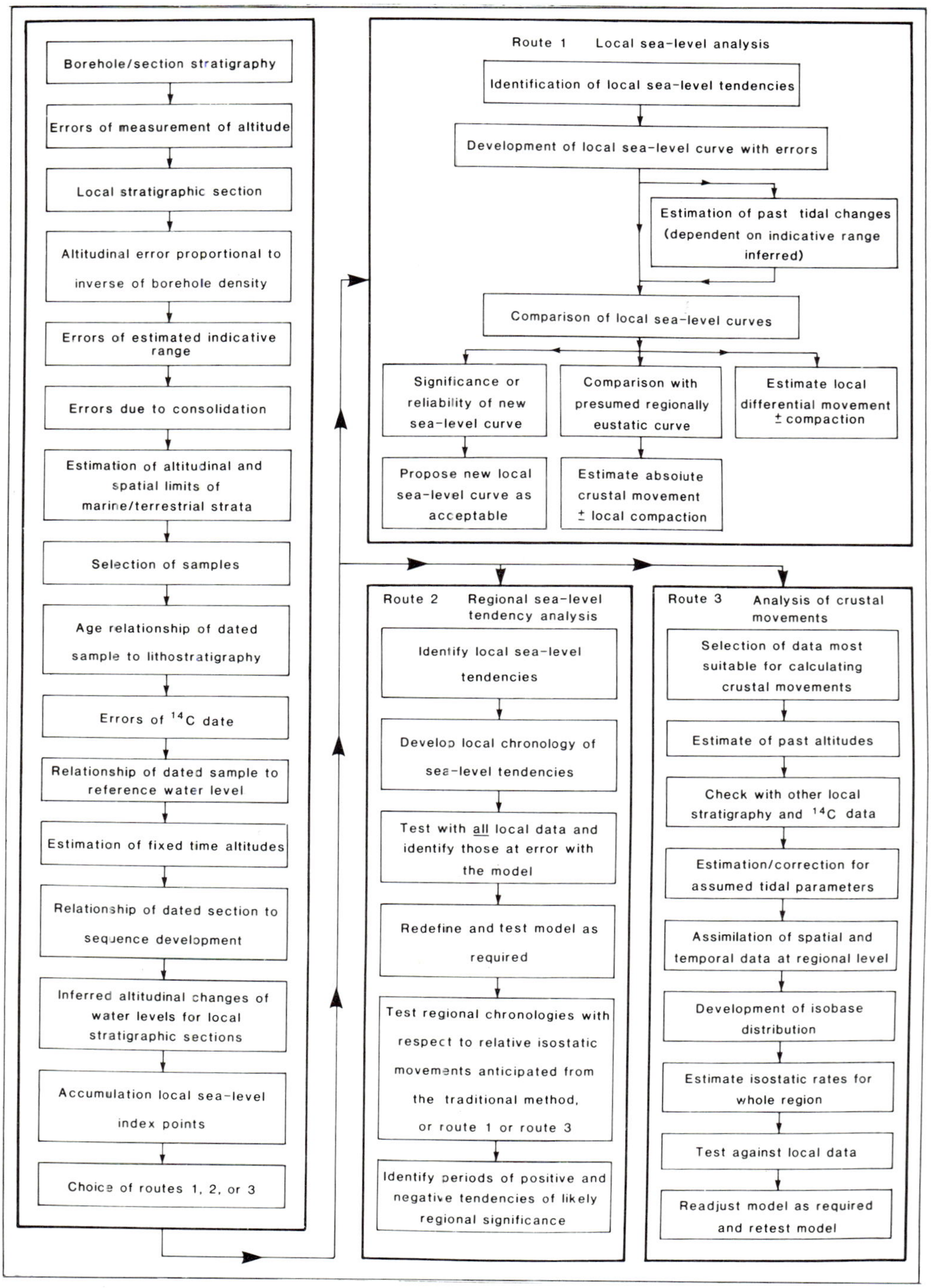

Figure 12. Expanded model of sea-level research methods with explicit allowances for errors.

Route 1, local sea-level analyses

The simplest option is to analyse only the local factors. This is no more than the development of a local sea-level curve or chronology but with an estimate of the errors of data collection and interpretation. Periods of different sea-level tendencies may be evaluated. Figure 3 would represent such an analysis showing errors and tendencies, Fig. 7 indicates the errors only, and Fig. 10 indicates the tendencies and only a broad impression of altitudes. This route of analysis is shown in Fig. 12 (Route 1) and represents how existing sea-level studies may be developed if the relevant information is available for each sea-level index point.

Route 2, regional sea-level tendency analysis

Shennan (1980, 1982a,b) and others (Shennan *et al.*, 1983) have shown that when the tendency of each sea-level index point can be evaluated a chronology of periods of positive sea-level tendencies can be developed for each area. Existing sea-level data can be manipulated within an integrated data analysis system such as the one being developed at Durham (Tooley, 1982a; Shennan *et al.*, 1983) which allows the use of various techniques of analysis ranging from those outlined above to histogram analysis (e.g. Geyh, 1971, 1980; Shennan, 1979, 1982a,b; Shennan *et al.*, 1983; Tooley, 1982b).

The accumulation of around 800 ^{14}C data points, some of which are not related to sea level but may be used to test various models, allows basic statistical analysis of sea-level data to be carried out. It is preferable, though, to keep the analyses exploratory at the moment due to the complexities of the original data. The approach used up to now has been described by Shennan (1982b) and applied by him and others (Shennan, 1982a,b; Shennan *et al.*, 1983; Tooley, 1982b).

It has been argued that a presumed subsiding region, such as the Fenland, will be dominated by periods of positive tendencies of sea-level movement while recorded negative tendencies in such areas are more likely to have regional significance. Similarly transgressive tendencies in uplifted zones are more likely to have regional significance. Such analyses have been successfully applied to the Fenland and north-west England (Fig. 13). The analysis has been extended to include data from the Tay estuary in Scotland and has indicated anomalies in the proposed chronology from north-west England which require further investigation. A major advantage of this approach is that it identifies data which do not agree with proposed models, therefore concentrating future research on problematic areas. It is intended that comparisons can be made between all areas of England, Wales and Scotland when the time-consuming tasks of data acquisition and verification have been completed. This methodology is shown in Fig. 12 (Route 2).

It will be noted that the correlation methods described above do not include an analysis of the altitude parameters. Such data are obviously considered in the identification of sea-level tendencies but the errors involved in altitudinal reconstructions are probably too great at the moment to warrant statistical analyses.

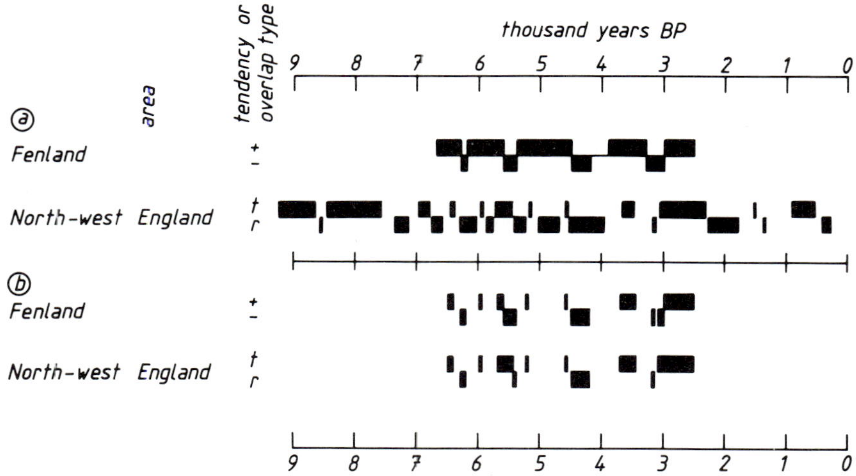

Figure 13. (a) "Wash" periods (positive tendencies of sea-level movement) and "Fenland" periods (negative tendencies of sea-level movement) from the Fenland compared to periods of transgressive and regressive overlap from north-west England (from Shennan, 1982b). (b) The periods of regionally significant tendencies as described in the text are extracted and the two chronologies compared for those periods only. The statistically significant agreement suggests non-random, therefore possibly regionally significant, tendencies of sea-level movement (Shennan, 1982a).

The method developed by Flemming (1982) can be readily applied to the data set and the evaluation of different models and associated residuals will be an obvious future application.

Other numerical methods of analysis (Devoy, 1982) offer less in terms of explanation since the statistical validity of the results obtained is questionable considering the nature of the data in terms of both geological and statistical reliability. The polynomial used is biased towards a smooth curve and appears statistically unstable. Neither the distribution of the residuals or the application of the model at different scales is considered, and overall a more exploratory approach is probably preferable at present.

Route 3, analysis of crustal movements

This approach is shown in Fig. 12 (Route 3). By testing the validity of a model produced by Flemming (1982) against all the available local stratigraphic and ^{14}C data its adequacy could be evaluated. Other numerical models should also be developed. The results of any analyses must also be tested by the application of the regional tendencies concept discussed above and the combined analyses should give consistent results.

Preliminary results and future prospects

The application of consistent operational definitions to sea-level data, the estimation of the errors of the data base and the utilization of the tendency concept have given convincing results in the situations in which they have been applied so far. For example, the sequences of transgressive and regressive overlap within the Fenlands, north-west England and the Tay estuary show a high degree of correlation suggesting that some controlling factor of regional significance exists. A problem that still exists is the reconstruction of the altitude parameters of both sea-level and crustal movement with any accuracy. The errors affecting each sea-level index point (Heyworth and Kidson, 1982; Shennan, 1982a) have been applied to data within the Fenland, the Humber estuary and at West Harlepool (Shennan, 1980). The "error boxes" of the Fenland and West Hartlepool data do not overlap, therefore indicating significant differences in relative sea-level rise, but the error limits in comparison with the Humber area overlap since at least 5240 B.P. Only the inferred differential movement between the Fenland and the Humber estuary can be identified for sediments at least as old as 6000 B.P., for which relative movement has been greater than present sampling and interpreted errors.

Nevertheless the significance of processes and phenomena at different spatial and temporal scales can be evaluated using the methods discussed. The status of different variables changes as the scale of study changes (Shennan, 1982b). Collection of data in a consistent manner allows an integrated, flexible approach to sea-level studies to be developed. At present the analyses are exploratory in nature but more complex multivariate methods can be developed from the existing base. For example, the synchroneity of periods of positive and negative tendencies of sea-level movement in areas of different isostatic history depends on the inferred rates of rise or fall of the sea level and crustal factors (Shennan, 1982b). Such rates have yet to be calculated. Discrepancies within such comparisons may indicate local or regional changes of sedimentation rates or palaeo-tidal changes. These different studies can all be combined using an integrated geographical information system and the approaches suggested clearly identify exceptions to proposed models, therefore those areas most suitable to study local and regional processes in the future.

CONCLUSIONS

While there exists a lot of excellent data relating to sea-level changes and crustal movements within England and Wales during the Flandrian, information for the Late Devensian is sparse. This is in contrast with the situation in Scotland for which Sissons (1981 and this volume) has shown that the Main Lateglacial Shoreline is a very clearly recorded feature and indeed he suggests that the submerged erosional features in Cardigan Bay (Dobson *et al.*, 1971) and Start Bay (Kelland, 1975) are of the same age (Loch Lomond Advance). If this is so then it may be possible to compare the sea-level and crustal movements of England, Wales and Scotland back to the Late Devensian. However, it has been argued in this chapter that it is not yet possible to integrate all the relevant information to give any reliable regional synthesis of either sea-level or crustal factors, particularly in the resolution of the altitude parameters. Alternative methods have been presented which require a re-evaluation of much of the existing data so that progress can be made beyond the development of local studies. Initial investigations, dealing mainly with the resolution of chronologies of sea-level tendencies (Shennan *et al.*, 1983), suggest that this is possible.

ACKNOWLEDGEMENTS

It is a pleasure for me to acknowledge the help and advice of Dr M. J. Tooley during the research and preparation of this chapter, for permission to reproduce Fig. 9 and to use some unpublished data. I am grateful to my colleagues Mr M. J. Davis and Dr B. A. Haggart for their lively discussion of many of the topics presented. My original research in the Fenland was greatly aided by the radiocarbon dates made available by Dr M. A. Geyh. I thank Professor C. Kidson and Mr A. Heyworth for permission to reproduce Figs 5, 6 and 7. Some of the maps were produced by the Cartographic Unit at the Department of Geography, and the photographic reductions were carried out by the Photographic Unit at the Department of Geography. Mrs K. Cowton typed the manuscript. My thanks go to all these people concerned in the production of the chapter, and finally I wish to thank Professor J. I. Clarke for the use of the facilities at the Department of Geography, University of Durham.

REFERENCES

Ackeroyd, A. V. (1972). *Phil. Trans. R. Soc.* **A272**, 151-169.

Andrews, J. T., Bowen, D. Q. and Kidson, C. (1979). *Nature* **281**, 556-558.

Beckett, S. C. and Hibbert, F. A. (1979). *New Phytol.* **83**, 577-600.

Bowen, D. Q. (1977). *In* "The Quaternary History of the Irish Sea" (C. Kidson and M. J. Tooley, eds), pp. 223-256. Seel House Press, Liverpool.

Brown, E. H. (1979). *Trans. Inst. Br. Geogr.* N.S. **4**, 449-462.

Carr, A. P. (1970). *Z. Geomorph.* N.S. **14**, 289-300.

Carr, A. P. and Baker, R. E. (1968). *Trans. Inst. Br. Geogr.* **45**, 107-123.

Cathles, L. M. (1975). "The Viscosity of the Earth's Mantle". Princeton University Press, Princeton.

Catt, J. A. (1977). "Yorkshire and Lincolnshire". Guidebook for Excursion C7 (D. Q. Bowen, ed.), X INQUA Congress, 1977. Geoabstracts, Norwich.

Churchill, D. M. (1965). *Quaternaria* **7**, 239-247.

Churchill, D. M. (1970). *In* "The Fenland in Roman Times: Studies of a Major Area of Peasant Colonization with a Gazetteer Covering All Known Sites and Finds" (C. W. Phillips, ed.), pp. 132-146. *Royal Geographical Society Research Series* **5**.

Coles, B. P. L. and Funnell, B. M. (1981). *Spec. Publs. int. Ass. Sediment* **5**, 123-131.

Culver, S. J. and Banner, F. T. (1978). *Marine Geol.* **29**, 187-207.

Cunliffe, B. W. (1980). *In* "Archaeology and Coastal Change" (F. H. Thompson, ed.), pp. 37-55. *The Society of Antiquaries Occasional Paper* (N.S.) **1**, London.

Devoy, R. J. N. (1977). *Nature* **270**, 712-715.

Devoy, R. J. N. (1979). *Phil. Trans. R. Soc.* **B285**, 355-407.

Devoy, R. J. N. (1980). *In* "Archaeology and Coastal Change" (F. H. Thompson, ed.), pp. 134-148. *The Society of Antiquaries Occasional Paper* (N.S.) **1**, London.
Devoy, R. J. N. (1982). *Proc. Geol. Ass.* **93**, 65-90.
Dobson, M. R., Evans, W. E. and James, K. H. (1971). *Marine Geol.* **11**, 27-69.
Dunham, K. C. (1972). *Phil. Trans. R. Soc.* **A272**, 81-86.
Eden, R. A., Holmes, R. and Fannin, N. G. T. (1978). *Rep. Inst. Geol. Sci.* No. 77/15.
Evans, D. J. and Thompson, M. S. (1979). *Proc. Geol. Assoc.* **90**, 1-14.
Fairbridge, R. W. (1961). *In* "Physics and Chemistry of the Earth 4" (L. H. Ahrens, F. Press, K. Rankama and S. K. Runcorn, eds), pp. 91-187. Pergamon Press, London.
Flemming, N. C. (1982). *Proc. Geol. Ass.* **93**, 113-125.
Gaunt, G. D., Bartley, D. D. and Harland, R. (1974). *Bull. geol. Surv. Gt. Br.* **48**, 1-24.
Gaunt, G. D. and Tooley, M. J. (1974). *Bull. geol. Surv. Gt. Br.* **48**, 25-41.
Godwin, H. (1940). *Phil. Trans. R. Soc.* **B230**, 239-303.
Godwin, H. (1943). *J. Ecol.* **31**, 199-247.
Godwin, H. (1978). "Fenland: its Ancient Past and Uncertain Future." Cambridge University Press, Cambridge.
Godwin, H. and Clifford, M. H. (1938). *Phil. Trans. R. Soc.* **B229**, 323-406.
Godwin, H. and Newton, L. (1938). *New. Phytol.* **37**, 333-344.
Godwin, H. and Vishnu-Mittre (1975). *Phil. Trans. R. Soc.* **B270**, 561-604.
Greensmith, J. T. and Tucker, E. V. (1973). *Geol. Mijnb.* **52**, 193-202.
Greensmith, J. T. and Tucker, E. V. (1980). *Proc. Geol. Ass.* **91**, 169-175.
Hallam, S. J. (1970). *In* "The Fenland in Roman Times: Studies of a Major Area of Peasant Colonisation with a Gazetteer Covering All Known Sites and Finds" (C. W. Phillips, ed.), pp. 22-113. *Royal Geographical Society Research Series* **5**.
Hawkins, A. B. (1971). *Quaternaria* **14**, 115-130.
Haynes, J. R., Kiteley, R. J., Whatley, R. C. and Wilks, P. J. (1977). *Geol. J.* **12**, 129-158.
Heyworth, A. (1978). *In* "Dendrochronology in Europe" (J. Fletcher, ed.), pp. 279-288. *Brit. Archaeol. Rep. Int. Ser.* **51**.
Heyworth, A. and Kidson, C. (1982). *Proc. Geol. Ass.* **93**, 91-111.
Hibbert, F. A. (1980). *In* "Archaeology and Coastal Change" (F. H. Thompson, ed.), pp. 103-105. *The Society of Antiquaries Occasional Paper* (N.S.) **1**. London.
Hillaire-Marcel, C. and Fairbridge, R. W. (1978). *Geology* **6**, 117-122.
Hodson, F. and West, I. M. (1972). *Proc. Geol. Ass.* **83**, 421-441.
Jardine, W. G. (1975). *In* "Quaternary Studies" (R. P. Suggate and M. M. Creswell, eds), pp. 163-168. The Royal Society of New Zealand, Wellington.
Jardine, W. G. (1979). *In* "The Quaternary History of the North Sea" (E. Oele, R. T. E. Schütterhelm and A. J. Wiggers, eds), pp. 159-174. *Acta. Univ. Symp. Univ. Ups. Annum Quingentesimum Celebrantis* **2**, Uppsala.
Jardine, W. G. (1981). *Geol. Mijnb.* **60**, 297-304.
Jardine, W. G. (1982). *Proc. Geol. Ass.* **93**, 25-41.
Jelgersma, S. (1961). *Meded. geol. Sticht.* Serie C. VI **7**, 1-100.
Jelgersma, S. (1966). *In* "World Climate 8000 to 0 B.C." (J. S. Sawyer, ed.), pp. 54-69. *Proceedings of the International Symposium held at Imperial College, London, April 1966.* Royal Meteorological Society, London.
Jelgersma, S. (1979). *In* "The Quaternary History of the North Sea" (E. Oele, R. T. E. Schütterhelm and A. J. Wiggers, eds), pp. 233-248. *Acta. Univ. Ups. Symp. Univ. Ups. Annum. Quingentesimum Celebrantis* **2**, Uppsala.

Jennings, J. N. (1952). *Royal Geographical Society Research Series* **2**, 1-66.
Kelland, N. C. (1975). *J. geol. Soc. Lond.* **131**, 7-17.
Kelsey, J. (1972). *Phil. Trans. R. Soc.* **A272**, 141-149.
Kenna, R. (1978). *J. Merseyside Archaeol. Soc.* **2**, 27-34.
Kidson, C. (1977). *In* "The Quaternary History of the Irish Sea" (C. Kidson and M. J. Tooley, eds), pp. 257-298. Seel House Press, Liverpool.
Kidson, C. and Heyworth, A. (1973). *Proc. Ussher Soc.* **2**, 565-584.
Kidson, C. and Heyworth, A. (1976). *Q. Jl. Engineering Geol.* **9**, 217-235.
Kidson, C. and Heyworth, A. (1978). *Nature* **273**, 748-750.
Kidson, C. and Heyworth, A. (1979). *Proceedings of the 1978 International Symposium on Coastal Evolution in the Quaternary*, pp. 1-28. *São Paulo, Brasil.*
Kidson, C., Gilbertson, D. D., Hayes, J. R., Heyworth, A., Hughes, C. E. and Whatley, R. C. (1978). *Boreas* **7**, 215-218.
Kolp, O. (1976). *Petermanns geogr. Mitt.* **120**, 1-23.
Lambert, J. M., Jennings, J. N., Smith, C. T., Green, C. and Hutchinson, J. N. (1960). *Royal Geographical Society Research Series* **3**.
Ludwig, G. G., Müller, H. and Streif, H. (1981). *Spec. Publs. int. Ass. Sediment* **5**, 211-219.
McGrail, S. and Switzur, R. (1975). *Int. J. Naut. Archaeol. and Underwater Explor.* **4**, 191-200.
Miller, G. H., Hollin, J. T. and Andrews, J. T. (1979). *Nature* **281**, 539-543.
Mörner, N.-A. (1971). *Geol. Mijnb.* **50**, 699-702.
Mörner, N.-A. (1973). *Palaeogeogr., Palaeoclimatol., Palaeoecol.* **13**, 1-14.
Mörner, N.-A. (1976). *Palaeogeogr., Palaeoclimatol., Palaeoecol.* **19**, 63-85.
Mörner, N.-A. (ed.) (1980a). "Earth Rheology, Isostasy and Eustasy." John Wiley, Chichester.
Mörner, N.-A. (1980b). *In* "Earth Rheology, Isostasy and Eustasy" (N.-A. Mörner, ed.), pp. 251-284. John Wiley, Chichester.
Mörner, N.-A. (1980c). *Palaeogeogr., Palaeoclimatol., Palaeoecol.* **29**, 281-300.
Morrison, I. A. (1976). *In* "Geoarchaeology" (D. A. Davidson and M. L. Shackley, eds.), pp. 159-175.
Mottershead, D. N. (1976). *In* "Portsmouth Geographical Essays, II" (D. N. Mottershead and R. C. Riley, eds), pp. 1-21. Portsmouth.
Mottershead, D. N. (1977). *In* "The Quaternary History of the Irish Sea", (C. Kidson and M. J. Tooley, eds), pp. 299-320. Seel House Press, Liverpool.
Rhind, D. W. (1972). *Trans. Nat. Hist. Soc. Northumb.* **41**, 159-164.
Rossiter, J. R. (1972). *Phil. Trans. R. Soc.* **A172**, 131-139.
Shennan, I. (1979). *Sea Level, Information Bulletin of I.G.C.P. Project No. 61* **I**, 6-11.
Shennan, I. (1980). "Flandrian sea-level changes in the Fenland." Unpublished Ph.D. Thesis, University of Durham.
Shennan, I. (1981). *In* "Florilegium Florinis Dedicatum" (L.-K. Königsson and Paabo, eds.), pp. 177-181. *Striae* **14**.
Shennan, I. (1982a). *Proc. Geol. Ass.* **93**, 53-63.
Shennan, I. (1982b). *In* "Climatic Change in Later Prehistory" (A. F. Harding, ed.), pp. 52-67. Edinburgh University Press, Edinburgh.
Shennan, I., Tooley, M. J., Davis, M. J. and Haggart, B. A. (1983). The analysis and interpretation of Holocene sea-level data. *Nature* **302**, 404-406.
Shephard-Thorn, E. R. (1975). *Proc. geol. Ass.* **86**, 537-547.
Simmons, B. B. (1980). *In* "Archaeology and Coastal Change" (F. H. Thompson, ed.), pp. 56-73. *The Society of Antiquaries Occasional Papers* (N.S.) **1**, London.

Sissons, J. B. (1981). *Nature* **291**, 473-475.
Skertchly, S. B. (1877). *Mem. Geol. Surv. U.K.* H.M.S.O., London.
Smalley, I. J. (1967). *Mercian Geol.* **2**, 267-278.
Smith, A. G. (1958). *New Phytol.* **57**, 19-49.
Smith, A. G. (1970). *In* "The Fenland in Roman Times: Studies of a Major Area of Peasant Colonisation with a Gazetteer Covering All Known Sites and Finds" (C. W. Phillips, ed.), pp. 147-164. *Royal Geographical Society Research Series* **5**.
Smith, D. B. and Francis, E. A. (1967). *Mem. Geol. Surv. U.K.*
Streif, H. (1979). *Oceanis* **5**, 303-306.
Swinnerton, H. H. (1931). *Q. Jl. geol. Soc. Lond.* **87**, 360-375.
Taylor, R. L. S. and Smalley, I. J. (1969). *Scient. Jl.* (July 1969), 54-59.
Thomas, G. S. P. (1977). *In* "The Quaternary of History of the Irish Sea" (C. Kidson and M. J. Tooley, eds), pp. 223-256. Seel House Press, Liverpool.
Tooley, M. J. (1974). *Geogrl. J.* **140**, 18-42.
Tooley, M. J. (1976). *Geol. J.* **11**, 137-152.
Tooley, M. J. (1977). "The Isle of Man, Lancashire Coast and Lake District." Guidebook for Excursion A4 (D. Q. Bowen, ed.), X INQUA Congress. Geoabstracts, Norwich.
Tooley, M. J. (1978a). "Sea-Level Changes in North-West England during the Flandrian Stage." Clarendon Press, Oxford.
Tooley, M. J. (1978b). *Int. J. Naut. Archaeol. and Underwater Explor.* **7**, 71-87.
Tooley, M. J. (1978c). *Geol. För. Stockh. Förh.* **100**, 203-212.
Tooley, M. J. (1979). *Proceedings of the 1978 International Symposium on Coastal Evolution in the Quaternary*, pp. 502-533. Sao Paulo, Brasil.
Tooley, M. J. (1980). *In* "INQUA Subcommision on Shorelines of Northwestern Europe Field Guide: Western Scotland and North Western England September 1980" (W. G. Jardine, ed.), pp. 71-110. Glasgow.
Tooley, M. J. (1982a). *Proc. Geol. Ass.* **93**, 3-6.
Tooley, M. J. (1982b). *Proc. Geol. Ass.* **93**, 43-51.
Valentin, H. (1953). *Geogrl. J.* **119**, 229-305.
Valentine, K. W. G. and Dalrymple, J. B. (1975). *Quaternary Research* **5**, 551-590.
van de Plassche, O. (1982). "Sea-level Change and Water Level Movements in the Netherlands during the Holocene." Free University, Amsterdam.
van de Plassche, O. and Preuss, H. (1978). "IGCP Project 61. Sea-level movements during the last deglacial hemicycle (ca 15.000 yr). Explanatory guidelines for completion of the computer form for sample documentation." Unpublished typescript. Amsterdam/Hannover.
Walcott, R. I. (1972). *Quaternary Research* **2**, 1-14.
West, R. G. (1972). *Phil. Trans. R. soc.* **A272**, 87-98.
West, R. G. (1980). "The Pre-glacial Pleistocene of the Norfolk and Suffolk Coasts." Cambridge University Press, Cambridge.
Wilks, P. J. (1977). "Flandrian sea-level change in Cardigan Bay." Unpublished Ph.D. thesis, University College of Wales, Aberystwyth.
Wilks, P. J. (1979). *Palaeogeogr., Palaeoclimatol., Palaeoecol.* **26**, 17-36.
Willis, E. H. (1961). *Ann. N.Y. Acad. Sci.* **95**, 368-376.

TWELVE

Glacial geology and glacial isostasy of the Hudson Bay region

W. R. PELTIER and J. T. ANDREWS

INTRODUCTION

The geometry of the central portions of the Laurentide Ice Sheet (i.e. its relative thickness) has traditionally been deduced by one of the following approaches:

(1) flow lines based on the distribution of erratics across wide, regional areas;
(2) deductive models of the pattern of growth of an ice sheet; and
(3) pattern of glacial unloading as measured by isobases or direction of tilting of glacial lake or marine shorelines.

In the last few years we can add to this list the additional methods of:

(4) realistic three-dimensional ice sheet growth models (not steady-state models); and
(5) reconstruction of probable ice thickness variations calculated from sophisticated models of earth rheology.

Because of the different approaches to the problem of the reconstruction of Pleistocene ice sheets (Andrews, 1982) there are indeed significant arguments as to the thickness of Late Wisconsin ice sheets; there are heated debates on the position(s) of ice divides; and there are underlying concerns about the rate of transportation of glacial debris during proposed shifts in the flow regimes of the Late Wisconsin ice sheets. Of particular concern to us in this chapter is the rôle of Hudson Bay in the glacial (Wisconsin) history of the North American Laurentide Ice Sheet (Fig. 1). These questions have been recently addressed in several papers (e.g. Andrews and Miller, 1979; Shilts *et al.*, 1979; Hillaire-Marcel *et al.*, 1980; Shilts, 1980; Denton and Hughes, 1981; Peltier, 1981a; Andrews, 1982; Peltier and Wu, 1982). These papers have discussed explicitly or implicitly the central problems of ice thickness, ice flow directions, and the locations of ice divides.

We propose to undertake the following:

(1) present a brief review of the historical, indeed philosophical, background to the problem which in essence examines the question: "what were the reasons for

SHORELINES AND ISOSTASY
ISBN 0 12 652960 4

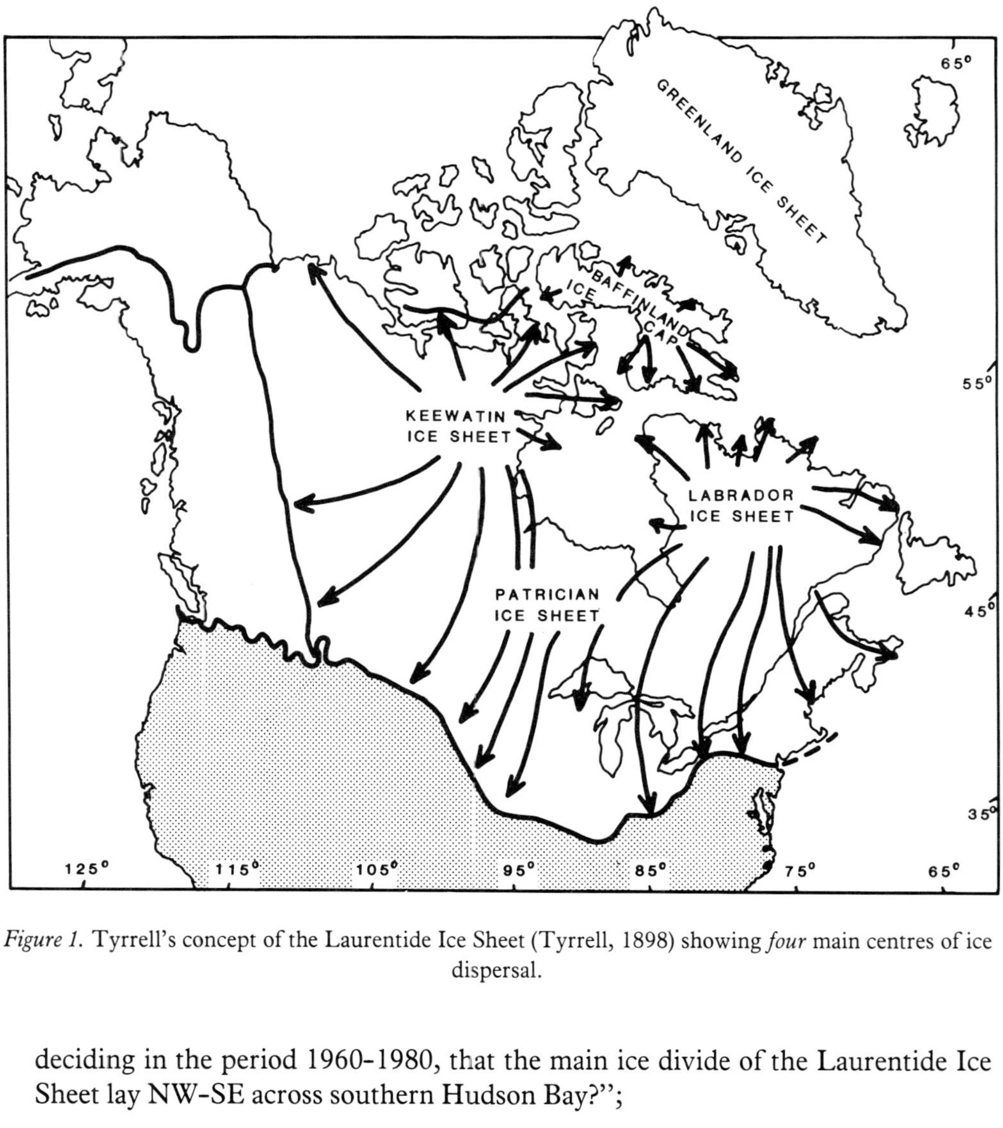

Figure 1. Tyrrell's concept of the Laurentide Ice Sheet (Tyrrell, 1898) showing *four* main centres of ice dispersal.

deciding in the period 1960–1980, that the main ice divide of the Laurentide Ice Sheet lay NW–SE across southern Hudson Bay?";

(2) present a review of new data which tends to confirm the original thinking of Tyrrell (1898) about the position of ice divides and the role of Hudson Bay (Fig. 1);

(3) briefly outline the evidence for successive marine invasions of Hudson Bay during the Wisconsin Glaciation (Shilts *et al.*, 1981; Andrews *et al.*, in press); and

(4) present the results of calculations on the thickness of the ice sheet over Hudson Bay based on geophysical models and examine the importance to earth rheological models to the chronology of deglaciation. Based on work on glacial geology and glacial chronology, we will argue that the Laurentide Ice Sheet over Hudson Bay was dynamic, with a major divide lying east of the eastern coast of

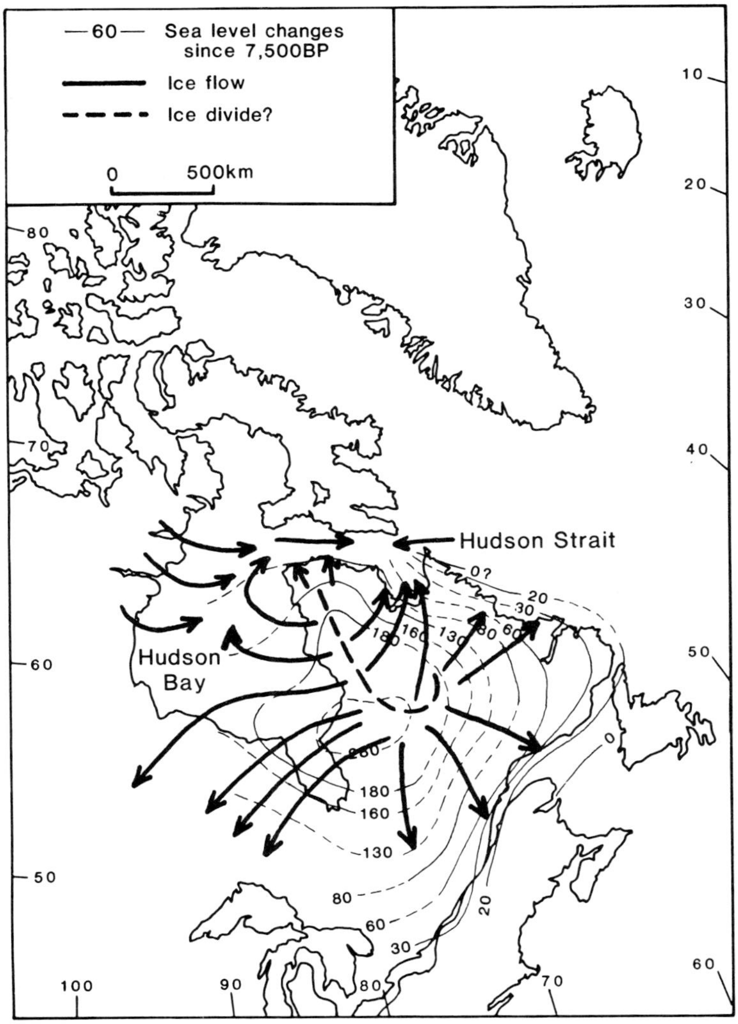

Figure 2. Isobase map of glacial recovery in the last 7500 years (from Andrews and Tyler, 1977 and Hillaire-Marcel and Occhietti, 1980) showing the relationship to inferred glacial flow lines (from Shilts, 1980).

Hudson Bay (Fig. 2). We will then evaluate whether this model is compatible with the evidence from relative sea-level curves and from free air gravity anomalies.

SURVEY OF THE IDEA OF A HUDSON BAY ICE DIVIDE

Shilts (in press) presented a stimulating account of the literature that led to the final development of the concept of a Hudson Bay ice divide (Flint, 1943). Flint's paper argued for a conceptual model of the development of the ice sheet which first developed in the mountains of Labrador with the divide migrating westward eventually to become established in a central, symmetrical position within the borders of the Laurentide Ice Sheet. In advancing this hypothesis, Flint (1943) undoubtedly made use of the growing body of observations on the marine limit in Hudson Bay which tended to show a dome of glacial isostatic rebound centred toward the southern half of the Bay extending southward to James Bay. This phase can be traced rather well in a series of marine limit maps published in the early half of the twentieth century (see Andrews, 1973 for review). Figures 2, 3 and 4 illustrate aspects of this work on the elevation of marine limits. However, these data ran counter to a longstanding concept of the Laurentide Ice Sheet which had been

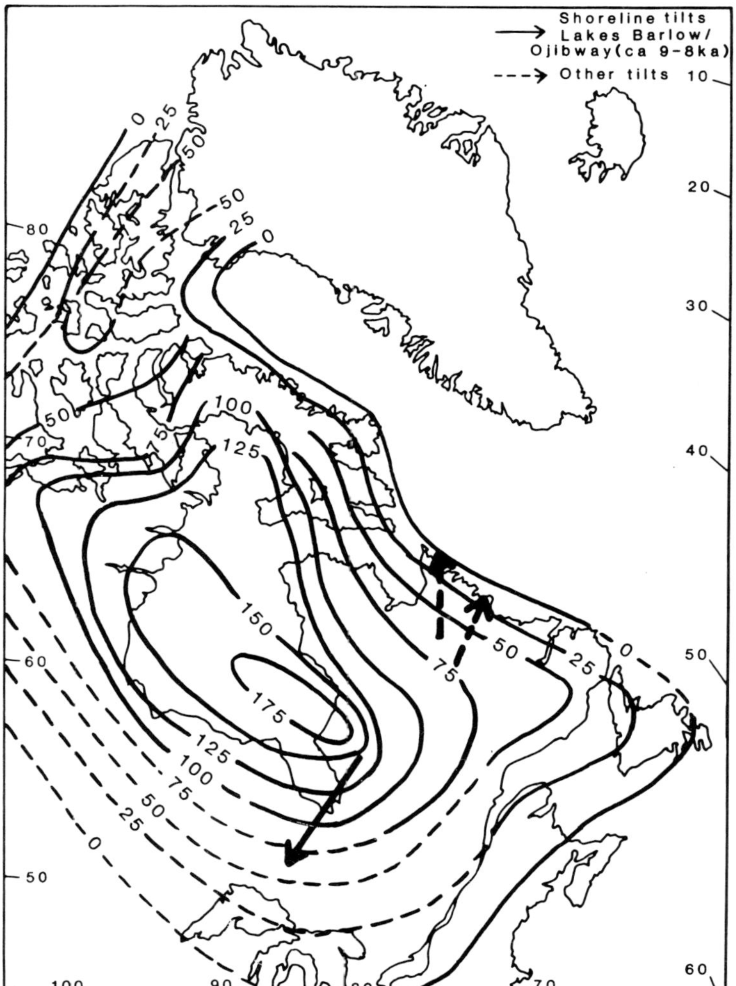

Figure 3. Walcott's isobase map showing recovery over the last 7000 years and superimposed over the directions of lake shoreline tilt (from Barnett and Peterson, 1964 and Vincent and Hardy, 1979).

developed by Canadian field geologists who based their conclusions on the mapping of ice movement indicators (see Fig. 1).

Since the publication of Flint's important paper in 1943, the idea of a single, massive ice divide lying NW–SE across southern Hudson Bay has been generally accepted. With the advent of ^{14}C dating and extensive geological investigations in Hudson Bay (e.g. Andrews and Falconer, 1969; Craig, 1969; McDonald, 1969) additional data were now available for the reconstruction of isobases or isolines on the post-glacial recovery of the Bay (e.g. Farrand and Gadja, 1962; Craig, 1969; Webber *et al.*, 1971). In retrospect, it is intriguing that Andrews (1970) reconstructed a single-domed ice sheet over Hudson Bay although when he attempted to map the glacial isostatic rebound over the last 6000–8000 years, the data showed two quite distinct centres of isostatic uplift centred over western Keewatin and the south-western sector of Hudson Bay (Fig. 4a). A key site in the pattern of the isobases is the Ottawa Islands at 59° 50′ N and 80° 00′ W where the marine limit of 158 m is closely dated by shells collected at 139 m and dated 7430 ± 180 B.P. This site necessitates or forces the pattern of relative sea-level change depicted in Fig. 4. Peltier and Andrews (1976) constructed a time thickness model for the Laurentide Ice Sheet (ICE -I) which embodied this single-dome concept at a presumed 18 000 B.P. glacial maximum.

In his review of the glacial isostatic rebound evidence from North America, Walcott (1972) drew on an enlarged body of ^{14}C-dated sites and presented a map of glacial rebound (Fig. 3) which differed in detail from that presented earlier by Andrews (1970) (see Fig. 4a). The main feature of Fig. 3 is the major centre of rebound located along the south-eastern coast of Hudson Bay.

These various reconstructions of the form of post-glacial sea-level change (Figs 2, 3 and 4) were implemented by several studies which determined the direction of dip of tilted or warped lake and marine shorelines (e.g. Goldthwait, 1910; Farrand, 1960; Barnett and Peterson, 1964). Andrews and Barnett (1972) reviewed this literature and presented a map which showed the directions of deformation for 30 sites in North America. Isobases should lie normal to the dip of these tilted surfaces but as Fig. 3 shows, they do not. Rather than indicating a rebound centre along the east coast of Hudson Bay, the intersection of the shoreline dips suggested several possible areas of updoming and implied a major zone of uplift toward the centre of the Labrador-Ungava Peninsula. In retrospect, Andrews and Barnett (1972) did not fully discuss the implication of their data to the model of a single-domed ice sheet, nor did they fully discuss the apparent contradictions between the strandline dips and inferred isobases (Fig. 3).

The difference between Laurentide Ice Sheet reconstructions may indeed be based on different concepts and data. However, we should be aware that reconstruction based on *ice flow indicators* (e.g. Shilts, 1980) and those based on glacial isostatic evidence (e.g. Peltier and Andrews, 1976) might be inherently different. This is because it is not necessary that an *ice divide* coincide with the *area of thickest ice* (cf. Andrews, 1982). Although we have not investigated the problem specifically, the location of an ice divide over a topographic high, and the location of maximum

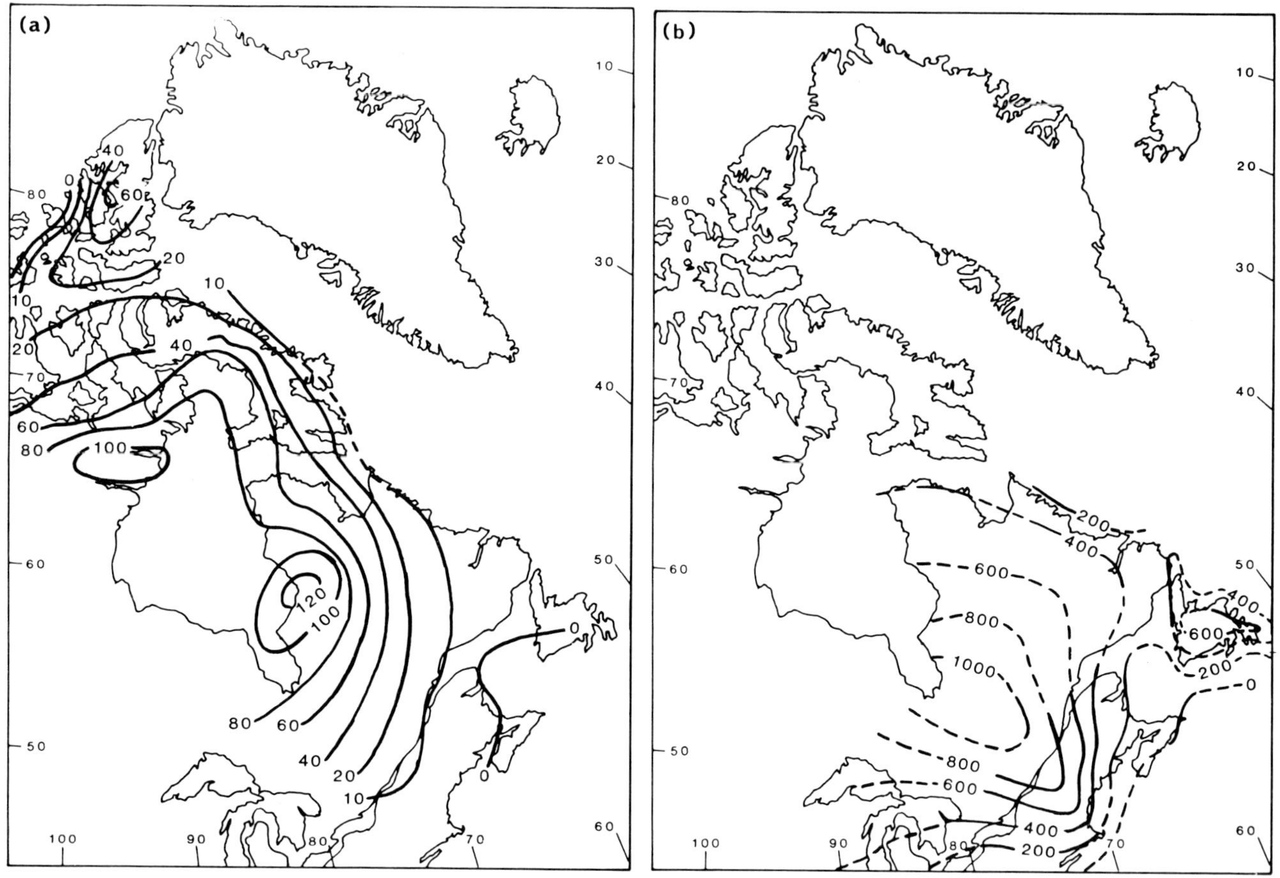

Figure 4. (a) Isobase map of postglacial uplift in the last 7000 years (from Andrews, 1970) compared with Fairchild's reconstruction (Fig. 4b). (b) Isobase map of postglacial uplift (from Fairchild, 1918).

glacial rebound close to a bathymetric low might be contained within the same glaciological reconstruction.

CURRENT DATA ON ICE FLOW AND GLACIAL ISOSTATIC REBOUND

In the last four to six years renewed research in Hudson Bay has resulted in a gradual change of emphasis, but has of late become crystallized into the concept of a multi-domed Laurentide Ice Sheet (Andrews and Miller, 1979; Shilts *et al.*, 1979; Hillaire-Marcel and Vincent, 1980; Hillaire-Marcel *et al.*, 1980; Shilts, 1980, and in press). A major impetus to research was the investigations associated with the proposed James Bay power facilities and the prospects of gas pipelines being routed along the western side of James Bay. Dionne (1975), Hillaire-Marcel (1979) and Shilts (1980, in press) discussed the evidence for ice flow indicators; Hillaire-Marcel (1979) and Hillaire-Marcel and Occhietti (1981) have discussed the form of the glacial isostatic rebound based on dated sea levels (Fig. 2) (see also Skinner (1974) and Hardy (1976)); and Vincent and Hardy (1979) have made a significant contribution in their study of the shorelines of the various phases of Glacial Lake Ojibway-Barlow as the lake expanded from an initial outlet near 46° 40′ N to finally extend northward to nearly 56° N along the eastern side of Hudson Bay (Hardy, 1976; Vincent and Hardy, 1979; Hillaire-Marcel and Vincent, 1980). At the same time, maps of ice thickness during different periods of glaciation and deglaciation were produced as a result of glaciological modelling (Paterson, 1972; Andrews and Mahaffy, 1976; Budd and Smith, 1981; Denton and Hughes, 1981) and from the consideration of relative sea-level variations and rheological models of the earth (Peltier, 1974, 1981a; Clark, 1980; Wu and Peltier, 1982, 1983). As depicted in Fig. 2, new isobase maps, based on data available up to the late 1970s has shown that isobases delimiting a major centre of rebound have been shifted significantly eastward (Andrews and Tyler, 1977; Hillaire-Marcel and Occhietti, 1980). These maps are more in agreement with postulated flow conditions during the Late Wisconsin glacial maximum than earlier efforts (Figs 2, 5), but it should be noted, that the isobase data and ice flow studies were essentially independent.

It is very important to stress that the available data on erratic and mineralogical provenances around southern Hudson Bay argue for a dominant *south-westerly* flow from Labrador and directed toward the Hudson Bay Lowlands and eventually toward the Lake Superior drainage. The evidence for erratics being carried *eastward* (Lee, 1960) is now discounted by the evidence of Dionne (1975) who demonstrated that all such erratics were located *below* the marine limit and were most probably derived from ice-rafting during deglaciation. The existence of a Patrician ice centre (Tyrrell, 1898) has not been advocated by Shilts (1980) but Dyke *et al.* (1982) have suggested that an additional ice dome was located over southern Hudson Bay/James Bay.

The pattern of deglaciation within Hudson Bay during the early Holocene is instructive, particularly since the rapid disintegration of the ice sheet about 8000 B.P.

has been discussed by several authors (Andrews and Falconer, 1969; Prest, 1970; Skinner, 1973; Hughes *et al.*, 1977; Denton and Hughes, 1981). In this regard the work of Hillaire-Marcel (1979), Vincent and Hardy (1979) and Hardy (1976) is extremely important in modifying the picture of events that must have occurred within Hudson Bay during this interval. These events are important in terms of the interpretation of the synchroneity of marine limits in parts of the Bay and the direction of post-glacial tilting.

Pictorially, we must now envisage the deglaciation of Hudson Bay as consisting of four major elements (Figs 5a,b):

(1) the growth and development of glacial lakes Barlow and Ojibway from the vicinity of an ice margin near Timiskaming (47° 30′ N) to the final glacial lake stage when glacial lake waters extended northward along the east coast of southern Hudson Bay to the vicinity of Great Whale River (Hillaire-Marcel, 1979; Vincent and Hardy, 1979; Hillaire-Marcel and Vincent, 1980). Concretions at Great Whale River date this event at close to 8000 B.P.

(2) To the north, ^{14}C dates from Hudson Strait (Blake, 1966; Matthews, 1967) and patterns of striations on the Ottawa Islands indicate that a calving bay developed in Hudson Strait and migrated southward down the Bay, *not* in the form of an extensive calving bay as incorrectly drawn by Hughes *et al.* (1977) and Denton and Hughes (1981), but as a rather confined entrant, possibly located along the line of contact between Labradorean and Keewatin ice (Figs 5a,b).

(3) Sometime toward 8000–8500 B.P. there was a south-eastward movement of ice from central Hudson Bay toward the James Bay Lowland during the Cochrane advances (probable surges) (Prest, 1970; Hardy, 1976; Vincent and Hardy, 1979). The ice-source for these late glacial surges may have been a residual mass of ice situated over west-central Hudson Bay (Hillaire-Marcel *et al.*, 1980; Dyke *et al.*, 1982) or it might have been an extension of the Keewatin ice flow regime (Shilts, 1980). Whether this represents some facet of Tyrrell's "Patrician Ice Sheet" is unknown.

(4) At some point in time close to 8000 B.P. the marine waters of the Tyrrell Sea broke through the ice dam in the south and replaced Glacial Lake Ojibway.

However, elsewhere around the Bay, the ice margins retreat moderately slowly despite being in water to depths of 250–300 m. Figure 5a lists the dates of deglaciation of various coastal segments and clearly demonstrates that the Hughes *et al.* (1977) calving Bay model does not fit the field evidence (Craig, 1969; Hillaire-Marcel and Vincent, 1980; Shilts, in press).

The important event for the study of the glacial isostasy of the Bay are the events in (4) above (see Fig. 5) as they indicate that the marine limit in southern and south-eastern James Bay is essentially isochronous and thus elevations on the marine limit can be used to extend the time period for which isochronous water plane solutions are available for the determination of rebound centres. These calculations are based on triangle solutions for tilts (Løken, 1962; Andrews, 1970) which give average

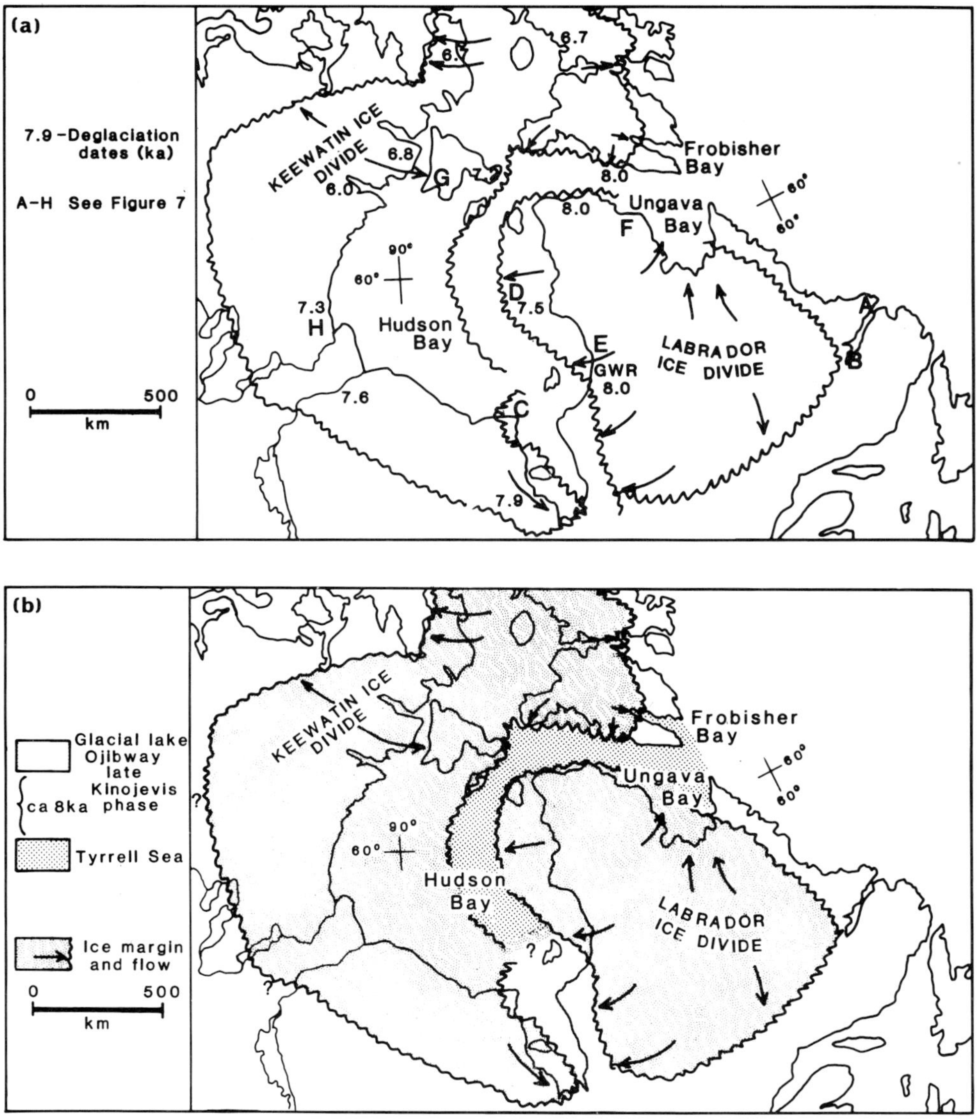

Figure 5. (a) Suggested ice margin at *c.* 8000 B.P. and dates on deglaciation of the coast of Hudson Bay. Location of sites with sea-level curves (Fig. 6) also shown. (b) Pattern of deglaciation *c.* 8000 B.P. in Hudson Bay prior to the drainage of Glacial Lake Ojibway into the encroaching Tyrrell Sea.

directions of tilt up to the north or north-east. The average solution is similar to the directions of tilt derived from the different shorelines of Glacial Lakes Barlow and Ojibway (Vincent and Hardy, 1979). These determinations are at variance with a calculation by Archer (1968) near Great Whale River which suggested a tilt up toward the west or south-west. Shoreline tilts for Labrador-Ungava are illustrated in Fig. 3 and suggest a centre(s) of rebound east of the Hudson Bay coast.

Over the course of the last 20 years, several reasonably dated relative sea-level curves have been constructed for various sections of the Hudson Bay and Labrador coast (see Hillaire-Marcel and Vincent, 1980 for review; also Walcott, 1980). A selection of these curves are shown as Fig. 6 and Table 1 lists estimated amounts of relative sea-level change (with respect to present) for the last 8000 years. The sites for the sea-level curves are shown in Fig. 5a. The distribution of the sites raises an obvious problem with the use of marine-oriented data for glacial isostatic studies. How do we take into account the massive regions where *no* relative sea-level data exists? A control on the orientation and gradient of the isobase maps is partly given by the directions and rates of tilt on glacial lake shorelines which were a distinctive feature of the deglaciation of Labrador-Ungava (Ives, 1960; Barnett and Peterson, 1964; Hughes, 1965; Vincent and Hardy, 1979). Of particular benefit for the understanding of the glacial history of Hudson Bay/Hudson Strait would be accurate determinations of directions of tilt of the large glacial lakes that existed south of the north-west coast of Labrador-Ungava (Ives, 1960; Matthews, 1967). To our knowledge, such data are not yet available. However, one of the most critical dates in the region is a ^{14}C assay on shells at 100–120 m south-west of Ungava Bay of only 6270 ± 100 B.P. (GSC-2311) (Blake, 1976). The age/elevation of this site is shown in Fig. 6 and demonstrates the high rate of isostatic recovery for this site even when compared with sites nearer the suggested former centre of the Laurentide Ice Sheet (e.g. site C, Fig. 5a). In addition, the date of 6015 ± 150 B.P. (I-1224) for deglaciation of Baker Lake, well inland from the coast in north-west Hudson Bay indicates emergence of approximately 120 m at the site (Figs 6 and 8) and may delimit a Keewatin centre of rebound. Comparison of the curves in Fig. 6 indicates

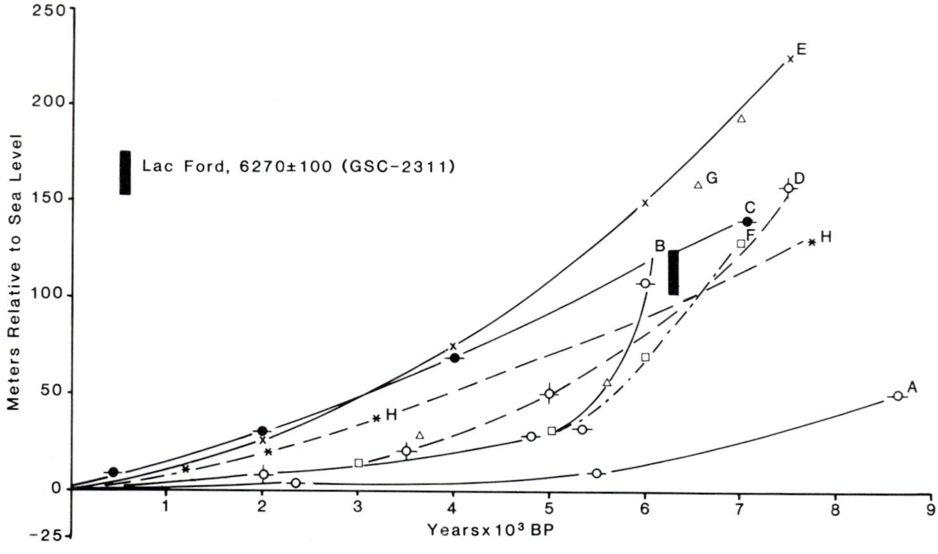

Figure 6. Relative sea-level curves from around Hudson Bay (see Fig. 5a) showing changes of sea level since deglaciation. A, outer Hamilton Inlet; B, inner Hamilton Inlet; C, Cape Henrietta Maria; D, Ottawa Islands; E, Richmond Gulf; F, Ungava Bay; G, Southampton Island (from original sources and Hillaire-Marcel, 1979).

TABLE 1
Sea level, above present (metres), at sites shown in Fig. 6

Time Ka	A Outer Hamilton	B Inner Inlet	C Cape He-Ma	D Ottawa Is.	E Richmond Gulf	F W. Ungava Bay	G South- Is.	H Churchill
1	1	3	17	3	12	3	3	10
2	2	9	30	7	25	9	9	20
3	5	13	47	12	47	12	12	38
4	6	20	70	30	75	20	39	50
5	8	30	92	50	110	30	45	95
6	13	110	120	85	150	70	95	115
7	25		140	120	200	130[b]	180	
8	40				(300)[a]			
9								
10								

[a]Inferred. [b]Inland date of 6270 ± 100 (GSC-2311) (1976) at 130 m.
A,B, Fitzhugh (personal communication); C, Webber *et al.* (1970) and Walcott (1972); D, Andrews and Falconer (1969); E, Hillaire-Marcel and Vincent (1980); F, Hillaire-Marcel and Vincent (1980); G, Bird (1970); H, Craig (1969).

that the majority show fairly rapid relaxations whereas the two curves from east and south-east Hudson Bay indicate a significantly longer relaxation. Walcott (1980) commented on the form of the curve from near Cape Henrietta-Maria (Webber *et al.*, 1970; Walcott, 1972) and suggested that in comparison to the Ottawa Islands relative sea-level curve (site D, Fig. 6) (Andrews and Falconer, 1969), the latter curve was probably in error. An attempt to examine this problem further has been made and two additional dates were obtained on samples from the Ottawa Islands. However, neither date really constrains the curve. However, it is worth noting that the curve with the total highest change in relative sea level over the last 7000 years (from Richmond Gulf, E, Fig. 6) relaxes faster than sites from the eastern side of the Bay. This may suggest a non-glacial isostatic control on at least part of the response possibly related to the structural high which is present along the west side of the Bay. The most important points that are brought out on the isobase map (Fig. 7, see Table 1 and Fig. 6) are:

(1) relative sea-level changes delimit at least two areas of sea-level change greater than 150 m in the last 7000 years;
(2) the site west of Ungava Bay suggests a major ice unloading zone between there and Richmond Gulf;
(3) the suggested sea level of *c.* 100 m for inner Hamilton Inlet (site B, Fig. 6) indicates that a significant ice mass lay close to the central-east sector of the Labrador-Ungava plateau.

Points (2) and (3) in particular, taken in conjunction with the pattern of shoreline deformations around the Peninsula (Fig. 3) map, indicate a broad zone of relatively

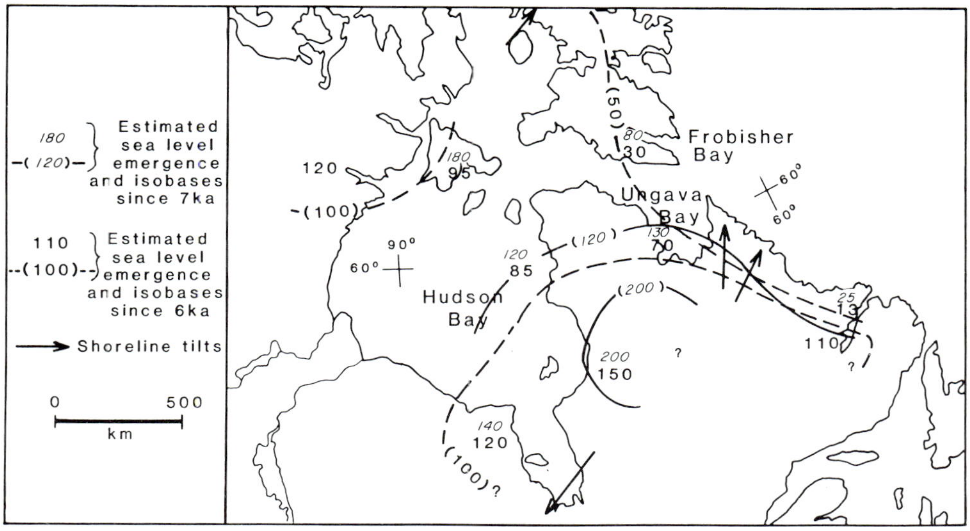

Figure 7. Isobases on emergence around Hudson Bay using the data in Fig. 6 and Table 1. Note the separation of the area of highest uplift into two discrete areas on either side of Hudson Bay.

high rebound which is delimited on Fig. 7 by the 100 m contour for 6000 B.P. A separate rebound centre over Keewatin appears to be required when the emergence of 120 m+ west of Southampton Island (Fig. 8; Craig, 1969) is compared to emergence of about 100 m for the same interval on Southampton Island (Bird, 1970; G, Fig. 6) and the figure of 85 m+ for the Ottawa Islands.

Hillaire-Marcel (1980) believes that the emergence data from Richmond Gulf can be interpreted as showing an initial history associated with the deglaciation of Hudson Bay followed by a subsequent period when the emergence was effected by the late deglaciation of residual ice located over central Labrador-Ungava. This hypothesis is difficult to test, but it is associated with the troublesome question of whether shifts in ice divides and delayed deglaciation cause perturbations in the emergence history of a region. Dyler (1974) argued that these effects could be observed in the post-glacial emergence of Baffin Island. It appears to us that the relatively long history of shoreline isobases striking NW-SE from the initial development of the Great Lakes through Glacial Lakes Barlow and Ojibway casts some doubts on the effectiveness of these responses. It is, however, a question that requires further study and possibly modelling (see Conclusions).

WISCONSIN GLACIAL AND DEGLACIAL HISTORY OF HUDSON BAY

In our initial models of the Laurentide Ice Sheet—which we termed ICE-1 (Andrews and Peltier, 1976; Peltier and Andrews, 1976)—the ice sheet was assumed to be in isostatic equilibrium at 18 000 B.P. Changes in extent and thickness were

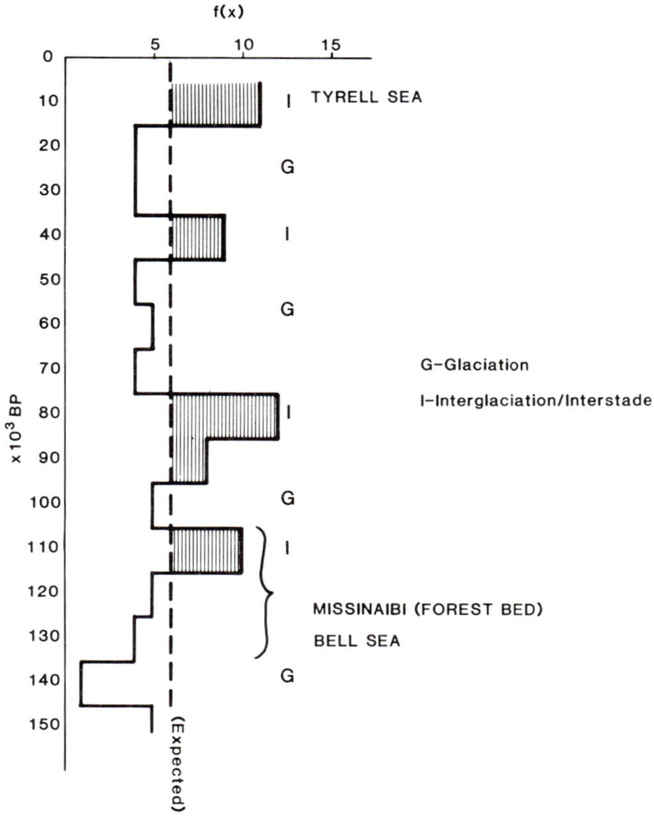

Figure 8. Estimated ages of marine shells collected from tilts and waterlain sediments in the Hudson Bay Lowlands. Age estimates are based on the total amino acid ratios for D-alloisoleucene: L-isoleucene (data from Andrews *et al.*, in press).

then forecast and the predicted changes in relative sea level compared with ^{14}C-dated sea-level curves. Further discussion of this and other models (ICE-2, -3, and -4) are contained in the next section (see Peltier, 1981a; Wu and Peltier, 1983). The importance of the equilibrium constraint at 18 000 B.P. was not explicitly examined in our initial studies. However, the global isotopic record from marine foraminifera (Shackleton and Opdyke, 1973; Shackleton, 1977) do indicate that during the last glacial/interglacial cycle there have been globally significant fluctuations in ice mass on time-scales of 10^4 years. Did these affect the central region of the Laurentide Ice Sheet? This is a profound question that requires thought, not an unqualified "yes" or "no". Andrews and Barry (1978) suggested that the Laurentide Ice Sheet had a stable core which was maintained on time-scales of 10^5 years and an outer periphery where rapid fluctuations of ice extent occurred. However, recent studies on the stratigraphy and amino acid geochronology of tills and non-glacial sediments from the Hudson Bay and James Bay Lowlands challenges this models (Shilts *et al.*, 1981; Andrews, 1982; Andrews *et al.*, in press; Shilts, in press). This challenge also comes at a time when physical models of ice sheet growth (Budd and Smith, 1981) have produced major intervals of deglaciation across Hudson Bay during two intervals at *c.* 40 000 B.P. and *c.* 80 000 B.P. (Budd and Smith, 1981; Andrews, 1982).

In our chapter we do not have the space to detail these investigations nor to elaborate the stratigraphy. However, Fig. 8 illustrates the main output from the amino acid analysis of over 140 shells and shell fragments from tills and non-glacial sediments within the Hudson Bay Lowlands (Shilts *et al.*, 1981; Shilts, in press; Andrews *et al.*, in press). The chronology of Fig. 8 is controlled by assuming that the marine sediments of the Bell Sea (Skinner, 1973; Shilts *et al.*, 1981) are of last interglacial age and date from about 130 000 B.P. These sediments are overlain by the forest and peats of the terrestrial members of the Missinaibi Formation (e.g. Prest, 1970) which have a ^{14}C date of > 76 000 B.P. (Stuiver *et al.*, 1978). Amino Acid kinetics then predict a long-term effective diagenetic temperature in the Lowlands of +0·6° for the last 130 000 years. With this assumption shells of the lateglacial Tyrrell Sea (Lee, 1960; Craig, 1969) give amino acid estimates of 8000–9000 B.P. compared with ^{14}C dates of 7000–8200 B.P. If this equation is then applied to the amino acid D/L ratios for the bivalve *Hiatella arctica* the distribution of age estimates of Fig. 8 results. This graph suggests that there have been marine episodes within the Hudson Bay Lowlands during the Wisconsin Glaciation. This further implies that the ice load disappeared from over Hudson Bay and probably from the western shores of Hudson Bay; however, our data do not deny the existence of a substantial ice cover across Labrador-Ungava during one or more of the major deglacial periods indicated in Fig. 8. We do emphasize that the amino acid data are also consistent with the stratigraphy within the Lowlands where non-glacial fluvial and lacustrine sediments occur stratigraphically between the sediments of the Bell and Tyrrell Seas (McDonald, 1969; Skinner, 1973; Shilts, in press).

In concluding this brief review of the observations we note that various lines of field evidence indicate that the *dominant* ice flow and hence thickness gradient over southern Hudson Bay was from the north-east. These data, in themselves, do not necessarily deny an interval (possibly during the Wisconsin 18 000 B.P. maximum, or even at earlier times) when a major ice divide(s) lay within or close to Hudson Bay (Dyke *et al.*, 1982). There is no single piece of data which can be used to firmly establish the existence of a major ice divide extending NW–SE across Hudson Bay. Isobases and strandline tilts appear to support the notion that a major centre of rebound was located east and north-east of James Bay. Evidence from multiple till/non-glacial sections along the major rivers draining the Hudson Bay/James Bay Lowlands suggest a complex Wisconsin glacial history that they may have some effect on the interpretation of glacial isostatic modelling experiments. In the next section we will proceed to enquire as to how geophysical models of the isostatic adjustment process may be employed to further constrain our interpretations.

ISOSTATIC ADJUSTMENT CONSTRAINTS ON MANTLE RHEOLOGY AND GLACIATION HISTORY: WHEN ISOSTATIC EQUILIBRIUM PREVAILS INITIALLY

As mentioned in the introduction to this chapter, there have been several recent advances made in the problem of modelling the Earth's mechanical response to Quaternary ice loads. Historically the subject has been primarily of concern to the geophysical community since studies of the rebound of the crust following deglaciation have provided the only means by which one could infer the effective viscosity of the planetary mantle. The importance of this parameter of course lies in the fact that it is a fundamental ingredient in thermal convection models of the process of continental drift (Peltier, 1980, 1981a; Jarvis and Peltier, 1982). The new theory which has been developed to describe the process of glacial isostatic adjustment, in papers which began with Peltier (1974), differs very much from previous elaborations in that explicit models of the history of glaciation are required to implement it. Predictions of the model therefore depend jointly upon the mantle viscosity profile and upon the glaciation history so that the model may be employed to infer the properties of both functionals directly from the observations. Fairly comprehensive reviews of the theory will be found in Peltier (1981b, 1982). We will confine the discussion here to an account of the most recent results which have been obtained with it rather than of the mathematical structure of the theory itself.

The results of this analysis for the Hudson Bay region have begun to play a particularly important role in geophysical theory. This is due to the fact that the Laurentide Ice Sheet which formed here had the largest spatial scale of those which existed at Würm-Wisconsin maximum, so that the local relative sea-level variations forced by its disintegration are more sensitive to the deep structure of the planetary mantle than are those from other localities. In particular, both the relative sea-level and free air gravity data from Hudson Bay are sensitive to the viscosity of the mantle below the phase transition at 670 km depth which marks the base of the upper mantle (Peltier, 1981b; Peltier and Wu, 1982; Wu and Peltier, 1983). This is important geophysically because it has been suggested that the mantle convective circulation which drives surface plate tectonics does not transfer material across this interface, a circumstance which would imply the existence of a thermal boundary layer at this depth and thus a marked change of viscosity (Peltier, 1981a).

Because predictions of the model depend jointly upon the mantle viscosity profile and the glaciation history we cannot expect to understand the extent to which the data constrain one of these imperfectly known inputs to the model without simultaneously understanding the contribution of the other. As discussed in Peltier (1976), the inverse problem for glacial isostasy is fundamentally non-linear and we are therefore obliged to solve it in an iterative fashion which requires that we must be able to provide a good first guess to the glaciation history in order to initialize the iteration sequence. If this first guess is "sufficiently close" in some sense to the true glaciation history then we may reasonably expect the sequence of iterations to converge. We first fix the glaciation history and refine the viscosity model $\nu(r)$ to achieve a "best" fit to the data. We next fix $\nu(r)$ and refine the glaciation history, continuing the process until convergence is achieved. A formal inverse theory whereby this process may be implemented was outlined in Peltier (1976) and the first step along the iterative path was described by Peltier and Andrews (1976) who constructed a deglaciation chronology called ICE-1 on the basis of *a-priori* information and used it to calculate preliminary relative sea-level curves for a number of sites from which ^{14}C-controlled relative sea-level histories were available. These calculations established that the viscosity of the mantle had to be essentially constant in order to fit the observed emergence around Hudson Bay.

Clark *et al.* (1978) and Peltier *et al.* (1978) employed a slightly more accurate version of the theory for post-glacial sea-level variations which included the self-gravitation of the oceans and showed that such a model was capable in principle of reconciling a large fraction of the global relative sea-level data set. In order to demonstrate this point, however, they were forced to modify the ICE-1 model of Peltier and Andrews (1976) quite drastically, essentially by delaying melting by 2×10^3 years which led to a marked discord with the deglaciation isochrones of Bryson *et al.* (1969). Clark (1980) has held to this assumption and employed the formal inverse theory suggested in Peltier (1976) to refine the glaciation history over Hudson Bay. This approach has turned out to be rather fruitless, partly because the time-shifted version of the ICE-1 chronology is not in fact a reasonable approximation to the melting history, and partly because the nature of the numerical calculation requires a large sacrifice in the spatial resolution with which the glaciation chronology is specified. For the new results which we will discuss here an alternative and computationally interactive approach has been adopted. Discussion will be restricted to points connected with the ability of the adjustment data to constrain the history of loading over Hudson Bay and the viscosity of the underlying mantle.

Wu and Peltier (1983) described an approach to the interactive adjustment of the loading history which seeks to minimize the misfits between predicted and observed relative sea-level data. For the most part this analysis has been based upon the assumption that isostatic equilibrium prevails initially and so does not provide information on the history of glaciation prior to glacial maximum which is the question which most concerns us here. With such analysis we are nevertheless able to investigate the extent to which the hypothesis of a single north-west trending ice divide over Hudson Bay at glacial maximum is compatible with the data and to demonstrate the ability of the new adjustment model to fit the observed relative sea level and free air gravity information. In Fig. 9 we show three time slices through the ICE-1 model for Hudson Bay and corresponding data for a model called ICE-2 which is a refinement of ICE-1 effected to minimize the misfits between predicted and observed relative sea-level data for the Hudson Bay region. It will be observed that the most characteristic feature of the ICE-1 model, which was the early disintegration of the ice complex over Hudson Bay itself, is *less* pronounced in the ICE-2 model. The separate Labrador and Keewatin ice divides are both *less* distinct

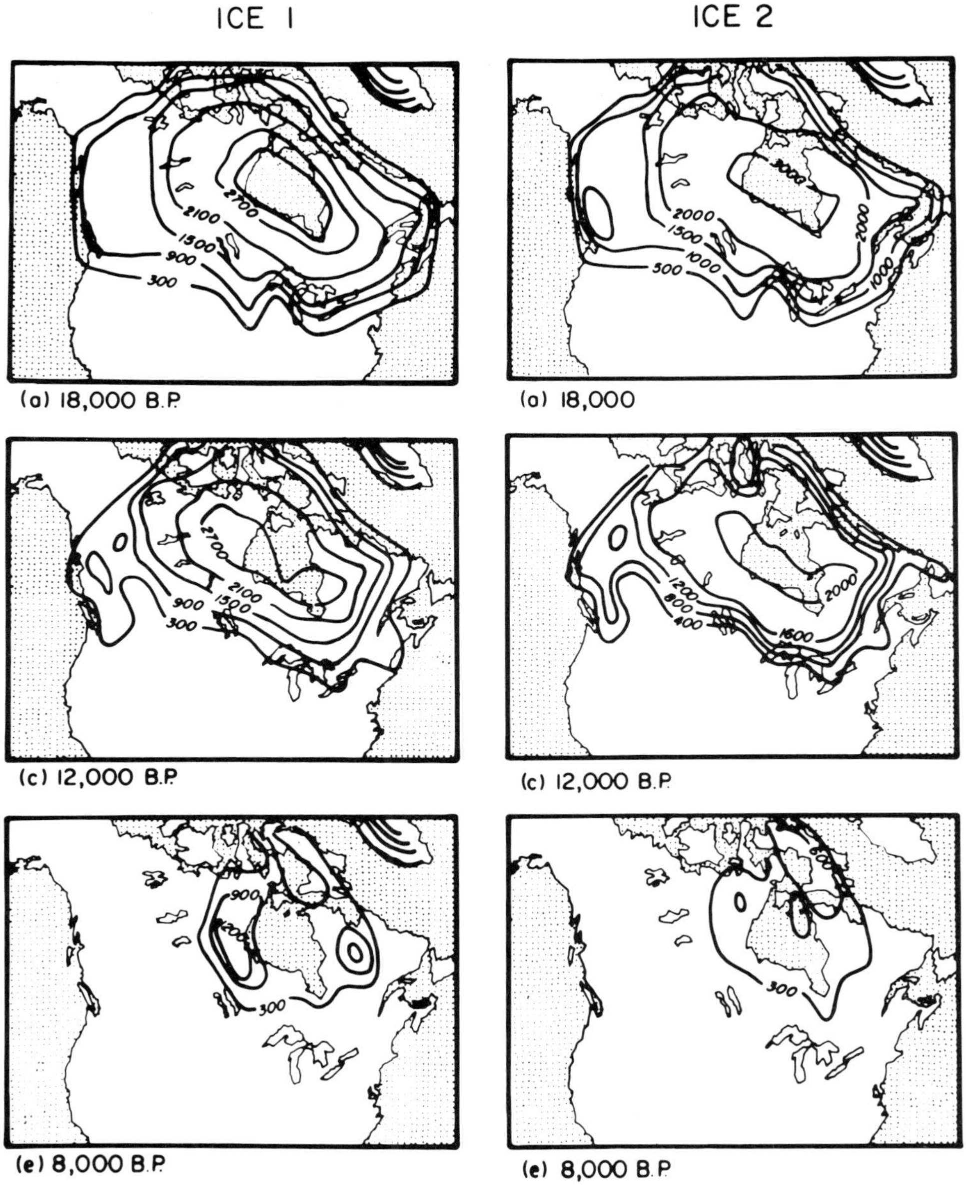

Figure 9. Ice thickness (m) for three times in the melting history of the Laurentide Ice Sheet.

in the refined ice history. Figures 10 and 11 show a selection of comparisons between observed and predicted relative sea-level data at several sites from North America and for both the ICE-1 and ICE-2 models. Calculations are shown for three realistic earth models which differ from one another only in the magnitude of the mantle viscosity beneath 670 km depth. Inspection of these figures shows that the uniform viscosity model fits the observed relative sea-level data best and that the ICE-2 model is a distinct improvement over ICE-1 at most sites. An exception to

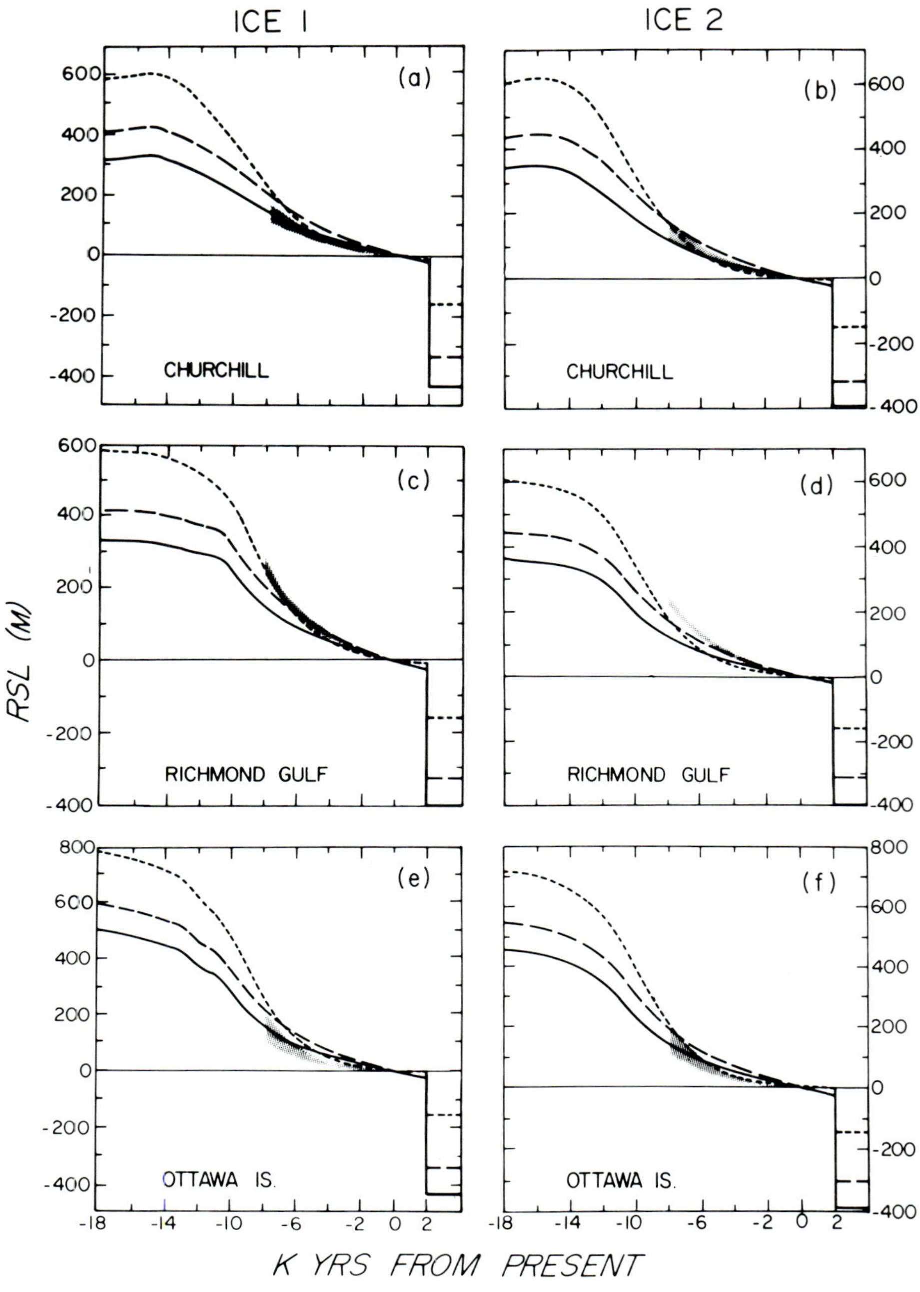

Figure 10. Comparisons of observations of relative sea level (hatched regions) with theoretical predictions. Theoretical curves are for three models which differ from one another only in the viscosity of their mantles beneath 670 km depth. The short dashed curve is for a model with a uniform mantle viscosity of 10^{21} Pa s (10^{22} Poise). The long dashed curve is for a model with lower mantle viscosity of 10^{22} Pa s while the solid curve is for a model with lower mantle viscosity of 5×10^{22} Pa s.

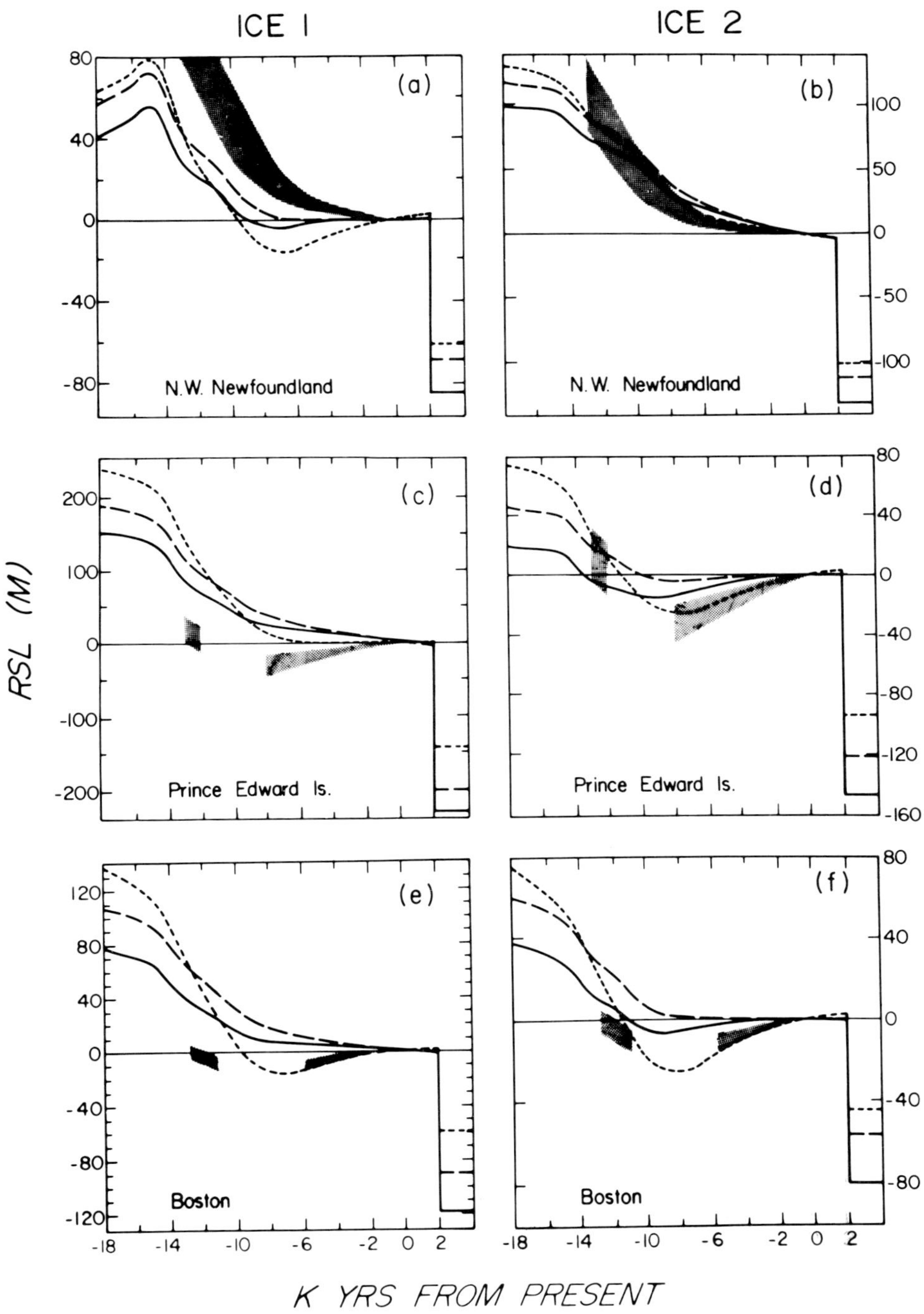

Figure 11. Same as Fig. 10 but for different geographical regions.

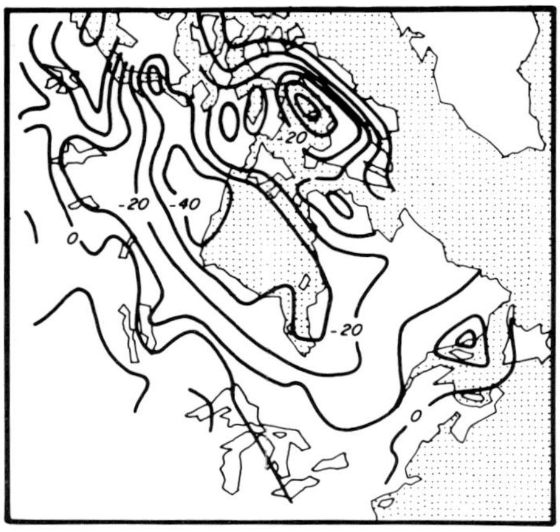

Figure 12. The free air gravity signal over Hudson Bay. Contours are in mgal.

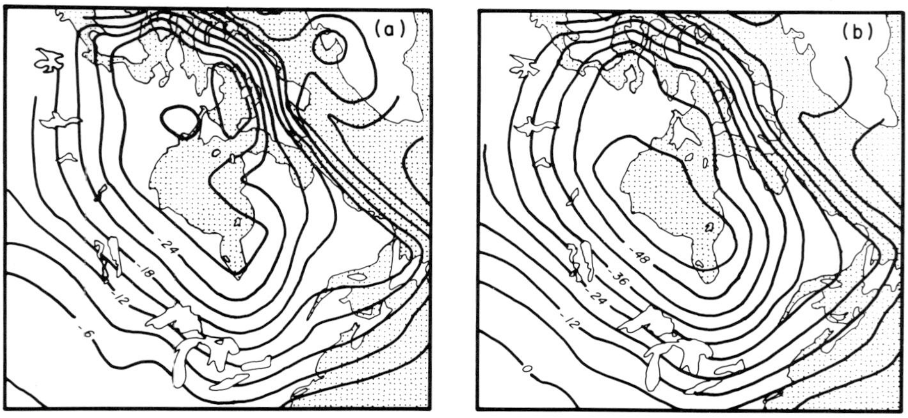

Figure 13. (a) Predicted present day free air gravity anomaly for an earth model with a uniform mantle viscosity of 10^{21} Pa s. (b) Same as (a) but the earth model has a lower mantle viscosity of 10^{22} Pa s.

the general truth of the latter statement obtains at Richmond Gulf where the relative sea-level data suggest that melting should be delayed somewhat at this site or that local ice thickness should be increased. The Richmond Gulf data were not employed to constrain the ICE-2 melting history, however, and it should be quite clear that further adjustment of this history would also reconcile these data.

A second constraint upon the mantle viscosity profile and the deglaciation history is provided by the observed free air gravity signal over Hudson Bay, which is shown in Fig. 12. This figure is based upon Walcott's (1970) analysis of the data of Innes *et al.* (1968) and shows a negative elliptical anomaly with long axis trending north-west and a magnitude of 35–40 mgal. Comparison of this anomaly with the model ice sheet topography at 18 000 B.P. shown in Fig. 10 reveals a strong spatial correlation, suggesting that the assumption in the ice sheet reconstruction of a single north-west trending ice divide at glacial maximum might not be too unreasonable. Figure 13 shows predicted present-day free air gravity maps for Hudson Bay for two different earth models using the ICE-2 deglaciation history. The first earth model (a) has a uniform mantle viscosity of 10^{21} Pa s and the second (b) is identical except for the viscosity of the mantle beneath 670 km depth where it is 10^{22} Pa s. Both calculations are based upon the assumption that isostatic equilibrium prevails initially. Inspection of this figure shows that the uniform viscosity model fits the observed peak gravity anomaly rather well (Fig. 13a). Equally clear is the fact that the model with high lower mantle viscosity predicts a peak free air anomaly which is near 60 mgal and this is completely ruled out by the data. The data seem to require a uniform mantle viscosity profile and an ice sheet at glacial maximum which consists of a single elliptical dome trending north-west and centred upon Hudson Bay. It must be kept clearly in mind, however, that this conclusion has been predicated upon the assumption that a state of isostatic equilibrium prevailed at 18 000 B.P. It should also be clear that there is strong indication in the gravity field for *local* maxima in ice thickness over Baffin Island and Keewatin. These data therefore suggest that the actual ice sheet topography may have contained some of the features of both of the models described previously.

ISOSTATIC ADJUSTMENT CONSTRAINTS ON MANTLE RHEOLOGY AND GLACIATION HISTORY: EFFECTS OF THE LOAD CYCLE

In order to investigate the sensitivity of our conclusions to the assumption of initial isostatic equilibrium we are obliged to introduce explicit models of the loading history prior to deglaciation. Some constraint on this important early phase of the glaciation chronology is provided by the oxygen isotope stratigraphy in sedimentary cores taken from the deep ocean basins. An example of such data, reproduced from Shackleton and Opdyke (1973), is shown in Fig. 14 where it is compared to a simple sawtooth waveform with period 10^5 years ago. It is well recognized that the $^{18}O/^{16}O$ time series is essentially proxy for the time dependence of continental ice volume (Shackleton and Opdyke, 1973) and it is also clear on the basis of spectral analyses of these times series that most of the variance in the ice coverage record is concentrated on a time-scale of 10^5 years, although there is also significant variance on the Milankovitch periods near 2×10^4 years (associated with orbital precession) and 4×10^4 years associated with the variation of orbital obliquity (e.g. Hays *et al.*, 1976). Since the solar insolation time series contains no power on the period of 10^5 years (e.g. Berger, 1978) it remains unclear as to why the record of cryospheric fluctuations is dominated by this time-scale. The importance of the suggestion from the field data around Hudson Bay, that at least the Bay itself may have been clear of ice at both 40 000 and 80 000 B.P., is that these periods are times of relative seasonal solar insolation maxima due to the variation of orbital obliquity. It could therefore be that the periodic disintegration of the central dome of the ice cap over Hudson Bay on this time-scale is a prominent source of the signal at this period which is

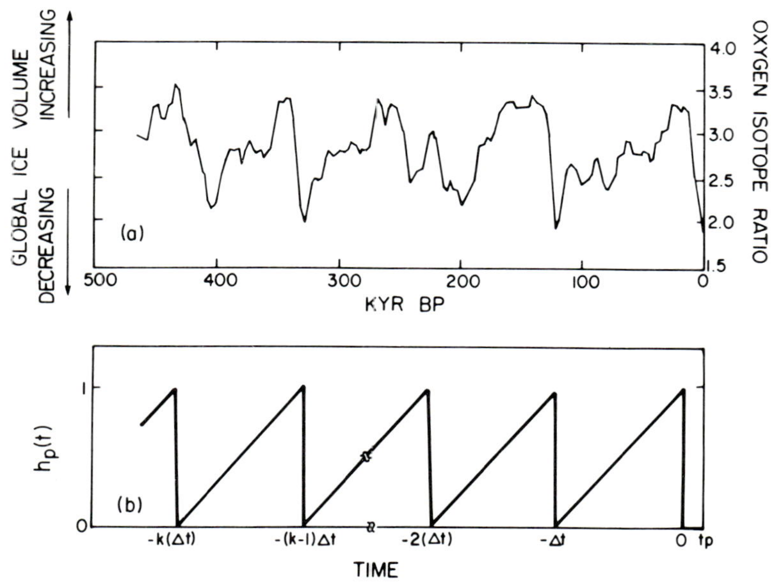

Figure 14. Oxygen isotope ratio $^{18}O/^{16}O$ as a function of time (depth) in a typical deep sea sedimentary core. The lower curve shows a sawtooth approximation to the implied history of loading.

found in deep sea sedimentary cores. Indeed some such phenomenon must account for this observed signal. The question which we wish to address here is whether one might be able to find additional evidence in the relative sea-level record for this quasi-periodic accretion and disintegration of the central Laurentide ice divide and perhaps also in the observed free air gravity anomaly.

The simplest model which can usefully be employed to understand the influence of the load cycle upon isostatic adjustment data is one in which the ice sheet is represented as a single spherical cap and the oceans are neglected entirely. Figure 15 compares relative displacement predictions for the centre of such a Laurentide model ice cap with three representative relative sea-level curves from central Hudson Bay for the sites noted in the figure caption. Theoretical predictions are shown for two viscosity models which are labelled L1 and L2 and which correspond to the model with uniform mantle viscosity of 10^{21} Pa s and that with lower mantle viscosity of 10^{22} Pa s respectively. Relative displacement predictions for each viscosity model are shown for calculations which assume initial isostatic equilibrium (solid curves) and for calculations which include the effect of the load cycle (dashed curves). In each case in which the load cycle is included it has been approximated by the simple sawtooth history shown in Fig. 14, which is based upon the oxygen isotope data. These calculations were performed on the basis of the assumption that the radius of the ice cap remains constant while its volume fluctuates, an approximation which does not introduce any substantial error at the centre of the spherical cap (Wu and Peltier, 1983). Inspection of Fig. 15 shows that the load cycle contributes negligibly to the relative sea-level signal in central Hudson Bay since the solid and dashed curves differ from one another insignificantly. The

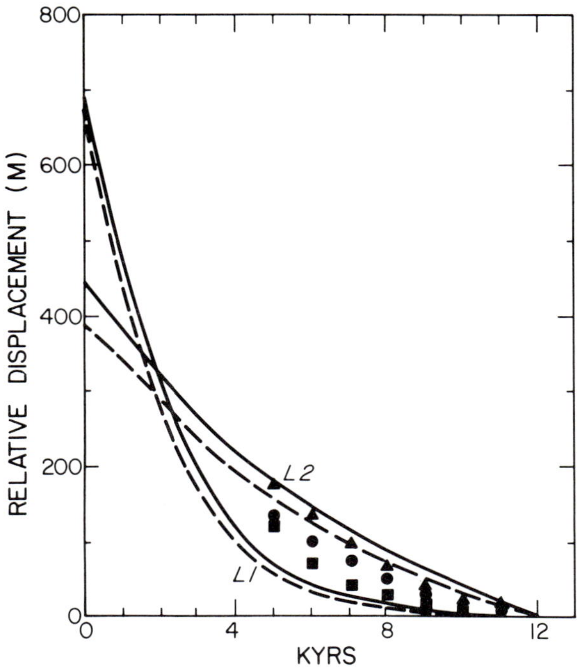

Figure 15. Relative displacement predictions for the centre of a Laurentide model disc load for two Earth models L1 and L2. These differ from one another only in their viscosities beneath 670 km depth where L1, which is uniform, has viscosity 10^{21} Pa s whereas L2 has viscosity 10^{22} Pa s. The solid curves are the predictions based upon the assumption of initial isostatic equilibrium whereas the dashed curves include the influence of initial departures from equilibrium. ■, data from Ottawa Islands; ●, data from Churchill; ▲, data from Castle Island.

mathematical explanation of this important point is provided in Wu and Peltier. Physically it is simply a consequence of the fact that relative sea-level data are measurements of the response relative to a present day zero, so that as long as the characteristic time-scale of the response is short compared to the time-scale on which significant fluctuations of the load occur, then no significant evidence of the load cycle will be seen in the relative sea-level record. This is a very important point which has not been made previously. It is quite clear on the basis of this characteristic behaviour that Earth models which are rejected on the basis of the observed relative sea-level record cannot easily be brought back into contention by invoking an effect due to the deviation from initial isostatic equilibrium. Figure 15 demonstrates that a firm upper bound on the viscosity beneath 670 km depth is $\sim 10^{22}$ Pa s (10^{23} Poise, cgs units). The actual viscosity of the lower mantle ν_{LM} would seem to be such that 10^{21} Pa s $\leq \nu_{LM} < 10^{22}$ Pa s.

Because the free air gravity anomaly is an absolute rather than a relative quantity it is not similarly insensitive to the loading history prior to glacial maximum. Figure 16 shows theoretical predictions of the present-day peak free air anomaly based upon the same disc load approximation to the Laurentide Ice Sheet which was used to compute the displacement response shown in Fig. 15. Peak anomalies are

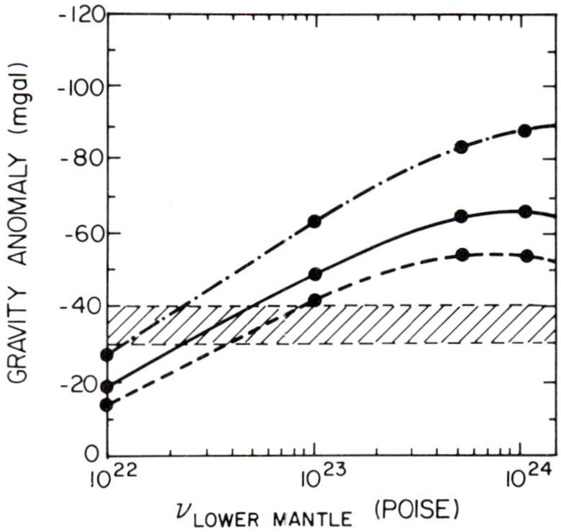

Figure 16. Predicted free air gravity anomaly at the centre of the model Laurentide disc load. The hatched region represents the observations (see text).

calculated as a function of the viscosity of the lower mantle beneath 670 km depth and compared to the observed peak anomaly from Fig. 12 which is shown as the hatched region. Three different theoretical calculations are superimposed on this figure. The dash-dotted curve is the theoretical prediction based upon the assumption that isostatic equilibrium prevailed at glacial maximum. The dashed curve includes the degree of initial isostatic disequilibrium implied by the specific viscosity model and the load cycle shown in Fig. 14. For this calculation the ice sheet radius was held fixed. The solid curve is the most accurate of the three calculations since it includes the variation of ice sheet radius which occurs as the volume fluctuates. The radius is determined on the basis of the assumption that the ice cap maintains an equilibrium plastic profile as the volume varies (Wu and Peltier, 1983).

On the basis of the comparisons shown in Fig. 16 it is quite clear that the gravity anomaly is influenced to a significant degree by the effect of initial isostatic disequilibrium. Neglecting this effect, no increase of mantle viscosity across the phase boundary at 670 km depth is required to explain the free air anomaly. Including this effect, on the basis of the assumption that the load history is sufficiently well approximated by that shown in Fig. 14, we *require* an increase of mantle viscosity across this boundary by a factor of about three. To the extent that higher frequency fluctuations of ice volume are important, for example on the Milankovitch periods of 2×10^4 and 4×10^4 years, this required increase must be considered a lower bound. As discussed in Wu and Peltier (in press), analysis of the data from Fennoscandia leads to precisely the same conclusion.

The fact that both the relative sea-level data and the free air gravity signal from Hudson Bay can be understood on the basis of an Earth model which has essentially constant viscosity has not been established previously. In fact all previous analyses have been predicated upon the assumption that it was essentially impossible to effect such a reconciliation. In particular, O'Connell's (1971) argument (based upon the observed non-tidal acceleration of rotation) that the viscosity of the mantle was constant was based upon the assumption that there was *no* free air gravity anomaly of glacial origin over Hudson Bay. Cathles (1975) basically accepted O'Connell's argument and believed that the anomaly described by Walcott was only fortuitously correlated with the geometry of rebound. Walcott (1980) continued to maintain that the observed anomaly was of glacial origin and that its existence required a very high value for the viscosity of the lower mantle. As discussed in Peltier (1981a) and Wu and Peltier (1983), the ability of the new isostatic adjustment model to explain both sets of data simultaneously in terms of a uniform viscosity model is due to the fact that it contains internal density discontinuities associated with the olivine-spinel and spinel-post-spinel phase transitions at 420 and 670 km depth.

The analyses presented here, of the effect of the load cycle upon relative sea-level and free air gravity data, is discouraging in so far as they imply that these data are probably not too sensitive to the history of loading prior to glacial maximum. It would therefore appear that they cannot be usefully employed to constrain the glaciation history. If it were not for the relative insensitivity of these data to the load cycle, however, they would not be very useful as constraints upon the mantle viscosity profile and therefore would not be nearly as useful as they are for geophysical purposes. It is perhaps unfortunate that the requirements of geophysics and of Quaternary geomorphology cannot be satisfied simultaneously. Nevertheless we have shown that the sea-level and gravity data exhibit some preference for the existence of a single north-west trending ice divide at glacial maximum. They simply cannot be invoked to argue either for or against the periodic disappearance of the ice load over Hudson Bay itself on the time-scale of 4×10^4 years suggested by the observations discussed in previous sections of this chapter. Radical alterations of the timing of the main deglaciation event which began at about 18 000 B.P. can, however, be ruled out completely by the combination of relative sea-level and free air gravity data. This is illustrated in the next subsection.

ISOSTATIC ADJUSTMENT CONSTRAINTS UPON THE TIMING OF THE LAST DEGLACIATION EVENT

The deglaciation histories ICE-1, 2 which have been employed in the calculation of relative sea-level and free air gravity data conform closely to the disintegration isochrones of Bryson *et al.* (1969). For relative sea-level data at sites located near and under the major ice sheets, the data can all be fit by a combination of this melting history and an Earth model in which the viscosity does not vary drastically as a function of depth. An important question which we should be concerned to answer is whether there might not be sufficient error in the deglaciation chronology to allow models with substantially different viscosity profiles to fit the data equally well. This raises the question of the uniqueness of the combined melting history and viscosity profile inferred previously.

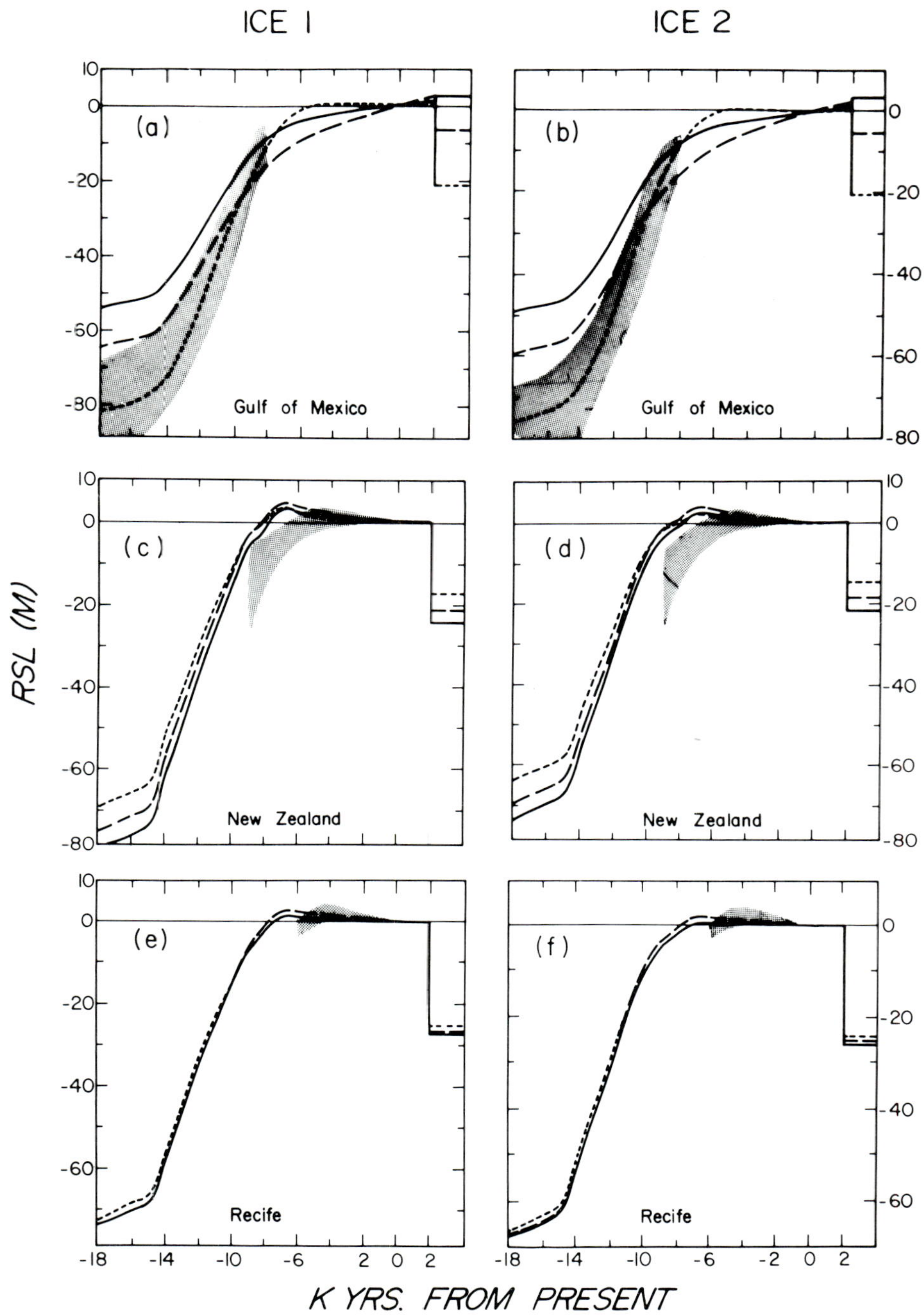

Figure 17. Predicted relative sea-level variations for the ICE-1 and ICE-2 models at sites remote from the Laurentide ice centre.

There *are* in fact some relative sea-level data which may be taken to imply a significant error in the basic melting chronology of the ICE-1, 2 models. These include virtually all relative sea-level data from sites sufficiently distant from the main ice centres and examples of these data are shown in Fig. 17 by the curves for South Island, New Zealand and Recife, Brazil. The theoretical predictions are for the same Earth models employed in the calculations shown in Figs 11 and 12 and calculations are shown both for ICE-1 and the refined ICE-2. Irrespective of the mantle viscosity profile the ICE-1, 2 models predict the first appearance of raised beaches at the two far field sites at 8000 B.P. rather than 6000 B.P. at which these features are actually observed. One way of remedying this misfit is clearly to delay the main northern hemisphere melting event by 2×10^3 years. This was the approach adopted in Clark *et al.* (1978), but as pointed out in Peltier *et al.* (1978) it leads to enormous errors at near field sites since the degree of emergence/submergence is now over-predicted by a factor of two. One way of remedying this new defect is to make the northern hemisphere ice sheets thinner than they are in the ICE-1,2 models to compensate for the errors produced by the delayed melting.

This procedure has been followed in Wu and Peltier (1983) and leads to the construction of a new deglaciation chronology called ICE-3,4 which is illustrated in Fig. 18 for three times during the melting event. ICE-4 was generated from the crude model ICE-3 by forcing it interactively to produce a best fit to a subset of the relative sea-level data from Canada. Comparing the ICE-4 and ICE-2 models we see that the separate Labrador and Keewatin ice divides are more pronounced in the latter than they are in the former. Figure 19 shows a few relative sea-level curves from sites beyond the Laurentide ice margin for the ICE-3,4 model, including Recife, Brazil and South Island, New Zealand. The misfits at these sites which were evident in the ICE-1,2 predictions have clearly been removed by delaying the global melting event by 2000 years. It must be kept in mind, however, that this improved fit to this subset of the data has been achieved at the expense of violating the ^{14}C controlled melting isochrones of Bryson *et al.* (1969) and this is clearly unsatisfactory.

Figure 20 shows predicted present-day free air gravity anomalies for calculations based upon the ICE-4 melting chronology and for three different earth models:

(a) is for the Earth model with uniform mantle viscosity of 10^{21} Pa s;
(b) is for an identical Earth model except that the viscosity of the lower mantle is 10^{22} Pa s; and
(c) is the result for an Earth model in which the lower mantle viscosity is 5×10^{22} Pa s.

All of these calculations have been done on the basis of the assumption that isostatic equilibrium prevails initially. Inspection of this figure demonstrates that for a thin ice sheet which melts late like ICE-3,4 we require an increase of mantle viscosity across 670 km depth of about a factor of 10 in order to explain the observed free air anomaly of -35 to -40 mgal. The anomaly predicted by the uniform mantle viscosity model shown in (a) is less than -20 mgal, which is completely unacceptable.

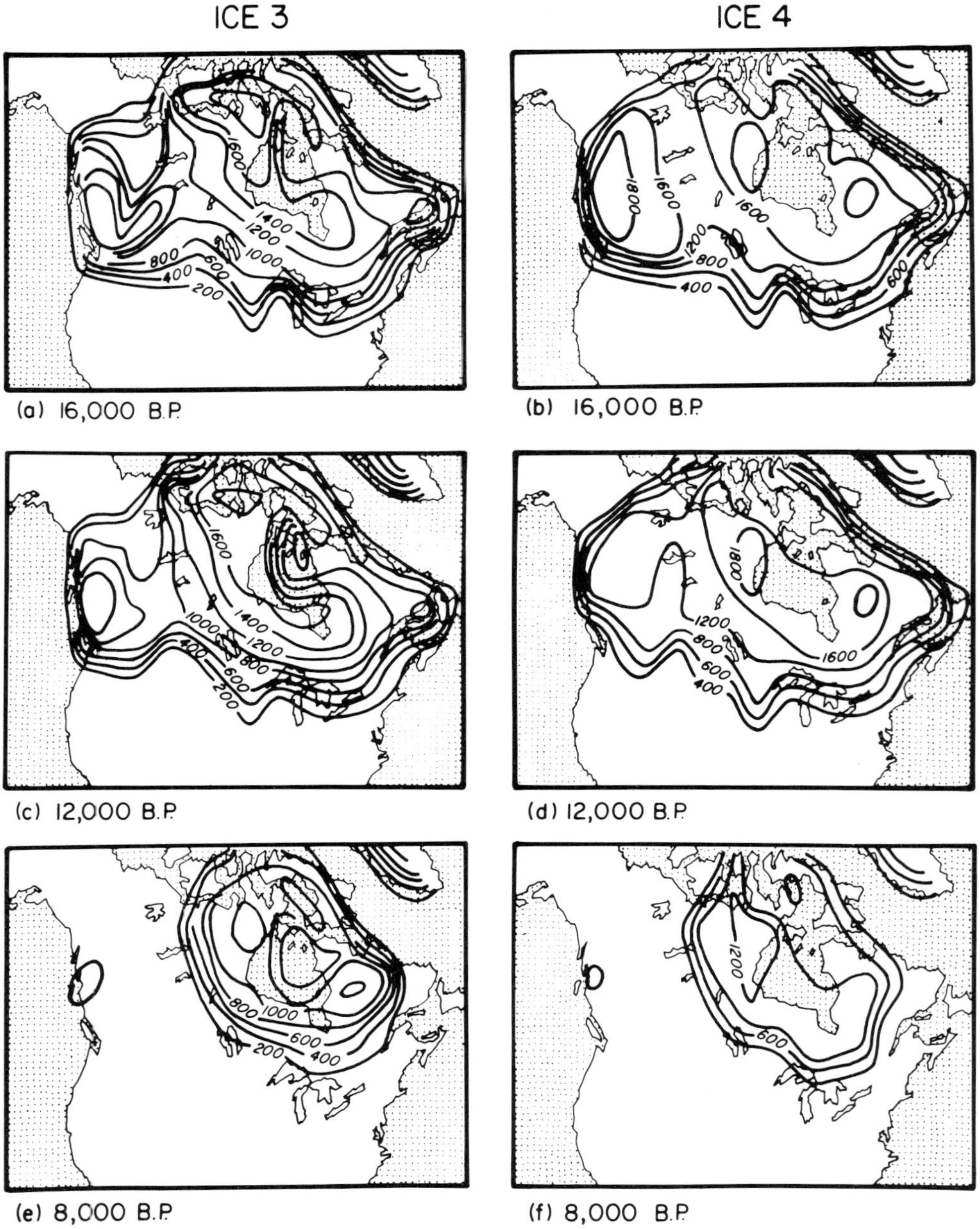

Figure 18. Three time-slices through the ICE-3 and ICE-4 melting chronologies.

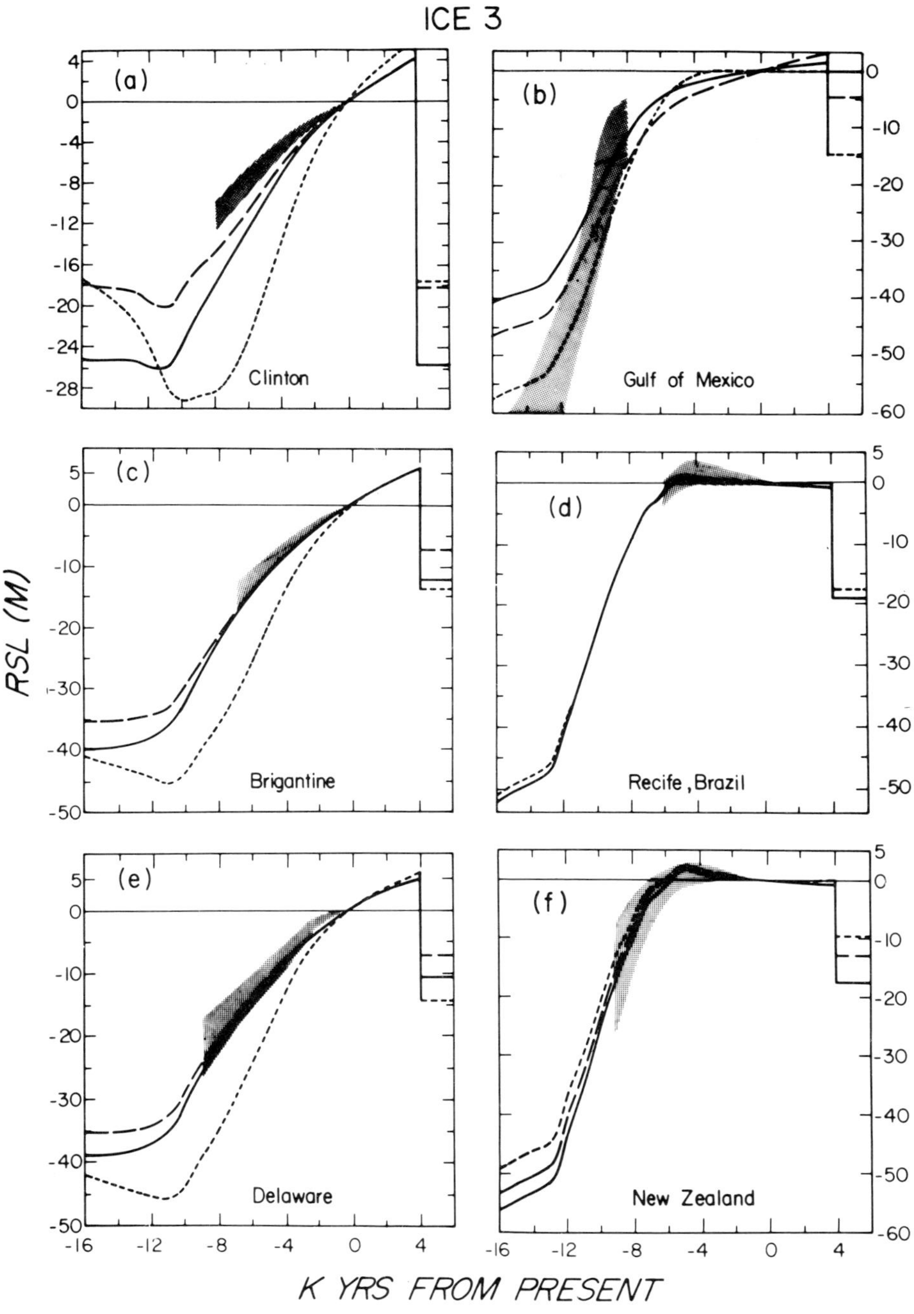

Figure 19. Relative sea-level predictions at peripheral and remote sites for the ICE-3 deglaciation chronology.

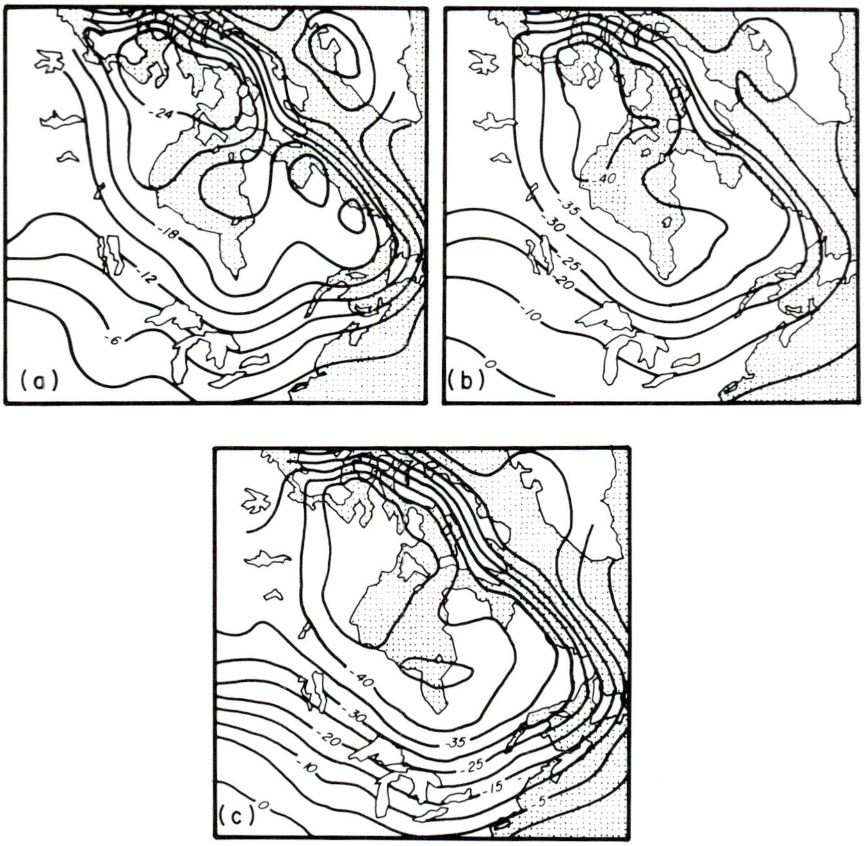

Figure 20. Free air anomaly predictions for Hudson Bay.

When the effect of initial isostatic disequilibrium is taken into account in the calculation it becomes abundantly clear that models of the ICE-3,4 type are completely rejected by the observations. Figure 21 shows the predicted present-day peak gravity anomaly as a function of the viscosity of the lower mantle for a suite of models which differ from one another only in this quantity. Calculations are based upon a disc load approximation to the ICE-3,4 deglaciation history of the Laurentide region and are shown for the same three versions of the glaciation history as were employed in Fig. 16. The dash-dotted curve is the calculation which assumes initial isostatic equilibrium, the dashed curve includes the effect of the load cycle estimated from the $^{16}O/^{18}O$ data assuming constant disc radius through the load cycle, and the solid curve includes the effect of the variation of load radius which accompanies the sawtooth fluctuation of volume. Comparison of Fig. 21 with Fig. 20a shows that the disc load calculation agrees with the realistic calculation in requiring a lower mantle viscosity of about 10^{22} Pa s in order to fit the observed peak free air anomaly when the effect of the load cycle is neglected. However, when the effect of the load cycle is included the solid curve shows that even a lower mantle viscosity of 10^{23} Pa s is only marginally capable of predicting the observed gravity

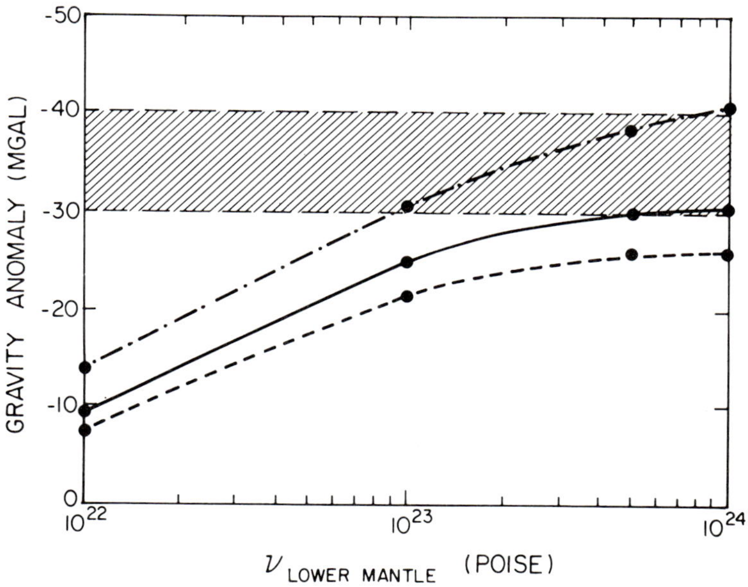

Figure 21. Free air gravity anomaly at the centre of a Laurentide model disc load as a function of the viscosity of the lower mantle. The meanings of the three predicted curves are as in Fig. 16.

signal and such a large increase of viscosity is completely incompatible with the relative sea level data. It is therefore clear that models of the ICE-3,4 type cannot simultaneously satisfy both types of data and must be rejected. Models which are capable of fitting the data must have disintegration isochrones which closely match those presented by Bryson *et al.* (1969) and therefore be of the ICE-1,2 type. As shown previously, such melting chronologies are compatible with the data only if the viscosity of the mantle is essentially uniform.

Since a significant delay in time of the main melting episode is not compatible with the vast majority of the observational data, the question remains as to how the differences between observations and far field predictions at sites like Recife, Brazil and South Island, New Zealand are to be reconciled. When it is recognized that the heights of the raised beaches which are predicted and observed at such locations are only a few metres at most, it is quite clear that the misfits between theory and observation can be explained easily. In the ICE-1,2 models melting is essentially complete by 8000 B.P. and it is at this time that raised beaches are predicted to form in the far field. Prior to this time the rate of addition of water to the ocean basins was more than sufficient to keep up with the local increase of planetary radius at such sites and so no raised beaches are cut. As soon as melting stops, however, raised beaches immediately appear. It is quite clear on this basis that if melting did not end abruptly at about 8000 B.P., as it does in the ICE-1,2 histories, but rather continued after this time as suggested for example by the "eustatic" sea-level curve of Shephard (1963), then the time of first appearance of raised beaches in the far field could easily be delayed by the observed amount. The source of the small amount of extra meltwater required could be Greenland or Antarctica, for example. On the basis of this argument we see that the small misfits at far field sites do not require any substantial modification of the deglaciation history in order to understand them. They certainly do not require the drastic delay of the melting event assumed in Clark *et al.* (1978) which has now been shown to be impossible to reconcile with the relative sea-level and free air gravity data from Hudson Bay.

CONCLUSIONS AND SUGGESTIONS FOR FURTHER RESEARCH

Recent research on the glacial geology of the Hudson Bay region of Canada appears to suggest that the early views of Tyrrell (1898) concerning the locations of major ice divides may be closer to the truth than those of Flint (1943) which have become dominant. However, geophysical models of glacio-isostatic adjustment seem to be better able to reconcile relative sea-level and free air gravity data when the glaciation-deglaciation chronology is close to that of Flint. It is clear that considerable further effort will be required to reconcile these competing views of the late Pleistocene and early Holocene glacial history of this region. In this effort we expect that a few areas of concentration may prove particularly fruitful in providing new insights and perspectives.

Among these areas, that concerned with the explicit numerical modelling of ice sheet accretion and flow may prove especially rewarding. Using these methods we may be able to ascertain whether the geometry of the ice sheet inferred by Shilts (1980) and Andrews (1982) is physically reasonable or whether additional ice divides (cf. Dyke *et al.*, 1982) were present which would affect materially the length of flow lines and hence glacier gradients.

Another avenue of investigation which should provide useful additional information concerns data on the magnitude and direction of the tilts of glacial lake shorelines. These have not yet been incorporated as constraints in the modelling of glacio-isostatic adjustment but should provide useful information since this field is much more sensitive to local variations in ice cover than either relative sea-level or free air gravity data. These data should also contain new information concerning the thickness of the continental lithosphere which is an important quantity in several different geodynamic applications. No quantitative analysis of the information content of these data has yet been performed.

Of particular importance in our analyses is temporal control on the evolution of the Laurentide Ice Sheet. We have to recognize that our several "pictures" of the Laurentide Ice Sheet (see figures) reflect not only different instantaneous forms of the ice sheet but also different reflections of the average conditions in our two basic data sets—glacial flow indicators and glacio-isostatic rebound. Further effort should help us to decide whether there is a significant spatial separation between ice divides and the area(s) of maximum ice thickness. These two concepts are not necessarily equivalent geographically.

REFERENCES

Andrews, J. T. (1970). "A geomorphological study of post-glacial uplift, with particular reference to arctic Canada". *Institute of British Geographers, Special Publication* No. 2, 156 pp.

Andrews, J. T. (1973). *Arctic and Alpine Research* **5**, 41-48.

Andrews, J. T. and Barnett, D. M. (1972). *Geog. Annlr.* **A54**, 1-11.

Andrews, J. T. and Barry, R. G. (1978). *Annual Review of Earth and Planetary Sciences* **6**, 205-228.

Andrews, J. T. and Falconer, G. (1969). *Canadian Journal of Earth Sciences* **6**, 1263-1276.

Andrews, J. T. and Mahaffy, M. A. (1976). *Quaternary Research* **6**, 167-183.

Andrews, J. T. and Miller, G. H. (1979). *Geology* **7**, 592-596.

Andrews, J. T. and Pelter, W. R. (1976). *Geology* **4**, 73-75.

Andrews, J. T. and Tyler, K. (1977). *Geographie Physique et Quaternaire* **31**, 389-400.

Andrews, J. T., Shilts, W. W. and Miller, G. H. (in press). Multiple deglaciation of the Hudson Bay Lowlands since deposition of the Missinaibi (last interglacial?) Formation. *Quaternary Research.*

Archer, D. R. (1968). *Arctic* **21**, 153-160.

Barnett, D. M. and Peterson, J. A. (1964). *Canadian Geographer* **8**, 173-181.

Berger, A. L. (1978). *Journal of Atmospheric Science* **35**, 2362-2367.

Bird, J. B. (1970). *Acta Geographica Lodziensia* **24**.

Blake, W., Jr. (1966). *Canadian Geological Surveyor Paper* 66-26.

Blake, W., Jr. (1976). *Geological Survey of Canada, Report of Activities, Part C* Paper 76-1C, 171-174.

Budd, W. F. and Smith, I. N. (1981). *In* "Sea Level, Ice, and Climatic Change", pp. 369-409. I.A.H.S. Publication No. 131.

Bryson, R. A., Wendland, M. W., Ives, J. D. and Andrews, J. T. (1969). *Arctic and Alphine Research* **1**, 1-14.

Cathles, L. M. (1975). "The Viscosity of the Earth's Mantle". Princeton University Press, Princeton.

Clark, J. A. (1980). *J. Geophys. Res.* **85**, 4307-4323.

Clark, J. A., Farrell, W. E. and Peltier, W. R. (1978). *Quaternary Research* **9**, 265-287.

Craig, B. G. (1969). *In* "Earth Science Symposium on the Hudson Bay" (P. J. Hood, ed.), pp. 63-77. *Geological Survey of Canada Paper* 68-53.

Denton, G. H. and Hughes, J. T. (eds) (1981). "The Last Great Ice Sheets". John Wiley and Sons, New York.

Dionne, J. C. (1975). *Revue Geographie Montreal* **31**, 453-457.

Dyler, A. S. (1974). *INSTAAR Occasional Paper* No. 12. University of Colorado, Boulder.

Dyke, A. S., Dredge, L. A. and Vincent, J. S. (in press). Configuration and dynamics of the Laurentide Ice Sheet during the Late Wisconsin maximum. *Geographie Physique et Quaternaire.*

Farrand, W. R. (1960). *Dissertation Abstracts* **21**(18).

Farrand, W. R. and Gaja, R. T. (1962). *Geographical Bulletin* **17**, 5-22.

Flint, R. F. (1943). *Bull. geol. Soc. Am.* **54**, 325-362.

Goldthwait, J. W. (1910). *J. Geol.* **16**, 459-476.

Hardy, L. (1976). "Contribution a l'étude géomorphologique de la portion québecoise des basses terres de la baie de James." Ph.D. thesis, McGill University, Montreal.

Hays, J. D., Imbrie, J., II and Shackleton, N. J. (1976). *Science* **194**, 1121-1132.
Hillaire-Marcel, C. (1979). "Les Mers post-glaciares du Québec: quelques aspects." Thèse de Doctorat d'Etat, L'Université Pierre et Marie Curie, Paris VI, France.
Hillaire-Marcel, C. (1980). *In* "Earth Rheology, Isostasy and Eustasy" (N. A. Mörner, ed.), pp. 215-230. John Wiley and Sons, New York.
Hillaire-Marcel, C. and Occhietti, S. (1980). *Z. Geomorph.* **24**, 373-392.
Hillaire-Marcel, C. and Vincent, J. S. (1980). *Paleo-Quebec* **11**, Montreal.
Hillaire-Marcel, C., Grant, D. G. and Vincent, J. S. (1980). *Geology* **8**, 466-467.
Hughes, O. L. (1965). *Geol. Soc. Amer. Spec. Pap.* **84**, 535-565.
Hughes, T., Denton, G. H. and Grosswald, M. G. (1977). *Nature* **266**, 596-602.
Innes, M. J. S., Goodacre, A. K., Weston, A. and Weber, J. R. (1969). "Science, History and Hudson Bay, Part V" (C. S. Beales, ed.). H.M.S.O., London.
Ives, J. D. (1960). *Cah. Géogr. Québ.* **4**, 323-343.
Jarvis, G. T. and Peltier, W. R. (1982). *Geophys. J.R. astr. Soc.* **68**, 389-424.
Lee, H. A. (1960). *Science* **313**, 1609-1611.
Løken, O. H. (1962). *Geological Bulletin* **17**, 23-56.
McDonald, B. G. (1969). *Geological Survey of Canada* Paper 68-53, 78-99.
Matthews, B. (1967). *Arctic* **20**, 176-202.
Paterson, W. S. B. (1972). *Review of Geophysics Space Physics* **10**, 885-917.
Peltier, W. R. (1974]). *Review of Geophysics Space Physics* **12**, 649-669.
Peltier, W. R. (1976). *Geophysical Journal Royal Astronomical Society* **46**, 669-706.
Peltier, W. R. (1980). *In* "Physics of the Earth's Interior" (A. M. Dziewonski and E. Boschi, eds), pp. 362-431. North Holland, Amsterdam.
Peltier, W. R. (1981a). *Annual Review Earth Planetary Science* **9**, 199-225.
Peltier, W. R. (1981b). *In* "Evolution of the Earth" (R. J. O'Connell and W. S. Fyfe, eds), pp. 229-248. Geophysical Union, Washington D.C.
Peltier, W. R. (1982). *Adv. Geophys.* **24**, 1-145.
Peltier, W. R. and Andrews, J. T. (1976). *Geophys. J.R. astr. Soc.* **46**, 605-646.
Peltier, W. R. and Wu, P. (1982). *Geophysical Research Letters* **9**, 731-734.
Peltier, W. R., Farrell, W. E. and Clark, J. A. (1978). *Tectonophysics* **50**, 81-110.
Prest, V. K. (1970). *In* "Geology and Economic Minerals in Canada", 5th edition (R. J. Douglas, ed.), pp. 676-674. Department of Energy, Mines and Resources, Ottawa.
Shackleton, N. J. (1977). *Phil. Trans. R. Soc.* **B20**, 169-182.
Shackleton, N. J. and Opdyke, N. D. (1973). *Quaternary Research* **3**, 39-54.
Shepard, F. P. (1963). *In* "Essays in Marine Geology", pp. 1-10. University of Southern California Press.
Shilts, W. W. (1980). *Nature* **285**, 213-218.
Shilts, W. W. (in press). Quaternary evolution of the Hudson/James Bay Region. *Guelph Symposium on Hudson Bay.*
Shilts, W. W., Cunningham, C. M. and Kaszki, C. A. (1979). *Geology* **7**, 537-541.
Shilts, W. W., Miller, G. H. and Andrews, J. T. (1981). *Bull. geol. Soc. Am. Abstracts* **13**, 553.
Skinner, R. G. (1973). *Geological Survey Canada Bulletin* **225**.
Stuiver, M., Heusser, C. J. and Che Yang, I. (1978). *Science* **200**, 16-21.
Tyrrell, J. B. (1898). *J. Geol.* **6**, 147-160.
Vincent, J. S. and Hardy, L. (1979). *Geological Survey of Canada Bulletin* **316**.

Walcott, R. I. (1970). *Canadian Journal of Earth Sciences* **7**, 716-727.

Walcott, R. I. (1972). *Review of Geophysics and Space Physics* **10**, 849-884.

Walcott, R. I. (1980). *In* "Earth Rheology, Isostasy and Eustasy" (N. A. Mörner, ed.), pp. 3-10. John Wiley and Sons, New York.

Webber, P. J., Richardson, J. W. and Andrews, J. T. (1970). *Canadian Journal of Earth Sciences* **7**, 317-325.

Wu, P. and Peltier, W. R. (1982). *Geophys. J.R. astr. Soc.* **70**, 435-486.

Wu, P. and Peltier, W. R. (1983). Glacial isostatic adjustment and the free air gravity anomaly as a constraint upon deep mantle viscosity. *Geophys. J.R. astr. Soc.* **74**, 377-449.

THIRTEEN

Glacio-isostatic effects of the Cordilleran Ice Sheet, British Columbia, Canada

J. J. CLAGUE

INTRODUCTION

The growth and decay of continental glaciers in British Columbia during the Pleistocene generated complex lithospheric deformations that were dominantly isostatic in nature (Mathews *et al.*, 1970; Clague, 1981; Clague *et al.*, 1982a). Vertical crustal movements caused by the melting of the late Wisconsin (Weichselian) Cordilleran Ice Sheet have been documented from studies of both deformed glacial lake strandlines (Fulton and Walcott, 1975) and landforms and sediments associated with sea levels differing from the present (Mathews *et al.*, 1970; Clague, 1975; Clague *et al.*, 1982a). Isostatic effects related to the growth of late Wisconsin Cordilleran glaciers are more poorly known, although some inferences can be drawn from observations on the late Wisconsin marine limit on the British Columbia coast.

The record of glacio-isostatic deformation in British Columbia is obscured to some extent by diastrophism related to plate interactions in the north-east Pacific Ocean (Riddihough, 1982). Western British Columbia is situated near the edge of the American lithospheric plate, and interactions of this plate with the Juan de Fuca, Explorer, and Pacific plates off the west coast of North America have deformed the crust throughout the Canadian Cordillera (Wheeler *et al.*, 1972; Riddihough and Hyndman, 1976; Riddihough, 1977, 1979; Souther, 1977; Monger and Price, 1979; Monger *et al.*, 1982). Vertical crustal movements resulting from these plate interactions cannot be easily separated from those of glacio-isostatic origin. However, tidal records and precise levelling in British Columbia suggest that rates of contemporary vertical tectonic movements are at least an order of magnitude lower than maximum rates of isostatic movements during and soon after deglaciation (Clague *et al.*, 1982a; Riddihough, 1982). Because late Pleistocene tectonic patterns and rates probably were not significantly different from those of the present, it is concluded that the large rapid crustal movements occurring during deglaciation of British Columbia were dominantly glacio-isostatic.

This chapter summarizes observed and inferred isostatic effects of Pleistocene ice sheets in British Columbia. Isostatic responses to the decay of the late Wisconsin ice

SHORELINES AND ISOSTASY
ISBN 0 12 652960 4

sheet in western British Columbia are emphasized because of the wealth of information on palaeo-sea levels in formerly glaciated coastal regions.

The organization of the chapter is as follows. A brief introduction to Cordilleran ice sheets is followed by sections dealing with the various isostatic effects of these ice sheets. Isostatic depression accompanying the growth of continental glaciers in British Columbia is considered first. Isostatic uplift during deglaciation is then evaluated by examining patterns of sea-level change that occurred as the late Wisconsin Cordilleran Ice Sheet disintegrated. Forebulge effects are documented in the next section by contrasting the sea-level records of the Queen Charlotte Islands at the periphery of the former ice sheet and the nearby northern mainland coast nearer the centre of isostatic uplift. The chapter closes with a short summary and synthesis.

CORDILLERAN ICE SHEETS

Repeatedly during the Pleistocene, global cooling triggered glacier growth in the high mountains of British Columbia. At the beginning of each major cold period, alpine glaciers advanced down mountain valleys. With continued cooling and perhaps increased precipitation, these glaciers coalesced to form piedmont complexes and small mountain ice sheets. During long-sustained cold periods, piedmont glaciers from separate mountain source areas joined to cover most of British Columbia (Davis and Mathews, 1944).

As the Cordilleran glacier complex expanded and thickened, its nourishment and surface flow patterns became less controlled by ground surface topography and more controlled by ice sheet morphology. At the climax of major glaciations, ice was sufficiently thick (*c.* 2500 m) over the interior of British Columbia for one or more ice domes to exist, with surface flow radially away from their centres (Fig. 1; Dawson, 1881; Kerr, 1934; Mathews, 1955; Wilson *et al.*, 1958; Fulton, 1967; Flint, 1971). At such times, ice streamed westward down fjords and river valleys and across coastal mountains to completely cover the British Columbia continental shelf (Prest, 1969). During lesser glaciations, the Cordilleran Ice Sheet probably remained a complex of coalescent piedmont glaciers that covered lowland and plateau areas but did not extend completely across the shelf (Fladmark, 1975, 1978, 1979).

Each Cordilleran ice sheet, including the last which formed during the late Wisconsin interval, thinned towards the west and south-west across the continental margin. Ice was thick over mainland fjords and coastal lowlands, somewhat thinner on the inner continental shelf, and very thin or absent on the outer shelf. For example, at the climax of late Wisconsin glaciation, the heads of large mainland fjords were buried by about 2000 m of glacier ice, whereas eastern Vancouver Island was covered by about 1000–1600 m of ice (Fig. 2; Wilson *et al.*, 1958). Parts of the Queen Charlotte Islands and western Vancouver Island probably remained ice-free throughout the late Wisconsin glacial interval.

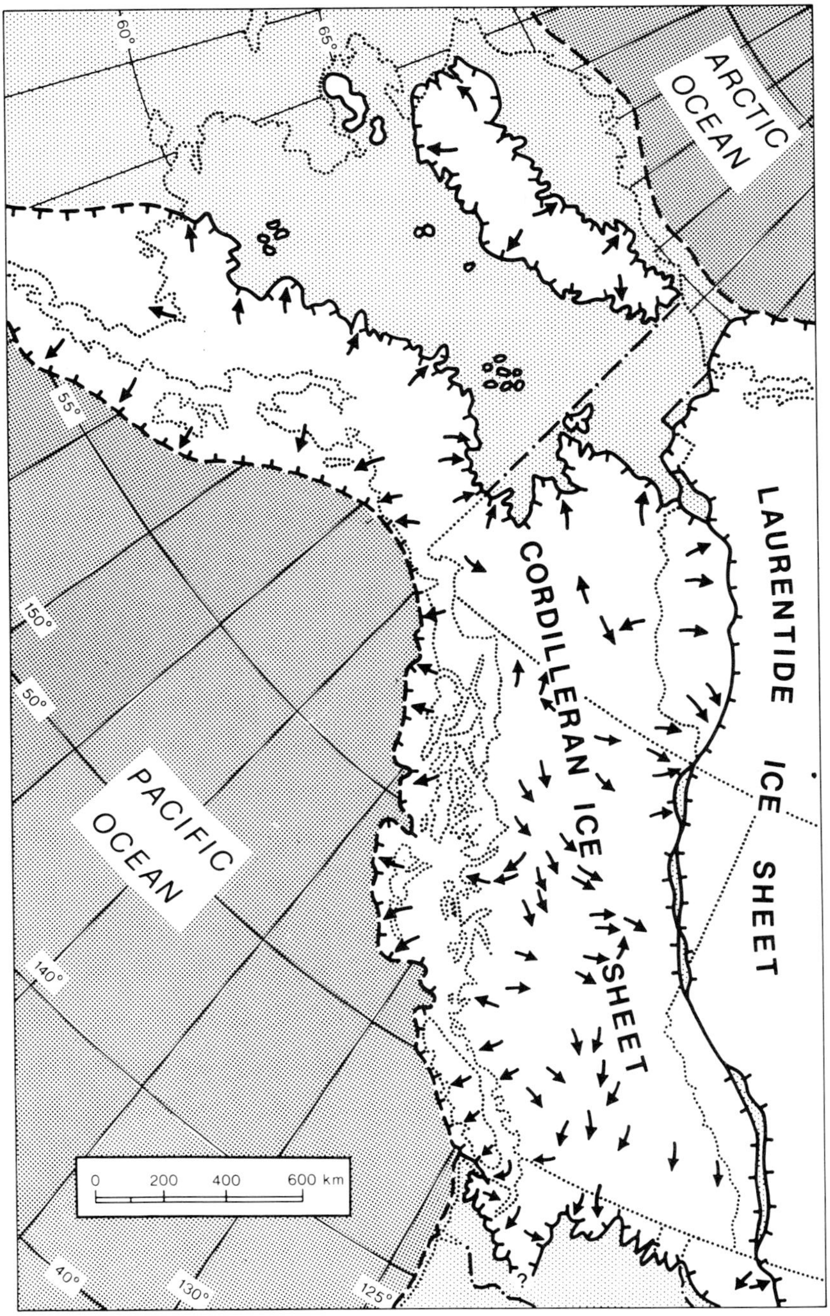

Figure 1. The Cordilleran Ice Sheet at its maximum extent. Arrows indicate generalized directions of flow. Non-glaciated coasts are drawn at the − 100 m isobath; glacier termini in oceans at the − 200 m isobath. Adapted from Flint (1971, Fig. 18-1) and reproduced with permission of John Wiley and Sons.

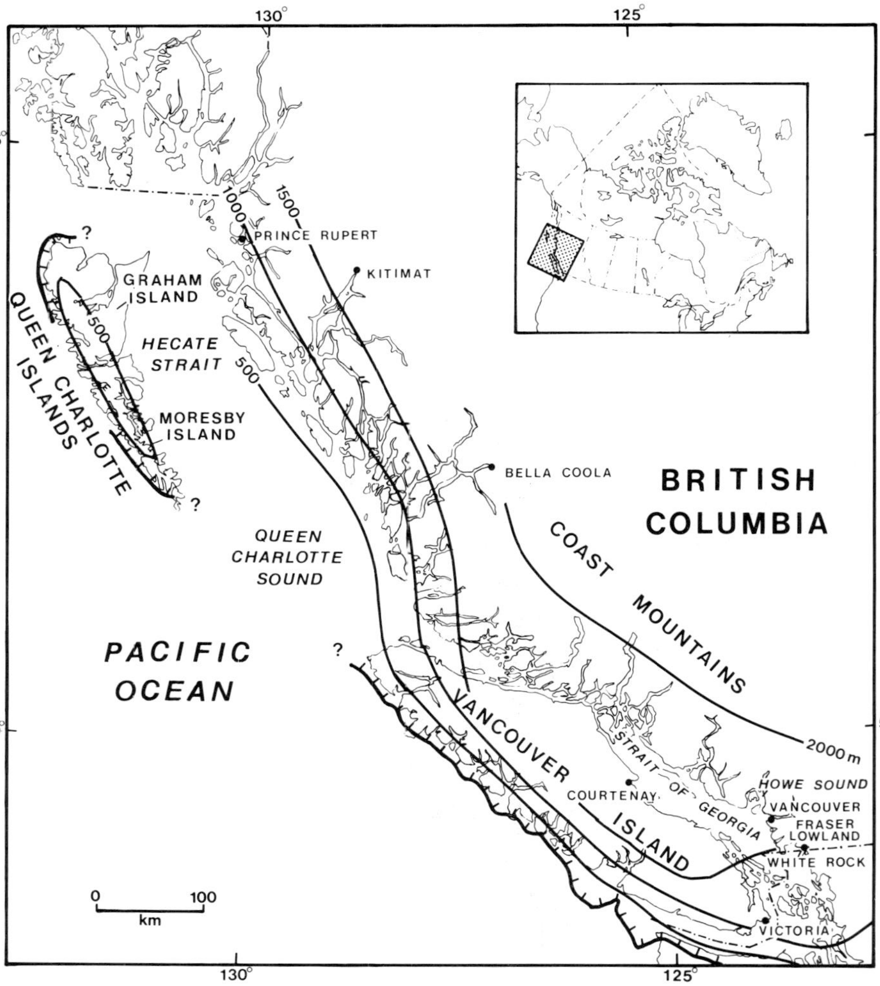

Figure 2. Map of British Columbia coastal region showing margin and surface contours of the Cordilleran Ice Sheet for the late Wisconsin glacial maximum (*c.* 14 500–15 000 B.P.). The surface contours have been adapted from Wilson *et al.* (1958) and are speculative.

GLACIO-ISOSTATIC EFFECTS OF CORDILLERAN ICE SHEETS

Isostatic depression

Loading of British Columbia by glacier ice during Pleistocene glaciations caused major isostatic adjustments in the crust and mantle of the region. Gradual thickening and expansion of glaciers led to progressive isostatic depression of the land surface. Initially, this depression was localized beneath the major mountain ranges that served as loci of glacier growth. As glaciers advanced out of the mountains and onto plateaux and lowlands, the area of crustal subsidence grew larger. Finally, at the climax of each major glaciation, most of British Columbia was displaced downward in response to loading by glacier ice. The magnitude of this depression depended primarily on the size of the ice sheet, the length of time during which it formed, and the structure and composition of the lithosphere. Presumably, isostatic depression was greatest beneath the centre of the ice sheet and decreased towards its periphery.

Although the magnitude of isostatic depression in British Columbia at the climaxes of major glaciations is unknown, some limits are provided by levels of shorelines that formed during deglaciation at the close of the Pleistocene. Relict shorelines occur on the mainland coast up to about 200 m above present sea level*

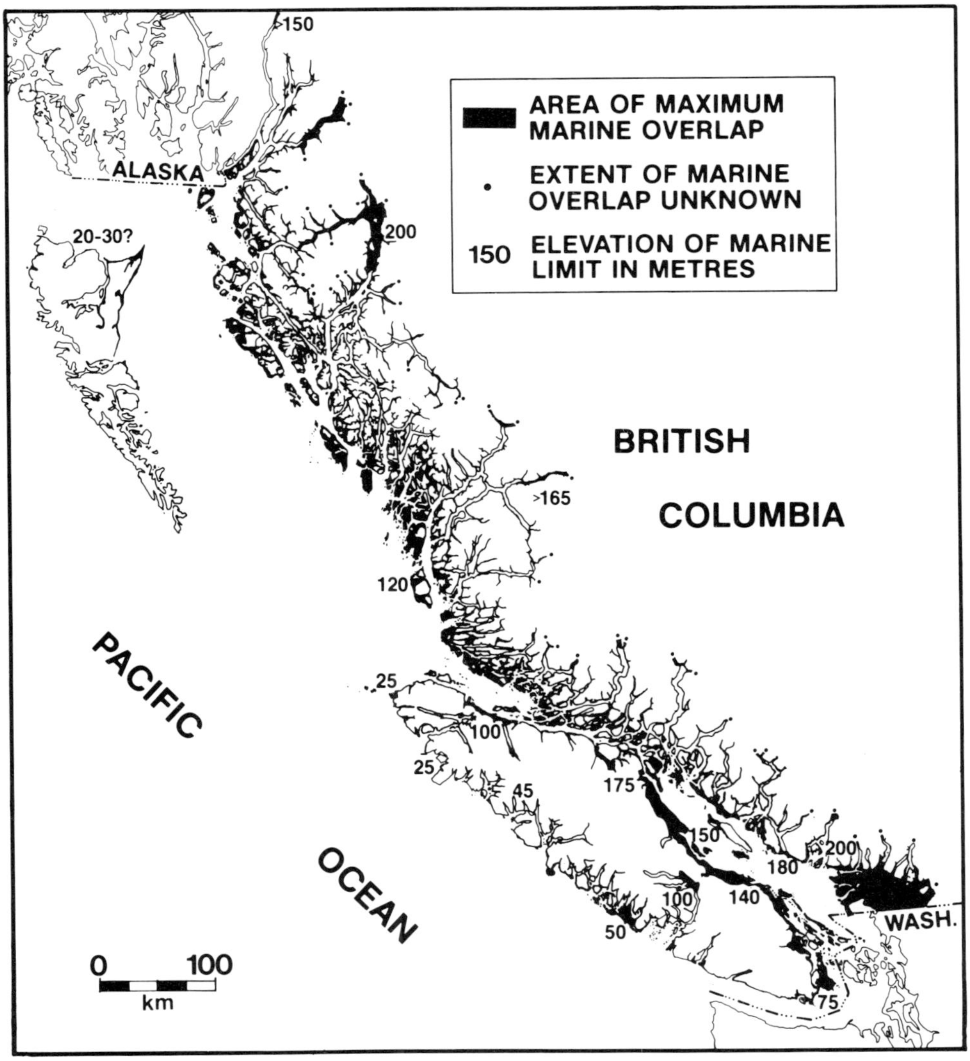

Figure 3. Maximum late Pleistocene-early Holocene marine overlap in British Columbia. Extent of overlap inland from the heads of most mainland fjords is unknown.

*Datum for all elevations in this chapter is mean sea level (m.s.l.).

(Fig. 3). Taking into account a eustatic sea-level lowering of perhaps 50–100 m at the time the highest shorelines formed, local glacio-isostatic depression of more than 250 m is indicated. In fact, this is a minimum value for isostatic depression in areas of thickest ice, because the Cordilleran Ice Sheet had decreased in size before the highest shorelines formed, and consequently isostatic rebound probably had commenced earlier.

The elevation of the late Pleistocene marine limit in coastal British Columbia varies in relation to distance from the main centres of ice accumulation and, to a lesser extent, to the timing of glacier retreat. In general, the marine limit is highest (*c.* 200 m) on the mainland coast and declines towards the west and south-west (Fig. 3; Mathews *et al.*, 1970; Clague, 1975). It is less than 50 m elevation on the west coast of Vancouver Island near the margin of the former Cordilleran Ice Sheet (Clague, 1981; Howes, 1981). Late Pleistocene shorelines were lower than present on the Queen Charlotte Islands, far from the centre of the former ice sheet, indicating that there was little or no glacio-isostatic depression of this region (Fladmark, 1975; Clague, 1981; Clague *et al.*, 1982a, in press). Some mainland areas below the regional marine limit (for example, the eastern Fraser Lowland) apparently were not transgressed by the sea at the close of the Pleistocene because they remained covered by ice until isostatic rebound was largely completed.

Isostatic uplift

Rapid deglaciation of British Columbia at the close of each major glacial cycle caused isostatic adjustments opposite in direction to those that occurred during ice sheet expansion. However, the pattern of isostatic uplift during deglaciation was not an exact mirror image of the pattern of antecedent isostatic depression. The main reason for this is that ice sheet decay did not duplicate in reverse ice sheet growth, as is shown by the following:

(1) Decay of the ice sheet was much more rapid than its growth. For example, at the close of the last glacial cycle, British Columbia was deglaciated within a period of only about 4500 years (*c.* 14 000–9500 B.P.); in contrast, the ice sheet grew to its late Wisconsin maximum size over a period of about 10 000 years or more (Clague, 1980, 1981).

(2) Whereas the ice sheet decayed by regional downwasting and widespread stagnation, it grew non-symmetrically by expansion from localized mountain source areas.

(3) Fluctuations of the margins of the Cordilleran Ice Sheet during deglaciation were different from those occurring during growth.

More detailed comparisons of isostatic adjustments preceding a glacial climax with those following the climax cannot be made at this time, in large part because the former are poorly known.

There is, however, a large amount of information bearing on isostatic responses to

Figure 4. Late Pleistocene raised delta deposits near Kitimat, British Columbia. Top, delta front; bottom, foreset-bedded gravel and sand. This delta was constructed 10 000-11 000 years ago by meltwater flowing into an arm of the sea that extended north of the present fjord head at Kitimat. The delta top is about 200 m above sea level.

deglaciation in British Columbia, especially for the period following the late Wisconsin glacial climax (Mathews *et al.*, 1970; Clague, 1975; Clague *et al.*, 1982a). For example, raised deltaic and glaciomarine sediments on the coastal lowlands of mainland British Columbia and Vancouver Island were deposited at the close of the last glaciation while these areas were being isostatically uplifted (Fig. 4). ^{14}C dates on wood and molluscan shells from these sediments provide information on the timing of this uplift. Similarly, ^{14}C-dated terrestrial sediments occurring at low elevations in coastal areas provide constraints on the level of the sea at the time of their deposition. From such data, patterns of late Pleistocene and Holocene sea-level change have been deduced for much of the British Columbia coast (Clague *et al.*, 1982a). The major glacio-isostatic implications of these data are summarized in the following sections.

Rate of uplift

Isostatic uplift was extremely rapid. For example, at Courtenay on eastern Vancouver Island (Fig. 2), the highest delta, 150 m above the present shore, has been ^{14}C-dated at 12 500 ± 450 B.P. (I(GSC)-9). In contrast, deltas 60–90 m below the marine limit about 65 km to the south-east have yielded dates ranging from 12 000 ± 450 to 12 400 ± 200 B.P. (I(GSC)-1 and GSC-1, respectively) (Mathews *et al.*, 1970, p. 695). Similarly, ^{14}C dates from a sequence of deltas near Kitimat on the northern mainland coast (Fig. 2) indicate that the shoreline was about 120 m above the present 10 100 ± 160 B.P. (GSC-2492), but had fallen to about 35 m by 9300 ± 90 B.P. (GSC-2425) (Clague *et al.*, 1982a). The error terms associated with these and other dates are sufficiently large and the dates themselves so closely spaced in time that it is not possible to assign precise figures to rates of uplift. Also, the amount of isostatic uplift that occurred while the ice sheet was thinning but before low-lying areas were transgressed by the sea is unknown. Nevertheless, most recorded uplift at each site occurred within a time interval of less than 2000 years. The "relaxation time" (defined by Heiskanen and Vening Meinesz (1958, p. 369) as the time for $1 - 1/e$, or about 63%, of the ultimate isostatic uplift to be achieved) may be less than 1000 years in many areas (Mathews *et al.*, 1970, p. 695). Such rates of uplift are broadly comparable to those from Greenland (Washburn and Stuiver, 1962; Lasca, 1966; Kelly, 1973; Ten Brink, 1974), for example, but are notably higher than those near the centre of the former Laurentide Ice Sheet (Andrews and Falconer, 1969; Webber *et al.*, 1970; Hillaire-Marcel and Fairbridge, 1978).

Timing of uplift

Isostatic uplift occurred at different times within British Columbia due to diachronous retreat of the late Wisconsin Cordilleran Ice Sheet (Table 1, Fig. 5). In general, regions that were deglaciated first rebounded earlier than those that were deglaciated at a late date. For example, isostatic uplift at Victoria on southern Vancouver Island (Fig. 2) was largely complete by 11 500 years ago, about 1500 years after initial deglaciation of the area. In contrast, the sea was at its upper limit at Kitimat about 11 000 years ago when the area was first deglaciated; most of the subsequent isostatic uplift there occurred between about 10 500 and 9000 B.P.

TABLE 1

Timing of late Pleistocene and Holocene sea-level changes caused by isostatic uplift, coastal British Columbia

Locality	Cross-over (years B.P.)[a]	Holocene low stand (years B.P.)
Victoria	*c.* 11 500	> 5500 < 9250
Courtenay	> 10 000 < 11 000	> 5700 < 8700
Vancouver	> 10 500 < 11 000	> 7300 < 8400
Bella Coola	> 9000 < 9500	< 9000
Kitimat	*c.* 9000	< 9000

[a]"Cross-over" occurred in response to isostatic uplift when the sea dropped below its present level relative to the land from its late glacial high stand.

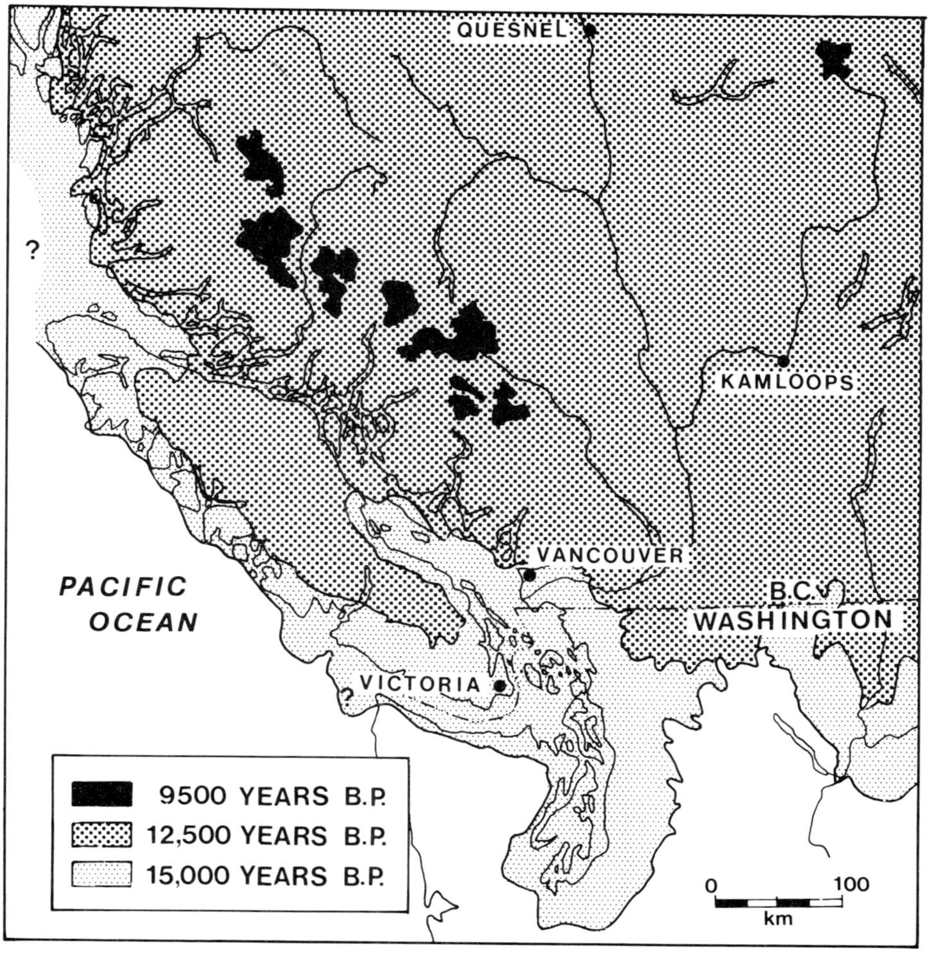

Figure 5. Decay of the Cordilleran Ice Sheet in southern British Columbia and northern Washington State at the close of the last glaciation. Approximate glacier margins at 15 000, 12 500, and 9500 B.P. (± 500 years) are depicted. Unglaciated areas within the confines of the ice sheet are not shown. There is a close relationship between the pattern of glacier retreat shown here and the pattern of isostatic uplift caused by deglaciation.

Such data indicate that there were substantial regional differences in glacio-isostatic response to deglaciation, and show that the crust deformed in a complex non-uniform fashion as the ice sheet decayed.

Termination of uplift

The rate of isostatic uplift in each region decreased exponentially as a function of time during and following late Wisconsin deglaciation. The land, upon reaching its present level with respect to the sea (Table 1), continued to rise, but at a much slower rate than during preceding centuries.

On much of eastern Vancouver Island and the mainland, emergence culminated

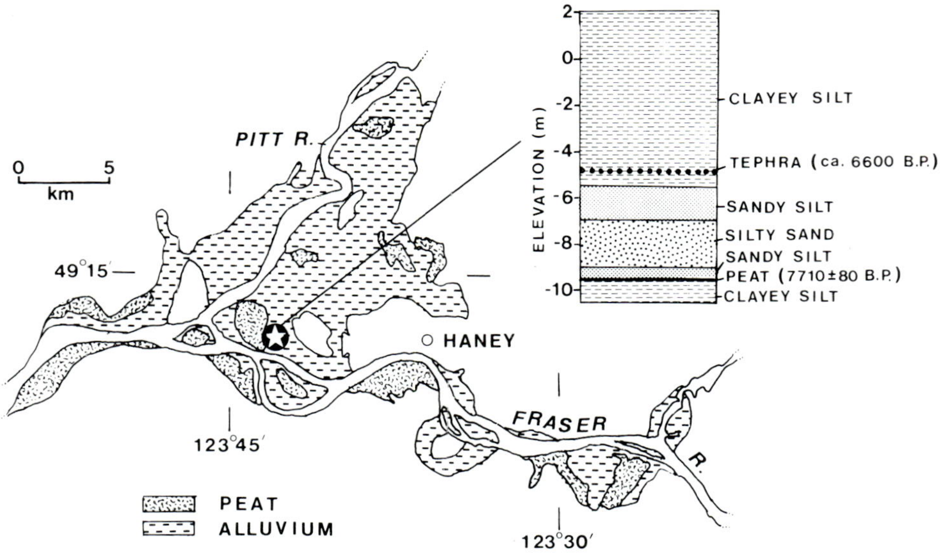

Figure 6. Stratigraphy of a core from the Fraser River floodplain 30 km east of Vancouver. Sediments above the peat are estuarine and fluvial, and were deposited in response to a rise in the level of the sea relative to the land.

during early to middle Holocene time with shorelines locally more than 10 m below the present. Near Vancouver, for example, terrestrial organic sediments occur widely at about − 10 to − 11 m elevation beneath the floodplain of the Fraser River and its tributaries (Fig. 6; Mathews *et al.*, 1970; Clague *et al.*, 1982a). These sediments delineate an alluvial surface graded to sea level in the Strait of Georgia at least 12 m below the present.* Several ^{14}C dates on peat from this buried alluvial surface range from 7300 ± 120 to 8360 ± 170 B.P. (S-99 and GSC-225, respectively). The level of the sea *relative to the land* was apparently stable in the Vancouver area during this period (Clague *et al.*, 1982a). However, beacuse the sea was still rising eustatically at this time due to final disintegration of the Laurentide Ice Sheet (Curray, 1965; Hopkins, 1967; Denton and Hughes, 1981), there probably was compensatory isostatic uplift at Vancouver in order to account for relative sea-level stability there.

A marine transgression in the Vancouver area, which began about 7000–7500 years ago, triggered aggradation of the Fraser River floodplain and probably caused a marine incursion onto part of the proto-delta of the Fraser River. Peat dated 7710 ± 80 B.P. (GSC-3099) and occurring at − 10 m elevation beneath the Fraser River floodplain 30 km east of Vancouver is overlain by 11·5 m of silt and sand deposited in an estuarine environment under aggradational conditions (Fig. 6). A 2 cm thick layer of *c.* 6600 year-old tephra is interbedded with silt 5 m above the peat, indicating that significant aggradation occurred between 7700 and 6600 B.P.,

*The difference between high tide, which is the controlling datum for peat growth, and mean sea level is 2·5 m at Vancouver.

probably in response to a rise in the level of the sea relative to the land. It is concluded from these data that isostatic uplift at Vancouver was essentially complete by the time the marine transgression began 7000–7500 years ago.

Data from the Victoria area support these conclusions. As at Vancouver, shorelines were lower than at present during early to middle Holocene time (Mathews *et al.*, 1970; Clague *et al.*, 1982a). Fibrous peat accumulated 1-2 m below present sea level in Portage Inlet at Victoria between 9250 ± 140 and 5470 ± 115 B.P. (I-3676 and I-3673, respectively) (Fig. 7). Portage Inlet is a shallow basin connected to the sea by a narrow, rock-floored tidal passage. Peat was deposited in the inlet when it ceased to be connected to the sea, that is when relative sea level was more than 4 m below its present position. At that time, Portage Inlet became a shallow pond or bog with only minor freshwater inflow. The peat is both underlain and

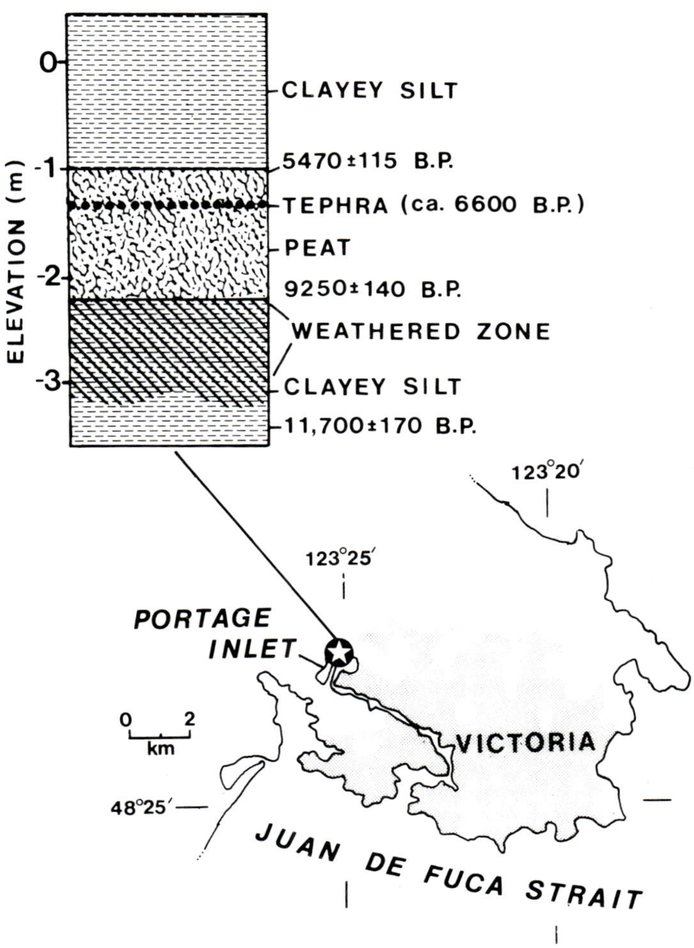

Figure 7. Stratigraphy of a core from Portage Inlet. This core provides evidence for a marine regression and a subsequent transgression on southern Vancouver Island (see text for details).

overlain by marine sediments, the older marine unit having yielded a ^{14}C date of 11 700 ± 170 B.P. (I-3675), and the younger unit presumably deposited after about 5500 B.P. (the age of the top of the peat) when the inlet was flooded as the sea rose to higher than −4 m elevation.

Sea level during deposition of the Portage Inlet peat was at least 11 m lower than at present, because shells in marine sediments have been leached and the sediments themselves oxidized and hardened by dessication down to at least −9 m elevation in the Victoria area (Mathews *et al.*, 1970; Crawford and Sutherland, 1971).* In addition, there is a widespread unconformity at about −9 to −15 m elevation within the Holocene sediment pile in several bays and inlets on southern Vancouver Island (Duffus *et al.*, 1978; Schurer *et al.*, 1979). This unconformity is likely to be an erosion surface formed by subaerial planation during a Holocene low stand of the sea (Clague *et al.*, 1982a).

In conclusion, in the Victoria area, as at Vancouver, isostatic uplift was virtually complete by the middle Holocene following a period of nearly stable sea level (i.e. eustatic rise approximately equalled residual isostatic uplift) when shorelines were below their present positions. Subsequent transgression of present shallow offshore areas resulted from a combination of diastrophism, continued eustatic sea-level rise, and possibly forebulge migration (see p. 333).

Limited data from the central and northern mainland coasts indicate a similar pattern of deglacial emergence succeeded by submergence. However, the timing and magnitude of the sea-level changes in these areas probably differed somewhat from those at Victoria and Vancouver (Table 1).

Local differences in glacio-isostatic response to deglaciation

Differences in the pattern and timing of isostatic uplift during deglaciation occurred not only between regions in British Columbia, but also locally within individual regions. In fact, there were major differences in the response of the crust to ice sheet decay over distances as short as a few tens of kilometres. Data from the southern mainland coast near Vancouver illustrate this concept.

Although the Vancouver area was deglaciated as early as about 13 000 B.P., piedmont and valley glaciers persisted in the nearby central and eastern Fraser Lowland and in Coast Mountain valleys and fjords directly north of Vancouver until about 11 000 B.P. Differences in the disposition of ice loads and timing of glacier retreat in this region caused local differences in isostatic uplift. For example, the western Fraser Lowland was probably undergoing major isostatic uplift before upper Howe Sound (Fig. 2), 50 km to the north. Near White Rock in the western Fraser Lowland, glaciomarine sediments dated 11 590 ± 280 and 12 625 ± 450 B.P. (GSC-226 and I(GSC)-6, respectively) were deposited after the sea had fallen more than 100 m from the local marine limit (Clague and Luternauer, 1982). In contrast, upper Howe Sound was probably still buried beneath glacier ice 11 500 years ago,

*Tidal amplitude at Victoria is 1·5 m.

and major isostatic uplift had not yet begun there. Wood from a Howe Sound raised delta, deposited when the sea was more than 41 m above the present shore, has been dated 10 690 ± 180 B.P. (GSC-227), indicating that substantial uplift occurred there after this date. In the western Fraser Lowland at the same time, sea level probably was within 10 m of its present position (Clague and Luternauer, 1982).

Howe Sound and the central and eastern Fraser Lowland experienced delayed isostatic uplift because they were deglaciated later than areas to the south and west. As a consequence, hypothetical late Pleistocene water planes in the Fraser Lowland should be tilted down towards the south and west due to the subsequent differential uplift. Unfortunately, this tilting cannot be conclusively demonstrated because deglacial shoreline features in the Fraser Lowland, in general, are poorly developed and discontinuous. However, the amount of tilting required to explain the seemingly anomalous pattern of sea-level change within this region is well within the range of glacio-isostatic deformation documented for late Pleistocene glacial lakes in south-central British Columbia (Fulton and Walcott, 1975). Strandlines associated with these lakes are tilted 1·6–2·5 m km^{-1}, and comparable deformation in the Fraser Lowland might account for present vertical differences of 100 m or more in the positions of coeval deglacial shorelines within that region.

Forebulge effects

The pattern of late Pleistocene and Holocene sea-level change on the mainland and eastern Vancouver Island is similar to that predicted by numerical modelling for marginal areas of glaciated North America (Clark *et al.*, 1978). Clark *et al.* (1978) attributed the pattern of initial rapid emergence followed by submergence to migration of a collapsing proglacial forebulge (Walcott, 1972). The forebulge is created by lateral transfer of matter within the Earth away from the centre of ice accumulation. As the ice sheet thins and retreats, this material flows back towards the centre of accumulation, resulting in uplift. At the same time, in peripheral areas, submergence occurs because subcrustal material is depleted as it flows towards the uplifting region. The forebulge does not simply collapse in place, but rather migrates towards the ice load as it collapses (Peltier, 1974). This explains the period of submergence following initial uplift of the land, a sequence both predicted in the model of Clark *et al.* (1978) and observed on parts of the eastern Vancouver Island and mainland coasts. As the forebulge migrated inland towards the British Columbia interior at the close of the last glaciation, coastal areas stopped rising relative to the sea, and instead began to fall. However, this simple pattern of forebulge migration was probably complicated in many areas by non-uniform glacier retreat and possibly by short-lived glacier readvances.

Migration of a collapsing forebulge may partly explain the anomalous pattern of sea-level change on the Queen Charlotte Islands, a region far from the centre of the Cordilleran Ice Sheet and one only weakly glaciated during the late Wisconsin (Fladmark, 1975; Clague, 1981; Clague *et al.*, 1982b; Mathewes and Clague, 1982). Shorelines on the Queen Charlotte Islands were lower than the present from

Figure 8. Late Pleistocene terrestrial peat exposed in the intertidal zone of eastern Graham Island. Sea level was lower than at present when this peat accumulated about 10 000-12 500 years ago.

Figure 9. Fossiliferous marine sediments of early Holocene age (m) overlie late Pleistocene terrestrial peat (p) in a sea cliff on eastern Graham Island. This section shows that there was a transgression of the Graham Island lowlands by the sea during the early Holocene. At the same time, the sea was falling relative to the land on the British Columbia mainland coast to the east.

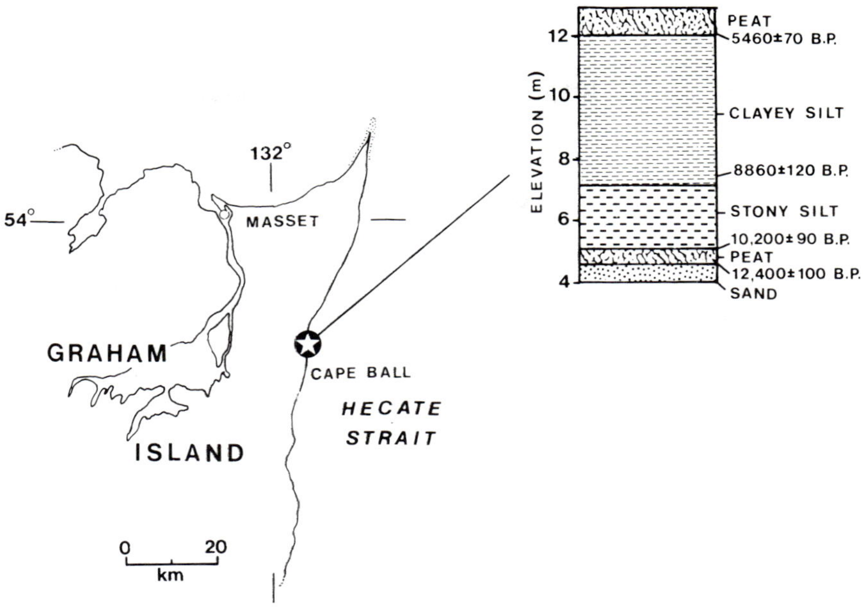

Figure 10. Stratigraphy of a low sea cliff at Cape Ball. The late Pleistocene peat near the base of the cliff was deposited when sea level was lower than at present. The overlying stony and clayey silts are marine and were deposited during a transgression that culminated during the early Holocene.

before 12 500 years ago until approximately 9500–10 000 years ago, in sharp contrast to the high shorelines on the mainland coast directly to the east (Clague *et al.*, 1982a,b). The evidence for low sea levels on the Charlottes is the occurrence of dated late Pleistocene peat containing *in situ* stumps in the intertidal zone and sea cliffs of eastern Graham Island (Figs 8, 9; Clague *et al.*, 1982b; Mathewes and Clague, 1982). At Cape Ball, for example, peat with bounding ^{14}C dates of 10 200 ± 90 and 12 400 ± 100 B.P. (GSC-3159 and GSC-3112, respectively) outcrops about 1 m above high tide level (Fig. 10). Other evidence for low sea levels on the Queen Charlotte Islands during late Pleistocene time includes: (1) drowned river mouths on Graham Island (Sutherland Brown, 1968); (2) possible submerged river valleys and associated wave-cut scarps up to a few tens of metres below present sea level in Hecate Strait off the north-east coast of Graham Island (Fladmark, 1975; Warner *et al.*, 1982); and (3) archaeological sites of late Pleistocene or early Holocene age in the present intertidal zone on Moresby Island (Hobler, 1978).

This period of low sea levels was followed by a marine transgression during which shell- and wood-bearing marine and estuarine mud and sand were deposited on top of older organic and glacial sediments (Figs 9, 10). The marine and estuarine sediments are best exposed in coastal bluffs of eastern Graham Island, where they occur up to about 12 m elevation and have yielded several radiocarbon dates ranging from 8610 ± 80 to 9350 ± 80 B.P. (GSC-2738 and GSC-3120, respectively) (Clague *et al.*, 1982a,b). At Cape Ball, for example, marine mud outcropping

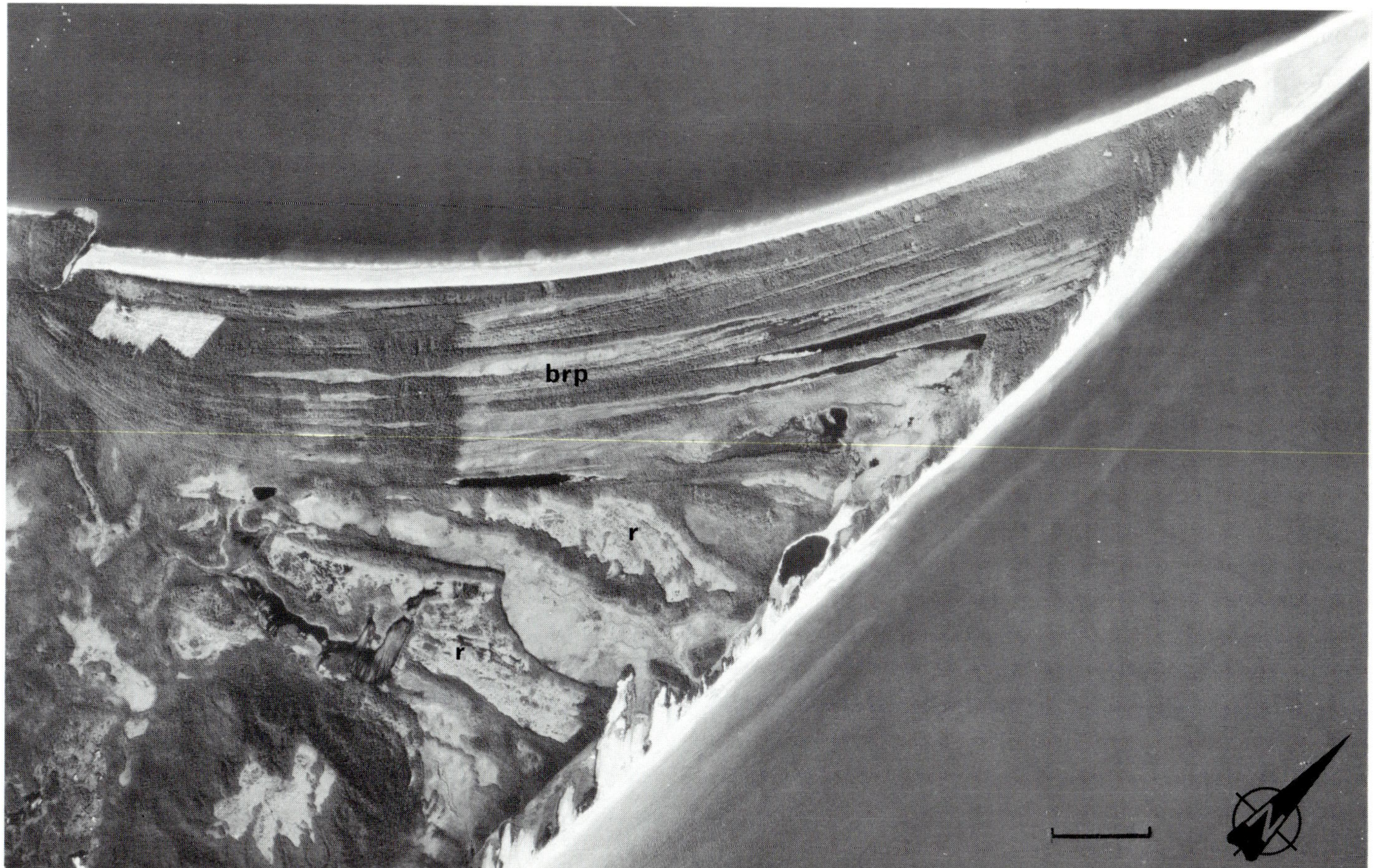

Figure 11. Elevated beach ridge plain (brp) and older recurves (r) at the north-eastern corner of Graham Island. These features formed during a middle and late Holocene marine regression. Province of British Columbia photo no. BC5630-198.

between 5 and 12 m elevation has been dated 8860 ± 120 B.P. (GSC-2734) (Fig. 10). This mud is both underlain and overlain by peats, the lower of which is of late Pleistocene age and the upper having commenced deposition about 5460 ± 70 B.P. (GSC-2963).

The transgression on the Queen Charlotte Islands culminated about 7500–8000 years ago when shorelines on north-eastern Graham Island were about 15 m higher than at present. This was followed by a fall of the sea relative to the land. Features that formed during this regression and are now elevated above sea level include wave-cut scarps, wave-cut benches overlain by littoral sand and gravel, bars, spits, and beach and dune ridges parallel to the modern shoreline (Fig. 11; Clague *et al.*, 1982a,b).

The pattern of sea-level change on the Queen Charlotte Islands contrasts sharply with that on the adjacent northern mainland coast. At Kitimat, 200 km east of Graham Island, the sea stood at about 120 m above the present shore 10 100 ± 160 B.P. (GSC-2492), whereas at the same time shorelines on the Charlottes were below present sea level. Between 10 100 ± 160 and 9300 ± 90 B.P. (GSC-2425), the sea fell rapidly relative to the land at Kitimat, whereas it rose during this same period on the Queen Charlotte Islands.

Sea-level changes on the mainland coast at Prince Rupert (Fig. 2), only 100 km east of Graham Island, have also differed from those on the Charlottes. Marine molluscs recovered from glaciomarine sediments at an approximate elevation of 11 m at Prince Rupert have been dated 12 700 ± 120 B.P. (GSC-2290), indicating that the shore there was higher than 11 m elevation at this time. Yet parts of the shallow offshore platform directly east of Graham Island probably were subaerially exposed at the same time (Warner *et al.*, 1982).

Isostatic uplift of the Prince Rupert area resulted in a marine regression during late Pleistocene and early Holocene time. Stratigraphic evidence of this sea-level fall is provided in exposures along McNeil River about 25 km east-south-east of Prince Rupert (Fig. 12). Here, marine pebbly, sandy mud containing abundant fossil wood and molluscan shells is sharply overlain by inclined parallel-bedded gravel, which in turn is overlain by organic-rich pebbly, silty sand grading upwards into peat. This stratigraphic succession is the product of progressive emergence of the site. The sandy mud of the lowest unit, which occurs below 3 m elevation and has yielded dates of 8580 ± 100 and 9180 ± 150 B.P. (GSC-2307 and GSC-2343, respectively), was deposited in a shallow marine environment. A driftwood mat at the contact between this unit and the overlying gravel has been dated 8460 ± 90 B.P. (GSC-2248). The driftwood probably accumulated in the upper intertidal zone, the modern counterpart of which is about 0–2 m lower at this site. The gravel was transported by the Skeena River north-west into the McNeil River embayment and deposited in shallow standing water. The upper surface of this unit, at slightly over 3 m elevation (less than 1 m below present high tide), probably approximates high tide level at the time of deposition of the unit. Although the gravel has not been dated directly, the base of the overlying pebbly, silty sand has yielded a date of 7710 ± 150 B.P. (GSC-2471). The latter unit is probably estuarine in origin

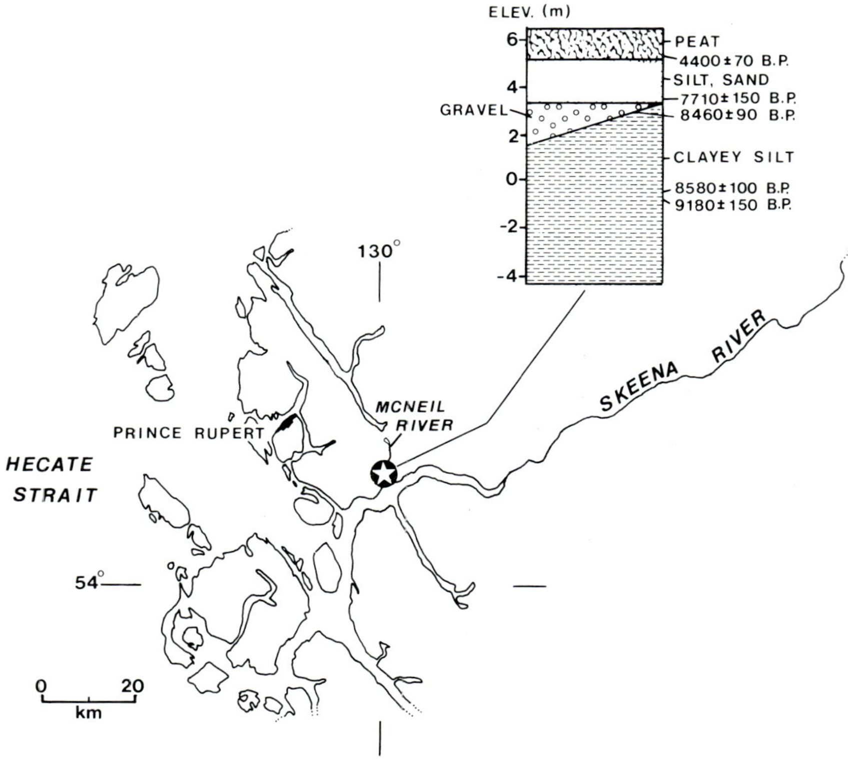

Figure 12. Stratigraphy of a section at McNeil River. This section provides information on the timing of isostatic uplift of the northern mainland coast (see text for details).

and was deposited when sea level was within a few metres of its present position. The peat at the top of the section was deposited when the sea was no higher, and possibly was lower, than at present. Wood collected from the transitional zone between this peat and underlying estuarine sediments yielded a radiocarbon date of 4400 ± 70 B.P. (GSC-2222). The above evidence indicates that there was a gradual fall in sea level in the McNeil River area between about 9200 and 7700 B.P., a period during which lowlands on the Queen Charlotte Islands were being transgressed by the sea.

In summary, late Pleistocene and Holocene sea-level fluctuations on the Queen Charlotte Islands have been opposite in direction to those on the adjacent mainland coast (Fig. 13). Submergence on the Charlottes at the close of the Pleistocene occurred at a time of emergence on the mainland. Middle-late Holocene emergence on the Charlottes took place when the sea was stable or rising at Prince Rupert.

These contrasting patterns are most easily explained by forebulge migration. Ice on the Queen Charlotte Islands at the late Wisconsin glacial maximum was thin, and some areas probably were ice-free. Transfer of matter in the Earth westward from beneath thicker parts of the Cordilleran Ice Sheet may have caused crustal bulging in the vicinity of the Queen Charlotte Islands. As a consequence of both this

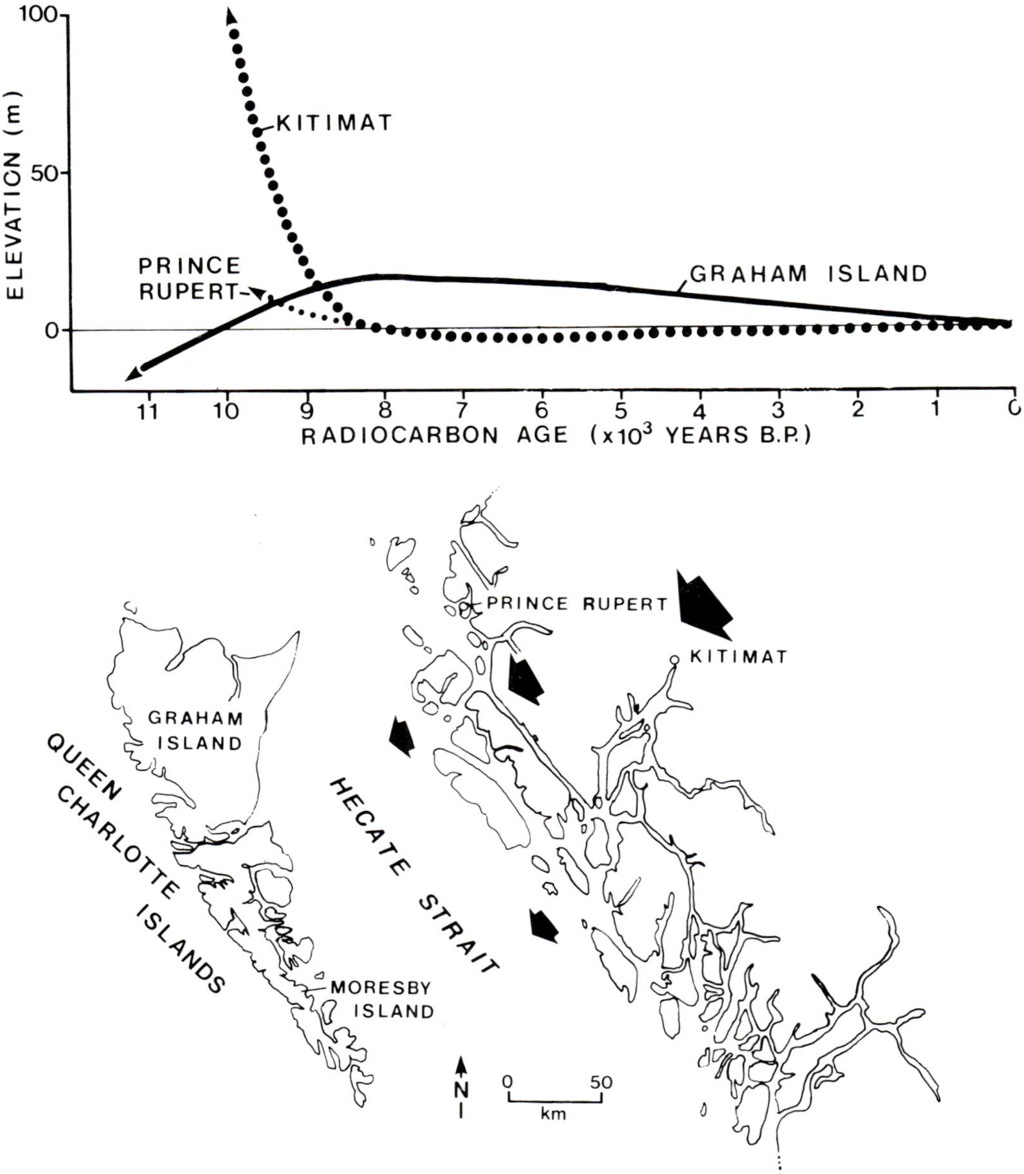

Figure 13. Opposing patterns of sea-level change on Graham Island and adjacent mainland British Columbia. The contrasting sea-level records may be explained, in part, by forebulge migration associated with decay of the late Wisconsin Cordilleran Ice Sheet. Arrows indicate generalized directions of glacier flow at the late Wisconsin maximum; their size is proportional to the thickness of ice.

crustal bulging and low eustatic sea levels, shorelines were lower than at present in this area during late Pleistocene time. The transgression that followed probably resulted from the combined effects of the eustatic rise in sea level accompanying deglaciation and forebulge migration towards the east. As mantle material flowed eastward towards the centre of the decaying ice sheet, the mainland was uplifted.

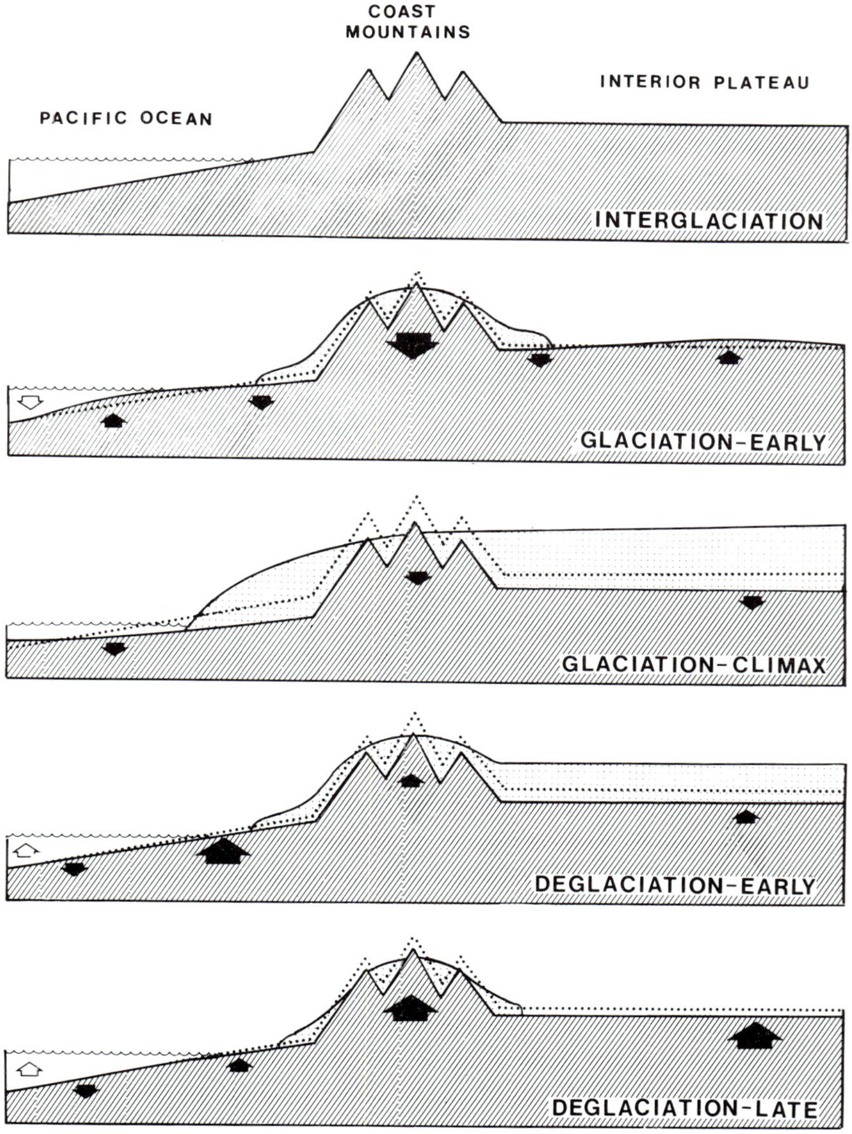

Figure 14. Summary of glacio-isostatic changes resulting from growth and decay of the Cordilleran Ice Sheet. Isostatic crustal deformation along a hypothetical profile extending across Queen Charlotte Sound into the British Columbia interior is shown for various stages of glaciation (dotted line is interglacial datum). Directions of vertical isostatic movements are shown by darkened arrows (continued downward movement is assumed at the climax of glaciation); directions of eustatic sea-level change are shown by open arrows. Lateral and vertical distances are not to scale.

SUMMARY AND SYNTHESIS

Complex vertical crustal movements accompanied the growth and decay of the Cordilleran Ice Sheet during the Pleistocene (Fig. 14). During the early phase of each major glaciation, valley glaciers and small ice fields grew in the high mountains of western Canada. Glacier expansion was accompanied by localized isostatic depression and lateral transfer of mantle material. Over a period of thousands of years, these glaciers thickened, coalesced, and advanced from the mountains onto adjacent plateaux and lowlands. Glacier growth was accompanied by an enlargement of the area of isostatic depression. At the climaxes of major glaciations, when British Columbia was almost completely enveloped by ice, all parts of the province were isostatically depressed, the amount of depression varying from region to region in relation to the thickness of the covering ice. Elevated glaciomarine sediments and shoreline features in coastal areas show that isostatic depression in British Columbia at the end of the last glaciation locally exceeded 250 m.

Major glaciations terminated with sharp climatic amelioration and rapid melting of glaciers. Rapid unloading triggered new isostatic adjustments as material moved laterally in the Earth from extraglacial regions towards the centre of the decaying ice sheet. Areas at the periphery of the ice sheet, which were deglaciated earliest, rebounded first. However, the total amount of uplift in these areas was less than in regions nearer the centre of the ice sheet where ice thicknesses were greater. As deglaciation progressed, the zone of rapid isostatic uplift migrated in step with receding glacier margins. Because the ice sheet thinned as it retreated, some uplift in each area probably occurred before that area was deglaciated.

Most of the British Columbia coastal zone experienced rapid isostatic uplift during late Pleistocene and early Holocene time. Uplift decreased in a non-linear fashion and was essentially complete by the middle of the Holocene. A mid-Holocene marine transgression on the mainland coast and on eastern Vancouver Island was caused either by forebulge migration or by a eustatic rise in sea level without compensatory isostatic uplift. Forebulge migration and rising seas were also responsible for a late Pleistocene-early Holocene transgression on the Queen Charlotte Islands. Isostatic uplift of the northern mainland and contemporaneous depression of the Charlottes probably resulted from migration of mantle material eastward from beneath the latter area towards the former as the late Wisconsin Cordilleran Ice Sheet decayed.

REFERENCES

Andrews, J. T. and Falconer, G. (1969). *Can. J. Earth Sci.* **6**, 1263-1276.

Clague, J. J. (1975). *Geol. Surv. Can. Paper* 75-1C, 17-21.

Clague, J. J. (1980). *Geol. Surv. Can. Paper* 80-13.

Clague, J. J. (1981). *Geol. Surv. Can. Paper* 80-35.

Clague, J. J. and Luternauer, J. L. (1982). "Excursion 30A: Late Quaternary Sedimentary Environments, Southwestern British Columbia". Field Excursion Guide Book, Eleventh Int. Congr. Sedimentol. *Int. Assoc. Sedimentol.*

Clague, J., Harper, J. R., Hebda, R. J. and Howes, D. E. (1982a). *Can. J. Earth Sci.* **19**, 597-618.

Clague, J. J., Mathewes, R. W. and Warner, B. G. (1982b). *Can. J. Earth Sci.* **19**, 1786-1795.

Clark, J. A., Farrell, W. E. and Peltier, W. R. (1978). *Quat. Res.* **9**, 265-287.

Crawford, C. B. and Sutherland, J. G. (1971). *Can. Geotech. J.* **8**, 77-93.

Curray, J. R. (1965). *In* "The Quaternary of the United States" (H. E. Wright, Jr. and D. G. Frey, eds), pp. 723-735. Princeton University Press, Princeton.

Davis, N. F. G. and Mathews, W. H. (1944). *J. Geol.* **52**, 403-413.

Dawson, G. M. (1881). *Q. Jl. Geol. Soc. Lond.* **37**, 272-285.

Denton, G. H. and Hughes, T. J. (eds) (1981). "The Last Great Ice Sheets". John Wiley and Sons, New York.

Duffus, H. J., Madill, J. W., MacFarlane, W. T. and Schurer, P. J. (1978). *R. Roads Military Coll., Coast. Mar. Sci. Lab. Manuscr. Rep.* No. 78-3.

Fladmark, K. R. (1975). *Natl. Mus. Man Mercury Ser., Archaeol. Surv. Can. Paper* No. 43.

Fladmark, K. R. (1978). *Univ. Alberta, Dep. Anthropol. Occas. Paper* No. 1.

Fladmark, K. R. (1979). *Am. Antiq.* **44**, 55-69.

Flint, R. F. (1971). "Glacial and Quaternary Geology". John Wiley and Sons, New York.

Fulton, R. J. (1967). *Geol. Surv. Can. Bull.* **154**.

Fulton, R. J. and Walcott, R. I. (1975). *Geol. Soc. Am. Mem.* **142**, 163-173.

Heiskanen, W. A. and Vening Meinesz, F. A. (1958). "The Earth and Its Gravity Field". McGraw-Hill Book Co., New York.

Hillaire-Marcel, C. and Fairbridge, R. W. (1978). *Geology* **6**, 112-122.

Hobler, P. M. (1978). *Can. J. Archaeol.* **2**, 1-13.

Hopkins, D. M. (1967). *In* "The Bering Land Bridge" (D. M. Hopkins, ed.), pp. 451-484. Stanford University Press, Stanford.

Howes, D. E. (1981). *B.C. Ministry Environ., APD Bull.* **5**.

Kelly, M. (1973). *Groenlands Geol. Unders. Rapp.* No. 59.

Kerr, F. A. (1934). *Trans. R. Soc. Can. Ser.* **3**, 28:4, 17-31.

Lasca, N. P. (1966). *Arctic* **19**, 349-353.

Mathewes, R. W. and Clague, J. J. (1982). *Can. J. Earth Sci.* **19**, 1185-1195.

Mathews, W. H. (1955). *Bull. Geol. Soc. Am.* **66**, 1657.

Mathews, W. H., Fyles, J. G. and Nasmith, H. W. (1970). *Can. J. Earth Sci.* **7**, 690-702.

Monger, J. W. H. and Price, R. A. (1979). *Can. J. Earth Sci.* **16**, 770-791.

Monger, J. W. H., Price, R. A. and Tempelman-Kluit, D. J. (1982). *Geology* **10**, 70-75.

Peltier, W. R. (1974). *Rev. Geophys. Space Phys.* **12**, 649-705.

Prest, V. K. (1969). *Geol. Surv. Can. Map* **1257A**.

Riddihough, R. P. (1977). *Can. J. Earth Sci.* **14**, 384-396.

Riddihough, R. P. (1979). *Can. J. Earth Sci.* **16**, 350-363.

Riddihough, R. P. (1982). *Tectonophysics* **86**, 319-341.

Riddihough, R. P. and Hyndman, R. D. (1976). *Geosci. Can.* **3**, 269-278.

Schurer, P. J., MacFarlane, W. T. and Duffus, H. J. (1979). *R. Roads Military Coll., Coast. Mar. Sci. Lab. Manuscr. Rep.* No. 79-8.

Souther, J. G. (1977). *In* "Volcanic Regimes in Canada" (W. R. A. Barager, L. C. Coleman and J. M. Hall, eds), pp. 3-24. Geological Association of Canada, Toronto.

Sutherland Brown, A. (1968). *B.C. Dep. Mines Pet. Resour. Bull.* No. 54.

Ten Brink, N. W. (1974). *Bull. Geol. Soc. Am.* **85**, 219-228.

Walcott, R. I. (1972). *Quat.* Res. **2**, 1-14.

Warner, B. G., Mathewes, R. W. and Clague, J. J. (1982). *Science* **218**, 675-677.

Washburn, A. L. and Stuiver, M. (1962). *Arctic* **15**, 66-73.

Webber, P. J., Richardson, J. W. and Andrews, J. T. (1970). *Can. J. Earth Sci.* **7**, 317-325.

Wheeler, J. O., Aitken, J. D., Berry, M. J., Gabrielse, H., Hutchison, W. W., Jacoby, W. R., Monger, J. W. H., Niblett, E. R., Norris, D. K., Price, R. A. and Stacey, R. A. (1972). *In* "Variations in Tectonic Styles in Canada" (R. A. Price and R. J. W. Douglas, eds), pp. 1-81. Geological Association of Canada, Toronto.

Wilson, J. T., Falconer, G., Mathews, W. H. and Prest, V. K. (compilers) (1958). "Glacial Map of Canada". Geological Association of Canada, Toronto.

PART FIVE

Shorelines and hydro-isostasy

FOURTEEN

Deformation of the North Queensland continental shelf in the Late Quaternary

D. HOPLEY

INTRODUCTION

Bloom (1967) revived some older speculation on coastal isostatic deformation by the post-glacial rise of sea level, suggesting that the load of water added to the continental margins by the transgression would produce subsidence of coastal areas in proportion to the average depth of water in the vicinity. Subsequently the concept of hydro-isostasy has been developed further by Walcott (1972), Mörner (1972), Chappell (1974) and Clark *et al.* (1978). Walcott (1972) showed how relative sea level at great distances from the glaciated regions was affected by the meltwater loading of the ocean floor. Chappell's (1974) work concentrated in particular on the critical hinge zone between oceans and continents. He suggested an average depression of ocean basins of about 8 m and mean upward movement of continents of about 16 m relative to the centre of the earth in the last 7000 years. Deflection in the ocean margin hinge zone was shown to vary with continental shelf geometry and rigidity of the underlying lithosphere. In particular, he suggested that for broad "box section" shelves the deflection can be significant. A more sophisticated spherical viscoelastic earth model was used by Clark *et al.* (1978), which they considered explained a large proportion of the observed variance in the global sea-level record. They suggested that six zones can be recognized with distinctive Holocene sea-level signatures. Their results reinforce the earlier work in emphasizing that variations should be most rapid across the continental margins of the world, with the potential for sites as little as 100 km apart having contrasting relative sea-level curves.

Unfortunately there are few locations where there is opportunity to confirm the rapid cross-shelf variation predicted by these models. Deep indentations into continental margins or extensions of peninsulas and continental islands out into the sea provide opportunity on only a very localized scale generally covering less than 30% of the width of continental shelves. One type of continental margin which does provide such opportunity is a dominantly carbonate tropical shelf with coral reefs extending across the full shelf width. Chappell (1974) recognized this possibility

SHORELINES AND ISOSTASY
ISBN 0 12 652960 4

and suggested that the margin of eastern Queensland over which the Great Barrier Reef has developed provides an excellent field laboratory for testing the hydro-isostatic models, stating

> undoubtedly this large shelf province, with its variations of geometry and possibly of upper Earth parameters, will yield very useful evidence for testing hydroisostatic predictions, from its archipelagoes of 'dipsticks'.

However, the "dipstick" analogy is not a perfect one, for the reefs far from providing a framework over which the sea has regressed and transgressed, have themselves responded to the sea-level changes. A simple relationship between sea level and reef growth does not exist. This chapter examines the results and implications of a research programme which has involved shallow drilling (to a maximum of 4·3 m) and ^{14}C dating of windward reef flats on the central Great Barrier Reef (Fig. 1) between 14° 39′ S and 20° 45′ S (Hopley, 1977, 1982; Hopley and Harvey, 1981, 1982) combined with seismic refraction investigations to determine the depth of pre-Holocene foundation from which the modern reef is growing (Harvey, 1977a,b; Harvey *et al.*, 1978, 1979; Harvey and Hopley, 1982; Searle and Harvey, 1982). More recently, since 1980, an extended programme has drilled into the reef tops to depths as great as 30 m and ^{14}C dating of these cores is now providing further information on the complex relationship between sea-level change and reef growth. Together these programmes suggest that shelf deformation is a major factor in the development of reefs of the central Great Barrier Reef.

ADVANTAGES AND COMPLEXITIES IN THE USE OF CORAL REEF EVIDENCE

Coral reefs extend to the very edge of the continental shelf of north Queensland with water depths of over 1000 m occurring within 1 km of the outermost reefs (Hopley, 1982). Recent echo-sounding off Townsville suggests that reefal terraces may also exist on the shelf slope down to depths of at least 160 m. A general relationship between coral reefs and sea level has long been recognized although individual species of coral generally have a wide range of depths over which they will grow. Decrease in illumination at depths below 50 m produces a decline in calcification rates to below that at which reef growth (as opposed to individual colony growth) is not possible largely through control of the functions of the symbiotic unicellular algae, the zooxanthellae, contained within the polyp tissue. An upper limit of growth is provided by emergence at low tide. In general this limit on the Great Barrier Reef approximates Mean Low Water Springs (M.L.W.S.) whatever the tidal range (2–6 m). A variation of up to 25 cm may occur due to year to year variation in tidal behaviour, varying exposures to wave action and varying tolerances to exposure of different species.

In some circumstances, water may be held back at low tide and not drain to the same level as in the open ocean. Moating may occur at any intertidal level and some

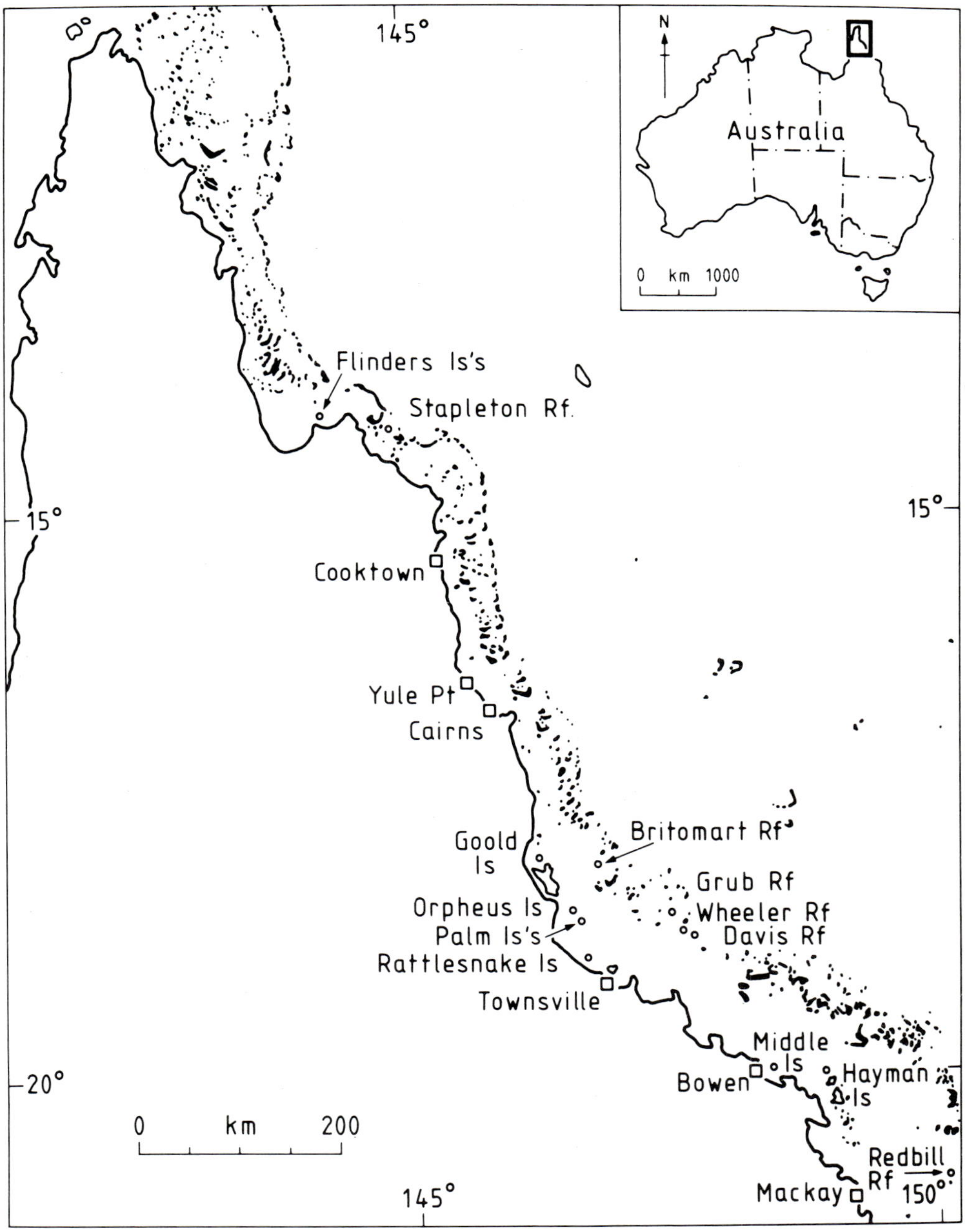

Figure 1. Central Great Barrier Reef, location of sites referred to in text.

lower-level moated pools may come into being only on the lowest Spring tides. Smaller moated pools form in depressions within the reef flat but much larger portions of reef flat are moated by a variety of damming agents. Most commonly these are rubble banks or ramparts of coral shingle deposited by storm waves around the reef flat margins. The efficiency of the dam may be enhanced by basal cementation of the shingle banks and/or filling of interstitial spaces by compaction and finer sediments. Water levels then drop to the level of the dam and show no further fall on the low tide. In the higher tidal range area of the central Great Barrier Reef, high algal rims commonly form dams around pools varying in size from a few m^2 to several km^2 (e.g. Ludington, 1979; Hopley, 1982).

The most important aspect of moating is that behind the dam the tide consistently falls to the same level unlike in the open water situation where variations between Spring and Neap tides may produce low water levels more than 2 m apart on the central Great Barrier Reef. The upper level of coral growth is thus firmly fixed and the result is the typical microatoll form consisting of flattened disc-shaped coral colonies with living vertical sides and dead centres often covered by crustose coralline algae (Scoffin and Stoddart, 1978; Stoddart and Scoffin, 1979). Microatolls obviously produce problems for determination of past sea levels, particularly in areas of large tidal range. However, whilst individual colonies may be found at levels up to Mean High Water Neaps (M.H.W.N.), such examples are very unusual and restricted. Large fields of moated microatolls do not normally grow to levels in excess of Mean Low Water Neaps (M.L.W.N.), and in emerged reef exposures where the former geometry of the reef top can be established with microatolls extending over 100 m^2 or more, their upper level can be taken conservatively as M.L.W.N. (for discussion see McLean *et al.*, 1978).

Larger problems exist in the interpretation of data from reef cores. There is little doubt that a reef borehole can indicate very clearly the nature of reef growth at a particular point on a reef (see below, Fig. 5). Sequential dating of carefully selected and prepared samples is necessary. Unfortunately, with the possible exception of some highly porous low energy Acroporid and Pocilloporid reefs from the Caribbean and eastern Pacific (Adey, 1975, 1978; Glynn and Macintyre, 1977) the upward growth rate of reefs appears to be slower than the rise of sea level, even during the latter part of the Holocene transgression when the foundations of the modern reefs were being inundated. The reef growth curve therefore is not a sea-level curve. Modern calcification studies (e.g. Kinsey and Davies, 1979) show that different zones on reefs may calcify at contrasting rates. The rapid transport of sediment in the reef system may also complicate the pattern of reef growth. Different parts of the reef may have accreted at highly contrasting rates. This has been confirmed by the drilling of reefs in the Bunker-Capricorn Group (Davies and Marshall, 1979, 1980). Davies and Marshall (1979) have prepared a hypothetical growth curve for a reef, which shows an early period of slow accretion when difficulties in colonizing the Pleistocene surface were encountered, a period of accelerated growth, and a period of decreasing growth as the reef approaches a stabilized sea level in which erosion in the surf zone would decrease actual accretion rates. Thus even the basal dates of the Holocene reef may not represent sea level at the time as a delay may have occurred between the inundation of the antecedent surface, when high turbidity may have existed, and coral colonization. However comparison of dated sequences from both Australia and the Caribbean with sea-level curves previously established by other means suggest that this delay may be minimal (e.g. Hopley *et al.*, 1978). Therefore it may be possible to use basal Holocene reef dates to establish a sea-level curve in a particular area in a way similar to the use of basal peat dating described by Bloom (1964). A relative relief of up to 14·5 m has been established on the pre-Holocene surface of individual reefs of the Great Barrier Reef (Harvey and Hopley, in press) whilst the total spread of depths ranges from surface exposures in the Torres Strait to in excess of −30 m in the central Reef. Theoretically at least, this should mean that dated sea level may be established for this total depth range. Fringing reefs growing from bedrock foundations may similarly provide a range of basal dates from different depths up the bedrock slope to establish a transgression rate.

Whilst corals and coral reefs are not the clear indicators of past relative sea levels that they are sometimes suggested to be, they do contain important information. Moreover, as large biogenic structures they provide a wealth of material for radiometric dating. However, the complex chemistry of the reef, especially when it is subaerially exposed, requires that great care and caution be applied to the choice, preparation and interpretation of materials and results (see reviews by Chappell *et al.*, 1974, 1978; Veeh and Green, 1977). ^{14}C results reported in this chapter are given in convential ^{14}C years except where otherwise stated. An environmental reservoir correction of -450 ± 35 years for these dates (Gillespie and Polach, 1979) and corrections for past variations in atmospheric ^{14}C concentration may be made using the calibration curve of Clark (1975) to 6500 B.P. and Barbetti's (1980) estimate for ages prior to 6500 B.P.

INTERPRETATION OF RESULTS FROM THE CENTRAL GREAT BARRIER REEF

Several lines of research over the past decade have all produced results which are strongly suggestive of a cross-shelf warping in central and northern Queensland but with significant regional variations in the nature and amount of movement.

Inner-shelf emergence *vs* outer-shelf submergence

Numerous authors over a period of 150 years of investigations have described features attributable to higher sea levels along the coastline and high islands adjacent to the Great Barrier Reef (see Hopley, 1974, 1978, 1982 for reviews). However, problems arise in the interpretation of much of the evidence, as discussed by Hopley (1982), largely as a result of the relatively common occurrence of high energy tropical cyclones and accompanying storm surges which may raise water levels by more than 3 m. Much depositional evidence is thus of dubious validity. High levels of cementation have also been equated with modern intertidal beach rock to substantiate emergence of up to 4 m (Hopley, 1968, 1971, 1975). However, problems remain as to the exact nature of the cementation process and whilst this author still believes that emerged beach rocks are valid evidence (possibly equated

TABLE 1
Emerged Holocene fringing reefs in North Queensland

Location	Emergence (m)	Age (^{14}C Ka)	Reference
Flinders Is's.	0·85–1·05	5·1–5·7	Chappell *et al.* (1982)
Yule Point	1·1–1·3	4·9	Bird (1971)
Goold Is.	1·0–1·4	5·7–6·3	Chappell *et al.* (1982)
Palm Is's.	0·9–1·1	5·3–5·5	Chappell *et al.* (1982)
Orpheus Is.	1·0	5·3[a]	Hopley *et al.* (in press b)
Rattlesnake Is.	0·8–1·2	5·3–5·5[b]	Hopley *et al.* (in press b)
Middle Is.	0·5–1·0	5·2[c]	Hopley (1975)

[a]SUA-1678, 5320 ± 100 B.P.; [b]GaK-7688, 5530 ± 130 B.P., GaK-9294 5380 ± 120 B.P.; [c]GaK-5215 5310 ± 115 B.P., GaK-4894 5290 ± 120 B.P.

to the highest astronomical tidal level rather than M.H.W.S. as quoted previously, which would reduce the amount of emergence by approximately 1 m), they are not used as primary evidence here.

Emerged reefs, however, are relatively common on the inner shelf and, even allowing for moating, provide strong evidence for emergence of between 1 and 2 m along large extents of the Queensland coast (Table 1). Nearly all are in the form of emerged microatolls and have been equated with the level of modern microatoll development. The period of highest relative sea level appears to have been about 5000 B.P., extending to approximately 3500 B.P. In these same areas modern sea level appears to have been achieved first prior to 6000 B.P. Where emerged beach rocks occur in close proximity the level of emergence is 1–2 m higher and the age of this emergence is 500–1000 years younger.

In contrast no site on the outer shelf has provided any evidence of a sea level higher than present. As discussed below there are many reefs which had reached the level of modern sea level and had commenced reef flat development by the time the emerged reefs of the inner shelf were being constructed. Former reports of emerged reefs in the outer barrier reefs (see for example Fairbridge, 1950, 1967) have all proved erroneous, the "emerged" reefs consisting of reef blocks thrown onto the reef flat by cyclonic storms. It also seems highly unlikely that any former reef developed to a higher relative sea level, has been entirely removed by erosion along the entire length of the Great Barrier Reef as has been implied by Maxwell (1968).

Thus there is the contrast between clear evidence for inner-shelf emergence and no such evidence on the outer shelf. Evidence for mid-shelf emergence from the reefs is minimal. The Great Barrier Reefs extend over 70% of the outer-shelf width north of Cairns but over a progressively smaller percentage of the shelf width to the south, although this is compensated for by an increase in absolute shelf width, the end result being an increase in the distance that the innermost reefs lie from the shoreline. North of Cairns the low wooded island reefs retain evidence of higher sea levels as the result of the inner-shelf location (McLean *et al.*, 1978). South of Cairns the innermost reefs of the outer barrier have a mid-shelf location and, like their outer-shelf counterparts, the majority do not show evidence for any emergence. One exception is Redbill Reef situated 70 km from the mainland on the 200 km wide shelf off Mackay. This reef is the innermost on this part of the Barrier Reef and its continental affinities are indicated by a small 1·6 ha outcrop of granite in Redbill Island. The reef has a massive algal rim reaching Mean Sea Level (m.s.l.). Drilling into this resulted in the recovery of *in situ* head corals 1·2–1·6 m above the level of modern open water coral growth and provided radiocarbon ages of 4570 ± 150 B.P. to 3920 ± 150 B.P. (see Hopley *et al.*, 1982, for discussion). No other mid-shelf or outer-shelf reefs of the central and northern Great Barrier Reef have provided similar evidence but on the southern reef Flood *et al.* (1979) describe evidence of 0·7 m of mid-Holocene emergence on Lady Elliott Island.

Reef top age and reef morphology

If contrasting relative sea-level curves apply to different parts of the Great Barrier Reef as the result of cross-shelf warping, then it is possible that some indication of this will be given by variations in reef top ages and in the degree of reef flat development which is dependent on the time the reef has been at modern sea level. However, the post-glacial sea-level rise has been at an average rate of *c.* 11 mm yr^{-1} whilst the recorded upward growth rates of reefs are generally about half this rate. Reefs have thus lagged behind the rise of sea level, even though the majority of modern reefs have developed from Pleistocene reefal foundations which in the Great Barrier Reef have been shown to have depths of between 5 and 25 m below modern sea level.

On the Great Barrier Reef drill cores have indicated mean accretion rates for a number of reefs of 2–7 mm yr^{-1} (Thom *et al.*, 1978; Hopley *et al.*, 1978; Davies and Marshall, 1979). If such accretion rates are taken and applied to the standard sea-level curve of eastern Australia of Thom and Chappell (1975), the depth of the Pleistocene foundation will determine the time at which the reef will reach modern sea level and when reef flat construction will be initiated. However, if the sea-level curve were more akin to that of Neumann (1971, 1972), which is generally accepted for the Caribbean, then growth from most antecedent platform levels will result in the reef catching up with the continued sea-level rise and subsequent upward growth will be at a rate determined by the slower sea-level rise.

In order to test the hypothesis implicit in both the hydro-isostatic models and the relationships between sea-level curves and reef growth, a programme was initiated in 1975 in the central Great Barrier Reef region which involved:

(i) shallow drilling of windward margins of 24 reefs to a maximum of 4·3 m and subsequent ^{14}C dating of *in situ* corals from the cores;

(ii) determination of the shallow depth of the pre-Holocene foundation in the area of the reef by seismic refraction techniques. This work was largely carried out by N. Harvey (1977a,b, 1980; Harvey *et al.*, 1979; Harvey and Hopley, 1982).

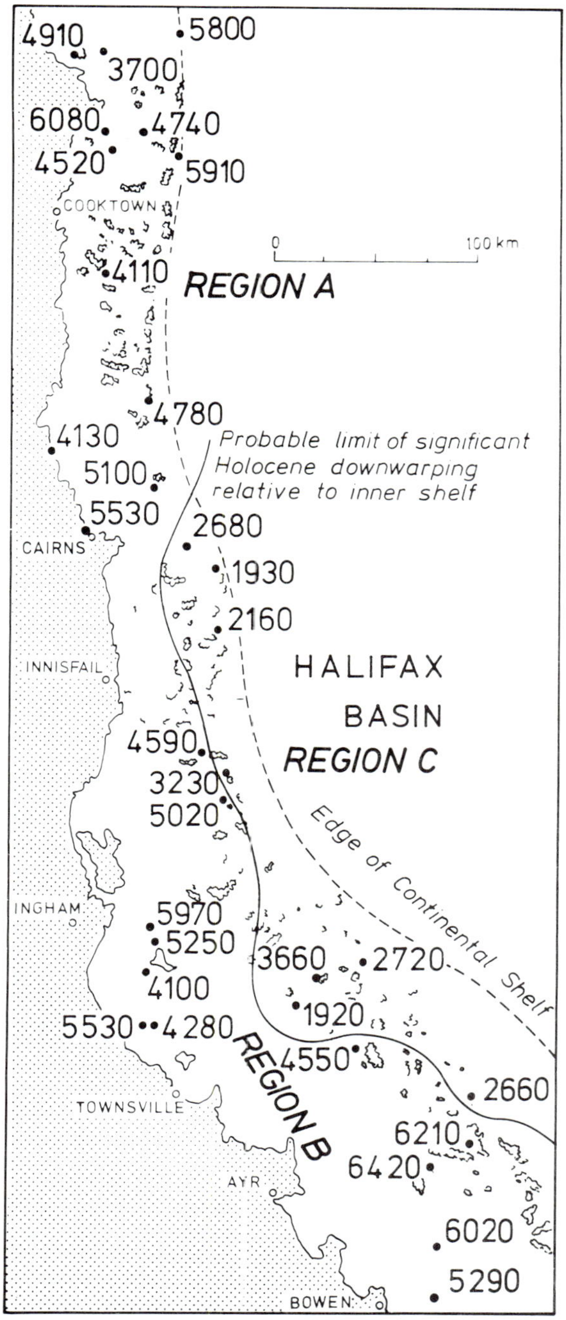

Figure 2. Oldest radiocarbon ages within 2 m of the surface of reefs on the central Great Barrier Reef and oldest age relating to modern sea level or higher on the adjacent mainland. The regions discussed in the text are indicated (based on Hopley, 1982, Fig. 6.11).

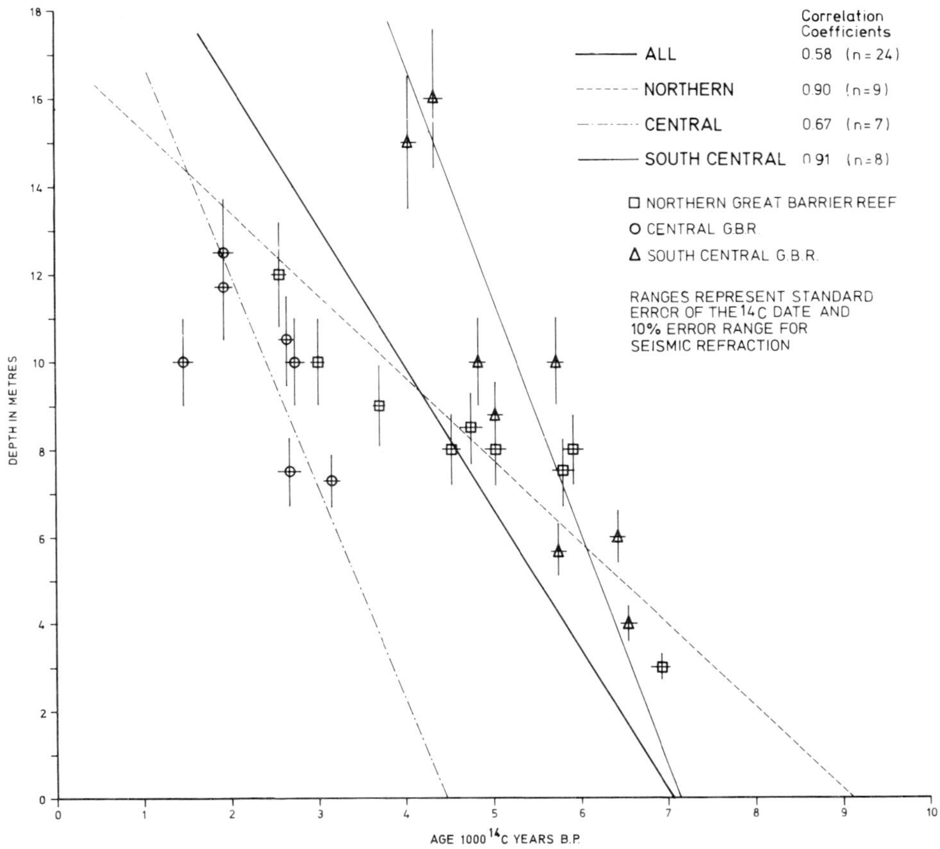

Figure 3. Plot of oldest radiocarbon age of windward reef cap against shallowest recorded depth to the pre-Holocene platform (from Hopley, 1982).

Figure 2 summarizes the data on reef top ages indicating the oldest ^{14}C age for the upper 1·5 m of the reef surveyed (with some inner-shelf dates added). Ages range from 1920 B.P. to 6420 B.P. with ages younger than 3800 B.P. coming exclusively from the outer shelf south of Cairns. If all reefs had the same accretion rate and if the same relative sea-level curve were to apply to the whole area then a plot of oldest age against shallowest depth to the pre-Holocene should produce a straight line plot, the slope of which should represent the accretion rate. The actual plot (Fig. 3) shows considerable scatter though there is sufficient linearity to produce a correlation coefficient of 0·58 (n = 24) significant at the 1% level with a regression line the slope of which represents a highly realistic accretion rate of 3·2 mm yr^{-1}.

The scatter of the plot may result from sampling problems but the presence of a regional variation in Holocene sea-level behaviour and/or reef accretion rates would produce a similar dispersal. To test for a spatial variation the data was sub-divided into three sets based on regional differences in shelf morphology, reef morphology and the previously recognized reef top ages. The regions are:

(A) the northern Reef north of Cairns with shallow and narrow (50 km) shelf, very steep continental slope and reef morphology characterized by a continuous line of outer ribbon reefs, large mid-shelf reefs and small inner reefs with complex low wooded islands (Stoddart *et al.*, 1978). Several reef top ages are in excess of 5500 B.P. in this region;

(B) the mid-shelf zone from Innisfail south with a similar range of reef top ages but with larger more irregular reefs and very little development of reef islands. The continental shelf widens to over 125 km in this area;

(C) the outer shelf south of Cairns characterized by younger reef top ages, smaller reefs and numerous submerged shoals.

Far better correlations (Fig. 3) are obtained for region A ($r = 0{\cdot}90$, $n = 9$) and region B ($R = 0{\cdot}91$, $n = 8$). Both are significant at the 1% level and represent highly contrasting accretion rates of 1·9 mm yr^{-1} and 5·5 mm yr^{-1} respectively. However, these rates are in accord with those for dated cores within each area, i.e. Stapleton, 2 mm yr^{-1} (Thom *et al.*, 1978) in region A and Britomart Reef, 4–6 mm yr^{-1} (Johnson *et al.*, in preparation) and Hayman, 4–5 mm yr^{-1} (Hopley *et al.*, 1978) in region B.

The correspondence of results from regions A and B highlights the discrepancy of region C, where the correlation coefficient is only 0·67 ($n = 7$), which is not significant but does suggest an accretion rate midway between those of adjacent regions of 4·7 mm yr^{-1}. A wide range of depths for the pre-Holocene antecedent reefal foundations have been recorded from this area which should result in a spread of reef top ages. Instead the oldest ^{14}C ages from these reefs cluster within the relatively narrow time span of 1920–2720 B.P. Such a result is exactly what would occur if a Caribbean type sea-level curve had operated in the region, i.e. reefs growing from a variety of platform depths would catch up with the continuously rising sea level at various times, but would all then accrete at a similar rate which is determined by the sea-level rise.

Figure 4 shows four transects across the shelf between 15 and 19°S, showing the highest level of *in situ* coral on each reef together with ^{14}C ages selected to indicate the highest levels of particular ages from the shallow cores. The seawards tilt is clearly seen with some inner-shelf emergence indicated. However, the degree of tilt in all except the northernmost transect is substantially increased if a *c.* 5000 B.P. isochrone is taken, the depression of this isochrone occurring only on the outermost shelf at 18°S, probably across the entire shelf at 18° 45′ S and again on the outer shelf only at 19°S. Such isochrones are indicative of levels reached by reef growth, not of sea level.

Reef morphology is consistent with major warping (see Hopley, 1982, for further discussion). The younger reef tops of the outer shelf between Cairns and Bowen have luxuriant living coral veneers, lack extensive algal pavement and are devoid of cays, a contrast to the older reefs especially to the north which are largely rubble covered and have wide algal zones and numerous vegetated and unvegetated islands.

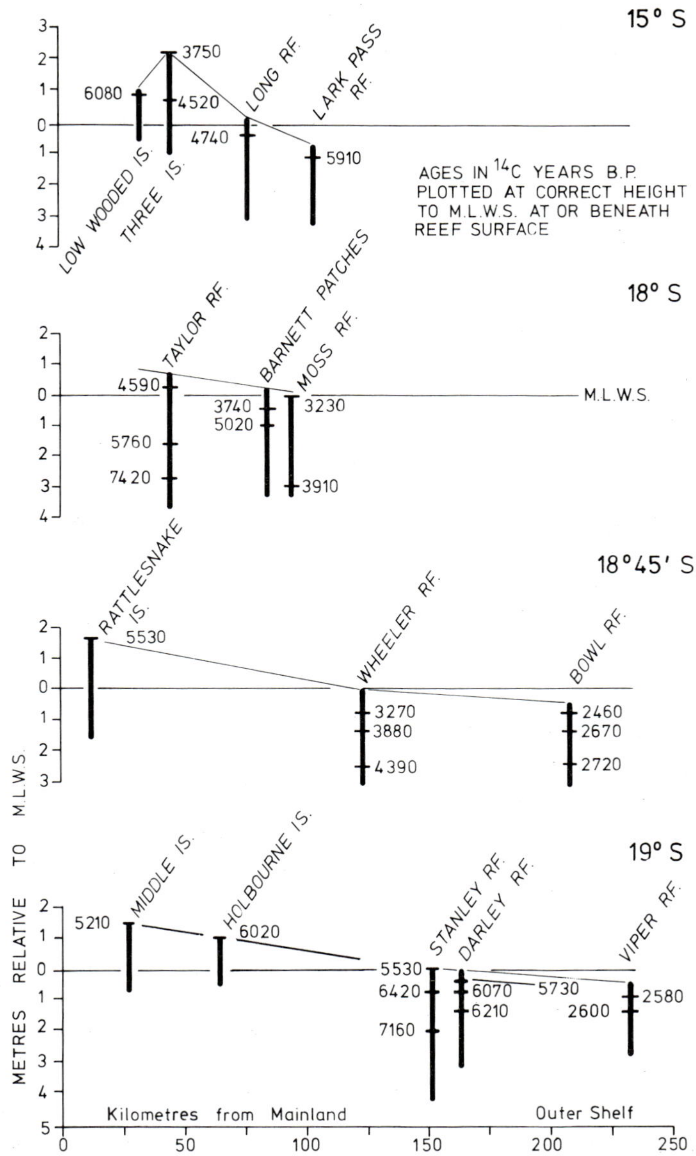

Figure 4. Maximum height of *in situ* corals across four shelf transects clearly showing decline in height seaward. Radiocarbon ages are shown (from Hopley, 1982).

The nature of the reef cap also varies: the older reefs, at modern sea level for up to 6000 years, being consolidated and cemented in contrast to the more open and porous framework of the younger reef tops (see Hopley, 1982, Fig. 8.7). The clearest illustration of the effects of the downwarping are seen in the gross morphology of the reefs. A reef classification scheme discussed by Hopley (1982) suggests that young reefs are patchily developed over pre-Holocene highs (cf. the resorbed reefs of Maxwell (1968) which he suggested in contrast were the oldest reef forms), and mature reefs are initially crescentic forms (with maximum reef flat development on the windward margin) and later with lagoonal morphology. Such reefs may be found in all areas of the Great Barrier Reef where they develop from deeper pre-Holocene foundations. However, the senile form, which is suggested as a planar reef in which former lagoons are infilled by the horizontal transfer of carbonate from the productive margins, resulting in a flat sediment-covered reef top, require sufficient time at a "stable" sea level to evolve. The absence of both mature lagoonal and senile planar reef types from the Reef from Cairns to south of Townsville is clearly in keeping with the patterns suggested above. Similarly, the presence of extensive submerged reefal shoals on the outer edge of the shelf in this region (Searle *et al.*, 1980, 1982), is also consistent with significant marginal downwarping.

Deeper cores and the depth of the pre-Holocene foundation

The deeper drilling programme initiated in 1980 has produced preliminary results which do not contradict the pattern of cross-shelf subsidence suggested by other evidence. Unfortunately no results are available from outer-shelf reefs but a comparison can be made between the growth patterns of inner-shelf fringing reefs (Orpheus and Rattlesnake Islands) and a mid-shelf reef (Wheeler) off Townsville (Fig. 5) where reef top ages in particular suggest that the cross-shelf subsidence is at its greatest.

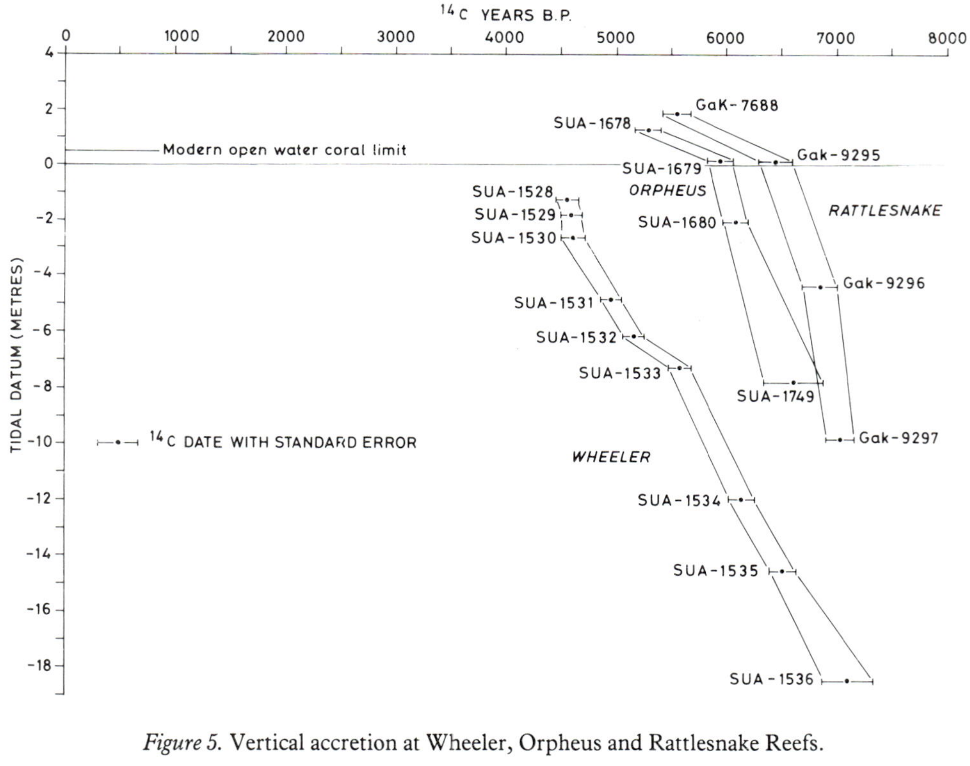

Figure 5. Vertical accretion at Wheeler, Orpheus and Rattlesnake Reefs.

Although the structure of the inner-shelf fringing reefs is quite different to that of the outer reefs, in particular containing extensive basal terrigenous sediments (see Hopley *et al.*, in press), the reef flats of both Orpheus and Rattlesnake were being constructed by at least 6000 B.P., the Pleistocene basement of the Holocene reef lying at 10–12 m depth and being inundated by at least 7000 B.P. In contrast, at about the same time growth was commencing on Wheeler Reef at a depth greater than 18 m and growth for any particular level has been 1000–1500 years later on Wheeler than on the inner fringing reefs. Further results are required to confirm this pattern and, because of problems outlined above regarding the possible delays in colonization of the reefal foundations and lag in the growth of reefs behind sea-level rise, caution is required in the interpretation. However, the discrepancy in age of reef tops of inner and outer shelves is now confirmed for greater depths and, in particular for the initiation of growth.

The deeper drilling programme has also provided confirmation of the depth of the pre-Holocene platform beneath modern reefs indicated by seismic survey. It has generally been presumed that the reef foundations have been formed by prior reef growth at past high sea levels, the most recent of which, prior to the Holocene, was the last interglacial *c.* 125 000 B.P. Worldwide evidence suggests that this was slightly higher than present sea level and north Queensland mainland coastal evidence is in agreement (see Hopley, 1982 for discussion). However, only in Torres Strait is the pre-Holocene at or near the surface of present reefs and over the area of the Great Barrier Reef discussed here it lies at depths generally between 5 and 30 m. Three possible causes for the depth can be discussed:

(i) slow last interglacial reefal growth did not allow reefs to fully develop and reach the sea level of that time. However, evidence suggests that sea water temperatures would have been at least as high as present (see, for example, Marshall and Thom, 1976) and the length of time of the last interglacial should have been sufficient for extensive reef flat development;

(ii) erosion has lowered the original level of the last interglacial reefs. Extrapolating from Trudgill's (1976) figures from Aldabra it is possible for there to have been up to 26 m of erosion in the 100 000 years of subaerial exposure (see Hopley, 1982, chapter 7 for discussion);

(iii) subsidence on the shelf has taken place which has lowered the level of the last interglacial reefal surface on the outer shelf as compared to the mainland and inner shelf where beach deposits related to sea level of the last interglacial are exposed.

It is possible that all these mechanisms are involved. However, the pattern indicated would support a degree of cross-shelf subsidence at least between Townsville and Cairns. North of Cairns the depth to the pre-Holocene on the outer shelf is approximately 10 m (Harvey, 1977). Near Townsville mid-shelf reefs including Grub, Wheeler, Davies (see Grimes, 1982) and Britomart (see Johnson *et al.*, in preparation), have depths recorded by drilling and/or seismic survey in the range of 18–24 m.

APPROXIMATION OF THE AMOUNT OF WARPING

The location of *in situ* coral does not accurately fix a contemporaneous sea level. At best it indicates that M.L.W.S. (the approximate upper limit of open water coral growth) was at or above that level. However, the reef flat dates obtained can act as a guide for the construction of sea-level curves for each of the three designated regions (Fig. 6). In region B the deeper cores from a number of reefs including Hayman Island (Hopley *et al.*, 1978) allow extrapolation of the curve to beyond 9000 years. However, as a first approximation it has been possible to extend the curve in all regions back to the early Holocene. This has been done via the following reasoning:

(i) mean reef accretion rates, verified by drilling have been established in each region;
(ii) the reef flat age and depth of antecedent platform are known;
(iii) by using the accretion rate to calculate the time needed for the reef thickness above the antecedent platform to accumulate, it is possible to calculate also the time that accretion commenced on the platform.

As this would necessarily follow the inundation of the platform the resulting date gives the earliest possible time for the attainment of sea level at the platform depth. Complications may arise from any delay in the colonization of the older reef foundation as discussed by Davies and Marshall (1979).

None the less the results appear to be reasonable, particularly when only reefs which are away from the margins of the designated regions are used (Fig. 6). Regions A and B have very similar curves which closely correspond to that of Thom and Chappell (1975). Modern sea level was achieved by about 6500 B.P. in both areas. In region B it is dated at 7400 B.P., less than 4 m below present. In both regions it was approximately 15 m below present 9000 years ago. The compatability of these results highlights the contrast with region C where modern sea level was first achieved by about 2700 B.P. and was 10–12 m below present 5000 B.P.

Although it is stressed that these results are first approximations (being tested by the deeper drilling programme commenced in 1980) they can be expressed regionally by contouring the dated or estimated level of the *c.* 5000 B.P. shoreline (Fig. 7). This suggests that only a minor amount of warping has taken place on the narrow shelf north of Cairns, but to the south the degree of outer shelf subsidence may be in excess of 6 m.

DISCUSSION

The idea of shelf warping with subsidence on the outer margin is not a new one. Hedley (1925) suggested that the reef between 17° and 18° was "recovering from a drowning movement" (p. 159) and Spender (1930) commented on the change in reef morphology off Cairns and noted that outer-shelf reefs were considerably lower

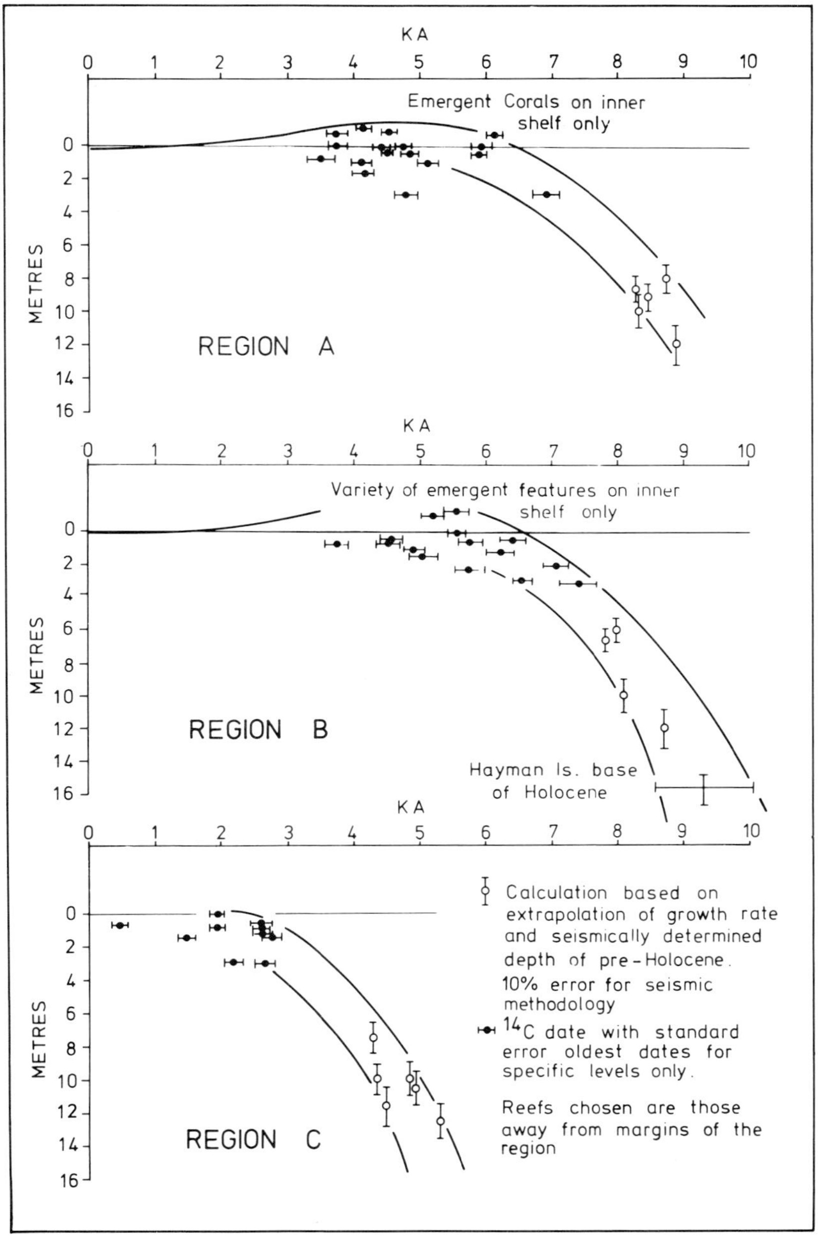

Figure 6. Possible sea-level curves for the three designated regions based on reef top ages, accretion rates and depth of pre-Holocene reefal foundations.

than those of the inner shelf. Fairbridge (1950) made similar observations, and, resurrecting the original comments of Jukes (1847), noted that the reef morphology between Cairns and Townsville "suggests very rapid subsidence locally" (p. 391) and that the reefs "are more juvenile than in the northern sectors" (p. 364).

The hydro-isostatic model of Chappell (1974) suggests that subsidence will take

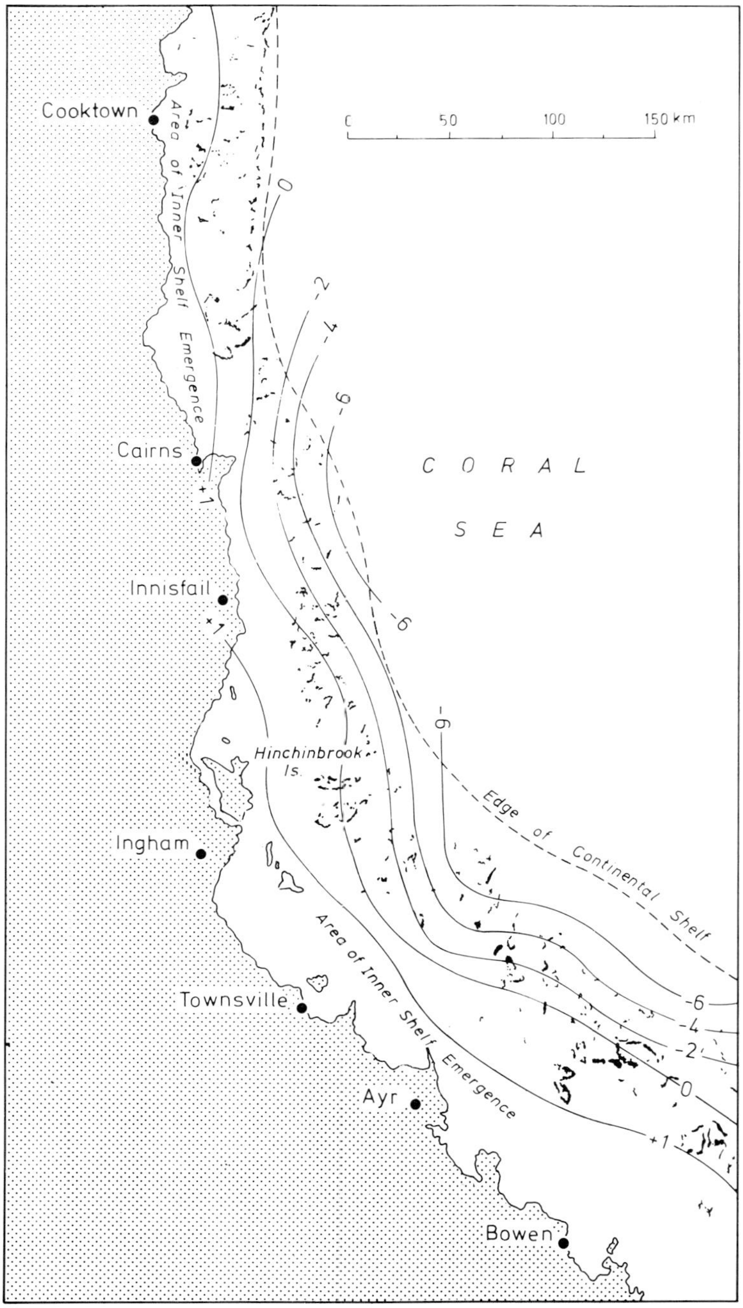

Figure 7. Interpreted pattern of Holocene shelf warping on the central Great Barrier Reef. Isobase values in metres.

place on the outer shelf with the possibility of inner-shelf uplift dependent on the location of the continental/oceanic hinge line. Thom and Chappell (1978) applied this model to the area of the Great Barrier Reef under discussion. Dependent on the nature of the lithosphere, they indicated a maximum of only 0·5 m subsidence of the 6000 B.P. shoreline on the outer shelf north of Cairns but up to 3 m opposite Townsville. Inner-shelf emergence ranged from 0·5 m in the north to 1·5 m near Townsville. More recently Chappell *et al.* (1982) have modelled the 5000 year isobase for north Queensland and compared it to evidence on the mainland and nearshore islands which closely fits the model. Along the entire coast north of Townsville the predicted pattern is one of *c.* 1 m emergence on the inner shelf compared to up to 1 m of submergence on the outer shelf. The northern Reef at 15°S closely fits the hydro-isostatic calculations but further south a considerably greater degree of warping is suggested. If the relative sea-level curves of Fig. 6 are accepted then the 5500 B.P. isobase may lie at depths of up to 14 m on the outer shelf opposite Townsville. This figure may be extreme but the degree of subsidence certainly appears to be greater than the 1–3 m suggested by the models. Vertical reef accretion of the order of 4·7 mm yr^{-1} as implied from Fig. 3 should have resulted in reefs of this area lagging behind their inner shelf and northern counterparts growing from similar antecedent platform depths by less than 560 years. Instead an age difference of up to 3000 years is indicated.

A contribution to the subsidence by regional tectonic downwarping may be indicated. Notably, the shelf edge area where maximum subsidence is suggested coincides closely with the known boundaries of the shelf marginal Halifax Basin (Fig. 2) which is known to have been active at least in Quaternary times (Benbow, 1980). Indeed, the reef top ages and related morphology may provide an accurate definition of the basin margins over the shelf.

Speculation on a tectonic element in the variation of height of mid-Holocene shorelines on the inner shelf has already been made (Hopley, 1978) with at least one major dislocation in the eastern side of Broad Sound identified (Cook and Mayo, 1977). It is possible that stresses produced beneath the inner shelf by hydro-isostatic movements may be released along old established fault lines, as has been shown for areas recovering from glacial loading (e.g. Sissons, 1972; Mörner, 1978). If the downwarping on the outer shelf were entirely due to regional tectonics then the rates of subsidence appear to be considerable and if extrapolated back through the late Pleistocene could place the level of the last interglacial far below the level of the pre-Holocene unconformity as presently suggested by drilling and seismic refraction survey. However, 125 000 years of subsidence may have allowed the addition of interstadial reef caps (see, for example, the sea-level curve of Bloom *et al.*, 1974), the most recent of which may have been only *c.* 30 000 B.P. Such an interpretation would be in keeping with the greater depths of the pre-Holocene platform beneath the modern reefs of the outer shelf but recovery of suitable materials for uranium series dating from the antecedent surface is required for confirmation. A complicating factor in the present depth of this surface beneath the reef which has achieved little attention to date, is the effect of unloading of the shelf during the period of sea-level regression. Only a more complete knowledge of the amounts of recoverable versus non-recoverable warping and the length of relaxation times will provide complete understanding of the vertical displacement of the last interglacial reefs of the Great Barrier Reef. Although the data presented here are largely circumstantial rather than conclusive, a degree of warping normal to the trend of the continental shelf of eastern Queensland appears to have taken place which may result from both tectonic and isostatic mechanisms. It appears to have had a significant influence on the development of the modern reefs of the Great Barrier Reef.

ACKNOWLEDGEMENTS

Research funding for this research project was provided by the Australian Research Grants Committee and the Australian Marine Sciences and Technologies Advisory Panel- Funding Advisory Panel. Radiocarbon dating was carried out by Dr M. Barbetti of Sydney University Radiocarbon Laboratory and Professor K. Kigoshi of Gakushuin University.

REFERENCES

Adey, W. A. (1975). *Atoll Res. Bull.* **187**, 1-67.
Adey, W. H. (1978). *Science* **202**, 4370, 831-837.
Barbetti, M. (1980). *Radiocarbon* **22**, 192-199.
Benbow, D. D. (1980). *APEAJ* **20**, 159-175.
Bird, E. C. F. (1971). *Australian Geographical Studies* **9**, 107-115.
Bloom, A. L. (1964). *J. Sedim. Petrol.* **34**, 599-603.
Bloom, A. L. (1967). *Bull. Geol. Soc. Am.* **78**, 1477-1494.
Bloom, A. L., Broeker, W. S., Chappell, J., Matthews, R. K. and Mesolella, A. J. (1974). *Quaternary Research* **4**, 185-205.
Chappell, J. (1974). *Quaternary Research* **4**, 429-440.
Chappell, J., Broeker, W. S., Polach, H. A. and Thom, B. G. (1974). *Proceedings, Second International Coral Reef Symposium.* **2**, 563-571.
Chappell, J., Thom, B. G. and Polach, H. A. (1978). *In* "Coral Reefs Research Methods" (D. R. Stoddart and R. E. Johannes, eds), pp. 81-91. UNESCO, Paris.
Chappell, J., Rhodes, E. G., Thom, B. G. and Wallensky, E. (1982). *Marine Geol.* **49**, 81-90.
Clark, J. A., Farrell, W. E. and Peltier, W. R. (1978). *Quaternary Research* **9**, 265-287.
Clark, R. M. (1975). *Antiquity* **49**, 251-266.
Cook, P. J. and Mayo, W. (1977). *Bur. Min. Res. Geol. Geophys. Bull.* **170**, 206 pp.
Davies, P. J. and Marshall, J. F. (1979). *Search* **10**, 276-279.
Davies, P. J. and Marshall, J. F. (1980). *Nature* **287**, 37-38.
Fairbridge, R. W. (1950). *J. Geol.* **58**, 330-401.
Fairbridge, R. W. (1967). *In* "Landform Studies from Australia and New Guinea" (J. N. Jennings and J. A. Mabbutt, eds), pp. 386-417. ANU press, Canberra.
Flood, P. G., Harjanto, S. and Orme, G. R. (1979). *Qd. Govt. Min. J.* Sept. 1979, 444-447.

Gillespie, R. and Polach, H. A. (1979). *Proceedings, 4th International Radiocarbon Conference* 404-421.

Glynn, P. and Macintyre, I. G. (1977). *Proceedings, Third International Coral Reef Symposium* **2**, 251-259.

Grimes, K. G. (1982). *Rec. Geol. Sur. Qd.* 1982/18.

Harvey, N. (1977a). *Proceedings, Third International Coral Reef Symposium* **2**, 45-51.

Harvey, N. (1976). *Catena* **4**, 333-339.

Harvey, N. (1980). Seismic Investigations of a Pre-Holocene Substrate beneath Modern Reefs in the Great Barrier Reef Province. Unpublished PhD thesis, James Cook University of North Queensland.

Harvey, N. and Hopley, D. (1982). *Proceedings, Fourth International Coral Reef Symposium* **1**, 549-554.

Harvey, N., Davies, P. J. and Marshall, J. F. (1978). *Bur. Min. Res. Geol. Geophys. Rec.* 1978/96, 9 pp.

Harvey, N., Davies, R. P. and Marshall, J. F. (1979). *Bur. Min. Res. J. Austr. Geol. and Geophys.* **4**, 141-147.

Hedley, C. (1925). *Reports, Great Barrier Reef Commissions* **1**, 157-160.

Hopley, D. (1968). *Aust. J. Sci.* **31**, 122-123.

Hopley, D. (1971). *Z. Geomorph.* **15**, 371-389.

Hopley, D. (1974). *Proceedings, Second International Coral Reef Symposium* **2**, 551-562.

Hopley, D. (1975). *In* "Geographical Essays in Honour of Gilbert J. Butland" (I. Douglas, J. E. Holmes and J. J. Pigram, eds), pp. 51-84. Department of Geography, University of New England, Armidale.

Hopley, D. (1977). *Proceedings, Third International Coral Reef Symposium* **2**, 23-28.

Hopley, D. (1978). *Phil. Trans. R. Soc.* **A291**, 159-166.

Hopley, D. (1982). "The Geomorphology of the Great Barrier Reef: Quaternary Development of Coral Reefs". John Wiley-Interscience, New York.

Hopley, D. and Harvey, N. (1981). *Proceedings, Conference on Environmental Engineering, Australian Institute of Engineering* **81/6**, 107-112.

Hopley, D. and Harvey, N. (1982). Radiocarbon ages and morphology of reef tops in the Great Barrier Reef between 140° 39′ S and 20° 45′ S: indicators of shelf neotectonics. *Proceedings, Fourth International Coral Reef Symposium* **1**, 523-530.

Hopley, D., McLean, R. F., Marshall, J. and Smith, A. S. (1978). *Search* **9**, 323-325.

Hopley, D., Davies, P. J., Harvey, N. and Isdale, P. J. (1982). The geomorphology of Redbill Reef, central Great Barrier Reef. *Fourth International Coral Reef Symposium* **1**, 541-547.

Hopley, D., Slocombe, A. M., Muir, F. and Grant, C. (in press). Nearshore fringing reefs in north Queensland. *Coral Reefs.*

Johnson, D. P., Cuff, C. and Rhodes, E. (in prep.). Reef sequences and geochemistry, Britomart Reef, Central Great Barrier Reef, Australia.

Jukes, J. B. (1847). "Narrative of the surveying voyage of HMS Fly, commanded by Capt. F. P. Blackwood in Torres Strait, New Guinea and other islands of the Eastern Archipelago, during the years 1842-1846 together with an excursion into the interior of the eastern part of Java", 2 vols. London, T and W Boone.

Kinsey, D. W. and Davies, P. J. (1979). *In* "Biogeochemistry of Mineral Forming Elements" (P. Trudingar and D. Swaine, eds), pp. 131-162. Elsevier Press, Oxford.

Ludington, C. A. (1979). *Aust. J. Mar. Fisheries Res.* **30**, 425-430.

McLean, R. F., Stoddart, D. R., Hopley, D. and Polach, H. (1978). *Phil. Trans. R. Soc.* **A291**, 167-186.

Marshall, J. F. and Thom, B. G. (1976). *Nature* **263**, 120-121.
Maxwell, W. G. H. (1968). "Atlas of the Great Barrier Reef". Elsevier, Amsterdam.
Mörner, N.-A. (1972). *Tellus* **42**, 586-592.
Mörner, N.-A. (1978). *Geology* **6**, 41-45.
Neumann, A. C. (1971). *Quaternaria* **14**, 41-43.
Neumann, A. C. (1972). *American Quaternary Association Second National Conference, Abstracts* 41-44.
Scoffin, T. P. and Stoddart, D. R. (1978). *Phil. Trans. R. Soc.* **B284**, 99-122.
Searle, D. E. and Harvey, N. (1982). *Marine Geol.* **46**, M9-M16.
Searle, D. E., Harvey, N. and Hopley, D. (1980). *Geol. Surv. Qld. Rec.* 1980/23.
Searle, D. E., Harvey, N., Hopley, D. and Johnson, D. P. (1982). Significance of results of shallow seismic research in the Great Barrier Reef province between 16° 10′ S and 20° 05′ S. *Proceedings, Fourth International Coral Reef Symposium* **1**, 531-539.
Sissons, J. B. (1972). *Trans. Inst. Br. Geogr.* **55**, 145-160.
Spender, M. R. (1930). *Geogrl. J.* **76**, 194-214, 273-297.
Stoddart, D. R. and Scoffin, T. P. (1979). *Atoll Res. Bull.* **224**, 1-17.
Stoddart, D. R., McLean, R. F. and Hopley, D. (1978). *Phil. Trans. R. Soc.* **B284**, 39-61.
Thom, B. G. and Chappell, J. (1975). *Search* **6**, 90-93.
Thom, B. G. and Chappell, J. (1978). *Phil. Trans. R. Soc.* **A291**, 187-194.
Thom, B. G., Orme, G. R. and Polach, H. A. (1978). *Phil. Trans. R. Soc.* **A291**, 37-54.
Trudgill, S. T. (1976). *Z. Geomorph. Suppl. Bd.* **26**, 164-200.
Veeh, H. H. and Green, D. C. (1977). *In* "Biology and Geology of Coral Reefs, IV, Geology" (O. A. Jones and R. Endean, eds), pp. 183-200. Academic Press, London and New York.
Walcott, R. I. (1972). *Quaternary Research* **2**, 1-14.

PART SIX

Concluding remarks

FIFTEEN

Shorelines and isostasy: retrospect and prospect

A. G. DAWSON and D. E. SMITH

INTRODUCTION

The study of relict shorelines has provided considerable information on the isostasy of the Earth's crust and mantle. The wealth of ideas and information in the chapters contained in this volume reflects how far our knowledge of the effects of loading and unloading of the Earth's crust has advanced in recent times. This chapter reviews the origins of the hypotheses of glacio-isostasy and hydro-isostasy, and considers briefly future prospects in studies of shorelines and isostasy.

RETROSPECT

The theory of isostasy developed, as is now well known, as a result of survey work in the eighteenth and early nineteenth centuries, culminating in the hypotheses of Pratt and Airy in 1855 and in the first use of "isostasy" by Dutton in 1889. Demonstrations of what might be termed the dynamic aspects of the theory, however, took place independently, as geologists seeking to explain local changes of level developed first the hypothesis of glacio-isostasy then later that of hydro-isostasy.

The hypothesis of glacio-isostasy can be seen as the culmination of many detailed and sometimes conflicting observations on relative sea-level changes made in Scotland during the eighteenth and early nineteenth centuries. During the eighteenth century, evidence was gradually accumulating to show that there had been widespread changes in sea level, some of them relatively recent, along the Scottish coastline. Much of this evidence consisted of organic remains in deposits above contemporary sea level (e.g. Sibbald, 1707; Nimmo, 1777; Statistical Account of Scotland, 1794) and it was at first supposed that there had been a widespread fall in sea level relative to the land. However, as time progressed, evidence of lower sea levels from "submerged forests" or "buried peat" began to be published (e.g. Blackadder, 1824). Faced with this conflicting information, writers began to postulate differential movement of the land (Anon., 1834; Martins, 1851; Geikie, 1861).

SHORELINES AND ISOSTASY
ISBN 0 12 652960 4

The conflicting evidence of both rises and falls of relative sea level was best displayed in the Forth valley, where raised estuarine sediments, containing marine fossils, could be seen to rest widely upon buried peat, beneath which lay marine silty sand. This and other similar sequences came to the notice of Thomas Jamieson, the factor of a large estate in north-east Scotland (Godsman, 1958), who had already begun to make significant contributions to geology. Following Martins (1851), who had suggested that the elevation of a bed of arctic marine shells near Edinburgh had been due to the land rising following the retreat of the ice, Jamieson proposed in 1865 that the occurrence of fossiliferous marine deposits now above sea level was due to the depression of the land beneath the ice and subsequent elevation. He wrote (p. 178):

> It has occurred to me that the enormous weight of ice thrown upon the land may have had something to do with this depression . . . and then the melting of the ice would account for the rising of the land, which seems to have followed upon the decrease of the glaciers.

In the case of the sequence in the Forth valley, he supposed that two such depressions and elevations had taken place.

This was apparently the first articulation of the hypothesis of glacio-isostasy, though Jamieson never used the term as far as is known. Shortly afterwards, in North America, first Whittlesey (1868) and later Shaler (1874), put forward similar explanations to explain evidence of emergence on the east coast. In 1882 and 1887 Jamieson elaborated his hypothesis, maintaining that the amount of depression and subsequent re-elevation would depend upon the thickness of the ice load. In 1882 and 1886, Gilbert extended the idea to include water loading, accounting for the varying elevations of Lake Bonneville terraces in terms of differential water loading in the lake basin, developing this further in his monograph of 1890. By the close of the 1880s, therefore, the hypotheses of glacio-isostasy and hydro-isostasy had been outlined.

The hypothesis of glacio-isostasy was thus developed as a result of research in several countries, but the further development of the hypothesis during the nineteenth century appears to have been very largely due to work in Scandinavia, ably summarized by Mörner (1979). The area had been the scene of numerous studies of changing sea levels dating back to before the time of Celsius, whose observations on the receding shoreline of the Gulf of Bothnia (e.g. Frisi, 1785) are well known. These observations led Playfair (1802) to suspect crustal movements, and in 1838 and 1840 Bravais (1842) demonstrated that marine terraces in northern Norway rose towards the head of the Altenfjord, indicating differential land movement. Later, similar observations were made in other fjord areas, and following work in 1882, De Geer was able to demonstrate differential crustal uplift in Scandinavia in several important contributions (e.g. 1888, 1890, 1896). De Geer produced maps of crustal displacement with lines of equal deformation, which he called isobases in 1890. De Geer's maps showed that the form of crustal uplift resembled a dome, and the area closely matched that of the former ice sheet. Other work followed, for example by Hansen in Norway in 1890, and in 1901 Brøgger, working in the Oslo area, deduced that as the front of the ice sheet had retreated, a wave of uplift moved from the periphery of the depressed area towards its centre. By the beginning of the present century, the pattern of crustal response to glacial unloading in Scandinavia was being reconstructed in impressive detail, providing a model for studies in other glacio-isostatically affected areas.

Ensuing decades have seen significant advances in many areas in understanding the mechanisms, patterns, time-scales, and side-effects of both glacio- and hydro-isostasy, as well as in identifying other forms of crustal loading. These advances have been made possible by increasingly detailed studies employing a wide range of techniques. In the process, the importance of isostatic processes in the development of the world's landscapes has appeared even greater than was perhaps at first thought.

PROSPECT

The many advances made in recent times in understanding the effects of crustal loading by ice and water will undoubtedly lead to further progress in the near future. Much of this progress will be made through the appreciation and resolution of problems of identification, interpretation and analysis which have been recognized in recent studies. Some of the main problems concerning isostatic effects in coastal marine environments lie in the following areas.

Measurement, correlation and interpretation of shoreline features

Measurement

Studies of glacio- and hydro-isostasy have been conducted in an impressively wide area of the globe, but in many areas studied the lack of adequate base maps and in particular datum networks (Gray, this volume) must inevitably make detailed study of shorelines extremely difficult. When mapping and altitude measurement are improved in these areas, further detail of crustal movement will become available and may entail revision of some present views.

Correlation

Correlation of shoreline features is also an area where the resolution of problems is awaited. Increasingly, studies designed to establish patterns of shoreline displacement have involved the detailed measurement of the altitudes of comparable features. Thus, for example, in some studies the inner margins of coastal terraces are measured, whilst in others it is the crests of beach ridges. Correlation even between similar features may be difficult, however, if the coastal environment has varied, for example in terms of sediment supply, climate or coastal processes.

A particular problem of correlation occurs in the case of some Lateglacial depositional terraces. In parts of Scandinavia (Eronen, this volume) and Scotland (Sissons, this volume), deltas and outwash terraces of former ice margins are related to impressive depositional terraces marking shorelines. In areas increasingly distant from the former ice margins, however, these terraces become fragmentary and frequently disappear. Thus shorelines which appear to be major levels in some areas cannot be traced in others, presumably due to sediment supply. This may pose problems of correlation which may only be resolved through detailed field investigation.

In 1934, Wright suggested in his "isokinetic theory" that relict shore features may have been produced during relative still-stands between the rising land and the rising sea level, thus implying that the shorelines should have been measurably diachronous. These depositional terraces, however, were probably produced during conditions of rapid sedimentation and may have been virtually synchronous.

Interpretation

Despite many years of the investigation of relict shorelines, there remain problems of interpretation. This can be illustrated in the case of emerged coastal rock platforms, which are well developed in some areas affected by glacio-isostasy. Until recently, such features on mid-latitude and polar coasts have been interpreted as having been produced during Pleistocene interglacials (e.g. Grant, 1980). However, the formation of rock platforms is now known to be highly complex (Trenhaile, this volume) and several conflicting views have been expressed. In particular, several workers (e.g. Rasmussen, 1981; Sissons, 1982) have attributed the origin of certain emerged rock platforms in Norway and Scotland to former periglacial shore erosion of rock. However, no detailed investigation of modern periglacial shore erosion of rock has been undertaken and the results of laboratory "freeze–thaw" experiments are ambiguous (McGreevy, 1981). The hypothesis of periglacial shore erosion of rock is therefore important to resolve, since other well-developed fossil rock platforms in middle and high latitudes may be open to reinterpretation.

Such problems of interpretation arise because first there is relatively little information on the nature of coastal processes in modern cold climate environments, and secondly there is almost no information on the altitudinal relationships between modern periglacial coastal landforms and sea level. These views are well illustrated in this volume by the contrasting interpretations of the beach ridge formations described by Fairbridge and by Taylor and McCann.

The interpretation of stratigraphic evidence for relative sea-level changes

In recent years, considerable progress has been made in reconstructing former sea-level changes in many areas of the world by using essentially stratigraphical approaches. The techniques used and their limitations are outlined in detail by Sutherland in this volume. In areas affected by glacio- and hydro-isostasy, local relative sea-level curves have been constructed (e.g. Cullingford *et al.*, 1981), many derived from studies of isolation basins (e.g. Stabell, 1980; Eronen, this volume; Hafsten, this volume). As more such curves are produced, detailed comparisons will become increasingly possible against the background of shoreline evidence, and further information on crustal movement may be gained.

Studies which combine both morphological and stratigraphical work may be particularly fruitful. Thus, where the evidence for a particular shoreline is virtually continuous over a wide area, detailed study of the related deposits at different locations (e.g. Smith *et al.*, in press) may indicate the degree of diachroneity of the shoreline and the manner in which displacement occurred.

The identification of changes in sea level independent of the local isostatic component is particularly important in studying isostatic movement. The stimulus of I.G.C.P. project 61 has led to the production of many Holocene sea-level curves in recent years (e.g. Devoy, 1979; Ota *et al.*, 1981; Shennan, 1982), some of which are from areas where glacio- or hydro-isostatic movement has been negligible over the time period involved. In addition, there has been publication of considerable information on earlier Late Pleistocene sea-level changes (e.g. Bloom *et al.*, 1974; Butzer, 1975). However, these advances have coincided with the discovery that a global "eustatic" sea level cannot be measured anywhere (Clark, 1980) and that regional differences in Holocene sea-level curves may be explained by ocean floor deformation and geoidal deformation of the ocean surface caused by changing ice and water loads. Thus separation of the isostatic component of movement (e.g. Mörner, 1971) will only be made where there is detailed regional information on sea-level change.

In the future, the establishment of the oxygen isotope record (Shackleton and Opdyke, 1973) may pose more questions than it may provide answers. For example, the inferred frequent and astonishingly rapid Pleistocene ocean volume changes imply repeated and relatively rapid growth and decay of Pleistocene ice sheets. As a result, patterns of interglacial sea-level changes may prove to have been as complex as those for the Holocene (cf. Bloom *et al.*, 1974; Butzer, 1975).

The importance of neotectonic activity

In the years since Sauramo (1939, 1955) first postulated the existence of hinge lines in the Scandinavian shoreline sequence, there has been a steady increase in interest in possible non-uniform uplift of the crust following the removal of an ice or water load. Studies by Sissons (1972), Lundquist and Lagerback (1976), Mörner (1980) and Sissons and Cornish (1982) have indicated that unloading by ice or water may be associated with neotectonic activity. Mörner has suggested that this may be the rule rather than the exception, and if this is so it seems unlikely that former shorelines in areas affected by glacio-isostatic uplift are uniformly tilted and/or gently warped (Sissons and Cornish). This would apply particularly to Lateglacial shorelines, where features would have been formed and displacement would have commenced during periods of relatively rapid isostatic uplift. Since it is these features which are commonly most fragmentary, the possibilities of non-uniform uplift and dislocation imply additional problems of correlation. This will have implications also for the construction of graphs of isostatic uplift (Mörner, 1971).

It has been suggested that rapid crustal unloading may be especially significant in inducing neotectonic activity (Sissons and Cornish). This may have implications for the interpretation of the histories of the large ice-marginal lakes of the Laurentide and Fennoscandian ice sheets. For example, we do not yet know whether the drainage of the Baltic ice lake (Bjorck, 1979) was a catastrophic or gradual event, but if the former is true, then widespread neotectonic activity may have taken place. As more detailed studies are undertaken, the extent of neotectonic activity will become clear and its significance will be assessed.

The effect of variations with time in the distribution of the load

Recent work on the distribution of raised shorelines in areas of glacio-isostasy has disclosed variations in the pattern of displacement between shorelines (e.g. Gray, this volume). Such variations imply changes in the distribution of the ice load, presumably in response to climatic changes during glaciation. Indirect evidence for such changes has come from the record of ocean sediments. Ruddiman *et al.* (1977) have inferred movements of the polar oceanic and polar atmospheric fronts in the North Atlantic during the last glaciation and have suggested that these changes would have affected the development of the Fennoscandian and British ice sheets. Their studies have contributed to the suggestion that the maximal limits of the last British ice sheet may not have been synchronous (Sissons, 1981) since, when the position of the fronts favoured the build-up of the more southerly parts of the ice-sheet, the more northerly parts would have been subject to diminished precipitation and may even have been retreating. Similar interpretations are applicable to the Laurentide and Fennoscandian ice sheets, the northern margins of which during the last glacial maximum may have been deprived of precipitation and thus subject to ice decay (cf. Boulton, 1979b). Similarly, Boulton (1979a) has argued that there was no major ice expansion in the Spitzbergen archipelago during the last glacial maximum at 18 000–20 000 B.P. Thus, there is reason to suggest that ice sheet growth and glacio-isostatic depression along the southern edges of mid-latitude ice sheets may have been accompanied by ice retreat and crustal rebound farther north.

As further studies of displaced shorelines are carried out, it seems likely that spatial variations in the growth and decay of the ice load will be further confirmed.

The extent of geophysical effects

Geophysical models of glacial unloading for the major ice sheets have now attained a high degree of sophistication. However, their accuracy remains subject to, amongst other factors, the accuracy of inferred deglaciation histories. The areas involved in these studies are vast, and further detailed morphological and stratigraphical studies could, in time, lead to revision of the pattern of shoreline displacement and consequent revision of geophysical models.

Andrews and Peltier (this volume) have discussed the methods by which relative sea-level data and free air gravity data may be reconciled with an earth of uniform mantle viscosity. Moreover, they have drawn attention to the Clark *et al.* (1978) model where Holocene sea-level data is partly explained by a delay in the melting of the main northern hemisphere ice sheets by 2×10^3 years, which conflicts with the ^{14}C-dated melting isochrones of the last Laurentide ice sheet (cf. Bryson *et al.*, 1969). The problem may be resolved partially by consideration of the melting histories of the Antarctic and Greenland ice sheets, since their contributions to Holocene sea-level changes remain largely unknown.

Certain geophysical models have been based upon the assumption that, during the last glacial maximum, the earth, its ice sheets and oceans were in isostatic equilibrium (cf. Andrews and Peltier, this volume). The possibility that isostatic disequilibrium existed during the last glacial maximum is presently being explored and modelled (ibid). However, our knowledge of environmental changes that took place between the last interglacial and the last glacial maximum is restricted despite recent progress through study of the oceanic sedimentary record (Ruddiman *et al.*, 1980) for this period.

CONCLUSION

Considerable progress has undoubtedly been made in studies of shorelines and isostasy in recent years. Inevitably, this has resulted in an appreciation of the complexity of both the features and the isostatic processes involved. Undoubtedly this will in time lead to more detailed reconstructions of former shorelines in areas affected by glacio-isostasy and hydro-isostasy, including areas for which apparently very detailed histories have already been proposed. The main need at present is for rigorous field investigations including detailed altitude measurement, supported by detailed sedimentological and microfossil studies and improved dating methods. The increasingly interdisciplinary approaches now involved in the study of shorelines and isostasy, perhaps partly illustrated by the different affiliations of the contributors to this volume, offer considerable promise for the future.

REFERENCES

Airy, G. B. (1855). *Proc. roy. Soc.* **7**, 240-242.

Anon, (1834). *The Scotsman* **18** (1546), 1 November, 1834.

Blackadder, A. (1824). *Memoirs of the Wernerian Society* **5**, 424.

Bjorck, S. (1979). *Fossils and Strata* **14**, 1-93.

Bloom, A., Broecker, W. S., Chappell, J. M. A., Matthews, R. K. and Mesolella, K. J. (1974). *Quaternary Research* **4**, 185-205.

Boulton, G. S. (1979a). *Boreas* **8**, 31-58.
Boulton, G. S. (1979b). *Boreas* **8**, 375-395.
Bravais, A. (1842). *In* "Voyages en Scandinavie, en Lapponie, au Spitzberg et au Faroe, pendant les années 1838, 1839 et 1840 sur la corvette La Recherche, 1" partie 1, 57-137.
Brøgger, W. C. (1901). *Norg. geol. Unders.* **31**, 731 pp.
Bryson, R. A., Wendland, M. W., Ives, J. D. and Andrews, J. T. (1969). *Arctic and Alpine Research* **1**, 1-14.
Butzer, K. (1975). *In* "After the Australopithecines" (K. W. Butzer and G. L. Isaac, eds), pp. 25-71. Mouton, The Hague, and Aldine, Chicago.
Clark, J. A. (1980). *In* "Earth Rheology, Isostasy and Eustasy" (N. A. Mörner, ed.), pp. 525-534. John Wiley and Sons, Chichester.
Clark, J. A., Farrell, W. E. and Peltier, W. R. (1978). *Quaternary Research* **9**, 265-287.
Cullingford, R. A., Caseldine, C. J. and Gotts, P. E. (1980). *Nature* **274**, 548-553.
De Geer, G. (1982). *Geol. Foren. Stockh. Förh.* **4**, 149-162.
De Geer, G. (1888). *Geol. Foren. Stockh. Förh.* **10**, 366-379.
De Geer, G. (1890). *Geol. Foren. Stockh. Förh.* **12**, 61-110.
De Geer, G. (1896). *Sver. geol. Unders. Afh.C.* **161**, 160 pp.
Devoy, R. J. N. (1979). *Phil. Trans. R. Soc.* **B285**, 355-407.
Dutton, C. E. (1889). *Bulletin of the Philosophical Society of Washington* **11**, 51-64.
Frisi, P. (1785). "Operum tomus tertius". Mediolani.
Geikie, A. (1861). *Edinburgh New Philosophical Journal* N.S. **14**, 102-112.
Gilbert, G. K. (1882). *United States Geological Survey, Second Annual Report* 169-200.
Gilbert, G. K. (1886). *Am. J. Sci.* **31** (April, 1886).
Gilbert, G. K. (1890). *United States Geological Survey Monograph* **1**, 438 pp.
Godsman, J. (1958). "A History of the Burgh and Parish of Ellon Aberdeenshire". W. and W. Lindsay, Aberdeen.
Grant, D. R. (1980). *In* "Earth Rheology, Isostasy and Eustasy" (N. A. Mörner, ed.), pp. 201-214. John Wiley and Sons, Chichester.
Hansen, A. M. (1890). *Arch. Math. Natur.* **14**, 254-343.
Jamieson, T. F. (1865). *Quart. J. geol. Soc. Lond.* **21**, 161-203.
Jamieson, T. F. (1882). *Geol. Mag.* **9**, 400-407.
Jamieson, T. F. (1887). *Geol. Mag.* Ser. 3, **4**, 344-348.
Lundqvist, J. and Lagerbäck, R. (1976). *Geol. For. Stockh. Förh.* **98**, 45-51.
McGreevy, J. P. (1981). *Progress in Physical Geography* **5**, 56-75.
Martins, C. (1851). *Edinburgh New Philosophical Journal* **50**, 301-318.
Mörner, N.-A. (1971). *Palaeogeography, Palaeoclimatology, Palaeoecology* **9**, 153-181.
Mörner, N.-A. (1979). *Geo Journal* **3** (3), 287-318.
Mörner, N.-A. (1980). *In* "Earth Rheology, Isostasy and Eustasy" (N. A. Mörner, ed.), pp. 251-284. John Wiley and Sons, Chichester.
Ota, Y., Matsushima, Y. and H. Moriwaki (1981). "Atlas of Holocene sea level records in Japan". Yokohama National University, Japan.
Petterssen, K. (1880). *Tromso Museums Aarshefter* **3**, 1-52.
Playfair, J. (1802). "Illustrations of the Huttonian Theory of the Earth". Edinburgh.
Pratt, J. H. (1855). *Proc. roy. Soc.* **7**, 176-182.
Rasmussen, A. (1981). *Norges, geol. Unders.* **369**, 1-31.
Ruddiman, W. F., Sancetta, C. D. and McIntyre, A. (1977). *Phil. Trans. R. Soc.* **B280**, 119-142.

Ruddiman, W. F., McIntyre, A., Niebler-Hunt, V. and Durazzi, J. T. (1980). *Quaternary Research* **13**, 33-64.

Sauramo, M. (1939). *Fennia* **66** (2), 1-26.

Sauramo, M. (1955). *Ann. Acad. Sci. Fennicae* A III **44**, 1-25.

Shackleton, N. J. and Opdyke, N. D. (1973). *Quaternary Research* **3**, 39-54.

Shaler, N. S. (1874). *Boston Society of Natural History, Memoirs* **2**, 320-340.

Shennan, I. (1982). *Proc. Geol. Ass.* **93**, 53-64.

Sibbald, R. (1707). "History and Development of Stirlingshire". Edinburgh.

Sissons, J. B. (1972). *Trans. Inst. Br. Geogr.* **55**, 145-159.

Sissons, J. B. (1981). *Boreas* **10**, 1-17.

Sissons, J. B. (1982). *Trans. Inst. Br. Geogr.* N.S. **7**, 205-216.

Sissons, J. B. and Cornish, R. (1982). *Quaternary Research* **18**, 268-288.

Smith, D. E., Cullingford, R. A. and Brooks, C. L. (in press). Flandrian relative sea-level change in the Ythan valley, North-East Scotland. *Earth Surface Processes and Landforms*.

Stabell, B. (1980). *Norsk. geol. Tidsskr.* **60**, 71-81.

Statistical Account of Scotland (1794). 20 vols, Edinburgh.

van de Plassche, O. (1982). *Mededelingen Rijks geologische dienst.* **36-1**, 1-93.

Whittlesey, D. (1868). *American Association for the Advancement of Science, Proceedings* **16**, 92-97.

Wright, W. B. (1934). *Comptes rendus du Congrès Internationale de Geographie Varsovie.* Section II, 534-553.

Subject Index

L

M

Special Publications of the Institute of British Geographers

No. 1 *Land Use and Resources: Studies in Applied Geography. A Memorial to Sir Dudley Stamp* (1968)

No. 2 *A Geomorphological Study of Post-Glacial Uplift with Particular Reference to Arctic Canada* (1970)
J. T. Andrews

No. 3 *Slopes: Form and Process* (1971)
Compiled for the British Geomorphological Research Group by D. Brunsden

No. 4 *Polar Geomorphology* (1972)
Compiled for the British Geomorphological Research Group by R. J. Price and D. E. Sudgen

No. 5 *Social Patterns in Cities* (1973)
Compiled for the Urban Study Group by B. D. Clark and M. B. Gleave

No. 6 A contribution to the International Hydrological Decade—*Fluvial Processes in Instrumented Watersheds: Studies of Small Watersheds in the British Isles* (1974)
Edited for the British Geomorphological Research Group by K. J. Gregory and D. E. Walling

No. 7 *Progress in Geomorphology: Papers in Honour of David L. Linton* (1974)
Compiled and edited for the British Geomorphological Research Group by E. H. Brown and R. S. Waters

No. 8 *Inter-Regional Migration in Tropical Africa* (1975)
I. Masser and W. T. S. Gould with the assistance of A. D. Goddard

No. 9 *Agrarian Landscape Terms: A Glossary for Historical Geography* (1976)
I. H. Adams

No. 10 *Change in the Countryside: Essays on Rural England, 1500–1900* (1979)
Compiled and edited for the Historical Geography Research Group by H. S. A. Fox and R. A. Butlin

No. 11 *The Shaping of Southern England* (1980)
Edited by D. K. C. Jones

No. 12 *Social Interaction and Ethnic Segregation* (1981)
Edited by P. Jackson and S. J. Smith

No. 13 The Urban Landscape: Historical Development and Management (1981)
Edited by J. W. R. Whitehand

No. 14 *The Future for the City Centre* (1983)
Edited by R. L. Davies and A. G. Champion

No. 15 *Redundant Spaces in Cities and Regions?* (1983)
Edited by J. Anderson, S. Duncan and R. Hudson

No. 16 *Shorelines and Isostasy* (1983)
Edited by D. E. Smith and A. G. Dawson

In preparation
No. 17 *Residential Segregation, The State and Constitutional Conflict in American Urban Areas* (1984)
R. J. Johnston